AF763154

Pelton and Crossflow Impulse Hydraulic Turbines

Zulcy de Souza • Geraldo Lúcio Tiago Filho

Pelton and Crossflow Impulse Hydraulic Turbines

Springer

Zulcy de Souza
Federal University of Itajubá
Itajubá, Brazil

Geraldo Lúcio Tiago Filho
Federal University of Itajubá
Itajubá, Brazil

ISBN 978-3-031-87085-9 ISBN 978-3-031-87086-6 (eBook)
https://doi.org/10.1007/978-3-031-87086-6

© The Editor(s) (if applicable) and The Author(s), under exclusive license to Springer Nature Switzerland AG 2025

This work is subject to copyright. All rights are solely and exclusively licensed by the Publisher, whether the whole or part of the material is concerned, specifically the rights of reprinting, reuse of illustrations, recitation, broadcasting, reproduction on microfilms or in any other physical way, and transmission or information storage and retrieval, electronic adaptation, computer software, or by similar or dissimilar methodology now known or hereafter developed.
The use of general descriptive names, registered names, trademarks, service marks, etc. in this publication does not imply, even in the absence of a specific statement, that such names are exempt from the relevant protective laws and regulations and therefore free for general use.
The publisher, the authors and the editors are safe to assume that the advice and information in this book are believed to be true and accurate at the date of publication. Neither the publisher nor the authors or the editors give a warranty, expressed or implied, with respect to the material contained herein or for any errors or omissions that may have been made. The publisher remains neutral with regard to jurisdictional claims in published maps and institutional affiliations.

This Springer imprint is published by the registered company Springer Nature Switzerland AG
The registered company address is: Gewerbestrasse 11, 6330 Cham, Switzerland

If disposing of this product, please recycle the paper.

Foreword

I have been presented with two challenges upon writing this introduction. The first challenge has been to summarize the importance of this work, given that it was written by two prominent professors who are remarkably capable of combining both theory and practical applications. The second challenge has been to position this book within our current context, i.e., the increased use of intermittent sources for power generation, e.g., solar and wind power.

I remember when Prof. Zulcy once told me "I feel the water passing through the turbine." This is what this book is about, the essence of a life-long project planner. Prof. Zulcy did not restrict himself to merely passing on the best international techniques but has strived to contribute his own experience and methods. Together with the master, the protégé, i.e., Prof. Tiago, has also followed suit by bringing together the best theories, combined with his own professional experience, to guarantee that this work will be truly excellent.

The impulse rotor hydraulic turbines discussed here, i.e., the Pelton and Bánki-Michell turbines, have varying applications. While Pelton turbines can be used in micro-power plant or large-scale power plant applications, Bánki-Michell turbines are used only in smaller projects. In contrast, the Pelton turbine is applied in conditions with relatively high heads, while Bánki-Michell turbines are normally used in medium to low heads. It is known that impulse turbines are not widely used in Brazil, while reaction turbines like the Francis and Kaplan turbines have been more widely applied. Furthermore, this work most surely will have impacts well beyond Brazil's borders, impacting Latin American countries, where rugged topographies make ideal settings for these machines.

Nonetheless, even in Brazil, there are many opportunities for exploring the use of these impulse turbines. The increased use of intermittent renewable resources has resulted in the need for changing operational criteria for hydroelectric power plants, since their reservoirs will no longer serve the function of regularizing energy over many years, but rather, will start to prioritize renewable energy cycles. Rapid variations in energy generation will require inverse responses from the so-called dispatchable generation units, e.g., hydroelectric power plants. Given this new scenario, many turbine characteristics will need to be reconsidered, since they will be subjected to large submergence variations.

Under these conditions, there is a remarkable opportunity for using Pelton turbines, since they operate within wide ranges, do not wear, and do not suffer performance losses or oscillations, contrary to reaction turbines (Francis and Kaplan turbines). These positive attributes can be combined with the fact that Brazil has large swaths of potential areas for installing these turbines to complement distributed generation. Furthermore, small power plants can use partial flows from large head waterways to promote operational flexibility. Small upstream and downstream reservoirs could, thus, mitigate environmental impacts. This is an opportunity for modernizing existing power plants given this new scenario.

Conversely, the robustness and the flexibility of Bánki-Michell turbines guarantee that they can be widely applied across Brazil, most notably now, given recent evolutions in automation and regulation technologies that allow them to be effectively integrated to electrical grids (where they exist) or to increase the quality of power supplied to isolated electrical systems.

This book serves as a guide for entrepreneurs, project designers, industries, and businesses, in selecting and manufacturing impulse (Pelton) turbines, using the best existing techniques, to strengthen national engineering practices, and in so doing, this book consolidates important knowledge that has been generated within Brazilian universities over many years.

Federal University of Itajubá, Itajubá, Brazil

Afonso Henrique Moreira Santos

Preface

This book is on hydraulic impulse turbines and was not initially intended to be included in the Turbomachinery Project series, precisely because of the complexity surrounding systematizing, calculating, and designing hydraulic impulse turbines (Pelton and Bánki-Michell turbines) using algorithms that can generate a computer program capable of providing viable results within the limitations of currently available market technology.

However, in light of the continued attention that has been given to the subject, which is available in the public domain, I, Prof. Zulcy de Souza, author of five previously published volumes on turbomechanics, have sought to collaborate with a colleague and friend, Prof. Geraldo Lucio Tiago Filho, to undertake the challenge of writing a new volume to this series, the result of which is this sixth volume, which addresses two main hydraulic impulse turbines, the Banki-Michell and Pelton turbines.

This sixth volume comprises three parts, namely:

Part I: The General Characteristics of Hydraulic Impulse Turbines (HITs)

Part I comprises two chapters.

- Chapter 1 describes the steps for designing rotor hydraulic impulse turbines. We then schematize and name their geometries.
- Chapter 2 studies and analyzes the behavior of stationary and moving water flows in stepped gradients, both with and without agitation, which can then be used to model components for either hydraulic or thermal turbomachines.

Part II: Banki-Michell Hydroelectric Impulse Turbines (BMHITs)

Part II comprises four chapters and is dedicated to studying the Banki-Michell hydraulic impulse turbine (BMHIT).

- Chapters 3 and 4 give a theoretical and hypothetical analysis of the real hydrodynamic water flows in a Banki-Michell hydraulic impulse turbine (BMHIT), from the inlet to the outlet at the rotor blades.
- Chapter 5 studies the geometry of the Banki-Michell hydraulic impulse turbine (BMHIT) and gives equations for the structural parts that make up this turbine, including vibrational parts.
- Chapter 6 compiles information from the previous chapters into an algorithm that can be used in a computer software program to calculate and design a Banki-Michell hydraulic impulse turbine (BMHIT).

Part III: Pelton Hydraulic Impulse Turbines (PHITs)

Part III comprises six chapters and is dedicated to studying the Pelton hydraulic impulse turbine (PHIT), with the following objectives.

- Chapter 7 establishes and theoretically equates the flow limits, i.e., the physical limits of the hydraulic flows, using a modeling hypotheses for flow geometries, starting at the input of the turbine cross-section, and going to the output at the rotors.
- Chapter 8 geometrically models the different types of pelton hydraulic impulse turbines (PHITs) that we have studied, characterizing each component by name, delimiting the space that each part occupies, and establishing criteria for the water flows as real fluid flows from the inlet to the outlet sections.
- Chapter 9 gives a procedure for obtaining the first approximation of the physical and geometric limits of the hydraulic path for real flows through the rotors of Pelton hydraulic impulse turbines (PHITs).
- Chapter 10 establishes and details a methodology for equating, geometrically tracing, and manufacturing the Pelton hydraulic impulse turbine (PHIT) rotors.
- Chapter 11 provides the basis for obtaining equations that preliminarily determine the thicknesses of each component that comprises the physical parts of a Pelton hydraulic impulse turbine (PHIT), via first approximation using hydrodynamic calculations and scaled designs for the hydraulic flow path, after defining the critical sections, the manufacturing materials, and the corresponding permissible tensions.

- Chapter 12 gives an algorithm to hydrologically and structurally calculate a Pelton hydraulic impulse turbine (PHIT) for preliminary designs and posterior manufacturing.

Itajubá, Brazil Zulcy de Souza
Itajubá, Brazil Geraldo Lúcio Tiago Filho

Symbology

	A
A	Cross-section area of the flow or component
A_g	Surface friction of the gasket
A_m	Average cross-section area of the flow
A_{t_i}	Cross-section area of the intermediate pipe flow
A_0	Cross-section area of the flow at the inlet
A_1	Cross-section area of the flow at the IN entrance
A_2	Cross-section area of the flow at the IN outlet covered by the angled aperture
A_δ	Area of the flow section at the IN outlet
A_3	Area of the flow section at the IN outlet
	B
b	Axial pad length or FW disc thickness
b_d	Length of the DF rods
b_{in}	Width of the IN
b_j	Bifurcation index 1; 2;...
B	Length of BMHIT rotor blades or inner shell width or shell thickness crown of the FW
B_a	Outer shell width
B_i	Height of the rotor shell on the inside of the PHIT housing
B_{in}	Length of the IN
B_{iu}	PHIT housing width in the DL
B_{io}	Width at the top of the PHIT housing
	C
c	Average absolute flow velocity
c_c	Average absolute flow velocity in the curves
c_{ca}	Average absolute flow velocity in the cylinder/needle cross-section
c_{ch}	Absolute shock velocity
c_g	Average velocity of rotation of the DF
c_m	Average absolute flow velocity projected in the meridional or radial directions
c_o	Maximum average absolute flow velocity in the jet

(continued)

c_{SA}	Average absolute flow velocity from the SA outlet
c_t	Average absolute flow velocity in the tubes
c_u	Average absolute flow velocity projected in the tangential direction
c_{m_4}	Average absolute projected flow velocity in the meridional direction at the RO inlet
c_{m_5}	Average absolute projected flow velocity in the meridional direction at the RO outlet
$c_{m_{4*}}$	Average absolute projected flow velocity in the meridional direction at the entry of the second passage through the RO
$c_{m_{5*}}$	Average absolute projected flow velocity in the meridional direction at the outlet of the first passage through the RO
c_{r_2}	Average absolute flow velocity in the tangential direction at the exit of the IN
c_{r_j}	Absolute velocity at general flow points in the radial direction
c_t	Average absolute flow velocity in the cylindrical tubes
c_{t_j}	Absolute velocity at general flow points in the tangential direction
$c_{r_{rj}}$	Resulting absolute velocity at general flow points
c_u	Average projected absolute velocity in the tangential direction
c_{u_4}	Average projected absolute flow velocity in the tangential direction at the RO inlet
c_{u_5}	Average projected absolute flow velocity in the tangential direction at the RO outlet
$c_{u_{4*}}$	Average absolute projected flow velocity in the tangential direction at the entry of the second passage through the RO
c_{u_5*}	Average absolute projected flow velocity in the tangential direction at the entry of the first passage through the RO
cv	Horsepower
c_0	Average absolute flow velocity at the inlet
c_1	Average absolute flow velocity at the IN entrance
c_2	Average absolute flow velocity at the IN exit
c_{2_t}	Average absolute flow velocity in the tangential direction
c_3	Average absolute flow velocity at the IN/RO interstice
c_4	Average absolute flow velocity at the RO inlet
c_{4_1}	Average absolute flow velocity at the minimum jet diameter under maximum load
c_5	Average absolute flow velocity at the RO exit
c_{4*}	Average absolute flow velocity at the entrance of the second passage through the RO
c_{5*}	Average absolute flow velocity at the entrance of the first passage through the RO
	D
dx	Element thickness
d_{o_1}	Jet diameter under maximum load
d_a	Diameter of the blade or needle
d_b	Internal diameter at the exit of the deviated bifurcation
d_{cf}	Bondschu formula or internal diameter of the penstock at flow entrance in the HT

(continued)

d_{cg}	Blade diameter
d_{ci}	Cylinder diameter
d_d	Disk diameter
d_e	Shaft diameter
d_{ed}	DF shaft diameter
d_{en}	Diameter of the DF camshaft
d_{ep}	Diameter of the pulley shaft on the DF command system
d_{ex}	External diameter of the RO shells
d_f	Diameter of the flow after exiting the brake
d_g	Gasket diameter
d_{in}	Diameter at the IN output
d_j	Pin diameter
d_m	Diameter of the trunnions on their supports or diameter of the DF drive shaft
d_{ti_f}	Inner diameter of the brake circuit tubes
d_{t_j}	General tube inner diameter
d_2	Diameter at the IN output
d_α	Element angle
$d_{\sigma r}$	Element stress in the radial direction
D	Reference diameter or pitch diameter
D_c	Inner diameter of the pipe
D_e	Outer diameter
D_{e_a}	Outer diameter of the shell edge
D_{e_c}	Outer diameter of the shell
D_i	Inner diameter of the RO
D_{i_a}	Inner diameter of the shell edge
D_{i_c}	Outer diameter of the shell edge
D_h	Hydraulic diameter
D_0	Inner diameter of the tube at the HT inlet
	E
e	Thickness
e_{c*}	Maximum shell thickness at the rib
e_{cg_a}	Thickness of the blade at its center of gravity
e_{cg_p}	Profile thickness at the center of gravity
e_{ci}	Cylindrical tube thickness
e_d	RO disk thickness or DF head
$e_{d_{min}}$	Minimum thickness of the RO axis
e_{fo}	Minimum thickness of the steel tube or walls
e_m	Root-mean-square error
e_h	Thickness of the sides of the DF
e_{min}	Minimum pipe or wall thickness
e_{blade}	Constant thickness of the RO blade
$e_{blade_{t_4}}$	Tangential thickness of the blade at the entrance of the RO
e_s	Supplementary thickness of tube
e_{su}	Supplementary thickness of the DF support plate

(continued)

e_{su}^{*}	Supplementary thickness
e_t	Tangential thickness of the RO blade
e_{tf}	Thickness of the pipe in the brake apparatus
e_{t_4}	Tangential thickness at the entrance of the RO
E	View of the distance in the cutaway of the shell
$E_{i_{PQf}}$	Internal specific energy lost to flow escapes
$E_{Q_{max}}$	Maximum shear strength
$E_{i_2}m$	Internal kinetic energy of the flow at the output of the IN
	F
f_e	Choke valve factor or profile thinning factor
$f_{e_{65-010}}$	NACA 65–010 choke valve factor or profile thinning factor
f_{se}	Safety factor for the formulas calculating energy losses
$f_{e_4}=f_{e_5}$	Choke factor of the flow at the entrance and output of the RO
$f_{e_{5*}}=f_{e_{4*}}$	Choke factor of the flow at the exit and in the entrance of the first and second passages through the RO
$f_{e_{4_t}}=f_{e_{5_t}}$	Tangential choke factor of flow at the entrance and output of the RO
$f_{e_{5*_t}}=f_{e_{4*_t}}$	Tangential choke factor of flow at the exit and entry of the first and second passages through the RO
$f_{d_{max}}$	Arrow on the maximum disk
$f_{e_{max}}$	Maximum arrow for the axis
$f_{e_{aux}}$	Auxiliary arrow
f_B	Relationship between internal shell widths
f_r	Electrical frequency for the consumer or grid
f_L	Relationship between internal shell lengths
f_T	Relationship between internal shell depths
f_{blade}	Arrow of the blade
$f_{blade_{exc}}$	Excitation frequency of the blade
f_{blade_n}	Vibrational frequency of blade
f_s	Safety factor
F	Force on the blade, or distance from the center of the jet to the inner diameter of the blade edge
F_a	Frictional force
F_{ae}	Frictional force on the axis of the DF
F_{c_h}	Active force of the flow on each shell half in the horizontal direction
F_j	Horizontal hydraulic force of the flow on the head of the DF
F_{r_d}	Force resulting from the flow and friction on the DF
F_h	Hydraulic force resulting from the flow on the DF
F_m	Driving/motor force
F_{max}	Maximum force
F_r	Force resulting from the flow and friction on the DF
F_R	Resulting force on the axis
$F_{R_{blade}}$	Resulting force on the blade
$F_{m_{4;5*}}=F_{m_{blade_1}}$	Force of the flow in the meridional direction in the first passage through the RO
$F_{m_{blade_2}}$	

(continued)

	Force of the flow in the meridional direction in the second passage through the RO
F_{r_δ}	Resulting force on the axis
$F_{u_{4;5*}} = F_{u_{blade_1}}$	Force of the flow in the tangential direction in the first passage through the RO
$F_{u_{blade_2}}$	Force of the flow in the tangential direction in the second passage through the RO
	G
g	Universal acceleration of gravity
g_s	Local acceleration of gravity
G_{cc}	Equivalent weight/mass of the crown blades
G_{co}	Equivalent weight/mass of the blades
G_d	Weight/mass of the disc
G_e	Weight/mass of the axis
G_r *or* G_{RO}	Weight/mass of the RO
G_{DE}	Weight/mass of the DF
GD^2_{GG}	Inertia of the GG
GD^2_{TH}	Inertia of the HT
GD^2_{VO}	Inertia of the FW
	H
h_{ga}	Water hammer
h_s	Height of the suction of the water hammer
h_{t_1}	Height of the energy plan in relation to the reference
h_{t_2}	Height of the reference plan for the output of the RO flow
H	Head of the water column at the inlet or direction/horizontal axis in meters
H_{cf}	Head of the forced conduit of the water column in meters
H_n	Nominal head of the water column at the inlet in meters
$H_{i_{cTS}}$	Internal head without a suction tube
$H_{i_{sTS}}$	Internal head without a suction tube
H_{i_0}	Head/internal energy of the water column at the entrance in meters
H_{max}	Head/maximum energy of the water column in meters
H_{min}	Head/minimum energy of the water column in meters
H_M	Head/energy including the water hammer of the water column in meters
	I
I_p	Polar moment of inertia
	J
J	Joule, SI unit of work
	K
k	Coefficient or sum of quantities
k_f	Coefficient of the welding efficiency or of tube defects
k_g	Gasket coefficient
k_m	Bearing loss coefficient
k_s	Bearing pad constant
k_{in}	Potential vortex in the injector
$k_{n_{qA}}$	Coefficient of the limitation range for specific rotation

(continued)

kW	Kilowatt, SI unit of power
	L
L	Internal height of the shell or equivalence in length
L_a	Length of the profile chord of the blade of the IN or the length of the needle in the IN or the length of the dividing edge of the shell
L_{ac}	Distance between the SCC input and point ea of the dividing edge of shell
L_{al}	Length of the directional flow blade
L_c	Length of the average curve arc or length of the circumference of the pad
L_{cci}	Length of the SCC
L_{ci}	Cylindrical length of the SCC
L_{co}	Length of the conical part of the IN
L_d	Deflector tube axle length or arc length of the DF
L_{di}	Length of the DI
L_{df}	Length of the DF in the direction of the flow axis
L_{d_1}	Internal diameter of the initial stretch of the flow diversions
L_{d_2}	Internal diameter of the diverted stretch of the flow
L_e	Equivalent length or distance from point EA to the jet axis at the entrance of the SA
L_g	Gasket length
L_h	Internal width or distance between the rods of the DF
L_{in}	Length of the IN
L_{Lb}	Diagonal length of the IN
L_m	Length corresponding to the spin of the DE or axial length of bearing
L_{m*}	Distance between the turning axis of the DE and the direction of the Fm
L_n	Projected flow length
L_p	Relative length of the chord
L_{blade}	Length of the average line of the bade
L^*_{blade}	Length of the average line of the bade for comparison
L_{t_o}	Length of the input tube in the SA
L_{t_i}	Length of the intermediate tube
L_{t_j}	General length of a straight axis tube
L_{tr}	Length of the circular to rectangular section transition piece
	M
mca	Water column in meters, unit of measurement for pressure in the TSU
M_a	Friction moment on the trunnion
M_f	Bending moment
$M_{f_{max}}$	Maximum bending moment
M_h	Hydraulic pressure moment on the blade
M_m	Motor moment of the SCC of the DF
M_T	Motor or driving moment
	N
n	Rotation of the HT or GG or number of coordinates considered in each section when calculating the experimental variance
$n_{e_{crit}}$	Critical axis rotation
n_G	EG rotation

(continued)

n_q	Specific rotation in the TSU relative to the flow
n_s	Specific rotation in the TSU relative to the power
n_{qA}	Specific rotation in SI units
n_{qA_1}	Specific rotation from the jet
n_{qA_R}	Specific rotation of the rotor
N_{QE}	Specific rotation of the IEC
n_T	Rotation of the HT
N_{Re}	Reynolds number
	P
p	Static pressure of the flow
p_c	The real step between the shells
p_{c*}	The theoretical step between the shells
p_{a_t}	Local atmospheric or barometric pressure
p_g	Static pressure of the flow at the input
p_{i_0}	Static pressure next to the inner wall in the input tube
p_{i_1}	Static pressure next to the inner wall at the input of IN
p_{max}	Static pressure from the maximum flow on the HFP walls
p_v	Static vapor pressure
p_s	Static suction pressure
p_o	Oil pressure at the pump outlet of the regulator system or pressure at the flow input n to the HT
p_6	Static pressure from the output flow
P	Power
P_e	Power on the axis
P_{gg}	GG power considering kinetic energy losses at the output of the RO
P_h	Hydraulic power
P_i	Internal power
P_{max}	Maximum power
P_{min}	Minimum power
P_n	Nominal power
P_{eM}	Power at the HT axis for head Hm
P_e	Power delivered to the HT axis
P_{e_s}	Power at the HT axis considering kinetic energy losses at the output of the RO
P_{ex}	Available external power from the flow at the output of the RO
$P_{i_{rg}}$	Internal power of the flow delivered to the RO considering only the geometry of its components
$P_{i_{rg1}}$	Internal power of flow delivered to the RO in the first passage considering only the blades geometry
$P_{i_{rg2}}$	Internal power of flow delivered to the RO in the second passage considering only the blade geometry
P_{i_0}	Internal power of flow at the input
P_{i_1}	Internal power of the flow at the input of the IN
P_{i_2}	Internal power of the flow at the output
P_{i_r}	Internal power of the flow delivered to the RO
$P_{i_{rg}}$	Internal power of the flow delivered to RO considering only the geometry

(continued)

P_{r_1}	Power of the flow delivered in the first passage through the RO
P_{r_2}	Power of the flow delivered in the second passage through the RO
P_{i_6}	Internal power of flow at the output
$P_{i_{P_{tr}}}$	Internal power of flow lost in the transition tube/in
P_{i_p}	Total internal power of the lost flow
$P_{i_{P_f}}$	Internal power of the flow lost to run
$P_{i_{P_{in}}}$	Internal power of the flow lost at the IN
$P_{i_{P_r}}$	Internal power of the flow lost at the RO
$P_{i_{P_{r1}}}$	Internal power of the flow lost in the first passage through the RO
$P_{i_{P_{r2}}}$	Internal power of the flow lost in the second passage through the RO
$P_{i_{P_{ch}}}$	Internal powers of the flow lost to shocks in the passages through the RO
$P_{i_{s_{TS}}}$	Internal power without a suction tube
P_{i_0}	Hydraulic internal power of the flow available at the input section of the HT
P_{i_r}	Internal power of flow at the rotor output
$P_{i_{r_s}}$	Internal power considering kinetic energy loss from the flow at the output of the RO
$P_{i_{P6}}$	Internal power lost at the flow outlet
$P_{P_{am}}$	Power lost to friction in smooth or the bearing housing pads
P_{P_d}	Power lost to ventilation on the discs installed on the axis
P_{P_g}	Power lost to ventilation
P_{P_m}	Total mechanical power lost
P_{P_v}	Power lost in the gaskets (seals)
	Q
q_p	Load distributed on the blade given its weight
q_{blade}	Hydrodynamic load distributed in the blade
q_r	Force distributed on the axis
q_t	Total load distributed on the blade
Q	Flow
Q_b	Flow diverted from the bifurcation
Q_f	Escape flow through the hydraulic brake circuit
Q_{max}	Maximum flow
Q_{min}	Minimum flow
Q_n	Nominal flow
Q_r	Flows through the RO or flow of the straight stretch of the bifurcation
Q_o	Oil flow in the servomotor
Q_1	IN flow
Q_{1*}	Injector flow under maximum load
$Q_{1/1}$	Flow with the flow controller totally open
Q_*	Flow through the RO blade channel
	R
r_D	Diameter and tangential velocities ratio
r_e	External radius of the FW
r_i	Internal radius of the FW

(continued)

r_j	General straight path
r_B	Ratio between the rotor length and IN
$r_{f_{blade}}$	Ratio of the arrows on the blade per meter of span
r_{in}	Initial radius of the outer spiral of the IN
r_{blade}	Radius of the average line of the RO blade
r_Q	Flow ratio
r_{ts}	Transmission ratio
r_v	Velocity ratio between u and c4
r_{v_t}	Velocity ratio between u and c4t
r_{ve}	Tangential and absolute velocity ratio at the entrance of the RO
R_{de}	Radius of the DF
R^2	Experimental variance
$R_{fe_{max}}$	Ratio of maximum arrows on the axis
R_i	Radius of the circumference of the vertical axis from the RO to the PHIT CA
R_m	Reaction force of the trunnion or the radius of the paraboloid of revolution
R_{io}	Radius of the circumference arc of the horizontal axis of the RO to the CA of the PHIT
	T
t	Theoretical step
t_{ag}	Water temperature
t_d	Closing time of the blade
t_d	Time to fully divert the jet
t_g	GG transitional time
t_h	Hydraulic time of forced conduit
t_T	Time that it takes a particle of water to travel through the blade
t_4	Theoretical step at the entrance of the first passage of the flow through the RO
t_{5*}	Theoretical step at the exit of the first passage of the flow through the RO
t_{4*}	Theoretical step at the entrance of the second passage of the flow through the RO
t_5	Theoretical step at the exit of the second passage of the flow through the RO
T_{cap}	Work capacity of the VR
	U
u	Average tangential velocity of the flow at the entrance of the RO
u_i	Average tangential velocity of the flow in diameter di
U_t	Wet perimeter for determining the hydraulic diameter
U_m	Wet perimeter of the internal cross-section of the transition part
U_o	Wet perimeter of the transverse section of the input pipe
U_1	Wet perimeter of the internal cross-section in 1 of the transition part
	V
v	Specific volume of the fluid or flow velocity
v_1	Average flow velocity at entry, as per NB 10280/1988
v_2	Average flow velocity at the BMHIT output, as per NB 10280/1988
V	Vertical direction or axis
V_c	Equivalent volume of a shell
V_{DE}	Volume of the DF

(continued)

V_{ev}	Volume of the flow axis between the rods of the DF
V_h	Volume of the DF rods
V_{pd}	Volume of the plate between the DF rods
	W
w	Average relative flow velocity
w_{ch}	Relative shock velocity
w_4	Average relative flow velocity at the RO entrance
w_5	Relative velocity of the flow at the output of the RO
w_{4*}	Average relative flow velocity at the entrance of the second passage through the RO
w_{5*}	Average relative flow velocity at the exit of the first passage through the RO
w_{∞}	Average velocity of the undisturbed attack flow on the profile
W	Moment of resistance
	X
x	Horizontal axis
x_a	Abscissa of the flap
x_i	General coordinate of a point
x_1	Distance from the disk to the trunnion
x_p	Abscissa relative to the profile
x_1	Horizontal distance from the RO disk to the trunnion
$\overline{x}$	Most probable value
	Y
y	Vertical axis
y_p	Ordinate of the corresponding profile
y_a	Ordinate of the flaps
y_{cg}	Ordinate of the profile in the blade's center of gravity
y_g	Distance from the RO axis to the blade's center of gravity
y_{max}	Maximum ordinate of the profile
Y_i	Specific internal work of the flow
Y_{i_r}	Specific internal work of the flow delivered to the RO
$Y_{i_{rg}}$	Specific internal work of the flow delivered to RO considering only the geometry of the components
$Y_{i_{rg1}}$	Specific internal work of the flow delivered to the RO in the first passage considering only the blade geometry
$Y_{i_{rg2}}$	Specific internal work of the flow delivered to the RO in the second passage considering only the blade geometry
$Y_{i_{cTS}}$	Specific internal work available at the output of the flow to the HT with a suction tube
$Y_{i_{sTS}}$	Specific internal work available at the outlet of flow to the HT without a suction tube
Y_{i_0}	Specific internal work at the entrance of the flow
Y_{i_2}	Specific internal work of the flow at the output
Y_{i_r}	Specific internal work of the flow delivered to the RO
Y_{i_4}	Specific internal work of the flow at the entrance of the RO
$Y_{i_{r_1}}$	Specific internal work delivered by the flow in the first passage through the RO

(continued)

$Y_{i_{r_2}}$	Specific internal work delivered by the flow in the second passage through the RO
Y_{i_p}	Total specific internal work lost by the flow
$Y_{i_{P_{br}}}$	Specific internal work lost by the flow in the bifurcation and straight stretch
$Y_{i_{P_{bd}}}$	Specific internal work lost by the flow in the bifurcation section from deflections
$Y_{i_{P_{in}}}$	Specific internal work lost by the flow in the IN
$Y_{i_{P_r}}$	Specific internal work lost by the flow in the RO
$Y_{i_{P_{r1}}}$	Specific internal work lost in the first passage of the flow through the RO
$Y_{i_{P_{r2}}}$	Specific internal work lost in the second passage of the flow through the RO
$Y_{i_{P_c}}$	Specific internal work lost from flows in curves
$Y_{i_{P_d}}$	Specific internal work lost from flow deflections
$Y_{i_{P_b}}$	Specific internal work lost from flows in bifurcations
$Y_{i_{P_t}}$	Specific internal work lost from flows in the pipes
$Y_{i_{P_{tr}}}$	Specific internal work lost from flows in the transition tube/IN
$Y_{i_{P_{ch}}}$	Specific internal work lost in flows from shocks in passages through the RO
$Y_{i_{P_{6*;3*}}}$	Specific internal work of the flow inside the RO
$Y_{i_{P_6}}$	Specific internal work lost from flows after departing the shells of the RO
$Y_{i_{rg}}$	Specific internal work of water as an ideal fluid flowing through a RO considering only the blade geometry
$Y_{i_{rg_1}}$	Specific internal work of water as an ideal fluid flowing in the first passage through RO considering the blade geometry
$Y_{i_{rg_2}}$	Specific internal work of water as an ideal fluid flowing in the second passage through RO considering the blade geometry
$Y_{i_{P_f}}$	Specific internal work lost to the flow run
Y_{i_0}	Specific internal work at the entrance of the flow to the PHIT
Y_{i_1}	Specific internal work at the entry of the flow to the IN
Y_{i_2}	Specific internal work at the output of the flow from IN
Y_{i_3}	Specific work and internal flow at the IN/RO interstice
Y_{i_4}	Specific internal work at the entrance of the flow at the RO shells
$Y_{i_{5*}}$	Specific internal work at the outlet of flow at the first passage through the RO
$Y_{i_{4*}}$	Specific internal work at the entrance of the flow in the second passage through RO
Y_{i_5}	Specific internal work at the output of the RO shells
Y_{i_6}	Specific internal work at the outlet of flow after exiting the RO shells
Y_{blade}	Specific internal work of the flow delivered to the blade
Y_p	Specific energy lost
Y_{blade}	Specific work from ideal fluid flows in a rotor with infinite infinitesimally thick blades
	Z
z_c	Number of RO shells
z_{c*}	Theoretical number of shells
z_j	Number of flow jets
z_p	Number of RO blades

(continued)

z_s	Number of bearing pads
z_{pp}	Number of EG pair poles
z_1	Vertical distance between the input and the output
z_2	Vertical distance from the reference plane to the RO exit
z_δ	Number of RO blades covered by the flow at the output from IN
	$\boldsymbol{\alpha}$
α	Machining angle on the back of a shell
α_a	Angle of the dividing edge of a shell
α_c	Real angular step between the shells
α_{c*}	Angular theoretical step between the shells
α_r	Angle that the direction of the flow without shocks forms with the direction with shocks at the rotor input
α_2	Angle that the direction of the absolute velocity of the flow forms with the horizontal/tangential flow at the output of the IN
α_4	Angle that the direction of the absolute velocity of the flow forms with the tangential flow at the input of the HIT rotor
α_5	Angle that the direction of the absolute velocity of the flow forms with the tangential flow at the output of the RO
α_{4*}	Angle that the direction of the absolute velocity of the flow forms with the tangential flow at the entrance of the second passage through the RO
α_{5*}	Angle that the direction of the absolute velocity of the flow forms with the tangential flow at the exit of the first passage through the RO
	$\boldsymbol{\beta}$
β	Complementary machining angle on the back of a shell
β_{blade}	Angle that the line direction of the blade at the input and at the HIT rotor output forms with the tangential direction
β_{blade*}	Angle that the direction of the blade AS forms at the exit and the entrance of the passages through the BMHIT with the tangential direction
β_{blade}	Angle that the line direction of the blade at the input and at the HIT rotor output forms with the tangential direction
β_{blade*}	Angle that the direction of the blade AS forms at the exit and the entrance of the passages through the BMHIT with the tangential direction
β_4	Angle of the blade/shell that the direction of the AS of the flow forms with the tangential flow at the entrance of the passage through the RO
β_5	Angle of the blade/shell that the direction of the AS of the flow forms with the tangential flow at the exit of the passage through the RO
β_{4*}	Angle that the direction of the relative velocity of the flow forms with the tangential flow at the exit of the first passage through the RO
β_{5*}	Angle that the direction of the relative velocity of the flow forms with the tangential flow at the exit of the second passage through the RO
	$\boldsymbol{\gamma}$
γ	Specific weight of water
γ_{steel}	Specific weight of the material used to manufacture the components
γ_{air}	Specific weight of air
	$\boldsymbol{\delta}$

(continued)

δ	Injector opening angle, or shock attack angle, or total diametrical span
δ_b	Deflection angle of the flow at the bifurcation
δ_e	Complement of the angle
	Δ
Δp_i	Internal power gains in the suction tube
Δp_g	Specific energy losses due to the RO geometry
	ε
ε; ε_t; ε_{ca}	Absolute average roughness of the material used to manufacture the component
ε_{tr}	Absolute average roughness of the material used to manufacture the transition piece
	η
η_{ap}	HPP yield
η_e	Yield on the axis
η_{el}	EG electrical yield
η_{ex}	External yield
η_{ex_s}	External yield considering kinetic energy losses at the output of the RO
η_f	Water flow yields between fixed and mobile HT parts
η_{gg}	GG yield
η_{gg_s}	GG yield considering kinetic energy losses at the output of the RO
η_i	Internal yield
η_{i_a}	Flow yield in the path
$\eta_{i_{in}}$	Internal yield of the IN
η_{i_e}	Internal entry yield
$\eta_{i_{tr}}$	Internal transition yield
$\eta_{i_{in_0}}$	Relative losses of available yield power in the injector relative to the input
$\eta_{i_{Qf}} = \eta_f$	Internal yield corresponding to power lost from leakage
$\eta_{i_{Qf_0}}$	Relative losses of available flow power corresponding to flows relative to the input
η_{i_r}	Internal flow yield in passing through the RO
η_{i_s}	Internal yield considering kinetic energy loss at the output of the RO
$\eta_{i_{r_0}}$	Relative available power losses of the corresponding flow in the passage through the RO
$\eta_{i_{r1}}$	Internal yield of the flow in the first passage through the RO
$\eta_{i_{r1_0}}$	Relative loss of available power of the corresponding flow in the first passage through the RO
$\eta_{i_{r1;1}}$	Part of the internal power of the flow received by the RO in the first passage
$\eta_{i_{r2}}$	Internal yield of the flow in the second passage through the RO
$\eta_{i_{r2_0}}$	Relative losses of available power of the corresponding flow in the second passage through the RO
$\eta_{i_{r1;2}}$	Part of the internal power of the flow received by the RO in the second passage
η_t	Total yield
η_{i*}	Internal yield without considering kinetic energy losses at the exit
η_{i_g}	Internal yield considering only the RO geometry
$\eta_{i_{gmax}}$	Maximum internal yield considering only the RO geometry

(continued)

$\eta_{i_{rg}}$	Internal yield considering only the RO geometry
$\eta_{i_{tp}}$	Internal yield considering the total losses
η_{i_3}	Internal yield from the IN/RO interstice
$\eta_{i_{3_0}}$	Relative losses of available power at the IN/RO interstice relative to the input
$\eta_{i_{6*3*}}$	Internal flow yield in passing through the RO
$\eta_{i_{6*;3*0}}$	Relative available power loss inside the rotor relative to the input
η_m	Mechanical yield
η_{m_s}	Mechanical yield considering kinetic energy loss at the RO output
η_t	Total HT yield
η_{blade}	Blade yield
η_0	Internal yield of the oil feeding system to the VR
	$\boldsymbol{\theta}$
θ_{co}	Angle of the needle of the IN
θ_d	Angle of jet diversion from the DF
θ_{d*}	Angle between direction Fh and Fj
θ_{dj}	General angles of flow deflections
θ_g	Angle of rotation of the DF
θ_{in}	Angle of the IN cone
θ_m	Average angle of the IN
θ_{blade}	Angle covered by the RO blade
θ_1	Angle that the water particles travel the AS in time t
$\theta_1; \theta_2; \theta_3$	Angles of the graph for calculating the RAT
	$\boldsymbol{\lambda}$
λ_p	Loss coefficient of the flow load with the wall of the component
λ_t	Loss coefficient of the flow load in tubes
λ_{tr}	Loss coefficient of the flow load in the transition piece
	$\boldsymbol{\mu}$
μ	Poisson modulus or dynamic oil viscosity
μ_{steel}	Poisson modulus for steel
μ_{ad}	Coefficient of friction for water/steel
	$\boldsymbol{\nu}$
ν_a	Cinematic viscosity of water
	$\boldsymbol{\Sigma}$
$\sum a; \sum_0^i a_j$	Sum of quantities and/or values
	$\boldsymbol{\rho}$
ρ	Specific mass of water
ρ_{air}	Specific mass of air
	$\boldsymbol{\sigma}$
$\sigma_{t_{adm}}$	Permissible material torsion stress
$\sigma_{f_{adm}}$	Permissible material flexion stress
$\sigma_{c_{adm}}$	Permissible material compression stress
σ_e	Permissible material flow stress
	$\boldsymbol{\tau}$

(continued)

τ_{cis}	Material shear stress
$\tau_{cis_{adm}}$	Permissible material shear stress
τ_{max}	Combined material flexion and shear stress
	ϕ
ϕ	Euler's dimensionless equation for velocity ratios
ϕ_i	Loss coefficient of the absolute average flow velocity
ϕ_{i_a}	Loss coefficient of the average absolute velocity in passageways
ϕ_{tg}	Theoretical velocity coefficient
$\phi_{0;\ 4}$	Loss coefficient of the flow velocity from PHIT input to the shells of the RO
ϕ_{tr}	Flow velocity coefficient in the transition piece
ϕ_2	Flow velocity coefficient for the IN
ϕ_3	Flow velocity coefficient for the IN/RO
ϕ_{3*}	Flow velocity coefficient inside the RO
	ψ
ψ	Energy coefficient from Euler's a-dimensional equation
ψ_{tg}	Theoretical energy coefficient
ψ_5	Velocity coefficient for the RO flow
	ω
ω	Angular velocity

Acronyms

3D	Three-dimensional
AutoCAD	Graphical Computational Software Program
ACL	Average current line
AFL	Average flow line
ARS	Average real streamline
AS	Average streamline
ATS	Average theoretical streamline
BM	Banki-Michell
BMHIT	Banki-Mitchell hydraulic impulse turbine
CAS	Casing
CFD	Computational fluid dynamics
COD	Coefficient of determination
CS	Control system
DCS	Dual command system
DF	Deflector
DL	Downstream water level
EG	Electric generator
ELETROBRAS	*Centrais Elétricas Brasileiras S.A.*
FDM	Fused deposition modeling
FF	Free flow
FO	Flow outlet
FW	Flywheel
GG	Generator group
HB	Hydraulic brake
HFP	Hydraulic flow path
HIT	Hydraulic impulse turbine
HFT	Hydraulic flow turbine
HPP	Hydroelectric power plant project
HT	Hydraulic turbine

HTM	Hydraulic turbomachine
IEC	International Electrotechnical Commission
IN	Injector
INMETRO	Brazilian National Meteorological Institute
IS	Injector system
LAF	Line of average flow
LHP	Large Hydropower Plant
LHPCH	Hydrodynamic Laboratory for Small Hydroelectric Power Plants
MI	Model one
MII	Model two
MIII	Model three
MPGG	Micro Pelton Generator Group
NACA	National Advisory Committee for Aeronautics
NBR	Brazilian Norm
OLADE	Latin American Energy Organization
PGG	Pelton Generator Group
PH	Pierre Henry
PHIT	Pelton hydraulic impulse turbine
PIL	Product Innovation Laboratory
PP	Power plant
PT	Project team
PVC	Polyvinyl chloride plastic
QCC	Quick control component
RB	Richard Bran
RHFP	Real hydraulic flow path
RLAF	Real line of average flow
RAT	Relative average trajectory
RO	Rotor
RS	Rotor shells
SA	Admission system
SCC	Slow control component
SHP	Small hydropower plant
SI	International System of Unit
SL	Streamline
ST	Suction tube
TC	Transition component
TLAF	Theoretical line of average flow
TM	Turbomachines
TSU	Technical System of Units
UNIFEI	Federal University of Itajuba
VR	Velocity regulator

Contents

About the Authors

Zulcy de Souza is a Full Professor at the Federal University of Itajubá, working in the field of Flow Machines, Hydraulic and Thermal Machines, Hydroelectric Power Plants, and Energy.

Geraldo Lúcio Tiago Filho is a Full Professor at the Federal University of Itajubá, working in the field of Hydraulic Flow Machines, Hydroelectric Power Plants, and Renewable Energy. He also serves as the Executive Secretary of the National Center for Small Hydropower Plants in Brazil.

Part I
The General Characteristics of Hydraulic Impulse Turbines (HITs)

Chapter 1
Introduction

As per the previous volumes of the Turbomachinery Project series, this chapter initially seeks to establish the stages that must be dealt with to project a Hydraulic Impulse Turbine (HIT), along with the various needs thereof, considering the many particularities of Turbomachines. This will extend from design studies to the operational and reliability studies. Once this has been completed, we will study the geometries of HITs, along with some of their particular characteristics.

1.1 Project Steps

Following the same steps described in Volume I, Projecting Turbomachines, and in general for Hydraulic Impulse Turbines with action rotors (HITs), we will transcribe some applicable steps here.

- We must preliminarily or initially make a design setup that seeks to establish the physical flow limits, which are called the Hydraulic Flow Path (HFP), as a first approximation,
- We must conduct a preliminary sizing, sketching out an HIT in both cutaway and in scale, and we must the establish the compatible and desired characteristics and the corresponding geometries with the Equations resulting from principles of mechanics, with the experimental coefficient limits, using calculation and design methodologies.
- We must design and manufacture a prototype or model that could allow an HIT to be manufactured, along with the circuitry for the tests, with all the necessary elements for carrying out the project and building the HIT model prototype.
- We must conduct tests and elaborate designs for manufacturing the HIT. This step will complement the previous step, and will be carried out in part in the test circuit, which will allow any Project Team (PT) to match expected performance with elements used in manufacturing the HIT. For this purpose, we will make

© The Author(s), under exclusive license to Springer Nature Switzerland AG 2025

Z. de Souza, G. L. Tiago Filho, *Pelton and Crossflow Impulse Hydraulic Turbines*,
https://doi.org/10.1007/978-3-031-87086-6_1

drawings of each component for casting, machining, and assembly, including the template design to control the manufacturing process, the finishing process, and we will offer specifications on the materials and components that will need to be used, along with any materials that will be necessary for assembling the HIT, including scripts and other complementary information, so that the project can proceed as planned.
- We must analyze the manufacturing stage, where all previous foreseen projections and recommendations from previous stages come together into physical fruition. At this stage, the HIT Installation, Operation, and Maintenance Manuals must be prepared and written by the Project Team (PT), along with the protocols and any final corrections to any project components.

To carry out these steps, the following information must be readily available. So, the Project Team (PT) must:

- Know the maximum, normal, and minimum values of the upstream and downstream water levels, the available headwaters, and the boosted flows for the product itself, i.e., the HIT.
- Know all maximum, minimum, and average annual levels of barometric pressure, temperature, relative humidity, physical and chemical air quality, the presence of any vibrations, electromagnetics, electrostatics, or other fields of nature for where the Hydroelectric Power Plant (HPP) site will be installed.
- Have a detailed understanding of access to the installation site, including specifications on, and characterizations of any roads or connected roadways, and/or any future projects that may be undertaken in the area, e.g., monuments.
- Understand the physical, chemical, and biological state of the water itself, e.g.:
  - Its specific mass and viscosity;
  - Its electrical, thermal, and sonic conductivity;
  - The presence of acids, bases, and/or other corrosive or noncorrosive chemical and/or biological substances;
  - Any suspended elements transported in the water, arising from erosion, sewage, or via other discharges;
  - Elements on the local ichthyofauna, including behavioral elements.
- Elaboratesized drawings of the blueprints, and of the section of the installation site, from the downstream to the upstream, and must mark the downstream and upstream operational levels.
- Elaborate the generated energy consumption system or the interconnected network grid, establishing the distances from places of use, or from transmission lines, along with the corresponding frequency and voltage characteristics.
- Establish the desired performance of the system.
- Outline the geometric limits of the installation, the flows, the drops, and brakes, from bench tests.
- Detail the equipment limitations, particularly the casting, welding, and machining limitations, along with any possible limitations associated with the assembly itself and tests conducted on the factory floor.

- Observe any specific contractual clauses, and meet the agreed-upon obligations among the parties, which must all be transcribed into specific documents.

The Project Team (PT) must also possess elements that can:

- Execute or monitor all the aforementioned steps within the respective scheduled timelines.
- Establish the design point characteristics of the HIT and the electric generator (EG), along with their operational limits, the control panel characteristics, the protection system, and all fixed, mobile, permanent, and/or future instrumentation that will be used.
- Supervise the manufacturing, assembly, and testing of factory components and model or prototypes for corresponding bench test, making sure that all specifications fall within contractual obligations.
- Supervise on-site installations, operational tests, and conduct operational monitoring during the warranty period.
- Develop and/or analyze installation tests.
- Execute, monitor, and supervise acceptance tests.
- Employ the Hydraulic Flow Turbine (HFT) fluid mechanics (CFD, CFX) computational packages to determine the most efficient Hydraulic Flow Path (HFP) in all aspects, particularly in terms of energy and the breadth of the operational field.
- Use computer packages to improve the structural part of the HFT in terms of weight, vibration, material fatigue, etc.
- Prepare installation, operation, and maintenance manuals.

1.2 Project Geometries

Of the various geometries used for HITs, only those geometries schematized and named in the figures will be considered and described in this volume.

1.2.1 Banki-Michell Hydraulic Impulse Turbine (BMHIT)

Figure 1.1 shows a cross-section view of a Banki-Michell Hydraulic Impulse Turbine (BMHIT), with flow control, Q in flow via a manual or automatic system using a speed regulator or, for small power, regulation via the load.

This Hydraulic Turbine (HT) is also known as a Cross-Flow, Transversal Flow, Michell, Michell-Ossberger, or Banki-Michell turbine, and here will be called a Banki-Michell Hydraulic Impulse Turbine (BMHIT).

This HIT operates with partial admissions, a transversal flow, and a double effect, i.e., the flow takes two passages through the rotor blades.

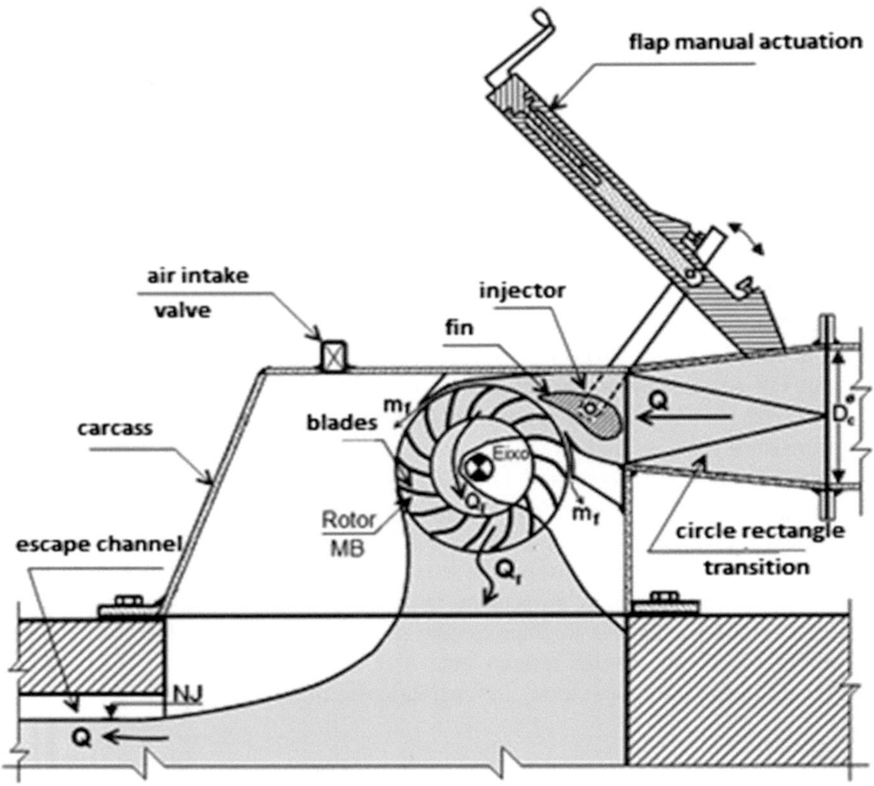

Fig. 1.1 Cross-section diagram of a BMHIT with the main components named

This turbine is constituted as a boundary-type turbine, between action and reaction turbines. A more complete classification would be a partial intake double effect action turbine.

As we will see, since it is easy to manufacture, it can be easily built in workshops with few mechanical and electrical resources, which can even be located in remote locations.

These considerations give BMHITs unique and inherent advantages, insofar as it:

- *Is easy to build*: Various components, like the rotor disks, the injector, and the housing can be manufactured from carbon steel sheets, and the blades can be calendered or cut from steel tubes. For small drops and low power, blades can be calendered or cut from galvanized steel tubes or from PVC.
- *Is self-cleaning*: Any objects that set on the blades during the first passage of water tend to be expelled in the second pass, after half a rotor rotation.
- *Is low-cost*: The start-up costs for a BMHIT are lower than for any other type of HT used in Low-Head Hydropower Stations, along with the operational and maintenance costs.

- *Has a low installation cost*: BMHIT installations requires little civil work.
- *Rotation versatility*: The rotation imposed on the BMHIT is suitable for use as a generator, for grain mills, sawmills, agricultural machinery, workshops and/or pumping stations.
- *Uses a suction tube*: The BMHIT power can be increased by installing a suction tube, depending on the characteristics of where it is installed.
- *Is cavitation-free*: In principle, it is not subject to the effect of cavitation, nor the effects arising from it.
- *Is ecologically suitable*: It has a positive ecological effect, since when in operation, it increases oxygen in water down the discharge channel.
- *Wide field of operation*: Its operational field, $\eta = f(Q)$, with H, and $n =$ constant is quite good compared to that of the Francis HT, for example, as will be shown in Volume III. Its reference field of application can be said to fall within the following limits:

$$2.0 < H\ (\mathrm{m}) < 250$$

$$0.04 < Q\ \left(\mathrm{m}^3/\mathrm{s}\right) < 15$$

$$0.5 < P_\mathrm{e}\ (\mathrm{kW}) < 3000$$

$$30 < n_{qA} = 3 \cdot n \cdot \frac{Q^{0.5}}{H^{0.75}} < 150$$

There are some disadvantages to this turbine, given its inherent operational characteristics, which include lower efficiency compared to other conventional turbines, a lack of compactness when operating under low heads, an inaccurate steering vane control system, mainly operating with a split injector, and noise pollution from the rotor blade ventilation, which operates partially empty.

1.2.2 Pelton Hydraulic Impulse Turbine (PHIT)

Figure 1.2 shows a cross-section view a Pelton Hydraulic Impulse Turbine (PHIT), on the horizontal axis, with the main components named.

A PHIT operates under high heads and small flows, with a horizontal axis with one or two jets, or a vertical axis for three to six jets, reaching powers greater than 300 (MW) per unit[1] and herds in the order of 1900 (m), which allow it to reach speeds greater than 180 (m/s) at the nozzle outlet.

This turbine has several advantages, in terms of its:

[1]Central Jostedal—Norway—1 PHIT of $P_n = 288$ (MW) and $P_{\max} = 301$ (MW)—5 injectors—$H_n = 1130$ (m)—$Q_n = 28.5$ (m^3/s)—$n = 428.6$ (rpm)—manufactured by Kvaerner Hydro—Oslo.

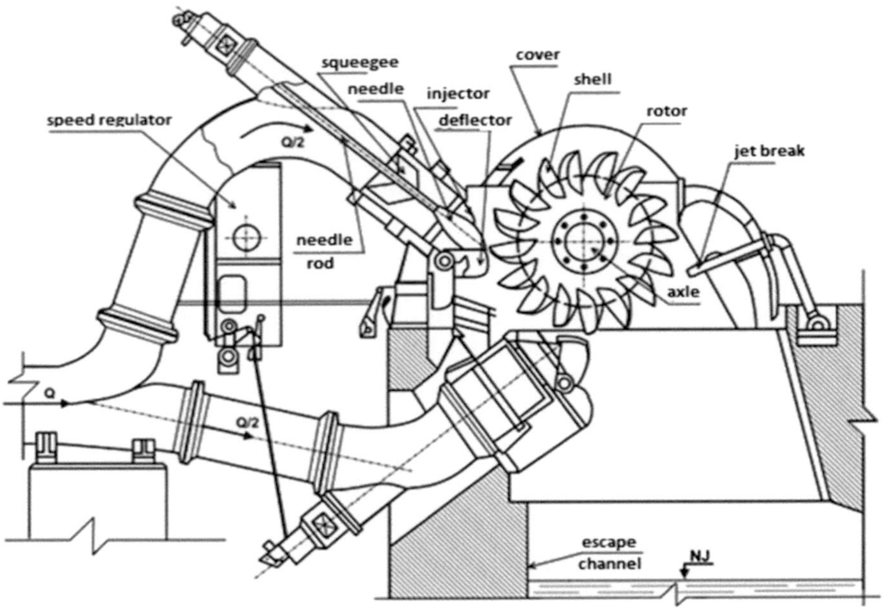

Fig. 1.2 Transversal section of a two-jet (PHIT), on the horizontal axis, with the main components named

- *Generator set arrangement*: turbine and generator, resulting in compact sets.
- *High efficiencies*: from 90 to 94%.
- *Flat efficiency curve*: more extensive than the Francis curve, especially if multiple injection nozzles are used, which allows the PHIT to operate under large flow variations.
- *Low manufacturing cost*: per unit of power, relative to similar turbines.
- *Simplified versions*: low-cost versions with moderate to high efficiencies at ~80 to 90%, making them attractive for microgeneration, where in some cases they can be used in heads of less than 20 m.
- *High flow rates*: the number of jets can be increased, and the rotor diameter can be reduced.

Figure 1.3 shows a blueprint of a PHIT with four jets, with the main component names, along with the hydraulic brake, which in this case is an injector. When in operation, it launches a jet onto the back of the blades to brake the rotor.

Figure 1.4 shows a longitudinal section of a Pelton Generator Group (PGG), on the vertical axis, with the main mechanical and electrical components named.

The PHIT uses a Velocity Regulator (VR) that allows it to be operated under load variations with the so-called double regulation. This adjustment allows variations in GG rotation to stay within established ranges when load variations occur, even if these load variations are sudden. To do this, first, the VR acts on the deflector,

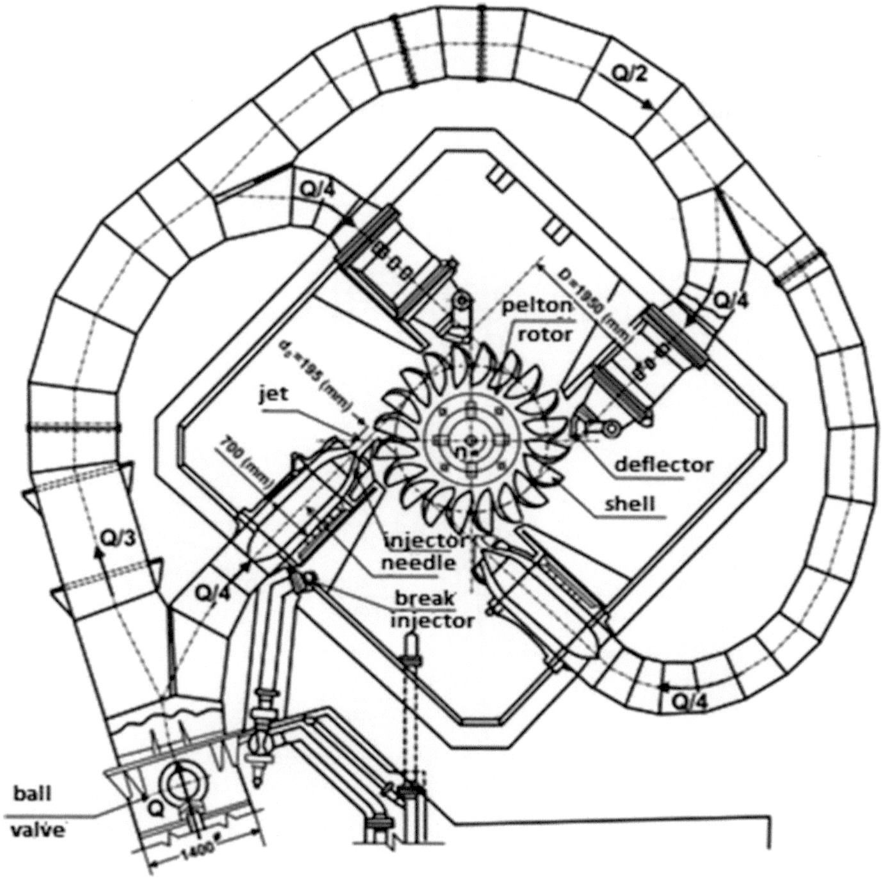

Fig. 1.3 PHIT blueprint with four jets, on the vertical axis, with the main components named

deflecting part of the flow, thereby adapting the hydraulic energy to the electrical energy required by the load with rotation variation within the range. Once this is done, the VR moves a needle within a predicted time, so that the pressure does not rise above predicted calculations (water hammer), and the flow supplied to the rotor supplies hydraulic energy to overcome losses and to supply the electrical energy required by the load. Figure 1.5 shows how the upper deflector acts before the needle moves under both reduced and increased load regimes.

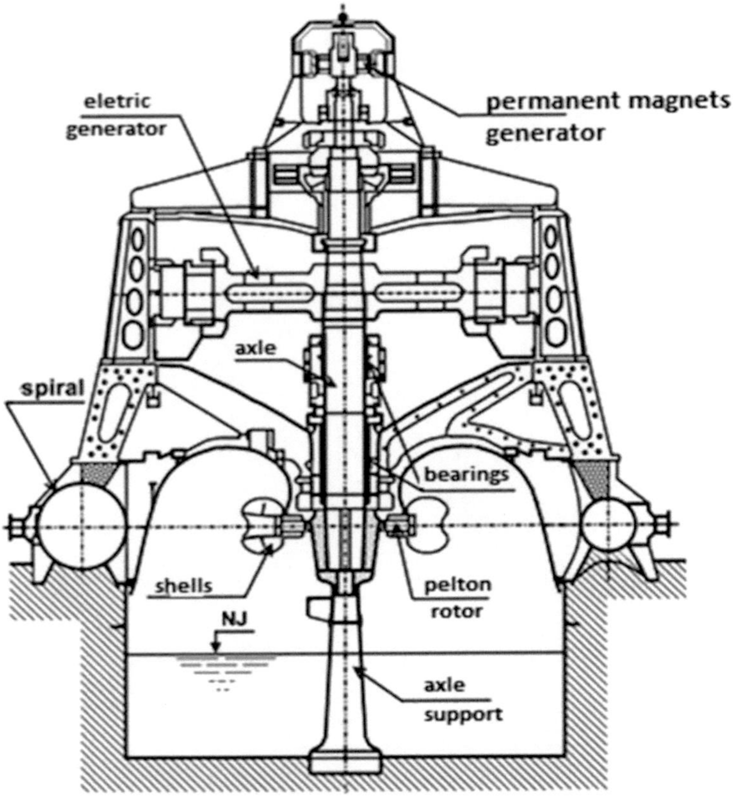

Fig. 1.4 Longitudinal section of a Generator Group (GG), on the vertical axis, of a PHIT, with the main components named

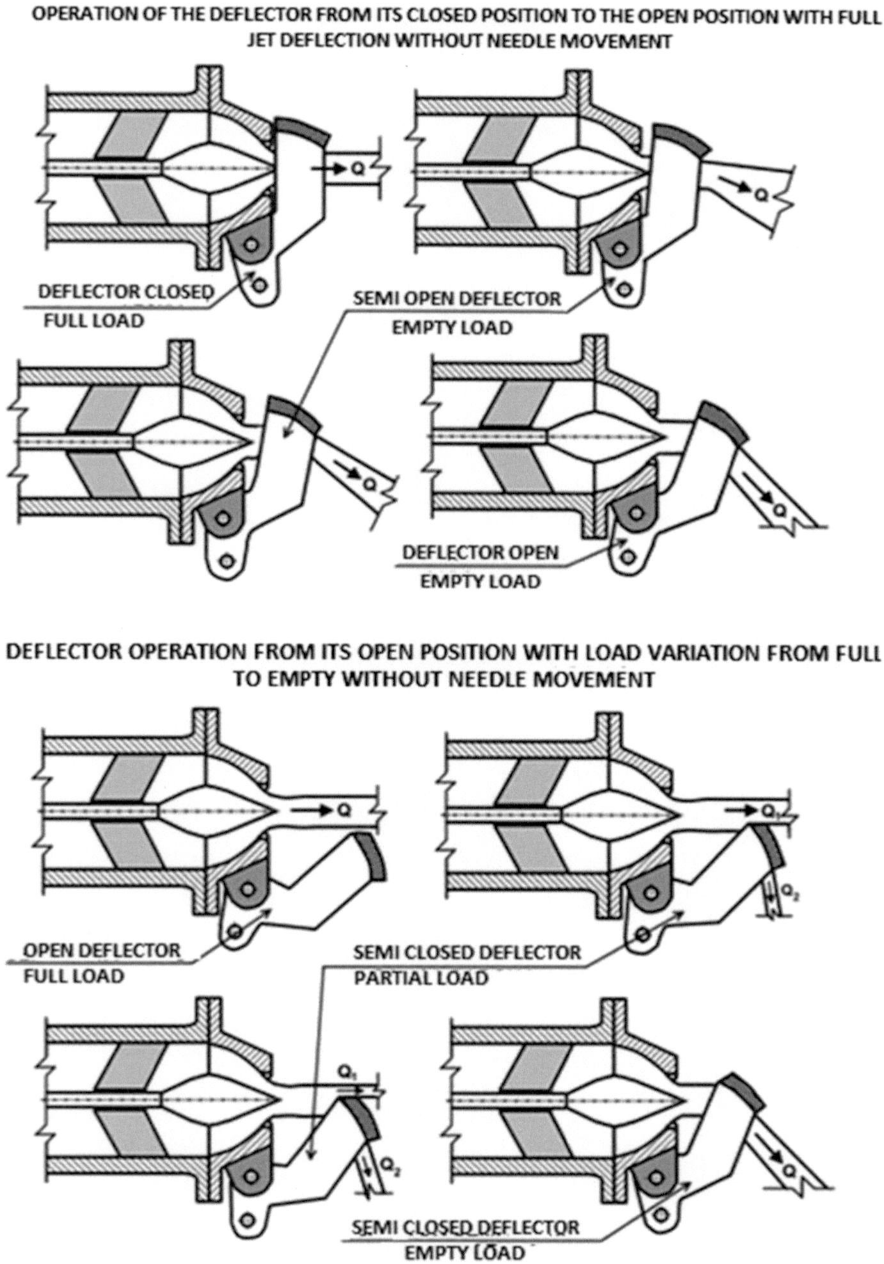

Fig. 1.5 Schematics showing the action of the upper deflector for reducing and increasing the PHIT load

Chapter 2
Considerations for Sizing Hydraulic Impulse Turbines (HITs)

A somewhat elementary knowledge of the Principles of Mechanics is essential when undertaking projects that employ Hydraulic Impulse Turbines (HITs), i.e., principles on the conservation of mass, energy, the volume of movements, the momentum of the volumes of movements, and energy degradation, since all these concepts are applied to studies on HITs.

The objective of this chapter is to synthesize knowledge on water flows, as ideal and real fluids acting on the two main components of HITs, i.e., the stator and the rotor. This water flow between the inlet and outlet of HITs is known as the Hydraulic Flow Path (HFP).

We will begin by studying the stator, i.e., a blade and rotor. The blade is generally represented using static two-dimensional geometry, while the rotor is usually represented using mobile two-dimensional geometry that moves horizontally at a constant speed.

2.1 Ideal Fluid Flows for an Infinitesimally Small Blade

Figure 2.1a shows a blade section, a component of an impulse turbine, that is fixed relative to the flows, which may or may not rotate around an axis.

At the entrance edge of the blade, an angle is formed by line α_1^o, tangent to the Line of Average Flow or better, the Average Streamline (AS) without shocks, corresponding to point 1 on the horizontal line, while angle α_2^o corresponds to the exit edge of the blade at *point* 2 on the corresponding horizontal line. Figure 2.1b shows an infinitesimally small blade coinciding with the AS, i.e., without shocks or friction between the fluid and the blade. In this case, the average flow speed c_1 is the same at the inlet and outlet, and only the direction of the flow changes i.e., α_1^o to α_2^o.

In Fig. 2.1c, a line r is made at point 1 with a horizontal direction $\alpha_r^o < \alpha_1^o$, where α_r^o is the angle formed by the path of the flow without shocks with the direction with

© The Author(s), under exclusive license to Springer Nature Switzerland AG 2025
Z. de Souza, G. L. Tiago Filho, *Pelton and Crossflow Impulse Hydraulic Turbines*,
https://doi.org/10.1007/978-3-031-87086-6_2

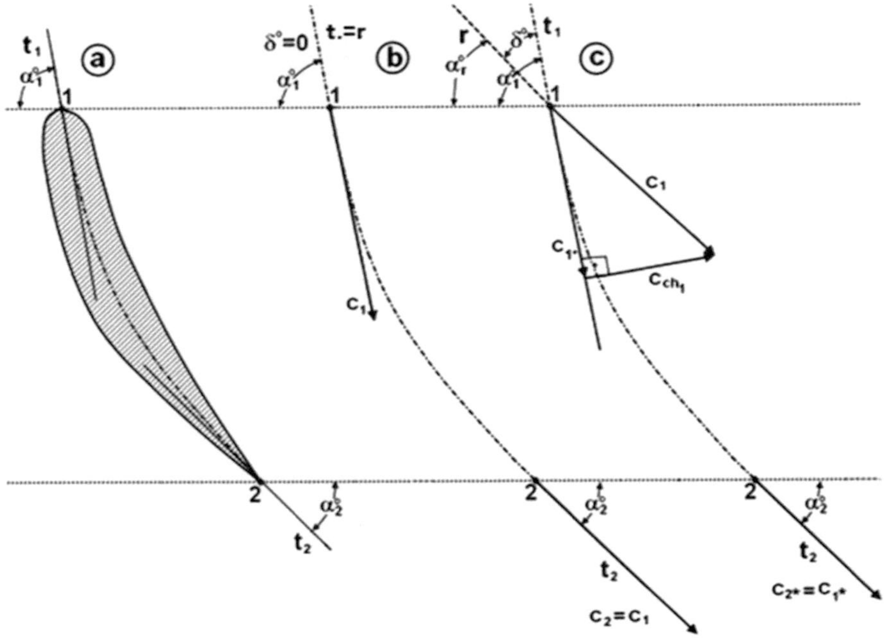

Fig. 2.1 Ideal fluid flows on a blade with and without inlet shocks

shocks at the rotor inlet, thereby giving δ^o, or the angle of the injector's reach, resulting in c, the velocity, that must flow along the AS, which will be projected in the direction of tangent t_1 at point 1, at the entrance edge of the blade. Therefore, the modulus is $c_{1*} < c_1$, resulting in c_{ch_1}, or the shock component.

Under the assumption of frictionless flow, the absolute velocity remains invariant in magnitude until the fluid exits the blade at point 2, undergoing solely a change in its direction. Consequently, Equation (2.1) provides a quantification of the specific work losses occurring at the blade inlet, which arise from kinetic energy shocks.

$$Y_{p_{ch_1}} = \frac{c_{ch_1}^2}{2} = \frac{c_1^2}{2} - \frac{c_{1*}^2}{2} \text{ (J/kg)} \tag{2.1}$$

2.2 Ideal Fluid Flows Without Shocks to an Infinitesimally Small Blade

2.2.1 Considerations

In general, when studying shockless flows for ideal fluids on infinitesimally small blades, the AS is known, at angle β_4^o at the inlet, and at angle β_5^o at the outlet, along

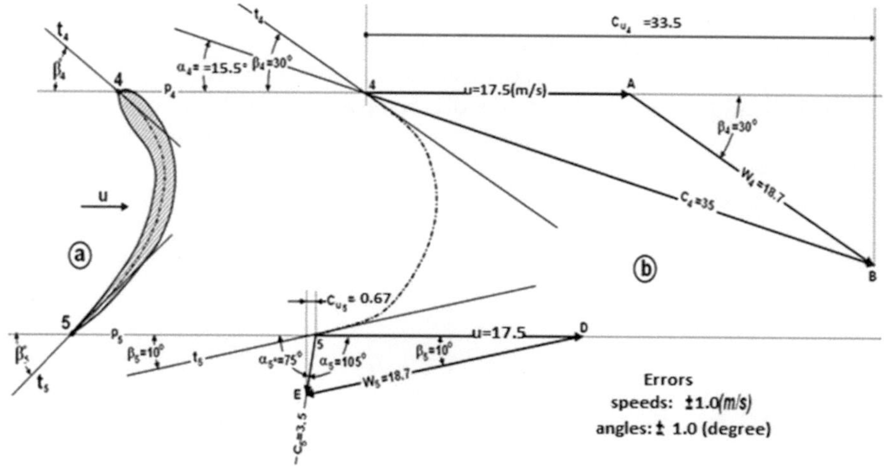

Fig. 2.2 Ideal fluid flows on a blade without shocks, geometrically calculating the velocity triangles

with the tangential or displacement blade speed u (m/s) and the absolute module speed at input C_4 m/s, as shown in Fig. 2.2.

Given these data, the objective then becomes determining the characteristics of the vectors that comprise the velocity triangles, and their angles at point 4 and point 5 (Fig. 2.2), and then to determine the specific theoretical work delivered to the blade, the corresponding internal yield, and its maximum value. These objectives can be achieved either geometrically, or analytically, as shown below.

2.2.2 *Geometrically*

One can draw the scaled velocity triangles at entrance edge (4), and at exit edge (5) of the AS (Fig. 2.2a), and the characteristic values (Fig. 2.2b), using the following AS data:

- $C_4 = 35$ (m/s)
- $u = 17.5$ (m/s)
- $\beta_4^o = 30(^\circ)$
- $\beta_5^o = 10(^\circ)$

… which can be obtained by following these steps:

1. Draw parallel lines p_4 and p_5, and then draw the AS to obtain input *point* 4, and output *point* 5.
2. Draw line t_4 and t_5 tangent to the AS, at point 4 and point 5, respectively, using angles β_4^o and β_5^o.

3. Draw the tangential velocity vectors u (m/s) horizontally, in scale, from point 4 and point 5, to obtain point A and point D, respectively.
4. Mark angle $\beta_4 = 30°$ from point A and horizontal line p_4 to the w_4 direction.
5. Obtain point B by placing the center of a compass at point 4, and with a velocity fixed scale, use the absolute velocity c_4 (m/s), to thereby obtain the velocity triangle 4; A; B; 4 for the AS at inlet 4 on the blade.
6. Using this scale, measure

$$w_4 = 18.7\ (\mathrm{m/s})$$

$$C_{u_4} = 33.5\ (\mathrm{m/s})$$
$$\alpha_4 = 15.5\,°$$

7. From the extremity of u over p_5, at point D, draw a line making angle $\beta_5 = 10^o$ with horizontal line p_5.
8. From point D on the straight line, mark $w_4 = w_5 = 18.7$ (m/s) in scale, thereby obtaining point E, and thus, velocity triangle 5; D; E; 5 for the AS at outlet 5 on the blade.
9. Using this scale, measure

$$c_5 = 3.5\ (\mathrm{m/s})$$

$$C_{u_5} = -0.65(\mathrm{m/s})$$
$$\alpha_5 = 105\,°$$
$$\alpha_5 = 75\,°$$

10. Using Euler's Equation for Hydraulic Turbines, calculate the theoretical specific work without shocks to the blade:

$$Y_{\mathrm{blade}_t} u \cdot (c_{u_4} - c_{u_5}) = 17.5 \cdot (33.5 + 0.65) = 597.63\ (\mathrm{J/kg})$$

11. Calculate the internal efficiency of the blade without shocks:

$$\eta_{\mathrm{blade}} = \frac{Y_{\mathrm{blade}_t}}{c_4^2/2} = \frac{2 \cdot 597.63}{35^2} = 0.9757 = 97.57\%$$

2.2.3 Analytically

To calculate an analytical solution, we will use the triangles from Fig. 2.3, and the data from the previous subitem, i.e.:

- $c_4 = 35$ (m/s)
- $u = 17.5$ (m/s)
- $\beta_5^o = 30\,°$
- $\beta_5^o = 10\,°$

Observing the following steps:

1. Calculate α_4 from triangles 4; D; B; 4 and A; D; B; A, where

$$u + w_4 \cdot \cos\beta_4 = c_4 \cdot \cos\alpha_4$$

and

$$w_4 \cdot \sin\beta_4 = c_4 \cdot \sin\alpha_4$$

thereby giving Equation,

$$\cos\alpha_4 - \frac{\sin\ \alpha_4}{tg\beta_4} - \frac{u}{c_4} = 0$$

Substituting the known values, and making iterations to Equation

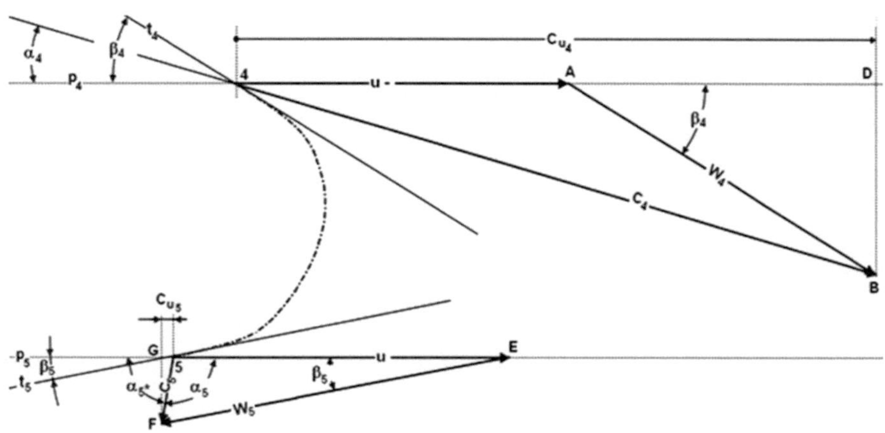

Fig. 2.3 Ideal fluid flows on a blade without shocks, analytically calculating the velocity triangles, the specific work, and the blade efficiency

$$\cos\alpha_4 - \frac{\sin\ \alpha_4}{tg30} - \frac{17.5}{35} = 0$$

will result in

$$\alpha_4 = 15^{\circ}$$

2. Calculate the relative velocity at the inlet of the blade using the expression obtained for triangle 4; *D*; *B*; 4;

$$w_4 = \sqrt{u^2 + c_4^2 - 2 \cdot u \cdot c_4 \cdot \cos\alpha_4}$$

   Substituting the values will result in

$$w_4 = \sqrt{17.5^2 + 35^2 - 2 \cdot 17.5 \cdot 35 \cdot \cos 15.54} = 18.74\ (\mathrm{m/s})$$

3. Since $w_4 = w_5$ calculate the absolute velocity at the blade outlet expressed by triangle 5; *E*; *F*; 5;

$$c_5 = \sqrt{u^2 + w_4^2 - 2 \cdot u \cdot w_4 \cdot \cos\beta_5}$$

   Substituting the values will result in

$$c_5 = \sqrt{17.5^2 + 18.74^2 - 2 \cdot 17.5 \cdot 18.74 \cdot \cos 10} = -3.39\ (\mathrm{m/s})$$

4. From triangles *G*; *E*; *F*; *G* and 5; *F*; *G*; 5; calculate

$$\alpha_{5^*} = \text{arc}\ \sin\left(\frac{W_4}{C_5} .\sin\ \beta_5\right) = \text{arc}\ \sin\left(\frac{18.74}{3.39} .\sin\ 10^{\circ}\right) = 73.63^{\circ}$$

   and

$$\alpha_5 = 180 - 73.73 = 106.27^{\circ}$$

5. Using Euler's Equation for Hydraulic Turbines, calculate the theoretical specific work, without shocks to the blade as:

$$Y_{\text{blade}_t} u \cdot (c_{u_4} - c_{u_5}) = u \cdot (c_4 \cdot \cos\alpha_4 + c_5 \cdot \cos\alpha_5^*) \rightarrow$$

$$Y_{\text{blade}_t} = 17.5 \cdot (35 \cdot \cos 15.54 + 3.39 \cdot \cos 73.73) = 606.73 \text{ (J/kg)}$$

6. Calculate the internal yield of the blade, without shocks, via Y_{blade_t},

$$\eta_{\text{blade}} = \frac{Y_{\text{blade}_t}}{c_4^2/2} = \frac{606.73}{35^2/2} = 0.9906 = 99.06\ (\%)$$

2.2.4 *Comments*

The specific theoretical work delivered to the blade by the fluid, as per Euler's Equation for hydraulic turbines, can be written as follows:

$$Y_{\text{blade}_t} = u \cdot (c_{4_u} - c_{5_u}) = u \cdot (c_4 \cdot \cos\alpha_4 + c_5 \cdot \cos\alpha_5) = u \cdot w_4 \cdot (\cos\beta_4 + \cos\beta_5) \text{ (J/kg)}.$$

This expression shows that the lower the angles β_4 and β_5, the higher the Y_{blade_t} for the chosen impulse blade geometry.

Thus, the theoretical efficiency of the flow across the blade η_{blade} depends on the geometry, and for the AS between inlet 4 and inlet 5 which is determined via:

$$\eta_{\text{blade}} = \frac{2 \cdot u \cdot w_4 \cdot (\cos\beta_4 + \cos\beta_5)}{c_4^2}$$

which shows us that, for η_{blade} to be maximum, angles β_4 and β_5 should be zero, i.e.,

$$c_4 = u + w_4$$

or

$$w_4 = c_4 - u$$

These considerations applied to η_{blade} allow us to write the following expression:

$$\eta_{\text{blade}} = 4 \cdot \frac{u}{c_4} \cdot \left(1 - \frac{u}{c_4}\right)$$

Taking the derivative of this expression, with respect to u/c_4, and setting the result to zero, we thereby obtain the following expressions:

$$\frac{\partial Y_{\text{blade}_{t\max}}}{\partial(u/c_4)} = 2 - 4 \cdot \frac{u}{c_4} = 0 \rightarrow \frac{u}{c_4} = \frac{1}{2} \rightarrow u = \frac{c_4}{2}$$

We will show that generally, and operationally, we should have $\beta_4 \geq 0$ and $\beta_5 > 0$, which will mean that we will not be able to obtain $\eta_{\text{blade}} = 1$, not even for ideal fluid flows without losses.

Regarding the differences between the geometric method and analytical method for obtaining the kinematic flow characteristics, we will see that the geometric method always results in greater uncertainties, as exemplified by the results already shown here.

2.3 Ideal Fluid Flows with Shocks to an Infinitesimally Small Blade

2.3.1 Considerations

When studying ideal fluid flows with shocks to an infinitesimally small blade, in principle, the LAF is known for both the inlet angle β_4^o, and the outlet angle β_5^o, along with the tangential or displacement velocity of the blade u (m/s), and the absolute inlet velocity c_4 (m/s) modulus and direction.

Using these data, we can then draw Fig. 2.4, starting from the AS line between horizontal lines p_4 and p_5. Then, at inlet 4, and outlet 5, we can draw lines t_4 and t_5 tangent to the AS, along with triangle 4; A; B; 4; with shocks, via u and c_4.

Since triangle 4; B; D; 4; is equal to triangle 4; A; B; 4; we can project $w_{4_{ch*}}$ in the direction without shocks, and from t_4 at inlet 4 of the LAF we can project w_{4*}, which is the relative velocity, and remains constant to outlet 5 of the LAF, without considering other losses.

The shock w_{ch}, is another component of velocity triangle 4; E; D; 4; that results in specific work losses, in function of shocks at the blade inlet $Y_{p_{\text{blade}}} = 0.5 \cdot w_{ch}^2$ (J/kg). Horizontal line segment $\bar{DB} = u$ can be drawn from D, thereby obtaining velocity triangle 4; A; F; 4; which is what would happen at the inlet if the flow was shockless at $c_{4*} = c_4$.

Since $w_{5*} = w_{4*}$ the velocity triangles at the outlet can be drawn at point 5 of the LAF.

2.3.2 Equations

Using input data from the previous subitem ($\beta_4^o, \beta_5^o, u, c_4, \alpha_{4*}$) we can then determine the vector characteristics comprising the velocity triangles, considering shocks at the inlet of the AS, and the angles at point 4 and point 5, and the theoretical specific work

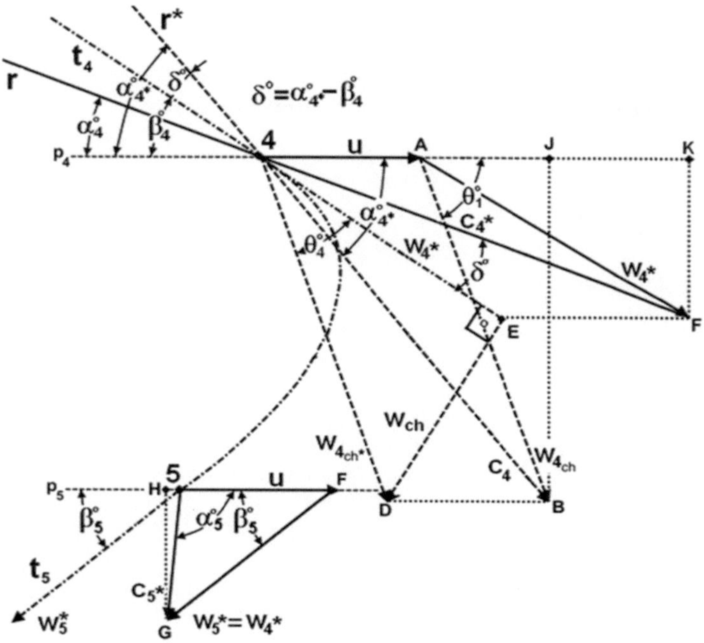

Fig. 2.4 Ideal fluid flows on the blade with shocks, analytically calculating the speed triangles, the specific work, and the blade efficiency

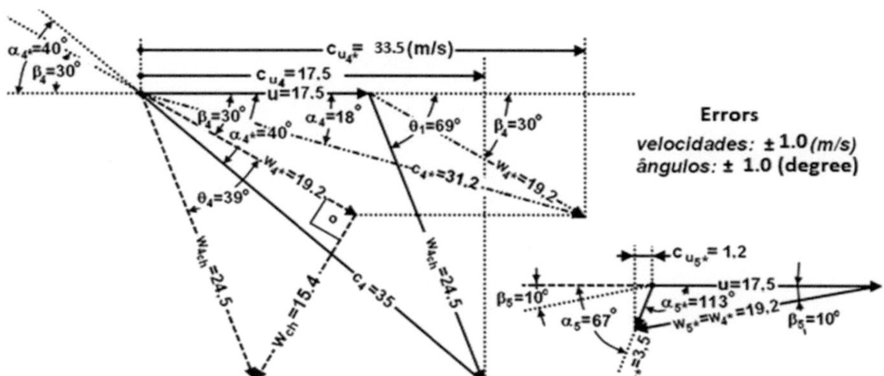

Fig. 2.5 Ideal fluid flows on the blade with shocks, drawing the speed triangles

delivered to the blade, and its corresponding internal efficiency. Similar to cases without shock, we will try to determine this either geometrically or analytically. Figure 2.5 shows the results for the geometric method. For practice's sake, one should try listing the required steps that would lead to these results, using Sect. 2.2.2 as a reference.

The study of—and consequent equations for—the calculations using the analytical method are based on Fig. 2.4.

The theoretical relative velocity with shocks at the blade inlet can be calculated using the data and speed triangle 4; *A*; *B*; *A*; from Fig. 2.4 as:

$$w_{4_{ch}} = \sqrt{c_4^2 + u^2 - 2 \cdot u \cdot c_4 \cdot \cos \alpha_{4*}} \tag{2.2}$$

Since $c_4 \cdot \sin\ \alpha_{4*} = w_{4_{ch}} \cdot \sin\ \theta_1$, the resulting angle between the directions of the velocity vectors c_4 and u is:

$$\theta_1 = \arcsin\left(\frac{c_4}{w_{4_{ch}}} \cdot \sin\ \alpha_{4*}\right)(\text{degrees}) \tag{2.3}$$

The angle θ_4 that the direction of the vectors w_{ch} and w_{4*} form among themselves is $\theta_1 = \theta_4 + \alpha_{4*} - \delta = \theta_4 + \alpha_{4*} - (\alpha_{4*} - \beta_4)$, resulting in:

$$\theta_4 = \theta_1 - \beta_4\ (\text{degrees}) \tag{2.4}$$

Considering shock w_4^*, and the respective shock component w_{ch}, the relative velocity can be calculated as follows:

$$w_{4*} = w_{4_{ch}} \cdot \cos \theta_4 \text{ and } w_{ch} = w_{4_{ch}} \cdot \sin\ \theta_4\ (\text{m/s}) \tag{2.5}$$

Considering shock, the absolute velocity can be calculated as follows:

$$c_{4*} = w_{4*} \cdot \frac{\sin\ \beta_4}{\sin\ \alpha_4} \tag{2.6}$$

The angle α_4 between the directions of velocity vectors c_4^* and u, considering $c_{4*} = c_4$, results in:

$$\alpha_4 = \arccos\left(\frac{c_4^2 + u^2 - w_{4*}^2}{2 \cdot u \cdot c_4}\right)\ (\text{degrees}) \tag{2.7}$$

The absolute velocity at the blade outlet is expressed as follows:

$$c_{5*} = \sqrt{w_{5*}^2 + u^2 - 2 \cdot u \cdot w_{5*} \cdot \cos \beta_5}\ (\text{m/s}) \tag{2.8}$$

The angle that the direction of velocity c_5^* with velocity u takes is calculated as:

$$\alpha_5 = \arccos\left(\frac{w_{5*}^2 - u^2 - c_{5*}^2}{2 \cdot u \cdot c_{5*}^2}\right) \text{ (degrees)} \tag{2.9}$$

The specific energy lost, per shock, at the blade inlet can be calculated as:

$$Y_{p_{ch}} = 0.5 \cdot w_{ch}^2 \text{ (J/kg)} \tag{2.10}$$

The specific work delivered from the ideal fluid to the blade, considering shock, can be calculated as:

$$Y_{\text{blade}_{t_{ch}}} = u \cdot [c_4 \cdot \cos \alpha_{4*} - c_{5*} \cdot \cos(180 - \alpha_5)] \text{ (J/kg)} \tag{2.11}$$

The yield with shocks at the blade inlet can be calculated as:

$$\eta_{\text{blade}_{cht}} = \frac{2 \cdot Y_{\text{blade}_{t_{ch}}}}{c_4^2} \tag{2.12}$$

The specific work delivered by the ideal fluid to the blade, considering shocks, can be calculated as:

$$Y_{\text{blade}_t} = u \cdot [c_4 \cdot \cos \alpha_4 - c_{5*} \cdot \cos(180 - \alpha_5)] \text{ (J/kg)} \tag{2.13}$$

The yield, without considering shocks at the blade inlet, can be calculated as:

$$\eta_{\text{blade}} = \frac{2 \cdot Y_{\text{blade}_4}}{c_4^2} \tag{2.14}$$

2.3.3 *Algorithm and Validation*

2.3.3.1 Input Data

The input data, in this case, will be the same input data from the previous subitem, i.e.:

- $c_4 = 35$ (m/s)
- $u = 17.5$ (m/s)
- $\beta_4^o = 30\,°$
- $\beta_5^o = 10\,°$ and $\alpha_{4*} = 40°$ added from the c_4 direction.

2.3.3.2 Step-By-Step Process with Validations

Calculate:

1. The theoretical relative velocity with shocks at the blade inlet as:

$$w_{4_{ch}} = \sqrt{c_4^2 + u^2 - 2 \cdot u \cdot c_4 \cdot \cos\alpha_{4*}} = \sqrt{35^2 + 17.5^2 - 2 \cdot 17.5 \cdot 35 \cdot \cos 40}$$
$$= 24.348\ (\mathrm{m/s})$$

2. The angle that the $w_{4_{ch}}$ direction forms with the u direction, as:

$$\theta_1 = \text{arc}\ \sin\left(\frac{c_4}{w_{4_{ch}}} \cdot \sin\alpha_{4*}\right) = \text{arc}\ \sin\left(\frac{35}{24.348} \cdot \sin 40\right) = 67.52\ (^\circ)$$

3. The angle that the $w_{4_{ch}}$ direction forms with the w_{4_*} direction, as:

$$\theta_4 = \theta_1 - \beta_4 = 67.52 - 30 = 37.52\ (^\circ)$$

4. The relative velocity, considering shock, as:

$$w_{4*} = w_{4_{ch}} \cdot \cos\theta_4 = 24.348 \cdot \cos 37.52 = 19.311\ (\mathrm{m/s})$$

5. The absolute velocity, considering shock, as:

$$c_{4*} = w_{4*} \cdot \frac{\sin\beta_4}{\sin\alpha_4} = 19.311 \cdot \frac{\sin 30}{\sin 18} = 31.246\ (\mathrm{m/s})$$

6. The shock component, as:

$$w_{ch} = w_{4_{ch}} \cdot \sin\theta_4 = 24.348 \cdot \sin 37.52 = 14.829$$

7. The angle between the directions of the vector velocities c_{4*} and u as:

$$\alpha_4 = \arccos\left(\frac{c_4^2 + u^2 - w_4^{*2}}{2 \cdot u \cdot c_4}\right) = \arccos\left(\frac{35^2 + 17.5^2 - 19.311^2}{2 \cdot 17.5 \cdot 35}\right) = 18.99\ (^\circ)$$

8. The absolute velocity at the blade outlet, as:

$$c_{5*} = \sqrt{w_{5*}^2 + u^2 - 2 \cdot u \cdot w_{5*} \cdot \cos\beta_5}$$
$$= \sqrt{19.311^2 + 17.5^2 - 2 \cdot 17.5 \cdot 19.311 \cdot \cos 10} = 3.681$$

9. The angle of the direction of velocity c_{5*} with the velocity u, as:

$$\alpha_5 = \arccos\left(\frac{w_{5*}^2 - u^2 - c_{5*}^2}{2 \cdot u \cdot c_{5*}^2}\right) = \arccos\left(\frac{19.311^2 - 17.5^2 - 3.681^2}{2 \cdot 17.5 \cdot 3.681}\right)$$
$$= 65.67 \text{ (degrees)}$$

10. The specific energy lost per shock at the blade inlet, as:

$$Y_{p_{ch}} = 0.5 \cdot w_{\mathrm{ch}}^2 \text{ or } Y_{p_{ch}} = 0.5 \cdot 14.829^2 = 109.950 \text{ (J/kg)}$$

11. The specific work delivered from the ideal fluid to the blade, considering shock, as:

$$Y_{\mathrm{blade}_{t_{ch}}} = u \cdot [c_4 \cdot \cos \alpha_{4*} - c_{5*} \cdot \cos(180 - \alpha_5)] = 17.5$$
$$\cdot [35 \cdot \cos 40 - 3.681 \cdot \cos(180 - 65.67)] = 495.742 \text{ (J/kg)}$$

12. The yield, with shocks at the entrance of the blade, as:

$$\eta_{\mathrm{blade}_{cht}} = \frac{2 \cdot Y_{\mathrm{blade}_{t_{ch}}}}{c_4^2} = \frac{2 \cdot 495.742}{35^2} = 0.8094 = 80.94\ (\%)$$

13. The specific work delivered from the ideal fluid to the blade, without shocks, as:

$$Y_{\mathrm{blade}_t} = u \cdot [c_4 \cdot \cos \alpha_4 - c_{5*} \cdot \cos(180 - \alpha_5)] \text{ or } Y_{\mathrm{blade}_t} = 17.5$$
$$\cdot [35 \cdot \cos 18.99 - 3.681 \cdot \cos(180 - 65.67)] = 605.704 \text{ (J/kg)}$$

14. The yield, without shocks, at the blade inlet, as:

$$\eta_{\mathrm{blade}} = \frac{2 \cdot Y_{\mathrm{blade}_4}}{c_4^2} = \frac{2 \cdot 605.704}{35^2} = 0.9889 = 98.89\ (\%)$$

2.3.4 Comments

- Upon comparing the specific theoretical work with and without shocks, $Y_{\mathrm{blade}_t} = 605.704$ and $Y_{\mathrm{blade}_{t_{ch}}} = 495.742$ (J/kg), we can see that, without shocks, the result was 22.18 (%) greater, with an equal increase in blade yield.
- To really drive home this concept, the reader should study the analogous case by imposing the following condition $\alpha_{4*} < \beta_4$, in Fig. 2.4, for example.
- One should observe that the shock, which is at its minimum at the projected point or even outside it, results in large working load variations with consistent

efficiency. This will also be true of Pelton Turbines. However, this is not true of other impulse turbines.

2.4 Graded Fluid Flows

Figure 2.6 shows a profile view of a segment of a flat set of impulse blades that move horizontally at speed u (m/s), under the force of a moving fluid without shocks.

This is called a moving graded geometry, given the finite pitch t between the blades, theoretically forcing the fluid to flow congruently with the channel they form, resulting in the AS, and the velocity triangles at inlet point 4 and outlet point 5, as per Fig. 2.6.

The finite thickness of the blades creates a bottleneck in the cross-section of the graded components, which can be determined for inlet 4 and outlet 5 using triangles a and b, as per Fig. 2.6, as follows:

$$f_{e_4} = \frac{t_4 - e_{t_4}}{t_4} = 1 - \frac{e_4}{t_4 \cdot \sin \beta_4}$$

and

$$f_{e_5} = \frac{t_5 - e_{t_5}}{t_5} = 1 - \frac{e_5}{t_5 \cdot \sin \beta_5} \tag{2.15}$$

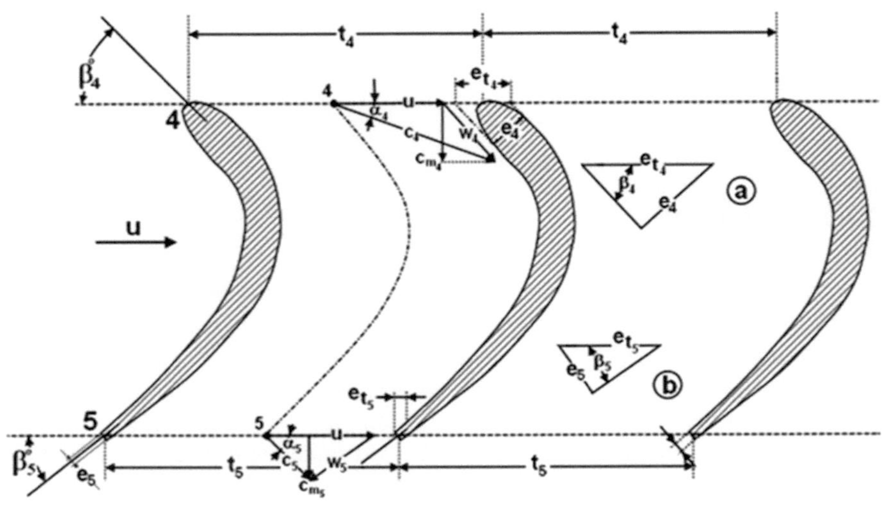

Fig. 2.6 Moving graded geometry flow

The symbols used in the expression, for inlet 4 and outlet 5, respectively, represent:

- t = the theoretical step between the blades;
- e = the blade thickness;
- e_t = the blade thickness projected in the tangential direction;
- β = the angle between the AS and the tangential directions (velocity u).

The variation of the narrowing factor between 4 and 5 can be said to be linear, as a first approximation, for simplified calculation.where:

- $Q_r(m^3/s)$ = is the water flow through the graded components;
- A (m^2) = is the cross-sectional area of the flow before and after the graded components; and
- c_m (m/s) = is the normal velocity in the section, called the meridional velocity.

According to the principle of conservation of mass, one can write the following equation:

$$Q_r = c_m \cdot A = \frac{c_m}{f_{e_4}} \cdot A \cdot f_{e_4} = \frac{c_m}{f_{e_5}} \cdot A \cdot f_{e_5} = c_{m_4} \cdot A_4 = c_{m_5} \cdot A_5 \tag{2.16}$$

2.5 Real Fluid Flows: Water

2.5.1 Considerations

Real fluid flows will be studied by modeling turbomachines comprising graded injector stators to transform pressure energy into kinetic energy. This is done in series with graded impulse motors with rotors that receive kinetic energy at constant pressure from a flowing fluid, and transform this into mechanical work. Figure 2.7 shows a flattened view of this geometric structure.

An AS was drawn in this figure, identifying characteristics at points 0 to 6, along with triangles with average speeds for rotor inlet 4 and outlet 5.

2.5.2 Conceptualizing Energy Losses and Yields

The specific energy losses that occur between 0 and 6, as per Fig. 2.7, at the projected maximum efficiency point, are as follows:

- $Y_{i_{p_0}} = Y_{i_{p_e}}$ (J/kg), internal specific work losses from turbulent flows and variations in cross-sections when passing through the empty space point (0), to the blade point (1). Average kinetic energy increases occur at this stator inlet region.

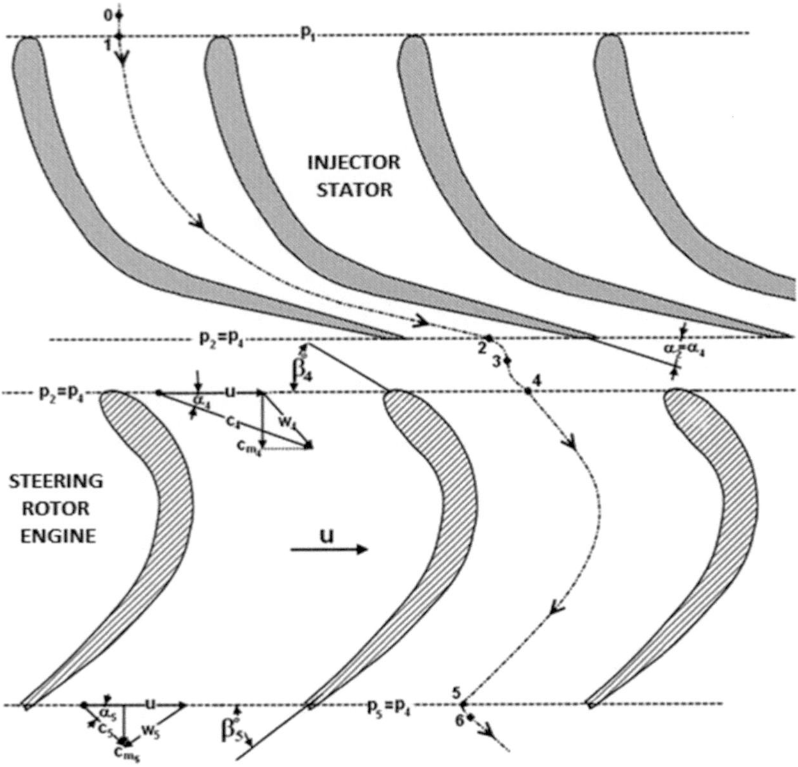

Fig. 2.7 Real fluid flows—water in graded injector stators and an impulse turbine rotor

- $Y_{i_{p_{1;2}}} = Y_{i_p}$ (J/kg), internal specific work losses from turbulent flows and friction with walls that define the Hydraulic Flow Path (HPF) in the stator (injector) channels between inlet 1 and outlet 2.
- $Y_{i_{p_i}}$ (J/kg), internal specific work losses from turbulence and variations in the transversal section of the flow in region 3, between the stator outlet point 2, and the rotor inlet at point 4. Initially, there are average kinetic energy losses in region 3, from point 2 to point 3, which then increases from point 3 to point 4.
- $Y_{i_{p_{4;5}}} = Y_{i_{p_r}}$ (J/kg), internal specific work losses from turbulent flows and friction with the walls that define the HFP in the stator (injector) channels between inlet point 4 and outlet *point* 5.
- $Y_{i_{p_6}} = Y_{p_s}$ (J/kg), internal specific work losses from turbulence and variations in the transversal section of the flow, and average kinetic energy losses at the rotor exit point at point 5 and region 6.

One must also consider flow mass losses in the spaces that define the stationary (housing) and rotating (rotor) systems, in addition to the specific work losses. These flow mass losses in impulse turbines are leakage flow volumes, Q_f(m^3/s), that do not delivery energy to the rotors, resulting in $Q_r = Q - Q_f$.

The available hydraulic power at the impulse turbine inlet can be obtained from the specific mass of water $\rho = 1000$ (kh/m^3), resulting in:

$$P = g \cdot H \cdot Q = Y \cdot Q \ (kW) \tag{2.17}$$

Taking P as the denominator, along with the conceptualized losses, the following expressions can be established for each of the respective yields:the internal entrance yield:

$$\eta_{i_e} = 1 - \frac{Y_{i_{p_e}}}{Y_{i_0}} \tag{2.18}$$

the internal injector yield:

$$\eta_{i_{in}} = 1 - \frac{Y_{i_{p_{in}}}}{Y_{i_0}} \tag{2.19}$$

interstitial internal yield:

$$\eta_{i_3} = 1 - \frac{Y_{i_{p_3}}}{Y_{i_0}} \tag{2.20}$$

the internal rotor yield:

$$\eta_{i_r} = 1 - \frac{Y_{i_{p_r}} \cdot Q_r}{Y_{i_0} \cdot Q} \tag{2.21}$$

the internal exit yield:

$$\eta_{i_s} = 1 - \frac{Y_{i_{p_6}}}{Y_{i_0}} \tag{2.22}$$

the outlet yield:

$$\eta_f = 1 - \frac{Q_f}{Q} \tag{2.23}$$

Thus, the internal yield will be:
the internal yield:

$$\eta_{\mathrm{f}} = 1 - \frac{Q_{\mathrm{f}}}{Q} \tag{2.24}$$

2.5.3 Determining the Energy Losses

One can only determine the most probable specific work losses and respective yields by testing models in the laboratory, or by testing installed impulse turbine prototypes. These results will give numerical coefficients and characteristic relationships that are, in principle, dimensionless, with minimum, average, and maximum values that can be used in studies on different impulse turbine geometries.

When flow channels i.e., injector and rotor channels, are geometrically well-defined, losses $Y_{p_{\mathrm{i}}}$ and $Y_{p_{\mathrm{r}}}$ can be established for studies and projects by using the universal head loss Formula:

$$Y_p = 0.8106 \cdot \lambda_p \cdot Q^{*^2} \cdot L \cdot D^{-5} \tag{2.25}$$

Where:

- $Y_p = g\,H_p$ (J/kg)—is the specific energy lost in the injector or rotor channel;
- L (m)—is the length of the *AS* through the injector or rotor channel;
- $D = \frac{4 \cdot A}{U}$ (m)—is the equivalent average diameter;
- A (m^2)—is the midsection drainage area;
- U (m)—is the wet perimeter of the midsection, calculated at the AS midpoint between the injector or rotor channel inlet and outlet;
- $Q^* = Q/z_{\mathrm{i}}$ (m^3/s) —is the flow rate into the injector;
- $Q^* = Q_{\mathrm{r}}/z_{\mathrm{r}}$ (m^3/s) —is the flow rate into rotor channel;
- $\lambda_p(1)$—is the energy loss coefficient, which can be determined using the P. K. Swamee Formula below:

$$\lambda_{\mathrm{p}} = \frac{1{,}325}{\left[\ln\left(\frac{\varepsilon}{3{,}7 \cdot D} + \frac{5{,}74}{Re^{0{,}9}}\right)\right]^2}$$

where:

$$N\frac{4 \cdot Q^*}{\pi \cdot \nu \cdot D}\;{}^{6}\frac{Q^*}{D}\,\mathrm{Re}$$

with

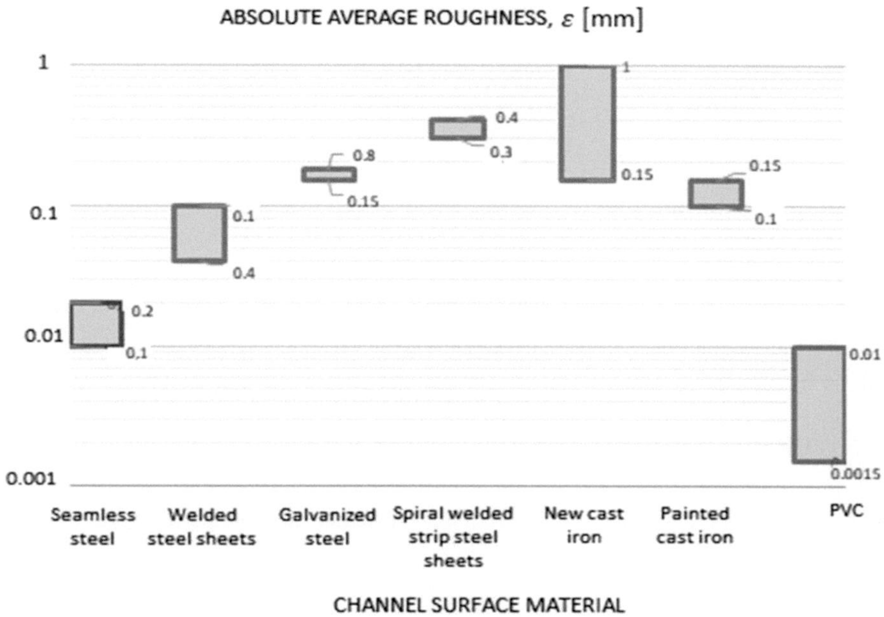

Fig. 2.8 Average Absolute Roughness, ε [mm], for different materials

$$5{\cdot}103 \le NRe \le 108$$

and

$$10^{-6} \le \frac{\varepsilon}{D} \le 10^{-2} \tag{2.26}$$

Where ε (m) is the average absolute roughness of the injector or rotor channel walls, obtained from Tables or Graphs, similar to Fig. 2.8, as a function of the channel wall material.

Also, coefficients that reduce the theoretical average speeds relative to real average speeds can be used for all specific energy losses. For example, the theoretical speed at the injector outlet (Fig. 2.6) can be determined using the following expression:

$$c_{2_t} = \sqrt{2 \cdot Y + c_0^2}\ (\mathrm{m/s}) \tag{2.27}$$

For coefficient ϕ_i, the average real speed would be:

$$c_2 = \phi_i \cdot c_{2_t} (\mathrm{m/s}) \tag{2.28}$$

The energy losses and injector yield would be, respectively:

$$Y_{p_i} = \frac{c_{2_t}^2 - c_2^2}{2} = \frac{c_{2_t}^2}{2} \cdot \left(1 - \phi_i^2\right) \rightarrow \eta_i = 1 - \frac{1 - \phi_i^2}{1 - \left(c_o / c_{2_t}\right)^2} \tag{2.29}$$

For $c_0 = 0$, the result would be:

$$\eta_i = \phi_i^2 \tag{2.30}$$

2.6 Final Comments and Recommendations

The energy losses, yields, and coefficients presented here, along with the general expressions for preliminarily determining them, set the foundations for studying impulse turbines in general, particularly the Banki-Michell (BMHIT) and Pelton (PHIT) turbines, as will be presented in the next chapters of this book.

The theoretical studies on shock, as shown here, are decisive in theoretically analyzing and testing the operational aspects of impulse turbines, particularly $\eta = f(Q)$, with H, and $n =$ *constants*, since heads have a strong influence on generated energy, and on impulse turbine lifespans outside the projected point.

Finally, as was shown here, the operationality of impulse turbines imposes limits to the HFP geometry, meaning that not all available energy can be converted into mechanical work, even under ideal conditions.

Part II
Banki-Michell Hydroelectric Impulse Turbines (BMHIT)

Chapter 3
Banki-Michell Hydroelectric Impulse Turbines (BMHIT), A Theoretical Hydrodynamic Analysis

Again, the reader should possess a somewhat elementary understanding of the Principles of Fluid Mechanics, i.e., the principles of conservation of mass, energy, the quantities of movements, the momentums of the quantities of movements, and energy degradation, to properly understand Banki-Michell Hydraulic Turbine (BMHIT) projects. These principles will allow the reader to establish mathematical expressions for the specific work, the internal theoretical power, and the flow transfers to the BMHIT rotor, and to determine the corresponding mathematics to express the maximized theoretical internal performance, the velocity ratios, and the angles of the velocity triangles.

This chapter lays out the steps for making calculations and numerical applications, and will conduct analyses, make comments, and recommendations on the content presented herein, which ultimately, will lay the foundation for studying real flow through BMHITs and allow us to establish algorithms to calculate and design them.

3.1 The Theoretical Trajectory of the Waterjet

This chapter studies water as an ideal fluid, i.e., a continuous frictionless system without turbulence, both in the interior, and at the limits of the areas bounded by fluid contact with the blades and the walls with the surface, for both absolute and relative flows.

According to the schematic in Fig. 3.1, which shows a cross-section of the BMHIT axis, it basically comprises two fixed components: the injector and its transition piece from the circular to the rectangular section; and one moving component: the rotor.

© The Author(s), under exclusive license to Springer Nature Switzerland AG 2025

Z. de Souza, G. L. Tiago Filho, *Pelton and Crossflow Impulse Hydraulic Turbines*,
https://doi.org/10.1007/978-3-031-87086-6_3

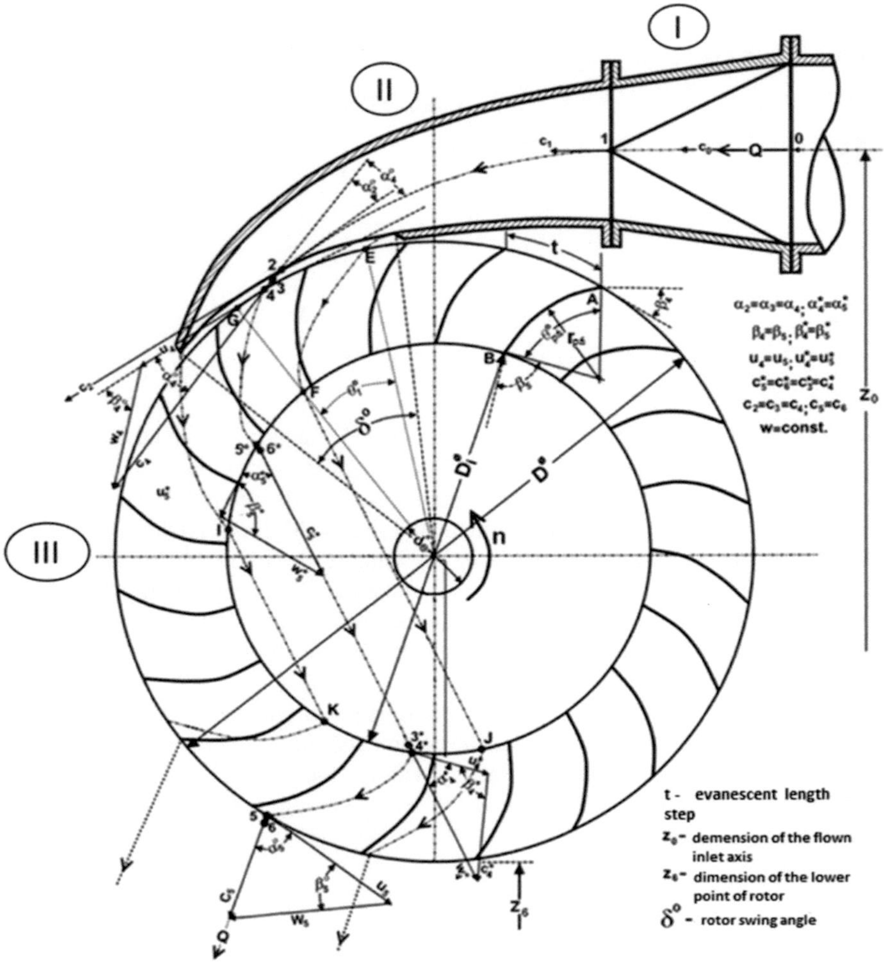

Fig. 3.1 Cross-section through the injector and rotor of a BMHIT, with the ATS drawing of the flow, and its limits

In most cases, a movable paddle is inserted into the injector for flow control, as shown in Fig. 4.1.

The schematic in Fig. 3.1 also shows the Theoretical Hydraulic Flow Path (THFP) from its inlet section 0, to the outlet 6 of the BMHIT. In this passageway, the flow is absolute:

(a) in the transition piece, 0; 1
(b) in the injector 1; 2
(c) in the interstice, 2; 3
(d) inside the rotor, 6 ∗ ; 3∗

After the rotor exit at outlet 6 the flow is relative. It delivers most of its energy to the rotor in its first pass, 4; 5∗, and delivers the rest in the second pass, 4 ∗ ; 5.

The Theoretical Line of Average Flow, or the Average Theoretical Streamline (ATS) of a material point of the flow, begins at 0, where flowing water energy is available, as:

$$Y_0 = g \cdot H \text{ (J/kg)}$$

where:

g (m/s^2) is the local acceleration of gravity;
H (mca) is the available head for the BMHIT.

This flowing energy Y_0, which is initially inside the transition piece from the circular section 0, to the rectangular section 1, then through the injector 1; 2 is thus transformed into kinetic energy:

$$\frac{c_0^2}{2} = Y_0$$

It then reaches absolute speed $c_2 = c_0$ (m/s) at the outlet of the injector forming angle α_2^o with the tangential direction of the rotor. This angle remains theoretically constant in the interstice from 3 to the rotor inlet 4, the midpoint of the angle of coverage of the injector outlet section δ^o.

To determine point 4 of the ATS at the entrance of the rotor, one needs to first draw the relative ATS between point 4 and point 5∗, which can be done by following the steps below, with reference to the drawing in Fig. 3.1.

1. Since the time t (s) that the water travels through arc $\bar{AB}$ of the blade axis, with the relative speed w (m/s), is the same as the time that the water travels through arc $\bar{EG}$, with the tangential speed u (m/s) of the rotor, this arc can be determined by:

   $$\bar{EG} = \frac{w}{u} \cdot \bar{AB}$$

   and the corresponding angle by:

   $$\theta_1 = 2 \cdot \frac{w}{u} \cdot \frac{r_{\text{blade}}}{D} \cdot \theta_{\text{blade}} \text{ (rad)}$$

   As time t (s) elapses, *point E* traverses the ATS relative to EF, which can be drawn with angles β_4^o and $\beta_5^o = \beta_5^*$, respectively at their points, from the entrance of rotor at E to its exit at F.

   $u = w$ was taken for the scaled drawing in Fig. 3.1.
2. Copy and transport EF to the ATS relative to 4; 5∗. In the Figure, the speed triangle was drawn at the entrance of the rotor at point 4 and at point 5∗ the corresponding speed triangle was drawn at the output.

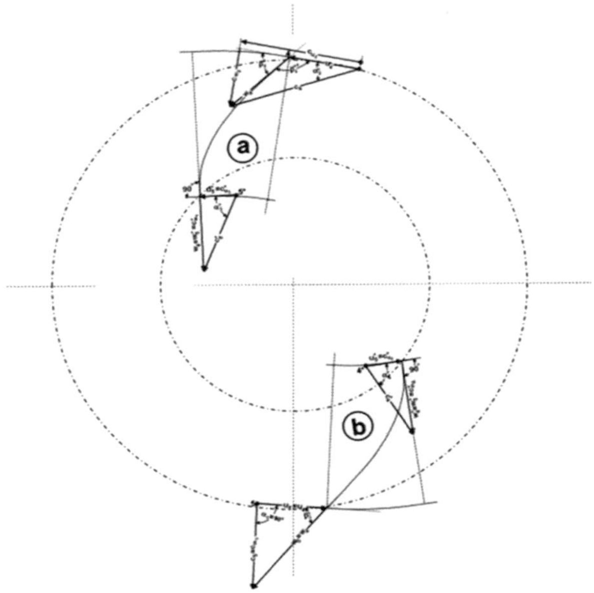

Fig. 3.2 Velocity triangles for the ATS. (**a**) the first passage of the flow through the rotor; and (**b**) the second passage

3. Since the flow is absolute in the interior of the rotor between 6∗ and 3∗ with $c_5^* = c_6^* = c_4^*$, straight lines $\overline{5^*6^*}; \overline{IK}; \overline{FJ}$, are drawn.
4. Copy, transport, and rotate *EF* to the relative ATS that start at 4∗; *K* and *J*.
5. Transport and rotate the speed triangle from 5∗ to 4∗.
6. Draw the velocity triangle at the rotor output 5, which will provide the corresponding flow output direction with the absolute velocity direction $c_5 = c_6$.

It is important to note that, in this case, the flow in passing through the interior of the rotor does not reach its axis, d_e. If it did, how would the flow in this region be drawn?

The speed triangles of the water flows through the rotor were reproduced, not-to-scale, in Fig. 3.2., for ease of understanding.

3.2 Internal Specific Work, Power, and Yield

This subitem will study and analyze the effects of the rotor geometry on performance under the hypotheses that the water flow is ideal, without internal friction, and with walls that delimit the flow, i.e., the Hydraulic Flow Path (HFP) between the inlet and outlet of the BMHIT.

Studying of the theoretical flow path between inlet and outlet of the BMHIT shows that this is only possible in the rotor if angle $\alpha_2 = \alpha_4$ were greater than zero, and consequently, also if β_4 were greater than zero at the entrance of the rotor blades.

Given these hypotheses, and considering only the effect of the geometry, as $Y_{i_{rg}}$ (J/kg), with α_2 and $\beta_4 > 0$, the theoretical internal specific work absorbed from the flow by the rotor can be determined by applying Euler's Equation for Turbomachines[1]:

$$Y_{i_{rg}} = u_4 \cdot c_{u_4} - u_5 \cdot c_{u_5}$$

Considering the velocity triangles in Fig. 3.1b:where:

$$\beta_4 = \beta_5 = \beta$$

$$u_4 = u_5 = u$$

$$c_{m_4} = c_{m_5} = c_m$$

and

$$c_2 = c_0$$

we have

$$c_{u_4} = \frac{c_m}{tg\beta_4} + u \rightarrow c_{u_5} = u - \frac{c_m}{tg\beta_4} \rightarrow c_m = c_4 \cdot \sin\alpha_4.$$

Substituting in Euler's Equation results in:

$$Y_{i_{rg}} = \frac{2 \cdot u \cdot c_m}{tg\beta}$$

or

$$Y_{i_{rg}} = \frac{2 \cdot u \cdot c_4 \cdot \sin\,\alpha_4}{tg\beta} \tag{3.1}$$

This Equation can be sized in terms of the theoretical energy and the velocity coefficients, as follows:

$$\frac{Y_{i_{rg}}}{u^2/2} = \frac{2 \cdot u \cdot c_m}{(u^2/2) \cdot tg\beta} = \frac{4}{tg\beta} \cdot \frac{c_m}{u}$$

or

---

[1] *Souza, Zulcy—Projeto de Máquinas de Fluxo—Tomo I—Base Teórica e Experimental—pg. 50—Editora Interciência—RJ—BR—2011.*

$$\psi_{tg} = \frac{4}{tg\beta} \cdot \phi_{tg} \tag{3.2}$$

In this expression the coefficients will be called:

$$\psi_{\text{tg}} = \frac{Y_{i_{\text{rg}}}}{u^2/2}$$

The theoretical geometric energy coefficient of the rotor related to the theoretical geometric specific work of the rotor and the specific kinetic work in inlet 4 of the BMHIT rotor.

$$\phi_{\text{tg}} = \frac{c_m}{u}$$

The theoretical geometric velocity coefficient, or the relationship between the average meridional velocity at the rotor inlet 4, with the corresponding speed tangential to the BMHIT.

These expressions show that, theoretically, there can be any flow and any direction at the exit of the blades in the first pass, which is why, for ease in manufacturing and for fixing the disks in the rotor, and to control manufacturing processes, this angle is normally fixed at $\beta_5^* = \beta_4^* = 90^{\text{o}}$.

The available internal power, P_{i_0} (kW) of the flowing water at the inlet 0 of the BMHIT, is obtained by multiplying the specific internal work available at its inlet, $Y_{i_0} = g \cdot H$ (J/kg), by the corresponding mass of flowing water, $\rho \cdot Q$ (kg/s), resulting in (kW):

$$P_{i_0} = 10^{-3} \cdot \rho \cdot Q \cdot Y_{i_0} = 10^{-3} \cdot \rho \cdot g \cdot Q \cdot H = 0,5 \cdot 10^{-3} \cdot \rho \cdot Q \cdot c_0^2 \text{ (kW)} \tag{3.3}$$

Considering only the blade geometry, and remembering the hypothetical assumptions already made, the theoretical power of the flowing water delivered to the BMHIT is

$$P_{i_{\text{rg}}} = 10^{-3} \cdot \rho \cdot Q \cdot Y_{i_{\text{rg}}} = 2 \cdot 10^{-3} \cdot \rho \cdot Q \cdot u \cdot c_4 \cdot \frac{\sin \alpha_4}{tg\beta_4} \text{ (kW)} \tag{3.4}$$

Considering only the rotor blade geometry, the theoretical internal performance of the BMHIT can be determined according to:

$$\eta_{i_{\text{tg}}} = \frac{P_{i_{\text{tg}}}}{P_{i_0}} = \frac{Y_{i_{rg}}}{Y_{i_0}} = \frac{2 \cdot u \cdot w(cos\,\beta_4 + cos\,\beta_5)}{c_0^2} \tag{3.5}$$

$Y_{i_{\text{rg}}}$ can also be obtained by summing the specific work of the flow in the two passages through the rotor channels, $Y_{i_{\text{rg}}} = Y_{i_{\text{rg}_1}} + Y_{i_{\text{rg}_2}}$, as:

$$Y_{i_{rg_1}} = u_4 \cdot c_{u_4} - u_{5*} \cdot c_{u_{5*}}$$

and

$$Y_{i_{rg_2}} = u_{5*} \cdot c_{u_{5*}} - u_5 \cdot c_{u_5} \tag{3.6}$$

To determine when $Y_{i_{rg}}$ is delivered in each pass through the rotor, one can merely establish the following relationships:

$$\frac{Y_{i_{rg_1}}}{Y_{i_{rg}}}$$

and

$$\frac{Y_{i_{rg_2}}}{Y_{i_{rg}}} \tag{3.7}$$

3.3 Geometric Limitations of the BMHIT

To be operationalized, The BMHIT, as shown in section form in Fig. 3.1, must geometrically have flow at the outlet of injector 2 and rotor entrance at angle $\alpha_4 > 0$, which limits the theoretical internal efficiency, $\eta_{i_{tg}}$, as shown below, with reference to expression (3.5).

This expression, $\eta_{i_{tg}} = \frac{2 \cdot u \cdot w \cdot (\cos \beta_4 + \cos \beta_5)}{c_0^2}$, considering the velocity triangles in Fig. 3.1, can be algebraically modified, using the following relationship:

$$c_4 \cdot \cos \alpha_4 = u + w_4 \cdot \cos \beta_4 \rightarrow w_4 = \frac{c_4 \cdot \cos \alpha_4 - u}{\cos \beta_4} \rightarrow w_4 = w \rightarrow \beta_4 = \beta_5 = \beta$$

After substitutions:

$$\eta_{i_{tg}} = 4 \cdot \left[\left(\frac{u}{c_4} \right) \cdot \cos \alpha_4 - \left(\frac{u}{c_4} \right)^2 \right] \tag{3.8}$$

For $\eta_{i_{tg\max}}$ one need merely calculate $\frac{\partial \eta_{i_t}}{\partial (u/c_4)} = 0$, and then u/c_4:

Table 3.1 Characteristics of geometric losses

α_4	u/c_4	$\eta_{\text{itg max}}$	Δpg
Degree	–	%	%
12	0.4890	95.7	4.3
14	0.4850	94.2	5.9
16	0.4810	92.4	7.6
18	0.4760	90.5	9.6

$$\frac{\partial \eta_{i_t}}{\partial (u/c_4)} = 0 = 4 \cdot \left[\cos \alpha_4 - 2 \cdot \left(\frac{u}{c_4} \right) \right]$$

or

$$\frac{u}{c_4} = \frac{\cos \alpha_4}{2} \tag{3.9}$$

Substituting in (3.8), gives:

$$\eta_{i_{\text{tgmax}}} = \cos^2 \alpha_4 \tag{3.10}$$

Practically, for the flow to enter and leave the rotor with minimal shock and leakage losses, one must have, constructively, $\alpha_4 \geq 12^{\circ}$, and for geometric losses reasons, the maximum value must be set at $\alpha_4 \leq 18^{\circ}$.

In BMHIT studies, it is recommended that, for each unit of theoretical internal specific work of the flow available at the entrance to the BMHIT, there be a geometric loss Δp_{g} at the injector outlet and rotor inlet, which is a direct function of the average angle α_4^{o} established for the flow at the entrance of the rotor. This angle will determine the β angle of the Average Streamline (AS) at the entrance and exit of the rotor blades.

Table 3.1 shows the corresponding values of u/c_4, $\eta_{i_{\text{tgmax}}}$ and the constant values of the respective geometric losses, $\Delta p_{\text{g}} = 100 - \eta_{i_{\text{tgmax}}}$ for the range of α_4.

Under these considerations, one can determine the expression for calculating the absolute speed at the input of the rotor, as follows:

$$Y_{i_4} = \eta_{i_{\text{tgmax}}} \cdot Y_{i_2} = u \cdot c_{u_4} = 0.5 \cdot c_4 \cdot \cos \alpha_4 \cdot c_4 \cdot \cos \alpha_4 = 0.5 \cdot (c_4 \cdot \cos \alpha_4)^2$$

which results in:

$$c_4 = \frac{\sqrt{2 \cdot Y_4}}{\cos \alpha_4} \ (\text{m/s}) \tag{3.11}$$

On the other hand, considering Fig. 3.2, where $\beta_{5*} = \beta_{4*} = \alpha_5 = 90^{\circ}$, gives

$$c_{m_{5*}} = w_{5*} = c_{m_{4*}} = w_{4*}$$
$$c_{u_{4*}} = c_{u_{5*}} = u_i$$
$$c_{u_5} = 0 \quad (3.12)$$

The ratio of diameters, r_D, should be introduced in this subitem to advance this study, which will be analyzed in the next chapter, which relates the following characteristics:

$$r_D = \frac{D_i}{D} = \frac{u_i}{u} \quad (3.13)$$

3.4 Application

3.4.1 Considerations

- The main objective of this application is to develop and validate an algorithm that presents a set of alternatives according to the previously established characteristics, for theoretically studying a BMHIT, to provide elements that allow for collaborative decision-making during the course of an executive project.
- Draw an energy diagram or a Sankey[2] diagram.

3.4.2 Algorithm Input Data

- Project data
  - $H = 40(\text{m})$
  - $Q = 1.0\ (\text{m}^3/\text{s})$
  - water as an ideal fluid.
- Preliminarily fixed data
  - $\alpha_4 = 14^o$
  - $\beta_{5*} = \beta_{4*} = \alpha_5 = 90^o$
  - $r_D = 0.5$

---

[2] *Matthew Henry Phineas Riall Sankey, Irish-born engineer and captain of the Royal Engineers, known as the creator of the Sankey diagram. He was a member of the board of directors of engineers Willans & Robinson (1889 to 1904). Consultant engineer in the area of steam and internal combustion (1904 to 1909). From 1909 until his death in 1926, he was a board member of Marconi's Wireless Telegraph Company. (Ref. Wikipedia).*

- $g = 9.81\ (\mathrm{m/s^2})$
- $\rho = 1000\ (\mathrm{kg/m^3})$.

3.4.3 Script and Validation

1. The specific internal work available from the flow at the entrance to the BMHIT:

$$Y_{i_0} = Y_2 = g \cdot H = 9.81 \cdot 40 = 3924\ (\mathrm{J/kg})$$

2. The internal power available from the flow at the entrance of the BMHIT:

$$P_{i_0} = Q \cdot Y_{i_0} = 1 \cdot 392.4 = 392.4\ (\mathrm{kW})$$

3. The theoretical absolute velocity of the flow at the entrance of the BMHIT is equal to that of the exit of the injector:

$$c_0 = c_2 = \sqrt{2 \cdot Y_0} = 28.014\ (\mathrm{m/s})$$

4. The maximum theoretical internal yield, considering the geometry of the flow at the entrance of the rotor at angle $\alpha_4 = 14^\circ > 0^\circ$, which results in a loss of energy due to the geometry:

$$\Delta_{p_g} = 1 - \cos^2\alpha_4 = 1 - \cos^2 14 = 0.0595 = 5.95\ (\%)$$

5. The internal specific work of the flow at the entrance of the rotor:

$$Y_{i_4} = Y_{i_2} = 392.4\ (\mathrm{J/kg})$$

6. The internal power of the flow at the entrance of the rotor:

$$P_{i_4} = Q \cdot Y_{i_4} = 1 \cdot 392.4 = 392.4\ (\mathrm{kW})$$

7. The absolute velocity of the flow at the entrance of the rotor:

$$c_4 = \frac{\sqrt{2 \cdot Y_{i_4}}}{\cos \alpha_4} = \frac{\sqrt{2 \cdot 392.4}}{\cos 14} = 28.872 \text{ (m/s)}$$

8. The tangential velocity at the entrance of the rotor:

$$u = 0.5 \cdot c_4 \cdot \cos \alpha_4 = 0.5 \cdot 28.972 \cdot \cos 14 = 14.056 \text{ (m/s)}$$

9. The absolute velocity component of the flow at the entrance of the rotor in the tangential direction:

$$c_{u_4} = c_4 \cdot \cos \alpha_4 = 28{,}872 \cdot \cos 14 = 28{,}014 \text{ (m/s)}$$

10. The absolute velocity component of the flow at the entrance and exit of the rotor in the meridional direction:

$$c_{m_4} = c_{m_5} = c_{m_6} = c_4 \cdot \sin \alpha_4 = 28.872 \cdot \sin 14 = 6.985 \text{ (m/s)}$$

11. The relative velocity of a constant flow in its first and second passage through the rotor, with water taken as an ideal fluid, resulting in $c_{u_5} = 0 \rightarrow \alpha_5 = 90^o$, and the velocity triangles at the rectangular exit resulting in:

$$w_4 = w_5 = w = \sqrt{u^2 + c_{m_5}^2} = \sqrt{14.056^2 + 6.985^2} = 15.695 \text{ (m/s)}$$

12. The average constructive angle at the entrance of the rotor blades, which makes the velocity tangential to the relative velocity:

$$\beta_4 = \beta_5 = \beta = \arcsin \frac{c_{m_4}}{w_4} = \arcsin \frac{6.985}{15.695} = 26.426^o$$

13. The absolute velocity at the exit of the rotor:

$$c_5 = \frac{u}{\cos \beta_5} = \frac{14.056}{\cos 26.426} = 15.696 \text{ (m/s)}$$

14. The internal power at the exit of the rotor:

$$P_{i_6} = Q \cdot 0,5 \cdot Y_{i_6}^2 = 1 \cdot 0.5 \cdot 6.985^2 = 24.395\ (\text{kW})$$

15. The internal power of the flow delivered to the rotor:

$$P_{i_r} = Q \cdot (P_{i_4} - P_{i_6}) = 1 \cdot (392.4 - 24.395) = 368.005\ (\text{kW})$$

16. The internal yield of the BMHIT, $\eta_{\mathrm{i_{rg}}} = \frac{\mathrm{P_{i_r}}}{\mathrm{P_{i_0}}} = \frac{368.005}{392.4} = 0.9378 = 93.78\ (\%)$. The difference of 0.4 (%) between this value, and the value obtained in step 4, is due to the approximation adopted in the calculations using up to three decimals.
17. The tangential velocity of the rotor in internal diameter:

$$u_i = r_D \cdot u = 0.5 \cdot 14.056 = 7.028\ (\text{m/s})$$

18. The absolute velocity at the exit of the flow in the first passage through the rotor:

$$c_{5*} = \sqrt{u_i^2 + w^2} = \sqrt{7.028^2 + 15.695^2} = 17.197\ (\text{m/s})$$

19. The angle between the directions of the velocities of the meridional and absolute flows in the first passage through the rotor, with $c_{m_{5*}} = w = 15.695\ (\text{m/s})$,

$$\alpha_{5*} = \arcsin \frac{c_{m_{5*}}}{c_{5*}} = \arcsin \frac{15.695}{17.197} = 65.875^o$$

20. The internal power of the flow delivered to the rotor in the first passage:

$$P_{i_{r_1}} = Q \cdot (Y_{i_{r4}} - u_i \cdot c_{u_{5*}}) = 1 \cdot \left(392.4 - 7.028^2\right) = 343.007 \quad (\text{kW})$$

21. The internal power of the flow delivered to the rotor in the second passage, as:

$$P_{i_{r_2}} = P_{i_{5*}} - P_{i_6} = 7.028^2 - 24.395 = 24.998 \quad (\text{kW})$$

22. The internal power of the flow delivered to the rotor from the sum of the two passages, $P_{i_r} = P_{i_{r_1}} + P_{i_{r_2}} = 343.007 + 24.998 = 368.005\ (\text{J/kg})$. The value is equal to the value obtained in step 14, demonstrating the veracity of the calculations up to this step.

23. The internal efficiency of the BMHIT:

$$\eta_i = \frac{P_{i_r}}{P_{i_0}} = \frac{368.005}{392.4} = 0.9378 = 93.78\ (\%)$$

24. The internal efficiency of the flow delivered to the rotor in the first passage, relative to the power available at the entrance of the BMHIT:

$$\eta_{i_{r_1}} = \frac{P_{i_{r_1}}}{P_{i_0}} = \frac{343.344}{392.4} = 0.8750 = 87.50\ (\%)$$

25. The internal efficiency of the flow delivered to the rotor in the second passage, relative to the power available at the entrance of the of the BMHIT:

$$\eta_{i_{r_2}} = \frac{P_{i_{r_2}}}{P_{i_0}} = \frac{24.661}{392.4} = 0.0629 = 6.29\ (\%)$$

26. The theoretical specific efficiency of the flow delivered to the rotor in the second passage, relative to the power available at entrance 4∗:

$$\eta^*_{i_{r_2}} = \frac{P_{i_{r_2}}}{P_{i_{4*}}} = \frac{24.661}{49.056} = 0.5027 = 50.27\ (\%)$$

27. Draw the velocity triangles for the entrance and exit of each of the flow passages through the rotor in Fig. 3.3.
28. Draw the Sankey diagram in Fig. 3.4.

3.4.4 Output Results

Table 3.2 summarizes the data and the results.

3.4.5 Analysis, Comments, and Recommendations

- Considering that the blade geometry of the rotor at the entrance and exit is established by fixing α^o_4, $\beta_{5*} = \beta_{4*} = 90^o$, $c_{m_4} = c_{m_5} = c_{m_6}$ is calculated in step 10, and β^o_4 is calculated in step 12. Considering further that the flow rate through

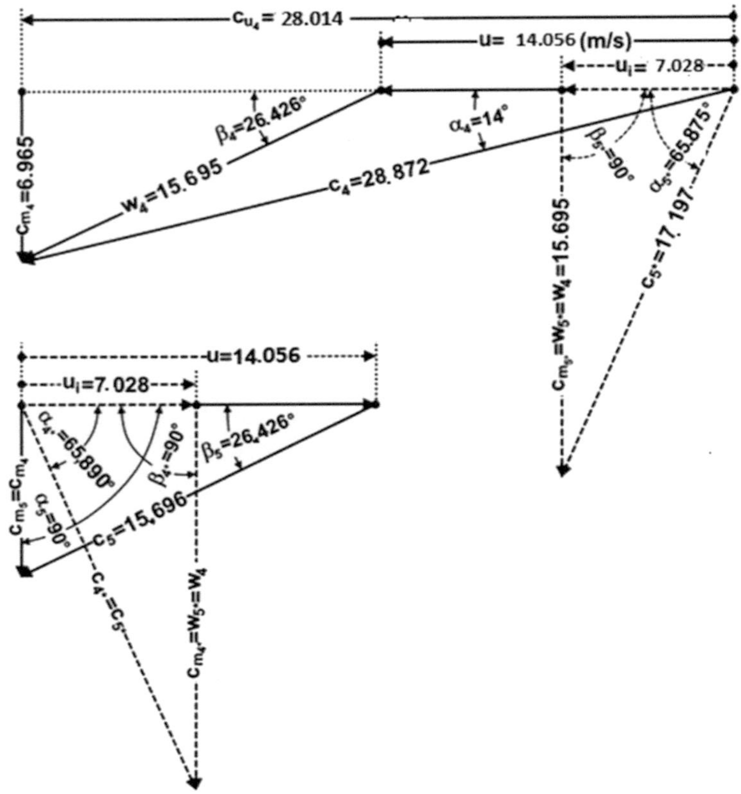

Fig. 3.3 Velocity triangles of the ATS for the two flow passages through the rotor of the BMHIT

the rotor is constant, the power of the flow at the rotor outlet can be determined as $P_{i_6} = Q \cdot 0.5 \cdot c_{m6}^2$ and consequently, the available power from the flow at the axis as $P_{i_4} = P_{i_3} - P_{i_6}$, which was completed in step 14.

- The calculations show that the water flow in the second passage through the rotor blades is small relative to the first passage, rarely exceeding 10 (%), for practically all cases. In the case studied, the results were $\eta_{i_{r_1}} = 87.50\ (\%)$ and $\eta_{i_{r_2}} = 6.29\ (\%)$ respectively.
- If we were to compare the respective energies available at the flow entrances in the rotor in the first and second passages, the result would be, $\eta_{i_{r_1}} = 87.50\ (\%)$ and $\eta_{i_{r_2}}^* = 50.27\ (\%)$. This result shows that the second flow passage through the rotor is much smaller than the first.

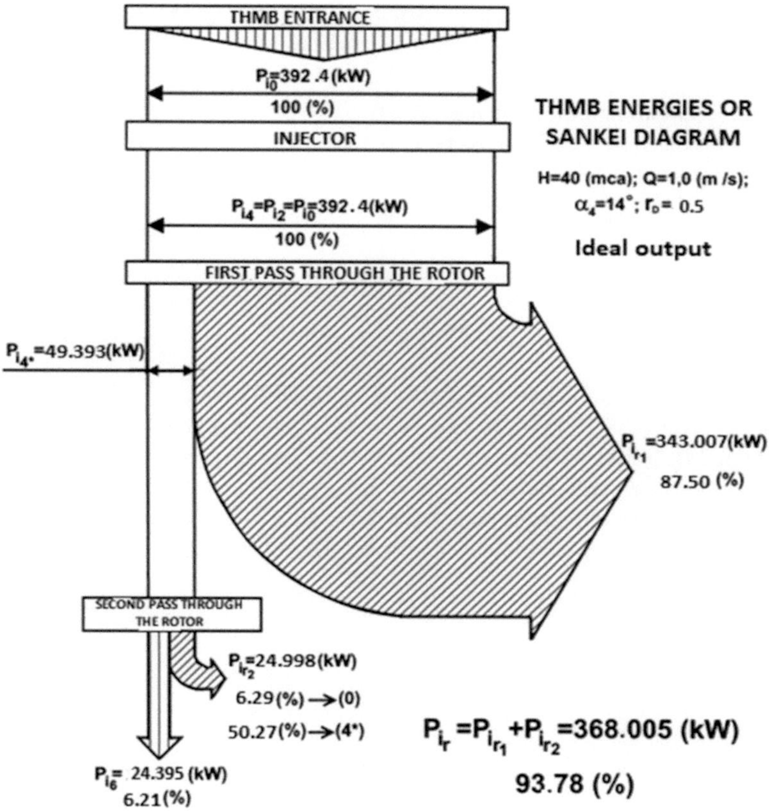

Fig. 3.4 Energy or Sankey diagrams of this application

Table 3.2 Data and results

H	Q	α_4	r_D	$\beta_{4*} = \beta_{5*}$	α_5	g	ρ	Υi0–Υi4	Pi0
mca	m^3/s	Degree	–	Degree	Degree	m/s^2	kg/m^3	J/kg	kW
40.00	1.00	14.00	0.50	90.00	90.00	9.81	1000.00	392.40	392.40

c0 = c2	Δpg	Υi4	Pi4	c_4	u	cu4	cm4 = cm5	c_5	cu5
m/s	%	J/kg	kW	m/s	m/s	m/s	m/s	m/s	m/s
28.01	5.95	392.40	392.40	28.87	14.06	28.01	6.99	15.70	0.00

w4 = w5	β4 = β5	Pi6	Pir	ηi rg	u_i	c_{5*}	cm5* = w
m/s	Degree	kW	kW	%	m/s	m/s	m/s
15.70	26.43	24.40	368.01	93.78	7.03	17.20	15.70

c_5	α_5*	Pir1	Pir2	η ir1	η ir2	η* ir2
m/s	Degree	kW	J/kg/kW	%	%	%
15.70	65.88	343.31	25.00	87.50	6.29	50.27

- Table 3.1 shows that the range of variation of the maximum internal performance, following our assumptions within the practical limits of $12 \le \alpha_2^o \le 18$, results in $90.45 \le \eta_{i_{g\max}} \le 95.68$. In this case, if the internal specific work losses represented by $\eta_i = 95$ (%) and the flow leakages by $\eta_f = 96$ (%) were considered, the yield limits on the *BMHIT* axis must be in the following range:

$$0.9045 \cdot 0.96 \cdot 0.95 = 0.8249 \le \eta_e = \eta_{i_{tg}} \cdot \eta_i \cdot \eta_f \le 0.9568 \cdot 0.96 \cdot 0.95 = 0.8726$$

or

$$82.49 \le \eta_e\ (\%) \le 87.26$$

Currently, manufacturers in Brazil offer BMHITs with $P_e < 500$ (kW) with η_e not exceeding 78 (%), while European and Asian manufacturers claim to have obtained $P_e \ge 500$ (kW) for $85 < \eta_i(\%) < 88$. The range within this analysis, and according to research carried out in Brazil, is $0.78 \le \eta_i < 0.88$

- Considering that the blades must have $\beta_{5*} = \beta_{4*} = 90$ (^{o}), the expressions for the energy and velocity coefficients can be expressed as follows

$$\frac{2 \cdot Y}{u^2} = \frac{2}{\eta_i} \cdot \frac{c_m}{u} \cdot ctg\beta_4$$

therefore:

$$\psi_{t_g} = \frac{2}{\eta_i} \cdot ctg\beta_4 \cdot \phi_{t_g} \tag{3.14}$$

- Considering the results of this application, the respective guiding values are:

$$\psi_{t_g} = \frac{2 \cdot Y_{i_r}}{u^2} = \frac{2 \cdot 368.005}{14.056^2} = 3.73 \rightarrow \phi_{t_g} = \frac{c_{m4}}{u} = \frac{6.985}{14.056} = 0.497$$

Chapter 4
The Real Hydraulic Flows Path (RHFP) in a Banki-Michell Hydraulic Impulse Turbine (BMHIT)

Based on the theoretical studies carried out in the previous chapters, particularly in Chap. 3, this chapter seeks to obtain a first approximation of the geometric physical limits of the real hydraulic flow path (RHFP) for the BMHIT to delimit the space occupied by the water flow as a real fluid, between the turbine inlet and outlet. This space comprises the physical limits of the injector, the director vane, the rotor, and the housing. Therefore, we will carry out a study on our preliminary needs, i.e., data, the limitations, and the results, particularly on the operational field, and the limits of the maximum hydrodynamic forces to which each component is submitted. This will allow us to establish an algorithm for preliminarily calculating the hydrodynamics of a BMHIT, and in the next chapter, to make the respective mechanical calculations.

4.1 Normalization and Flows

In principle, the head values are fixed for hydroelectric power plant projects (HPPs) as[1]:

- $H_{\max}$
- $H_n = H$
- $H_{\min}$ (mca)

and the flows as:

- $Q_{\max}$
- $Q_n = Q$
- $Q_{\min}$ (m^3/s)

[1] (mca) - designação de metro de coluna de água de pressão sendo: 1 (mca) = 9.81 (N/m^2).

© The Author(s), under exclusive license to Springer Nature Switzerland AG 2025

Z. de Souza, G. L. Tiago Filho, *Pelton and Crossflow Impulse Hydraulic Turbines*,
https://doi.org/10.1007/978-3-031-87086-6_4

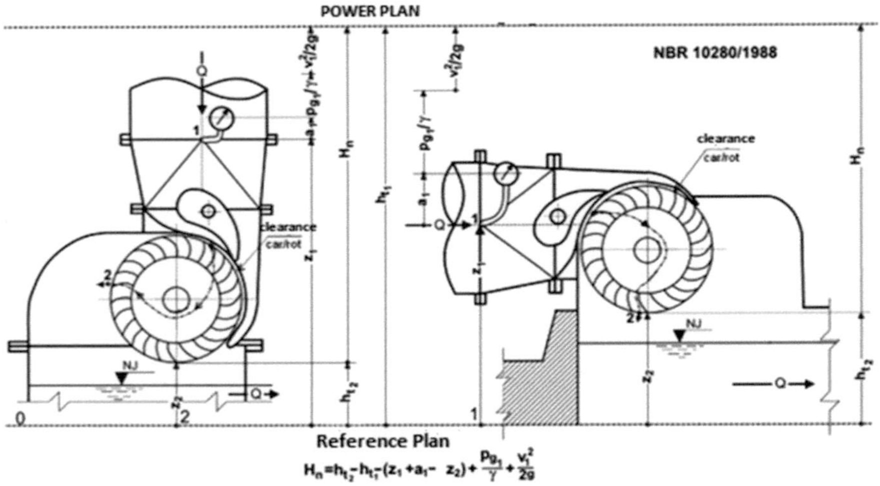

Fig. 4.1 Nominal head H_n for a BMHIT according to NBR 10280/1988

… along with their likely average annual durations. The definition of "head" has been established by Brazilian Standard—NBR 10280/1988, considering a nominal head, H_n, as shown in Fig. 4.1.

The figures show that, for BMHITs, the vertical distance z_2 must as low as possible, so that the downstream water level does not drown the rotor, so that the head can be maximized. Regarding the positioning of the engine room floor, and the minimum and maximum river levels, it may be technically and economically interesting to install a suction tube at the outlet of the BMHIT housing, equipped with a pressure control valve, sized to prevent the water level from rising and drowning the rotor during turbine operation. Also, as we will analyze in due course, this suction tube can recover a part of the gap, h_{t_2}, between the rotor and the downstream water level (DL), in the escape channel, by reducing the atmospheric pressure, and consequently, increasing the power.

We should note that the Brazilian standard does not consider kinetic energy at the rotor output as a specific energy loss:

$$2 - E_2 = 0.5 \cdot v_2^2 \left(\frac{\text{J}}{\text{kg}}\right)$$

Figure 4.2 shows the dimensionless curves of relative yield versus the relative flow:

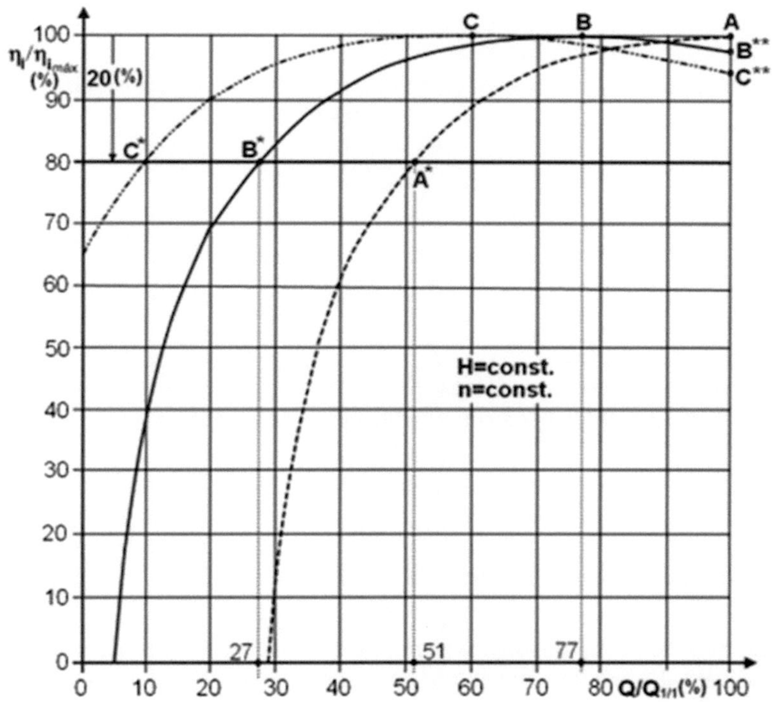

Fig. 4.2 Diagram of function $\frac{\eta_i}{\eta_{i_{\frac{1}{1}}}} = f\left(\frac{Q}{Q_{\frac{1}{1}}}\right)$ for a BMHIT

$$\frac{\eta_i}{\eta_{i_{\max}}} = f\left(\frac{Q}{Q_{\frac{1}{1}}}\right) = f(r_Q)$$

with r_Q referring to the flow ratios.

These curves relate the instantaneous internal BMHIT yield of sector η_i with their maximum value $\eta_{i_{max}}$ as a function of the corresponding instantaneous flow ratio Q, with that of the distributor in the fully open position, represented here by:

$$Q_{\frac{1}{1}} = Q_{\max}$$

in tests carried out with $H = const$ and $n = const$.

The curved black line, connecting points B^{**}, B^{*}, and B, obtained from laboratory tests, indicates that the internal efficiency (evaluated at point B) decreases by 20% when the flow ratio is reduced to 77%. As a result, the *BMHIT* operates within a flow ratio range of $27\% < r_Q\ (\%) \leq 100\%$, corresponding to points B^{*} and B^{**}, respectively.

This indicates that the BMHIT must be designed for a flow at:

$$Q_r = 0.77 \cdot Q_{\frac{1}{1}}$$

In the same figure, the following curves were drawn by transposition:

$$A **, A, A*$$

$$C *, C **, C, C*$$

… with flow ratios and operational fields for the same drop in internal efficiency at 20 (%) at

$A, A * \rightarrow Q_r = Q_{\frac{1}{1}}$ field $51 < r_Q$ (%) ≤ 100 and $C *, C, C ** \rightarrow Q_r = 0.6 \cdot Q_{\frac{1}{1}}$ field $10 < r_Q$ (%) ≤ 100.

We should note that, in any $Q/Q_{\frac{1}{1}}$ established for the system comprising transition parts, the injector and blade must be designed considering the maximum flow, i.e.:

$$Q_{\max} = Q_{\frac{1}{1}}$$

4.2 The Trajectory of the Waterjet

One can begin drawing the approximate limits of the waterjet trajectory, as a real fluid, between the entrance and exit of the rotor, by fixing the working angle of the injector δ (Fig. 4.3), which coincides with the corresponding angle at the rotor entrance, which can be drawn as follows:

The time it takes for a particle with relative speed w to travel through a blade arc $\bar{BC}$ between diameters D and D_i is the same time as it takes for the rotor to rotate through angle θ_6, causing the particle to travel through arc $\bar{AK}$. Thus, the trajectory of the particle is arc $\bar{AE}$.

Figure 4.3 allows us to write:

$$\theta_5 = \frac{\overline{BC}}{r} = \frac{\omega \cdot t_T}{r}$$

and

$$\theta_6 = \frac{\overline{AK}}{\frac{D}{2}} = \frac{2 \cdot u \cdot t_T}{D} = \frac{\pi \cdot n \cdot t_T}{30}$$

Eliminating t_T gives the angle by which the particle rotates, covering the distance between D and D_i, resulting in:

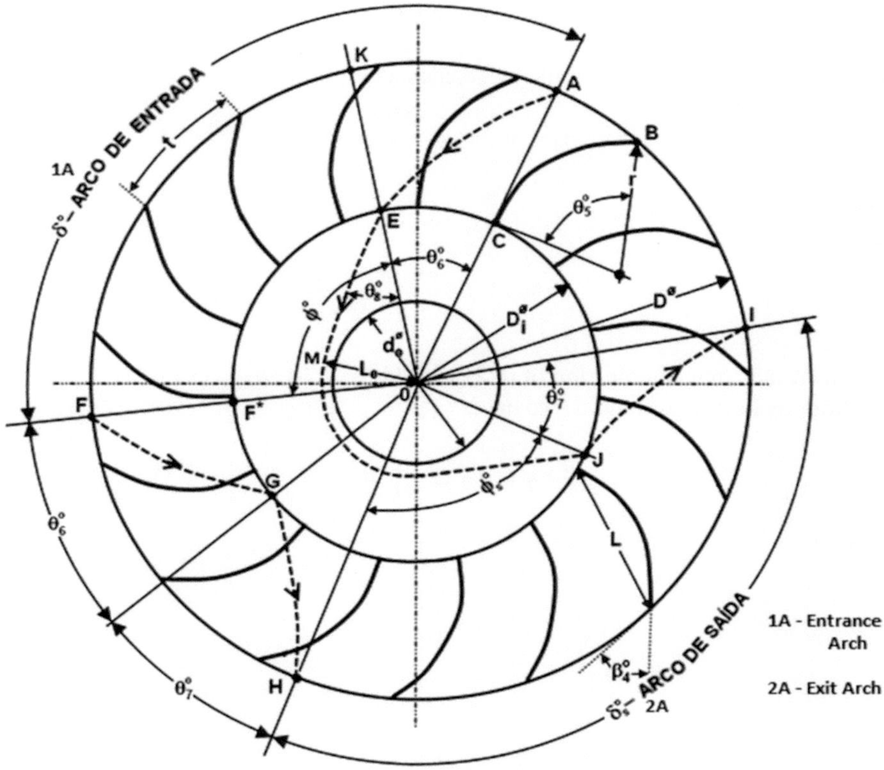

Fig. 4.3 Approximate drawing of the water flow in the BMHIT rotor, in a cross-section view to its axis, with named and designated geometric features

$$\theta_6 = \frac{\pi \cdot n \cdot r}{30 \cdot \omega} \cdot \theta_5 \tag{4.1}$$

Similarly, we can show that the angle by which the particle rotates over distance D to D_i on the second passage through the rotor is:

$$\theta_7 = \frac{\pi \cdot n \cdot r}{30 \cdot \omega_i} \cdot \theta_5 \tag{4.2}$$

Likewise, a particle at A, i.e., the initial jet entry angle δ point, must travel AE to JI. At the same time, another particle at F, the extreme point of this angle, must travel path FGH. Arc AF corresponds to the water jet inlet, and HI to the outlet, thus:

$$\varphi = \delta + \theta_6$$

and

$$\varphi_s = \delta_s + \theta_7 \tag{4.3}$$

Considering velocity losses in the blade channel as:

$$w_i = w_{4*} = \psi \cdot w = \psi \cdot w_4$$

and given that:

- $$w = w_4 = u = \frac{\pi \cdot D \cdot n}{60}$$

… this results in a combination of (4.2) with (4.3).

After making substitutions, we have:

$$\varphi = \delta + 2 \cdot \frac{r}{D} \cdot \theta_5$$

and

$$\varphi_s = \delta_s + \frac{2}{\psi} \cdot \frac{r}{D} \cdot \theta_5$$

as

$$\delta \cong \delta_s \rightarrow \varphi_s > \varphi \tag{4.4}$$

We should note that the real hydraulic flow path (RHFP) established and shown in Fig. 4.3 is only an first approximation of a real situation, since fluctuations from turbulence and centrifugal forces, or even from the BMHIT axis itself, can cause instantaneous changes to the shape.

4.3 Energy Losses

4.3.1 Considerations

With reference to Fig. 4.4, the internal energy losses in a BMHIT, from the flow section at entrance 0, to the rotor exit at section 6, can be preliminarily determined by conceptualizing the characteristic dimensionless relationships, whether of the flow, or of the geometry, which are called coefficients, ratios, yields, or efficiencies. The average values of these characteristics are obtained from laboratory model tests or from on-site prototype installations, and can be instrumentally monitored within standardized specifications.

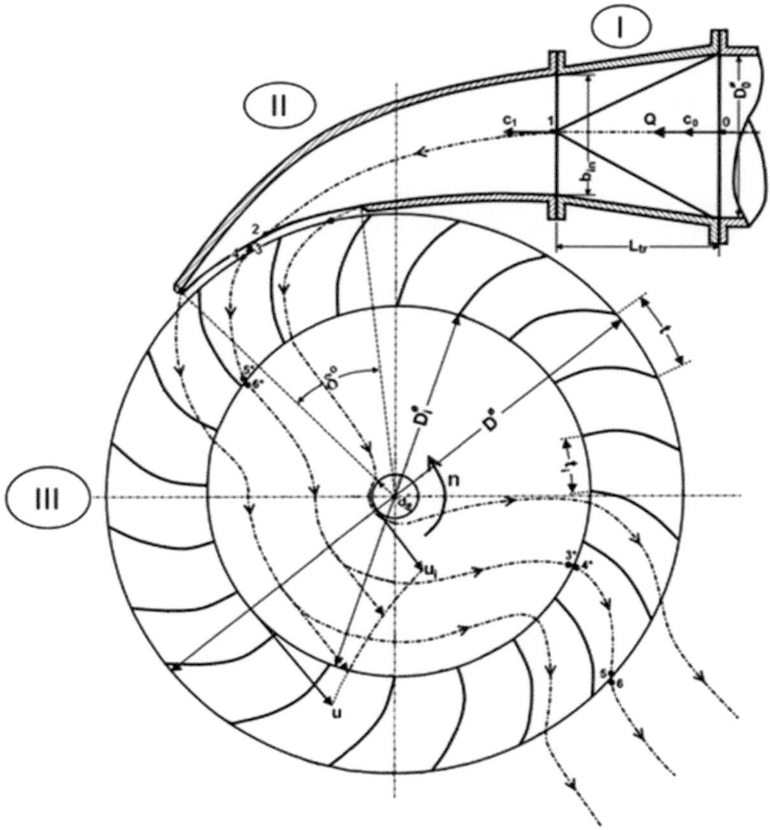

Fig. 4.4 Schematic cross section of the BMHIT axis for modeling the flat flow in the interior for the equating effect

The most commonly used dimensionless and dimensional characteristics will be conceptualized, and the recommended technical limits for their variations will be established, so that a BMHIT hydroelectric power plant (HPP) can be implemented.

4.3.2 The Available Energy at Inlet Section 0

Since the force penstock is sized based on technical and economic considerations, we can introduce the coefficient of the available kinetic energy at the entrance of the BMHIT, conceptualized according to the following expression:

$$\phi_0 = \frac{c_0}{\sqrt{2 \cdot Y_{i_0}}} \rightarrow 0.15 \leq \phi_0 \leq 0.25 \tag{4.5}$$

Since the objective at this point is to provide as much hydraulic energy as possible, the lowest possible values should be set for the highest available internal heads and vice versa. Once the value of this coefficient has been established, the average flow velocity in the penstock at the entrance of the transition can be calculated, followed by the kinetic and pressure energy that must be transformed into kinetic energy at the outlet of the BMHIT injector.

To obtain a preliminary value for the penstock diameter, the *Eletrobrás* company for small hydroelectric plants (SHPs), recommends using the Bondshu Formula:

$$D_0 = 1.27 \cdot \frac{Q_{\frac{1}{1}}^{0.4286}}{\left(H_{i_0} + h_{ga}\right)^{0.1429}} \ (\text{m}) \tag{4.6}$$

Q (m^3/s)
H_{i_0} (mca)
h_{ga} (mca)
water hammer

In preliminary studies, the value adopted for this shock is 30 (%) of the internal net head of the flow

$$h_{\text{ga}} = 0.3 \cdot H_{i_0} \ (\text{mca})$$

4.3.3 Transition I of Circular Section 0 to Rectangular Section 1

Between circular section 0, and rectangular section 1 of transition I, the specific internal energy loss or specific internal work lost from the flow, $Y_{i_{\text{Ptr}}} \left(\frac{\text{J}}{\text{kg}}\right)$, can be evaluated with good approximation, using the so-called universal pressure loss equation:

$$Y_{i_{\text{Ptr}}} = \lambda_{\text{tr}} \cdot \frac{c^2}{2} \cdot \frac{L}{D_{\text{h}}} = 0.8106 \cdot \lambda_{\text{tr}} \cdot Q^2 \cdot \frac{L}{D_{\text{h}}^5} \left(\frac{\text{J}}{\text{kg}}\right) \tag{4.7}$$

The quantities involved in this Equation, for the transition circular section 0, and rectangular section 1, as per Fig. 4.5, can be expressed as:

- The hydraulic diameter—

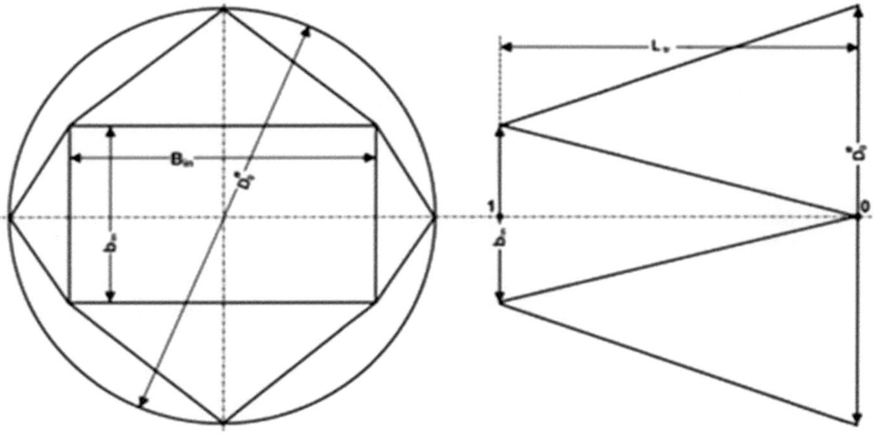

Fig. 4.5 Transition piece, *I*, with the dimensions named

$$D_{\mathrm{h}} = \frac{A_m}{U_m} = \frac{A_0 + A_1}{U_0 + U_1}\ (\mathrm{m}) \tag{4.8}$$

- The area of transversal sections—

$$A_0 = \frac{\pi \cdot D_0^2}{4} \rightarrow A_1 = B_{\mathrm{in}} \cdot b_{\mathrm{in}}\ \left(\mathrm{m}^2\right) \tag{4.9}$$

- The wet perimeters of the cross sections—

$$U_0 = \pi \cdot D_0 \rightarrow U_1 = 2 \cdot (B_{\mathrm{in}} + b_{\mathrm{in}})\ (\mathrm{m}) \tag{4.10}$$

- The length of the transition piece—

$$L = L_{\mathrm{tr}}\ (\mathrm{m}) \tag{4.11}$$

- The coefficient of internal energy—λ_{tr} (1)—P. K. Swamee Formula

$$\lambda_{\mathrm{tr}} = \frac{1.325}{\left[\ln\left(\frac{\varepsilon_{\mathrm{tr}}}{3.7 \cdot D_h} + \frac{5.74}{N_{\mathrm{Re\,tr}}^{0.9}}\right)\right]^2} \rightarrow$$

$$N_{\mathrm{Re}_{\mathrm{tr}}} \cong 1.26 \cdot 10^6 \cdot \frac{Q_{\frac{1}{1}}}{D_{\mathrm{h}}} \rightarrow$$

$$10^{-6} \leq \frac{\varepsilon_{\text{tr}}}{D_{\text{h}}} \leq 10^{-2} \rightarrow 5{\cdot}103 \leq N_{\text{Re}_{\text{tr}}} \leq 108$$

The cast, forged, and in-house machined transition parts with rounded corners

$$\varepsilon_{\text{tr}}\ 0.02\ (\text{mm})$$

The transition piece in welded steel sheet metal, with finished internal welds, image in Fig. 2.8.

Considering the energy loss, the specific internal work of the flow available at the exit of the transition piece, Y_{i_1} (J/kg), the average absolute velocity loss coefficient, ϕ_{tr}, and the internal efficiency or average internal energy efficiency of this component of the BMHIT, $\eta_{i_{\text{tr}}}$, can be determined by the following expressions

$$Y_{i_1} = Y_{i_0} = Y_{i_{\text{ptr}}}(\text{J/kg}) \rightarrow \varphi_{\text{tr}} = \sqrt{\frac{2 \cdot Y_{i_1}}{2 \cdot Y_{i_0}}} = \sqrt{\frac{Y_{i_1}}{Y_{i_0}}} = \frac{c_1}{c_0} \rightarrow \eta_{i_{\text{tr}}} = \frac{Y_{i_1}}{Y_{i_0}} = \varphi_{\text{tr}}^2$$
$$\rightarrow 0.97 \leq \varphi_{\text{tr}} \leq 0.99 \tag{4.12}$$

Coefficient φ_{tr} can be determined experimentally.

Observing normative provisions for installing the measurement system that allows the flows and static pressures in the respective standardized sections to be measured, we can calculate the respective specific work:

$$Y_{i_0} = \frac{p_{i_0}}{\rho} + 0.5 \cdot \left(\frac{Q}{A_0}\right)^2 \left(\frac{\text{J}}{\text{kg}}\right) \rightarrow Y_{i_i} = \frac{p_{i_1}}{\rho} + 0.5 \cdot \left(\frac{Q}{A_1}\right)^2 \left(\frac{\text{J}}{\text{kg}}\right) \tag{4.13}$$

… and the errors and uncertainties, for the permanent water flows in the installed BMHIT, either in the laboratory or on-site.

When elaborating a Basic BMHIT Project, one may need to install an intermediate tube with a rectangular cross section, of area $A_{\text{ti}} = b_{\text{in}} \cdot B_{\text{in}}$, and length L_{ti}, between the exit of the transition piece and the entrance of the injector. In so doing, the specific energy loss $Y_{\text{p}_{\text{Ptr}}}$ of this tube can be evaluated using the universal pressure loss equation, and then added to the transition piece, $Y_{\text{p}_{\text{Ptr}}}$.

The preliminary sizing of the transition part, in some cases, can be done based on the assumption that the cross section at the injector inlet is square, therefore $b_{\text{in}} = B_{\text{in}}$ (m), which allows the first approximation to be calculated as:

$$B_{\text{in}} = b_{\text{in}} = \sqrt{\frac{Q_{\frac{1}{1}}}{c_1}}\ (\text{m}) \tag{4.14}$$

The length of the transition piece, L_{tr} as a first approximation, is equal to the largest dimensions between D_0 and $B_{\text{in}} = b_{\text{in}}$.

4.3.4 The Injector of Rectangular Section II at Input 1 and at the Circular Section at Output 2

Similar to the previous item, here we can also establish an absolute velocity loss coefficient of the real flow of water through the injector, $\phi_{\text{in}} = \phi_2$, as the ratio of the square roots of the specific internal energies available from the flow at the inlet section of injector 1, and output 2, as well as the internal performance or average internal energy efficiency, $\eta_{i_{\text{in}}}$, and the corresponding specific internal energy loss in this component, as:

$$\varphi_2 = \sqrt{\frac{Y_{i_2}}{Y_{i_1}}} = \frac{c_2}{c_1} \rightarrow \eta_{i_2} = \frac{Y_{i_2}}{Y_{i_1}} = \varphi_2^2 \rightarrow Y_{i_{pin}} = Y_{i_{p2}} = \left(1 - \varphi_2^2\right) \cdot Y_{i_1} \text{ (J/kg)} \quad (4.15)$$

Once the injector geometry is fixed, the model can be built, installed, and set up, and through specific tests with certain fixed constant characteristics, we can determine the specific work and the corresponding average absolute velocities of the flow, and via (4.15) the corresponding values of φ_2. The technical range for φ_2 currently available is as follows:

$$0.96 \leq \phi_2 \leq 0.98 \quad (4.16)$$

The lowest values are used for welded, unmachined, or only partially machined injectors, or when they include transition losses. They are also used in the clearance area of the injector/rotor, 3. The highest values should be used for cast or welded injectors, but they should be machined so that the roughness on the internal walls is less than 0.001 (mm), and do not encompass transition losses or the injector/rotor clearance area.

4.3.5 The Clearance Area 3, Between the Section of Injector Outlet 2, and Rotor Inlet 4

The clearance area, interstice, or labyrinth, 3 in Fig. 4.6, between the injector outlet section, 2, and the rotor inlet, 4, of the BMHIT, is essential for operation, causes energy losses in actual water flow due to turbulence and direction change in average absolute velocity at the outlet of the injector.

This causes specific energy losses that are difficult to determine experimentally, but which can be estimated by introducing a loss coefficient for the average absolute velocity at the injector/rotor labyrinth $\phi_{la} = \phi_3$ and the internal efficiency of the water flow in this region, as follows:

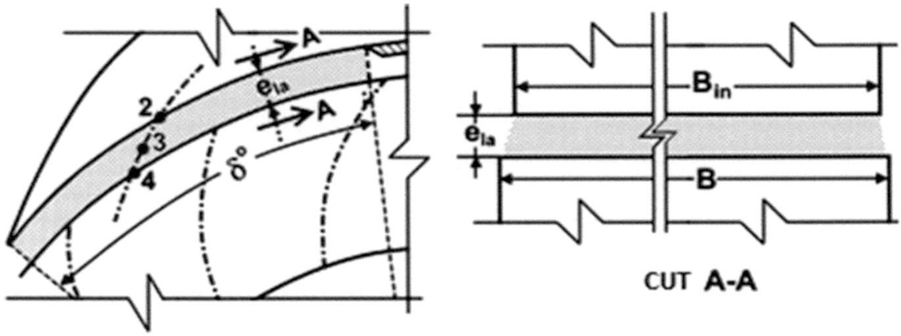

Fig. 4.6 Detail showing that the clearance is between the injector, II, and the rotor, III

$$\varphi_2 = \sqrt{\frac{Y_{i_3}}{Y_{i_2}}} = \frac{c_3}{c_2} \rightarrow \eta_{i_{la}} = \frac{Y_{i_3}}{Y_{i_2}} = \varphi_3^2 \rightarrow Y_{i_{\text{blade}_{la}}} = Y_{i_{p2}} = \left(1 - \varphi_2^2\right) \cdot Y_{i_1}\ (\text{J/kg}) \quad (4.17)$$

The losses from this coefficient are small, and in preliminary studies are included in the injector for clearances that do not exceed 2 (mm). Otherwise, preliminary studies can be adopted, $\varphi_2 \geq 0.98$.

4.3.6 The Variation of the Cross-Section Flow in Rotor III

If not fully compensated for by an increased rotor width, B, relative to the injector, B_{in}, the finite thickness of the rotor blades, III, in the tangential direction, e_{blade}, which appears in the corresponding cross section of the rotor axis, Fig. 4.7, produces throttling in the water flow as it passes through the rotor, which will cause variations in the modulus of the average meridional velocity cm. In this case, as shown in the diagram in Fig. 4.7, for the first pass, there is a reduction in the cross section from 3 to 4, and an increase from 5∗ to 6∗.

If not compensated for, these variations in the average meridional flow velocity can be determined by introducing throttling coefficients, as follows:

$$f_{e_4} = f_{e_5} = \frac{t_4 - e_{\text{blade}_{t4}}}{t_4} = 1 - \frac{e_{\text{blade}}}{t_4 \cdot \sin\beta_4} \rightarrow f_{e_{5*}} = f_{e_{4*}} = \frac{t_{5*} - e_{\text{blade}_{t5*}}}{t_{5*}} = 1 - \frac{e_{\text{blade}}}{t_{5*} \cdot \sin\beta_{5^*}} \quad (4.18)$$

To verify how these coefficients vary, we will give an example considering the following data:

- $e_{\text{blade}_{min}} = 3$ (mm)
- $e_{\text{blade}_{max}} = 20$ (mm)

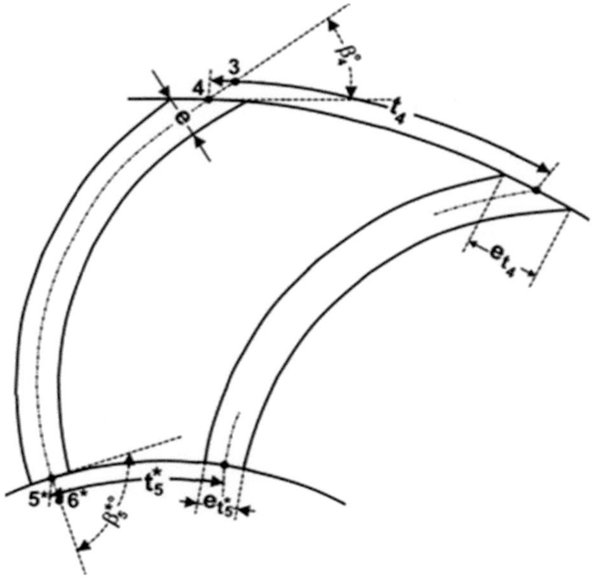

Fig. 4.7 Rotor blades, III, with their geometric characteristics

Table 4.1 Strangle FACTORS

$e_{pá}$	$\beta_4 = 24°$		$\beta_4 = 30°$		$\beta_4 = 36°$		$\beta_{5*} = 90°$
	$e_{pá\ t4}$	f_{e4}	$e_{pá\ t4}$	f_{e4}	$e_{pá\ t4}$	f_{e4}	f_{e5*}
mm	mm	–	mm	–	mm	–	mm
3	7.38	0.94	6.0	0.95	5.10	0.96	0.95
5	12.29	0.90	10.0	0.92	8.51	0.93	0.92
7	17.21	0.86	14.0	0.89	11.91	0.91	0.89
9	22.13	0.82	18.0	0.86	15.31	0.88	0.86
12	29.51	0.76	24.0	0.81	20.42	0.84	0.81
16	39.34	0.69	32.0	0.74	27.22	0.78	0.74
20	29.18	0.61	40.0	0.68	34.03	0.73	0.68
Eq.	$f_{e4} = 1–0.020\ e_{pa}$		$f_{e4} = 1–0.016\ e_{pá}$		$f_{e4} = 1–0.014\ e_{pá}$		$f_{e5*} = 1–0.008\ e_{pá}$
Faixas	$0.607 \leq f_{e4} \leq 0.959$						$0.680 \leq f_{e5*} \leq 0.9652$

- $D = 500$ (mm)
- $Di = 250$ (mm)
- $t_4 = 0.5 \cdot (D - D_i) = 125$ (mm)
- $t_{5*} = 62.5$ (mm)
- $\beta_4 = \beta_5 = 24^o \rightarrow 30^o \rightarrow 36^o$
- $\beta_{5*} = \beta_{4*} = 90^o$

The results of the calculations for $f_{e_4} = f_{e_5}$ and $f_{e_{4*}} = f_{e_{5*}}$ as well as the respective linear equations have been transcribed in Table 4.1.

Table 4.2 Spiral radius

q_j	r_{qj}	q_j	r_{qj}
degrees	m	degrees	m
0	0.21	75.00	0.29
15	0.23	90.00	0.31
30	0.24	97.50	0.32
45	0.26	105.00	0.33
60	0.27	120.00	0.36

The f_{e_4} can also be obtained from the relationship between the normal areas, or flows considering the arcs in the cross sections of Fig. 4.4 $\bar{a_{3_\delta}} = \bar{a_{4_\delta}}$, which results in:

$$Q_3 = Q_4 = \eta_f \cdot Q = c_{m_3} \cdot \overline{a_{3_\delta}} \cdot B_{\text{in}} = c_{m_4} \cdot \overline{a_{4_\delta}} \cdot B \cdot f_{e_4}$$

Therefore:

$$f_{e_4} = \frac{c_{m_3}}{c_{m_4}} \cdot \frac{B_{\text{in}}}{B} \tag{4.19}$$

Combining expressions (4.18) and (4.19) results in an expression that allows us to estimate a first approximation value for the constant thickness of the rotor blades, as:

$$e_{\text{blade}} = \left(1 - \frac{c_{m_3}}{c_{m_4}} \cdot \frac{B_{\text{in}}}{B}\right) \cdot t_4 \cdot \sin \beta_4 \tag{4.20}$$

4.3.7 *The Flow Through the Rotor in Spaces with Blades, Sections 4;5* and 4*;5*

As was seen in the previous chapter, the flow must deliver the maximum available energy during the two passages through the channels between the rotor blades. As water flows have losses, both in the first and second passage at average relative flow velocities, w at the respective inlets are greater than at the outlets. These reductions at these velocities can be evaluated by introducing a loss coefficient in the corresponding average relative velocity $\psi_{4;5^*} \cong \psi_{4^*;5} = \psi_5$. This coefficient can be obtained once the respective average relative flow velocities in the channels formed by the rotor blades are known, as shown by the velocity triangle averages in Fig. 4.8.

$$\psi_5 = \frac{w_{5*}}{w_4} \cong \frac{w_5}{w_{4*}} \rightarrow 0.95 \leq \psi_{4;5*} \cong \psi_{4*;5} = \psi_5 \leq 0.98 \tag{4.21}$$

The lowest possible constant values in the range should be used in preliminary studies for long blades, large numbers, high meridional and relative velocities, and rough blade walls, and the highest possible values should be used for an inverse situation.

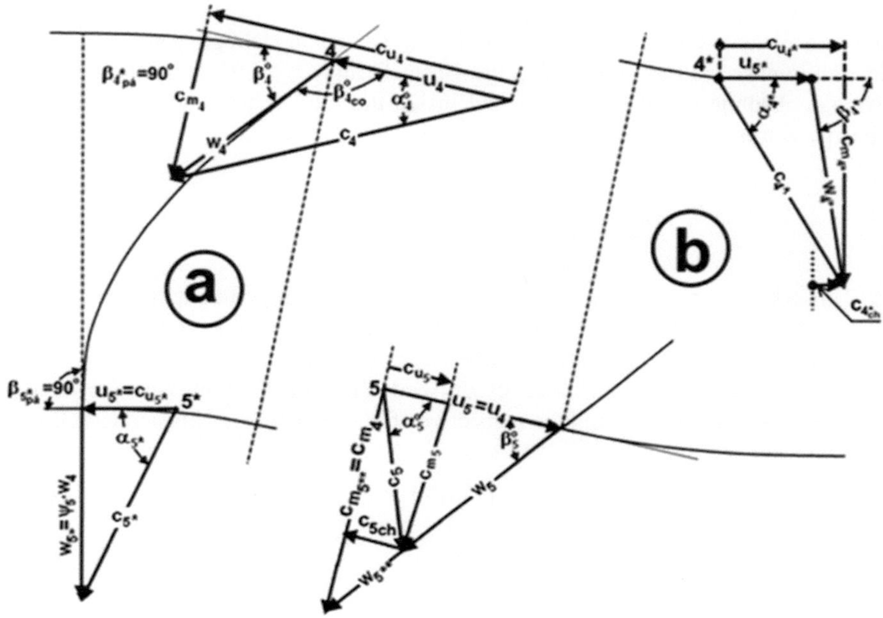

Fig. 4.8 Velocity triangles for real flows

We can determine the specific internal work lost from flows passing through the rotor, considering the triangles of the average flow velocities drawn in Fig. 4.8, resulting in:

- **First passage, 4:5***

$$Y_{i_{p_{r_1}}} = 0.5 \cdot \left(w_4^2 - w_{5*}^2\right) = \left(1 - \psi_5^2\right) \cdot w_4^2 \left(\frac{\mathrm{J}}{\mathrm{kg}}\right) \rightarrow \eta_{i_{r_1}} = \psi_5^2 \tag{4.22}$$

- **Second passage, 4*:5**

$$Y_{i_{p_{r2}}} = 0,5 \cdot \left[\left(w_{4*}^2 - w_5^2\right) + w_{4*\mathrm{ch}}^2\right] \left(\frac{\mathrm{J}}{\mathrm{kg}}\right) \rightarrow \eta_{i_{r_2}} = 1 - \frac{Y_{p_{i_{r2}}}}{Y_{i_{4*}}} \tag{4.23}$$

4.3.8 The Flows Inside the Rotor in Spaces Without Blades 6;3**

The actual flow of water inside the rotor in spaces without blades, which starts at the exit of the first passage, i.e., section 6∗, and at the entrance of the second one, i.e., 3∗, is highly turbulent, and due to internal friction and shaft rotation, it is rotational.

Therefore, given that energy losses are difficult to determine experimentally, they must be evaluated using computational fluid dynamics modeling resources, or CFD. In preliminary studies and during the Basic Design of a BMHIT, the so-called average absolute speed loss coefficient inside the rotor is used, which relates the specific works available in 3∗ and 6∗

$$\phi_{6^*3^*} = \phi_{3^*} = \frac{Y_{3*}}{Y_{6*}} = \frac{c_{3*}^2}{c_{6*}^2} \rightarrow 0.96 \leq \phi_{6^*;3^*} = \phi_{3^*} \leq 0.99 \tag{4.24}$$

This coefficient, which is very difficult to determine experimentally, reduces c_{6*} after exiting the first passage flowing by the rotor to c_{3*}, before entering the second passage, which is why the indicated interval must be used carefully.

Once this coefficient has been established, the corresponding specific energy internal losses of the flow and the respective efficiency can be evaluated as follows:

$$Y_{i_{p6*;3*}} = Y_{6*} - Y_{3*} = \left(1 - \phi_{3*}^2\right) \cdot Y_{6*} \left(\frac{\mathrm{J}}{\mathrm{kg}}\right) \rightarrow \eta_{i_{ir}} = \phi_{3*}^2 \tag{4.25}$$

4.3.9 Leakage Flows

Flow leaks Q (m^3/s) occur in the essential spaces between the injector and the rotor in a BMHIT, as mentioned above (gap, interstice, or labyrinth), and shown in the diagrams in Fig. 4.9, which are represented and called here:

Q_{f_1} and Q_{f_2} $\left(\frac{\mathrm{m}^3}{\mathrm{s}}\right)$ are the lateral leakage flows, where: $Q_{f_1} \cong Q_{f_2}$ $\left(\frac{\mathrm{m}^3}{\mathrm{s}}\right)$, Q_{f_3}, and Q_{f_4} $\left(\frac{\mathrm{m}^3}{\mathrm{s}}\right)$ are leakage flow rates at the injector outlets.

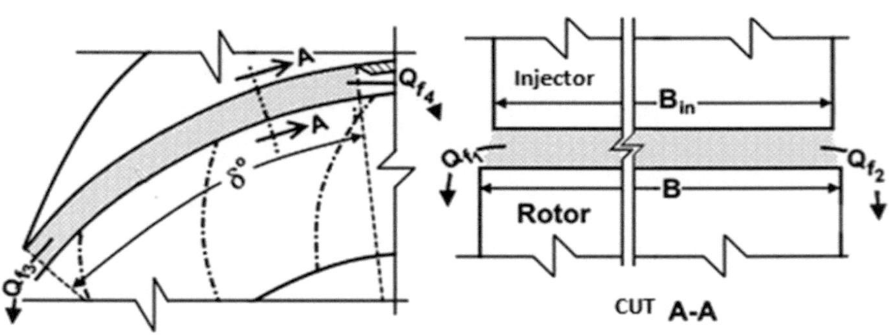

Fig. 4.9 Schematics showing the locations where flow leakage occurs in the gap between the injector and the BMHIT rotor

In this case, $Q_{f_4} \cong 0$ since the fluid flows in the opposite direction of the rotor's rotation.

The sum of these flows, or:

$$Q_f = Q_{f_1} + Q_{f_2} + Q_{f_e} \left(\frac{\mathrm{m}^3}{\mathrm{s}}\right)$$

… is called the THMB leakage flow and causes hydraulic energy loss, from flow in 3 to the rotor, E_{r_3}, which must be subtracted, since the flow of the rotor is:

$$Q_r = Q - Q_f$$

… which allows us to introduce the dimensionless characteristic, the leak yield, η_f, and the specific internal work of the leakage $Y_{p_{i_{Q_f}}}$ $\left(\frac{\mathrm{J}}{\mathrm{kg}}\right)$ as follows:

$$E_{i_{p_{Q_f}}} = \rho \cdot Q_f \cdot Y_{i_4} \rightarrow \eta_f = \frac{Q_r}{Q} = 1 - \frac{Q_f}{Q} \rightarrow E_{i_{p_{Q_f}}} = \rho \cdot Q \cdot Y_{i_3}$$

Since $Y_{i_3} \cong Y_{i_4}$, we have:

$$\frac{E_{i_{i_{Q_f}}}}{\rho \cdot Q} = (1 - \eta_f) \cdot Y_{i_4} \rightarrow Y_{i_{p_{Qf}}} = (1 - \eta_f) \cdot Y_{i_4} \left(\frac{\mathrm{J}}{\mathrm{kg}}\right) \tag{4.26}$$

4.4 The Speed Ratio

A dimensionless characteristic of fundamental importance, which must be established when projecting the BMHIT calculations, is the relationship between the tangential velocity, u, of the flow at the entrance and exit of the rotor, $u_4 = u_5 = u$ (m/s), with the absolute velocity at the entrance, c_4 (m/s). Here, this relationship will be called the speed relationship, $r_{ve} = \frac{u}{c_4}$, which is established by imposing the maximum condition for the internal efficiency of the BMHIT $\eta_{i_{\max}}$. Thus, based on average speed triangles for the input and output of the rotor, shown in Fig. 4.8, the following equation can be developed:

$$Y_{i_r} = u \cdot (c_{u_4} - c_{u_5})(\mathrm{J/kg})$$

which is the internal specific work delivered by the rotor to its shaft.

$$Y_{i_0} = \frac{c_0^2}{2} \text{ (J/kg)}$$

which is the available specific internal work at the entrance of the BMHIT.

•

$$\eta_i = \frac{2 \cdot u \cdot (c_{u_4} - c_{u_5})}{c_0^2} \cdot \frac{Q_r}{Q}$$

which is the internal yield.

Given the velocity triangles *a* and *b*, *f* can be written as:

$$c_{u_4} = w_4 \cdot \cos\ \beta_4 + u \rightarrow c_{u_5} = u - \psi_5^2 \cdot w_4 \cdot \cos\ \beta_4 \rightarrow w_4 \cdot \cos\ \beta_4 = c_4 \cdot \cos\ \alpha_4$$
$$\rightarrow c_4 \rightarrow \eta_4 = \frac{Q_r}{Q}$$

Taking these relationships into the internal yield expression yields:

$$\eta_i = \frac{2 \cdot \eta_f \cdot \phi_2^2 \cdot (1 + \psi_5^2)}{f_{e_4}^2} \cdot \left[\frac{u}{c_4} \cdot \cos\alpha_4 - \left(\frac{u}{c_4}\right)^2\right] \tag{4.27}$$

For η_i to be maximum with α_4, and for the other coefficients to be constant, one must have:

$$\frac{\partial\eta_i}{\partial\left(\frac{u}{c_4}\right)} = 0 \rightarrow \cos\alpha_4 - 2 \cdot \frac{u}{c_4} 0$$

so …

$$\frac{u}{c_4} = \frac{\cos\alpha_4}{2}$$

and …

$$\eta_{i_{max}} = \frac{(1 + \psi_5^2) \cdot \phi_2^2 \cdot \eta_f}{2 \cdot f_{e_4}^2} \cdot cos^2\alpha_4 \tag{4.28}$$

Considering the minimum values for these characteristics:

- $\phi_2 = 0.95 \rightarrow \psi_5 = 0.95$
- maximum leakage $\eta_f = 0.9$

- average $f_{e_4} = 0.95$
- and maximum $\alpha_4 = 18^o$

… the minimum internal yield, $\eta_{i_{min}}$, is given as:

$$\eta_{i_{\text{min}}} = \frac{(1 + 0.95^2) \cdot 0.95^2 \cdot 0.9}{2 \cdot 0.95^2} \cdot \cos^2 18 = 0.7744 = 77.47\ (\%)$$

Considering the maximum values of the characteristics:

- $\phi_2 = 0.97 \rightarrow \psi_5 = 0.99$
- minimum leakage $\eta_f = 0.98$
- average $f_{e_4} = 0.95$
- and minimum $\alpha_4 = 12^o$

… the maximum internal yield, $\eta_{i_{\text{max}}}$, is given as:

$$\eta_{i_{\text{max}}} = \frac{(1 + 0.99^2) \cdot 0.97^2 \cdot 0.97}{2 \cdot 0.95^2} \cdot \cos^2 12 = 0.9579 = 95.79\ (\%)$$

With the limits of α_4 (values given in Table 3.1), we can also evaluate the respective values for the maximum and minimum yields as follows:

$$\begin{aligned} \alpha_{4_{\text{max}}} &= 18^o \rightarrow \frac{u}{c_4} = 0.5 \cdot \cos 18 = 0.476 \rightarrow \eta_{i_{\text{max}}} = 77.47\ (\%) \rightarrow \alpha_{4_{\text{max}}} \\ &= 18^{\circ} \rightarrow \frac{u}{c_4} 0.5 \cdot \cos 18 = 0.476 \rightarrow \eta_{i_{\text{max}}} = 77.47\ (\%) \rightarrow \end{aligned} \tag{4.29}$$

… and when compared to the previously calculated values, they prove to be equal.

Bear in mind that the energy losses of the flow passage through the interior of the rotor were not considered, ϕ_{3*}, when obtaining expression (4.27). Having done this, the following expression for determining the internal efficiency of the BMHIT rotor results in:

$$\begin{aligned} \eta_i &= 1 - \phi_{3*}^2 \\ &\cdot \left[\frac{1 - \phi_{3*}^2}{\phi_{3*}^2} + \frac{\eta_f \cdot (1 - \psi_5^4)}{4 \cdot (f_{e_4} \cdot \cos \alpha_4)^2} + \frac{\eta_f \cdot f_{e_{5*}}^2 \cdot (1 - \phi_{3*}^2) \cdot (\psi_5^2 + r_D^2)}{4 \cdot (f_{e_4} \cdot \cos \alpha_4)^2} \right] \end{aligned} \tag{4.30}$$

4.5 BMHIT Rotor Geometry

The following characteristics will be addressed in this sub-item, geometrically, to obtain the real hydraulic flow path (RHFP) in the BMHIT.

4.5.1 The Injector Coverage Angle

Designers and manufacturers have adopted this angle in the following range:

$$60 \leq \delta^{o} \leq 120 \tag{4.31}$$

4.5.2 The Flow Angle at the Rotor Inlet

This angle, with the range studied and analyzed in Sect. 3.3 in Chap. 3, is transcribed below as follows:

$$12 \leq \alpha_4^{o} \leq 18 \tag{4.32}$$

4.5.3 The Blade Angle at the Exit of the First Flow Passage Equal to the Entrance of the Second Passage

The angle $\beta_{5*\text{blade}} = \beta_{4*\text{blade}}$ was already studied, and its value was fixed in Sect. 3.2 in Chap. as:

$$\beta_{\text{blade}} = 90^{o}$$

4.5.4 The Coefficient, or Ratio of Diameters, and Tangential Velocities

This ratio, $r_D = \frac{D_i}{D} = \frac{n_i}{n}$, corresponds to the ratio of the diameter (tangential velocity u_i) D_i of the rotor, corresponding to the exit of the first flow passage with diameter D (tangential velocity u) at the entrance of the rotor. r_D must be preliminarily fixed accounting for two opposing conditions, from a geometric standpoint, since the greater the average equivalent length of the channel between the rotor blades, the greater the losses, but the better the flow conditions (greater linearity of the respective average velocities inside it, thus, less turbulence). Normally, the following range is used in practice, considering the ratio of speeds:

$$0.42 < r_D < 0.65 \tag{4.33}$$

4.5.5 *The Blade and Its Features*

4.5.5.1 The Geometric Layout and Graphical Features

The geometric drawing of the blade axis via a circular arc, and the chord length of the blade's midline measurement, L_{blade}, in Fig. 4.10a, was done by fixing the following characteristics:

- $D = 450$ (mm)
- $D_i = 225$ (mm)
- $\beta_4 = 32^{\circ}$
- $\beta_{5*} = 90^{\circ}$

Given these initial data, the geometric drawing of the BMHIT rotor can be obtained using the following steps:

1. Calculate

$$r_{\text{blade}} = 0.25 \cdot \frac{D^2 - D_i^2}{D \cdot \cos\beta_4 - D_i \cdot \cos\beta_5} = 0{,}25 \cdot \frac{450^2 - 225^2}{450 \cdot \cos 32 - 225 \cdot \cos 90} = 99.5\ (\text{mm})$$

2. Draw two concentric circles to scale with diameters:

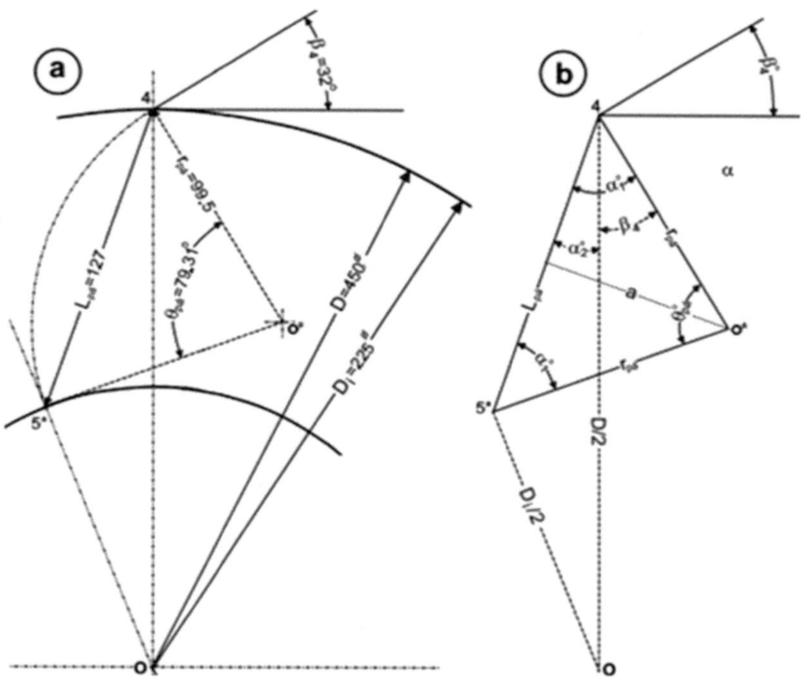

Fig. 4.10 Drawing the blade axis as a circumference arc and equation

$$D = 450\ (\text{mm})$$

and

$$D_i = 225\ (\text{mm})$$

respectively.

3. From point 4 on the vertical axis in the figure, dmark angle:

$$\beta_4 = 32\ (^o)$$

4. In scale, mark the following value from 4, normally next to angle β_4:

$$r_{\text{blade}} = 99.5\ (\text{mm})$$

… thereby obtaining point $O*$

5. With the center at $O*$ and radius $\bar{O * 4}$, draw an arc of circle 4; 5∗ that represents the axis of the blade in the projection in the figure.
6. Measure the following values for L_{blade} and θ_{blade} *to* scale, or measure one and calculate the other, i.e.:

$$L_{\text{blade}} = 127(mm)$$

and

$$\theta_{\text{blade}} = 79.31^{\text{o}}$$

4.5.5.2 Equation and Analytical Characteristics

Referencing Fig. 4.10b, we can analytically determine the blade length, L_{blade}, and the blade coverage angle, θ_{blade}, as follows:

- Triangle ***O*** ∗ ; **4**; **5**∗

$$2 \cdot \alpha_1 + \theta_{\text{blade}} = 180 \rightarrow \alpha_1 = 90 - 0.5 \cdot \theta_{\text{blade}}$$

as

$$\alpha_2 = \alpha_1 - \beta_4 = 90 - 0.5 \cdot \theta_{\text{blade}} - \beta_4$$

or:

$$\alpha_2 = 90 - (0,5 \cdot \theta_{\text{blade}} + \beta_4) \tag{4.34}$$

$$(0.5 \cdot L_{\text{blade}})^2 = r_{\text{blade}}^2 - a^2 = r_{\text{blade}}^2 - r_{\text{blade}}^2 \cdot \cos^2(0.5 \cdot \theta_{\text{blade}})$$

or:

$$L_{\text{blade}} = 2 \cdot r_{\text{blade}} \cdot \sqrt{1 - \cos^2(0.5 \cdot \theta_{\text{blade}})} \tag{4.35}$$

- Trainagle ***O***; **4**; **5**∗

$$\left(0.5 \cdot D_i^2\right) = L_{\text{blade}}^2 + (0.5 \cdot D)^2 - L_{\text{blade}} \cdot D \cdot \cos\alpha_2 \rightarrow L_{\text{blade}}^2 - (D \cdot \cos\alpha_2) \cdot L_{\text{blade}} + 0.25 \cdot \left(D^2 - D_i^2\right)$$

which results in:

$$L_{\text{blade}} = L_{\text{blade}}^* = 0.5 \cdot D \cdot \cos\alpha_2 \pm 0.5 \cdot \sqrt{(D \cdot \cos\alpha_2)^2 - \left(D^2 - D_i^2\right)} \tag{4.36}$$

Since we have 3 Equations and 3 variables, i.e., θ_{blade}, L_{blade} and α_2, after solving the system, we obtain the variables. This can be done by developing a software program from the algorithm steps in the following section. In practical terms, since the calculated values differ from 1 to 2 (%), take the mean integer value of the length L (mm), recalculating the angle $\theta_{\text{blade}}^{\text{o}}$.

4.5.5.3 Algorithm for Calculating the Length of the Rotor Blades

- **Input Data**
  - D
  - D_i
  - r_{blade} (mm)
  - β_4^{o}

An initial value is fixed for angle $\theta_{\text{blade}}^{\text{o}}$, which can be obtained from a scaled drawing of the axis of the blade. For validation, we will use the constant values from Fig. 4.10a and from the scaled drawing, i.e.:

- $D = 550$
- $D_i, = 225$
- $r_{\text{blade}} = 99.5$ (mm)
- $\beta_4 = 32^{\text{o}}$
- $\theta_{\text{blade}} = 79.\ 3^{\text{o}}$
- **Steps**

Calculate

1. The angle:

$$\alpha_2 = 90 - (0.5 \cdot \theta_{\text{blade}} + \beta_4) = 90 - (0.5 \cdot 79.3 + 32) = 18.35^{\circ}$$

2. The length of the reference blade:

$$L_{\text{blade}} = 2 \cdot r_{\text{blade}} \cdot \sqrt{1 - \cos^2(0.5 \cdot \theta_{\text{blade}})}$$
$$= 2 \cdot 99.55 \cdot \sqrt{1 - \cos^2(0.5 \cdot 79.3)} = 127.1 \text{ (mm)}$$

3. The length of the comparison blade:

$$L^*_{\text{blade}} = 0.5 \cdot D \cdot \cos\alpha_2 \pm 0.5 \cdot \sqrt{(D \cdot \cos\alpha_2)^2 - \left(D^2 - D_i^2\right)} = 0.5 \cdot 450 \cdot \cos 18.35$$
$$- 0.5 \cdot \sqrt{(450 \cdot \cos 18.35)^2 - \left(450^2 - 225^2\right)} = 125.8 \text{ (mm)}$$

Since

$$L_{\text{blade}} / L^*_{\text{blade}} = 127.1/125.8 = 1.01$$

then 1 (%)

$$L_{\text{blade}} = 0.5 \cdot (127.1 + 125.8) \cong 126 \text{ (mm)}$$

and

$$\theta_{\text{blade}} = 2 \cdot \arccos \sqrt{1 - \left(\frac{L_{\text{blade}}}{2 \cdot r_{\text{blade}}}\right)^2} = 2 \cdot \arccos \sqrt{1 - \left(\frac{126}{2 \cdot 99.5}\right)^2} = 78.57^{\circ}$$

4.5.6 *The Specific Rotation*

This feature relates two hydrodynamic quantities of the flow Q, Qr (m/s), $H = Y_{i_0}/g$, $H_{i_4} = Y_{i_4}/g$ (mca), with the rotation of the BMHIT rotor n (rpm), and is defined as follows:

$$n_{qA_{r0}} = 3 \cdot n \cdot \frac{Q^{0.5}}{H^{0.75}} \rightarrow n_{qA_r} = 3 \cdot n \cdot \frac{Q_r^{0.5}}{H_{i_4}^{0.75}} \tag{4.37}$$

Since $Q_r = \eta_f \cdot Q$, and $H_{i_4} = \phi_2 \cdot \phi_3 \cdot H$, we have $n_{qA_{r0}} \cong n_{qA_r}$, therby providing the recommended range for using the BMHIT, as:

$$50 \leq n_{qA_{r0}} \cong n_{qA_r} \leq 180 \qquad (4.38)$$

Expressions (4.37) and (4.38) allow us to establish limitations for n (rpm) for the BMHIT as a function of the established characteristics, H (mca) and Q (m3/s), which, after some algebra, results in:

$$n = k_{n_{qa}} \cdot \frac{H^{0.75}}{Q^{0.5}} \text{ (rpm) } 16 \leq k_{n_{qa}} \leq 6 \qquad (4.39)$$

4.5.7 *The Transmission Ratio*

This relationship, r_{ts}, relates the synchronous rotation n_{si}(rpm) of the 60 or 50 (Hz) EG, with what we established for the BMHIT, n (rpm). This ratio must be an integer, since it represents the number of EG pole pairs, z_{pp}. The most commonly used values for r_{ts} for direct coupling between BMHIT/EG are as follows:

EG = 60 (Hz)

$$r_{ts} = \frac{n_{si}}{n} = \frac{600}{n}; \frac{720}{n}; \frac{900}{n}; \frac{1200}{n}; \frac{1800}{n}$$

EG = 50 (Hz)

$$r_{ts} = \frac{n_{si}}{n} = \frac{500}{n}; \frac{600}{n}; \frac{750}{n}; \frac{1000}{n}; \frac{1500}{n} \qquad (4.40)$$

4.5.8 *Rotation and Rotor Diameters*

Given that:

$$u = \frac{\pi \cdot D \cdot n}{60} \rightarrow n \cdot D = \frac{60 \cdot u}{\pi \cdot n}$$

then:

$$D = \frac{60 \cdot u}{\pi \cdot n} \ (m) \rightarrow n = \frac{60 \cdot u}{\pi \cdot D} \text{ (rpm)} \qquad (4.41)$$

Considering expressions (4.39) and (4.41), and observing the maximum internal yield shown in Chap. 3, Sect. 3.3, where $u = 0,5 \cdot \sqrt{2 \cdot Y_{i_4}}$, the following recommended range for the rotation of the BMHIT rotor can be established as:

$$n = \frac{n_{qA_r}}{3} \cdot \frac{(Y_{i_4}/g)^{0.75}}{(\eta_f \cdot Q)^{0.5}} = \frac{n_{qA_r}}{3} \cdot \left(\frac{2}{g}\right)^{0.75} \cdot \frac{u^{1.5}}{(\eta_f \cdot Q)^{0.5}} = (50\ a\ 180) \cdot \frac{1}{3} \cdot \left(\frac{2}{9.81}\right)^{0.75}$$

or

$$n = (5\ a\ 18) \cdot \frac{u^{1.5}}{(\eta_f \cdot Q)^{0.5}}\ (\text{rpm}) \tag{4.42}$$

The following example aims to show how we can establish preliminarily values for n, D,and D_i, given the studies carried out thus far. For this, data on the following values will be used:

- $Q = 1.0\ (\text{m}^3/\text{s})$
- $\eta_f = 0.96$
- 60 (Hz) is the frequency of electric system, and
- $u = 13.169$ (m/s)

Initially, the rotation interval n must be determined via expression (4.42):

$$n = (5\ to\ 18) \cdot \frac{13.169^{1.5}}{(0.96 \cdot 1)^{0.5}} \cong 243 \text{ to } 878\ (\text{rpm})$$

In this range, EG = 60 (Hz):

$$z_{\text{pp}} = \frac{3600}{243} = 14.81$$

for the highest number of pole pairs so, $z_{\text{pp}} = 14$ poles, and $n = 257.143$ (rpm). While

$$z_{\text{pp}} = \frac{3600}{878} = 4.1$$

for the lowest number of pole pairs, so, $z_{\text{pp}} = 5$ poles and $n = 720$ (rpm).

Thus, (4.41) results in:

$$n \cdot D = \frac{60 \cdot u}{\pi} = \frac{60 \cdot 13.169}{\pi} = 251.509\ (\text{rpm} \cdot \text{m})$$

Thereby resulting in:

$$n_{\min} = 257.143\ (\text{rpm}) \rightarrow D_{\max} = 0.978\ (\text{m}) \rightarrow z_{\text{pp}} = 14 \text{ poles}$$

and

$$n_{max} = 720\ (\text{rpm}) \rightarrow D_{min} = 0{,}349\ (\text{m}) \rightarrow z_{pp} = 5\ \text{poles}$$

From a theoretical hydrodynamic standpoint, any set of these values will suffice, but choices will be limited by structural and cost-related issues of the Generator Group (GG).

With D fixed, the inner diameter will be calculated as:

$$D_i = r_D \cdot D\ (\text{m}) \tag{4.43}$$

4.5.9 The Number of Blades

The number of rotor blades, z_r, can be established considering the theoretical step at the rotor inlet:

$$t_4 \cong 0.5 \cdot (D - D_i)$$

so …

$$t_4 = 0.5 \cdot (D - D_i) \rightarrow z_r = \frac{\pi \cdot D}{t_4} \tag{4.44}$$

If z_r is a fraction smaller than 0.5, consider only the integer part, otherwise consider the immediately superior integer.

The number of blades, z_δ, covered by the central angle δ^o can be obtained as a function of z_r from the following expression:

$$z_\delta = \frac{\delta^o}{360} \cdot z_r \tag{4.45}$$

An application can clarify the use of the expressions in this sub-item. For example, with:

- $D = 450(\text{mm})$
- $D_i = 225\ (\text{mm})$
- $\delta = 90^o$

… we have

$$t_4 = 0.5 \cdot (D - D_1) = 0.5 \cdot (450 - 225) = 112.5\ (\text{mm}) \rightarrow$$

$$z_r - = \frac{\pi \cdot D}{t_4} = \frac{\pi \cdot 450}{112.5} = 12.5\ \text{blades}$$

Since z_r must be an integer, $z_r = 12$ blades will be adopted in this case, resulting in the pitch and number of blades covered by the distributor, respectively, as:

$$t_4 = \frac{\pi \cdot D}{z_r} = 12 = 117.8 \cong 118 \text{ (mm)}$$

and

$$z_{\delta=90} = \frac{\delta^o}{360} \cdot z_r = \frac{90}{360} \cdot 12 = 3 \text{ blades}$$

4.5.10 The Maximum Diameter of the Rotor Shaft

For reasons of rigidity and ease of construction, the BMHIT rotor is equipped with a thru-axle. It is important that, whenever possible, the diameter of this shaft does not interact with the water flow when passing through the interior of the rotor.

Figure 4.11, arising from Figs. 4.3 and 4.8a, shows this flow, where L_e is the distance which must always be greater than d_e, so the flow inside the rotor does not meet the BMHIT axis:

$$L_d > 0.5 \cdot d_e \rightarrow d_e < 2 \cdot L_d \tag{4.46}$$

The right triangle 0; M; E in Fig. 4.11 yields:

$$L_e = 0.5 \cdot D_i \cdot \cos \alpha_{5*} = 0.5 \cdot r_D \cdot D \cdot \cos \alpha_{5*} \tag{4.47}$$

We can also determine angle α^o_{5*} from Fig. 4.8a, as follows:

$$u + w_4 \cdot \cos \beta_4 = c_4 \cdot \cos \alpha_4$$

and

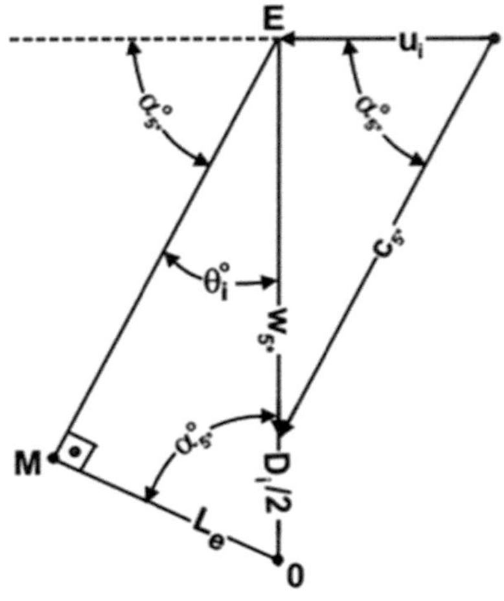

Fig. 4.11 Triangles for determining L_e and α_{5*}

$$w_4 \cdot \sin\beta_4 = c_4 \cdot \sin\alpha_4$$

thus

$$\frac{u + w_4 \cdot \cos\beta_4}{w_4 \cdot \sin\beta_4} = \frac{\cos\alpha_4}{\sin\alpha_4} \rightarrow$$

$$u + w_4 \cdot \cos\beta_4 = w_4 \cdot \frac{\cos\alpha_4 \cdot \sin\beta_4}{\sin\alpha_4} \rightarrow$$

$$u = w_4 \cdot \left(\frac{\cos\alpha_4 \cdot \sin\beta_4}{\sin\alpha_4} - \cos\beta_4\right) \rightarrow$$

$$u = w_4 \cdot \frac{\sin\beta_4 \cdot \cos\alpha_4 - \sin\alpha_4 \cdot \cos\beta_4}{\sin\alpha_4} \rightarrow$$

$$u_i = r_D \cdot w_4 \cdot \frac{\sin(\beta_4 - \alpha_4)}{\sin\alpha_4}$$

or finally

$$\alpha^{\text{o}}_{5*} = \arcsin \frac{\psi_5}{\sqrt{\left[\frac{r_D \cdot \sin(\beta_4 - \alpha_4)}{\sin\alpha_4}\right]^{2} + \psi_5^2}} \tag{4.48}$$

For a first evaluation, the minimum diameter of the rotor shaft, d_e, can be preliminary calculated from the torsion for steel, $\tau_{\text{adm}_t} = 29.43 \cdot 10^6$ (N/m^2), as:

$$d_e = 36.502 \cdot \sqrt[3]{\frac{P_{\max}}{\tau_{\text{adm}_t} \cdot n}} = 118 \cdot \sqrt[3]{\frac{P_{\max}}{n}}\ (\text{mm}) \rightarrow P_{\max}\ (\text{kW}) \rightarrow n\ (\text{rpm}) \tag{4.49}$$

4.6 The Hydrodynamic Sizing of the Injector

4.6.1 The Longitudinal Length of the Injector and Rotor

After first sizing the BMHIT rotor, one must then establish its transversal dimension, B, called the longitudinal length of the injector, which can be obtained by determining the corresponding dimension of the injector, B_{in}.

In addition to the known values for $Q_{1/1}$ (m^3/s) and H (mca), one must first fix $\alpha_2 \cong \alpha_4^o$, φ_2, and the angle of exit of the injector δ^{o} as being equal to the rotor entrance. Then, the following equation can be developed, Fig. 4.12:

Since …

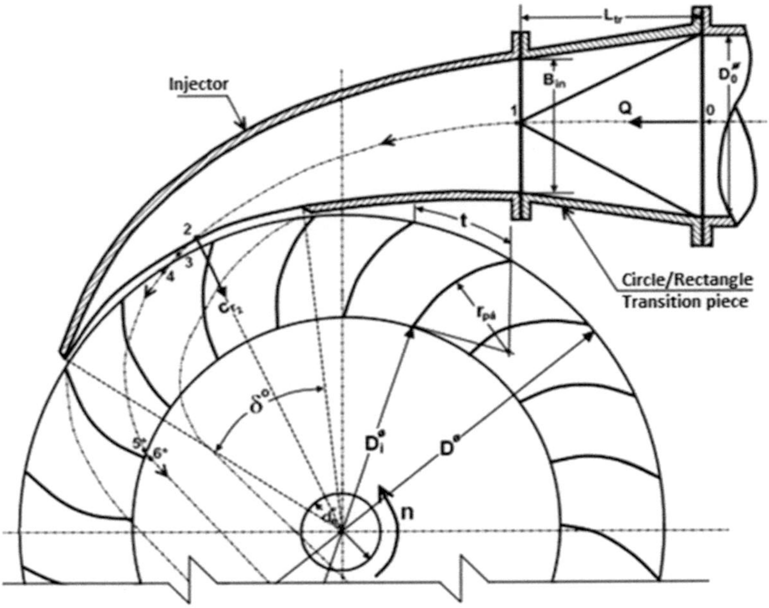

Fig. 4.12 BMHIT cross section with the characteristics of the transition piece/injector system

$$Q_{1/1} = A_\delta \cdot c_{r_2} \rightarrow A_\delta = \frac{\pi \cdot \delta \cdot D}{360} \cdot B_{\text{in}}$$

… we can make substitutions and isolate B_{in}, resulting in:

$$B_{\text{in}} = \frac{360 \cdot Q_{1/1}}{\pi \cdot \delta \cdot D \cdot c_{r_2}} \quad (4.50)$$

- $Q_1/1$ (m^3/s)
- D(m)
- c_{r_2}(m/s)
- α_4 (degrees)
- δ (degrees)

We will see that the structural and flow problems limit the values of B_{in}.

In practice, designers and manufacturers have adopted values between the interval limits of $60 \leq \delta^{\circ} \leq 120^{\circ}$.

Issues related to two-sector BMHITs, leakage losses, and blades and disk thicknesses mean that the rotor length, B, should be greater than the length of the injector, B_{in}, in the following range:

$$1.1 \leq r_B = \frac{B}{B_{\text{in}}} \leq 1.4 \text{ (m)} \tag{4.51}$$

For BMHITs with more than one sector, with rotors equipped with intermediate disks, this must be taken into account so that the throttling caused by the thickness of the disk can be compensated for, and so it comes out in the same order as:

?

4.6.2 *Injector Width*

The width of the B_{in} injector can be calculated by applying the continuity equation, resulting in:

$$Q_{1/1} = c_{2r} \cdot B_{\text{in}} \cdot b_{\text{in}} \rightarrow b_{\text{in}} = \frac{Q_{1/1}}{c_{2r} \cdot B_{\text{in}}} \text{ (m)} \tag{4.52}$$

As was already explained, a square width injector in the cross section of the inlet has been widely used, whenever possible, i.e., $b_{\text{in}} = B_{\text{in}}$.

4.6.3 *The Hydraulic Flow Path (HFP) of the Flow in the Injector*

Figure 4.13 gives a schematic of an injector, which in this case has plates that have been welded together.

The cross section of the BMHIT schematic in Fig. 4.14 shows the theoretical flow in the injector as the result of the superposition of a circular potential flow with sink considering a constant potential vortex $c_t \cdot r = k_{\text{in}} = \text{const}$, which allows us to make the following algebraic calculations to obtain an equation for the external surface with a constant width, B_{in}, represented in this section by the line between s, e.

Considering a line of average flow (LAF), r between the flow boundaries, starting from dq_{θ_j} and flowing through the elementary section $B_{in} \cdot dr$, where c_t is the tangential velocity, we have:

$$dq_{\theta_j} = c_t \cdot B_{in} \cdot dr = k_{in} \cdot B_{in} \cdot \frac{dr}{r}$$

Since this section is at θ_j (degrees) of radius r_s, the moving flow, or flow across the cross section is:

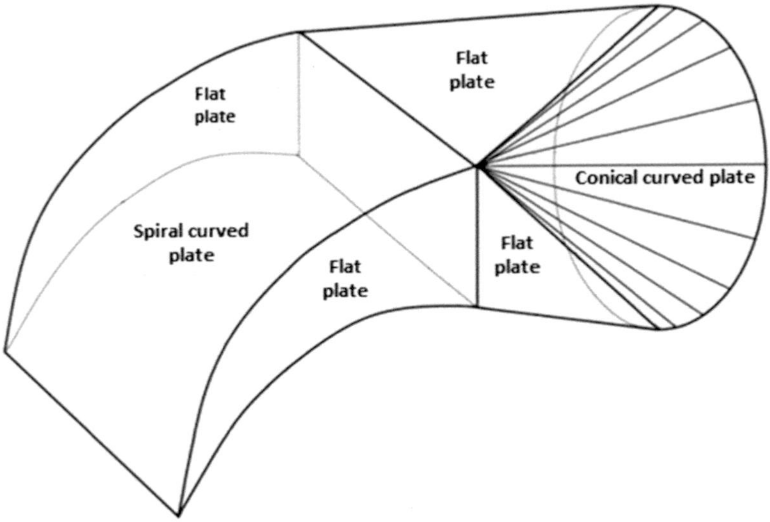

Fig. 4.13 Schematic of a BMHIT injector with the corresponding parts named

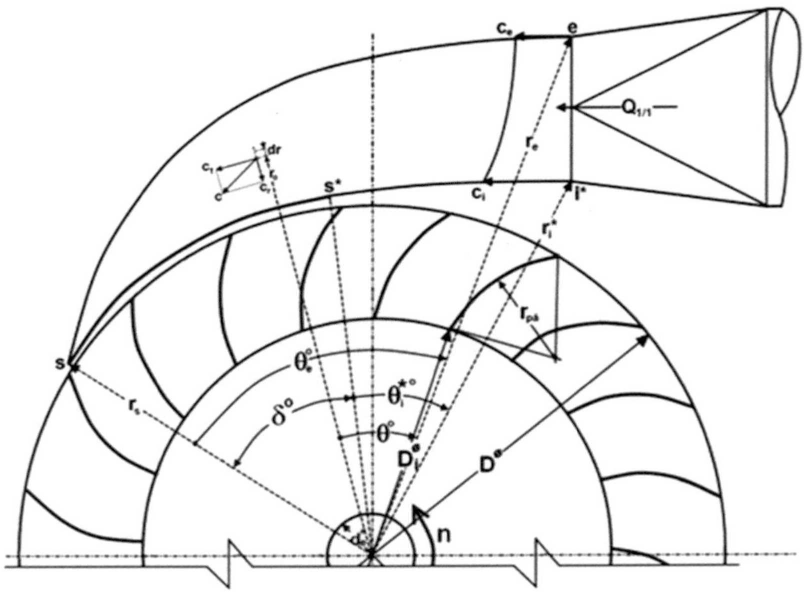

Fig. 4.14 Cross section of a BMHIT schematic showing the theoretical flows in the injector

$$Q_j = \int_{r_i}^{r_j} c_t \cdot B_{in} \cdot dr = k_{in} \cdot \int_{r_i}^{r_j} B_{in} \cdot \frac{dr}{r}$$

Considering further that Q_j (m^3/s) passes in θ_j^o to δ^o, then $Q_{1/1} = \frac{\delta}{\theta_j} \cdot Q_j$ will pass as:

$$\theta_j = \frac{\delta \cdot k_{\text{in}} \cdot B_{\text{in}}}{Q_{1/1}} \cdot \int_{r_1}^{r_{\theta_j}} \frac{dr}{r} = \frac{\delta \cdot k_{\text{in}} \cdot B_{\text{in}}}{Q_{1/1}} \cdot \ln \frac{r_{\theta_j}}{r_1} \rightarrow r_{\theta_j} = 0.5 \cdot D \cdot \exp\left(\frac{Q_{1/1} \cdot \theta_j}{\delta \cdot k_{\text{in}} \cdot B_{\text{in}}}\right) \quad (4.53)$$

It is important that c_2 and angle $\alpha_2 \cong \alpha_4$ be constant throughout the established coverage angle δ, due to the potential vortex, so the absolute velocity at the outlet of the injector depends only on the Head $H =$ const. for flow rates between $Q_{\min}$ and $Q_{1/1}$, and the angles at the beginning of the spirals are constant and equal to $\alpha_2 \cong \alpha_4$.

The expression for determining r_θ can also be expressed, considering that:

$$B_{\text{in}} = \frac{360 \cdot Q_{1/1}}{\pi \cdot \delta \cdot D \cdot c_{m_2}} = \frac{360 \cdot Q_{1/1}}{\pi \cdot \delta \cdot D \cdot c_2 \cdot \sin \alpha_2} \rightarrow k_{\text{in}} = 0.5 \cdot D \cdot c_{2_t} = 0.5 \cdot D \cdot c_2 \cdot \cos \alpha_2$$

Substituting these expressions in (4.53) gives:

$$r_{\theta_j} = 0.5 \cdot D \cdot \exp\left(\frac{\pi \cdot tag\alpha_2}{180} \cdot \theta_j^o\right) + e_{fo} \rightarrow 1 \leq e_{fo}(\text{mm}) \leq 4 \quad (4.54)$$

In this expression e_{fo} is the radial distance or clearance between the outlet of the injector and the inlet of the rotor, that is needed for the BMHIT to operate. This expression also shows that once a spiral has been drawn for a calculated BMHIT, all the other spirals can be drawn via transposition, the spiral relies only on D, $\alpha_2 = \alpha_4$, and θ_j^o being fixed.

4.7 Algorithm and Validation

4.7.1 Input Data

1. Established nominal characteristics

- $H0 = H = 40$ (m)
- $Q = Q_{1/1} = 1.0$ (m^3/s)
- 60 (Hz)

2. Preliminarily fixed characteristics

- $\phi_0 = 0, 2$
- $c_1 = 2 \cdot c_o$
- $\phi_{tr} = 0.98$
- $\varepsilon_{tr} = 0.03$ (mm)
- $\phi_{\text{in}} = \phi_2 = 0.97$
- $\phi_3 = 0.99$
- $\eta_f = 0.96$
- $\alpha_2 = \alpha_4 = 14^o$
- $\psi_5 = 0.97$
- $e_{fo} = 2$ (mm)
- $f_{e_{5*}} = f_{e_{4*}} = 0.80$
- $\phi_{3*} = 0.98$
- $r_D = 0.5$
- $B = 1.15 \cdot B_{in}$
- $\beta_{5*} = \beta_{5*_{\text{blade}}} = \beta_{4*_{\text{blade}}} = 90^o$
- $\rho = 1000$ (kg/m^3)
- $g = 9.81$ (m/s^2)
- Considering that the flow through the rotor is constant at $Q_r = \eta_f \cdot Q = 0.96 \cdot 1 = 0.96$ (m^3/s),we can establish that $c_{m_5} = c_{m_4}$ and $c_{m_{5*}} = c_{m_{4*}}$, implying that $\alpha_5 = 90^\circ$.
- Develop the computer program that will at least result in:
  - $\delta = 60^\circ$
  - 90°
  - 120°

4.7.2 Step-by-Step Script

Calculate to the nearest third decimal place:

1. The specific internal work available in the inlet flow:

$$Y_{i_0} = g \cdot H = 9,81 \cdot 40 = 392{,}4 \text{ (J/kg)}$$

2. Internal power available at the entrance of the flow:

$$P_{i_0} = Q \cdot Y_{i_0} = 1 \cdot 392.4 = 392.4 \text{ (kW)}$$

3. The absolute average velocity of the flow at the inlet of the *BMHIT*:

$$c_0 = \phi_0 \cdot \sqrt{2 \cdot Y_0} = 0.2 \cdot \sqrt{2 \cdot 392.4} = 5.603 \text{ (m/s)}$$

4. The loss of available internal work in the flow as it passes through the transition:

$$Y_{i_{P_{tr}}} = 0.5 \cdot \left(1 - \phi_{tr}^2\right) \cdot c_0^2 = 0.5 \cdot \left(1 - 0.98^2\right) \cdot 5.603^2 = 0.621\ (\mathrm{J/kg})$$

5. The loss of available internal power of the flow and its passage through the transition:

$$P_{p_{i_{tr}}} = Q \cdot Y_{p_{i_{tr}}} = 1 \cdot 0.621 = 0.621\ (\mathrm{kW})$$

6. The internal performance of the transmission:

$$\eta_{i_{tr}} = 1 - \frac{P_{p_{i_{tr}}}}{P_0} = 1 - \frac{0.621}{392.4} = 0.9984 = 99.84\ (\%)$$

7. The specific internal work available in the flow at the inlet of the injector:

$$Y_{i_{in}} = Y_{i_1} = Y_{i_0} - Y_{i_{P_{tr}}} = 392.4 - 0.621 = 391.779\ (\mathrm{J/kg})$$

8. The internal power available from the flow at the inlet of the injector:

$$P_{i_1} = Q \cdot Y_{i_1} = 1 \cdot 391.779 = 391.779\ (\mathrm{kW})$$

9. The theoretical average absolute velocity of the flow at the outlet of the injector:

$$c_{2_t} = \sqrt{2 \cdot Y_{i_1}} = \sqrt{2 \cdot 391.779} = 27.992\ (\mathrm{m/s})$$

10. The actual average absolute velocity of the flow at the outlet of the injector:

$$c_2 = \phi_2 \cdot c_{2_t} = 0.97 \cdot 27.992 = 27.152\ (\mathrm{m/s})$$

11. The specific internal work of the flow lost in the injector:

$$Y_{i_{P_{in}}} = 0.5 \cdot \left(c_{2_t}^2 - c_2^2\right) = 0.5 \cdot \left(27.992^2 - 27.152^2\right) = 23.160\ (\mathrm{J/kg})$$

12. The internal power lost from the flow as it passes through the injector:

$$P_{\mathrm{i}_{\mathrm{P_{in}}}} = Q \cdot Y_{i_{\mathrm{P_{in}}}} = 1 \cdot 23.160 = 23.160\ (\mathrm{kW})$$

13. The internal efficiency of the injector:

$$\eta_{\mathrm{i_{in}}} = 1 - \frac{P_{\mathrm{i_{P_{in}}}}}{P_{i_1}} = 1 - \frac{23.160}{391.779} = 0.9409 = 94.09\ (\%)$$

14. The relative loss of the available power of the flow in the injector in relation to P_{i_0}:

$$\eta_{\mathrm{i_{in_0}}} = \frac{P_{\mathrm{i_{P_{in}}}}}{P_{i_0}} = \frac{23.160}{392.4} = 0.0590 = 5.90\ (\%)$$

15. The available internal specific work of the flow at the outlet of the injector:

$$Y_{i_2} = Y_{i_1} - Y_{\mathrm{i_{p_i}}} = 391.779 - 23.160 = 368.619\ (\mathrm{J/kg})$$

16. The internal power available in the flow at the outlet of the injector:

$$P_{i_2} = Q \cdot Y_{i_2} = 1 \cdot 368.619 = 368.619\ (\mathrm{kW})$$

17. The actual average absolute velocity of the flow in the interstice, injector/rotor:

$$c_3 = \phi_3 \cdot c_2 = 0.99 \cdot 27.152 = 26.880\ (\mathrm{m/s})$$

18. The true average meridional velocity of the flow in the interstice, injector/rotor:

$$c_{m_3} = c_3 \cdot \ sin\, \alpha_2 = 26.880 \cdot \ sin\, 14 = 6.503\ \ (\mathrm{m/s})$$

19. The specific internal work of the flow lost in the injector/rotor interstice:

$$Y_{i_{p_3}} = 0.5 \cdot \left(c_2^2 - c_3^2\right) = 0.5 \cdot \left(27.152^2 - 26.880^2\right) = 7.348\ (\mathrm{J/kg})$$

20. The internal power lost from the flow as it passes through the injector/rotor interstice:

$$P_{i_{p_3}} = Q \cdot Y_{i_{p_3}} = 1 \cdot 7.348 = 7.348\ (\mathrm{kW})$$

21. The internal efficiency of the injector/rotor interstice:

$$\eta_{i_3} = 1 - \frac{P_{i_{p_3}}}{P_{i_2}} = 1 - \frac{7.348}{368.619} = 0.9801 = 98.01\ (\%)$$

22. The relative loss of available flow power in the injector/rotor interstice with respect to P_{i_0}:

$$\eta_{i_{3_0}} = \frac{P_{p_3}}{P_{i_0}} = \frac{7.348}{392.4} = 0.0187 = 1.87\ (\%)$$

23. The available internal flow specific work in the injector/rotor interstice:

$$Y_{i_3} = Y_{i_2} - Y_{i_{p3}} = 368.619 - 7.348 = 361.271\ (\mathrm{J/kg})$$

24. The internal power available in the flow in the injector/rotor interstice:

$$P_{i_3} = Q \cdot Y_{i_3} = 1 \cdot 361.271 = 361.271\ (\mathrm{kW})$$

25. The loss of specific energy from leakage equivalent to specific work on the rotor:

$$Y_{i_{pQf}} = (1 - \eta_f) \cdot Y_{i_3} = (1 - 0.96) \cdot 361.271 = 14.451\ (\mathrm{J/kg})$$

26. The internal power of the flow lost to leakage:

$$P_{i_{pQf}} = Q \cdot Y_{i_{pQf}} = 1 \cdot 14.451 = 14.451\ (\mathrm{kW})$$

27. The internal efficiency corresponding to the power lost to leakage:

$$\eta_{i_{Qf}} = 1 - \frac{P_{i_{pQf}}}{P_{i_3}} = ?$$

28. The relative loss of available power from the flow corresponding to the leakage in relation to P_{i_0}:

$$\eta_{i_{Qf0}} = \frac{P_{i_{pQf}}}{P_{i_0}} = \frac{14.451}{392.4} = 0.0368 = 3.68\ (\%)$$

29. The internal specific work available from the flow at the rotor inlet:

$$Y_{i_4} = Y_{i_3} - Y_{i_{pQf}} = 361.271 - 14.451 = 346.820\ (\mathrm{kJ/kg})$$

30. The internal power available in the flow at the rotor inlet:

$$P_{i_4} = Q \cdot Y_{i_4} = 1 \cdot 346.820 = 346.820\ (\mathrm{kW})$$

31. The absolute average velocity at the rotor inlet:

$$c_4 = \frac{\sqrt{2 \cdot Y_{i_4}}}{cos\,\alpha_4} = \frac{\sqrt{2 \cdot 346.820}}{cos\,14} = 27.143\ (\text{m/s})$$

32. The tangential velocity at the rotor inlet:

$$u = 0.5 \cdot c_4 \cdot \cos\alpha_4 = 0.5 \cdot 27.143 \cdot \cos 14 = ?$$

33. The component of the actual average absolute velocity at the rotor input in the tangential direction:

$$c_{u_4} = c_4 \cdot cos\,\alpha_4 = 27.143 \cdot cos\,14 = 26.337\ (\text{m/s})$$

34. The tangential velocity at the inside diameter of the rotor:

$$u_i = r_D \cdot u = 0.5 \cdot 13.168 = 6.584\ (\text{m/s})$$

35. The component of the absolute average velocity of the flow at the rotor inlet in the meridional direction:

$$c_{m_4} = c_4 \cdot \sin\alpha_4 = 27.143 \cdot \sin 14 = 6.566\ (\text{m/s})$$

36. The actual average relative velocity of the flow at the rotor inlet:

$$w_4 = \sqrt{c_{m_4}^2 + (c_{u_4} - u)^2} = \sqrt{6.566^2 + (26.337 - 13.168)^2} = 14.715\ (\text{m/s})$$

37. The average physical angle of the rotor blade in the tangential direction with the direction of the true average relative velocity of the flow at the entrance and exit of the rotor:

$$\beta_4 = \beta_5 = \beta_{\text{blade}} = \arcsin\frac{c_{m4}}{w_4} = \arcsin\frac{6.566}{14.715} = 26.501^{\text{o}}$$

38. The actual average relative velocity of the flow at the exit of the first passage through the rotor, equal to the corresponding average meridional velocity, for the physical blade angle $\beta_5 = 90^o$, thus, equal to $c_{m_{5*}}$:

$$w_{5*} = c_{m_{5*}} = \psi_5 \cdot w_4 = 0.97 \cdot 14.715 = 14.274\ (\text{m/s})$$

39. The specific internal work of the flow lost on the first passage through the rotor:

$$Y_{i_{p_{r1}}} = 0.5 \cdot \left(w_4^2 - w_{5*}^2\right) = 0.5 \cdot \left(14.715^2 - 14.274^2\right) = 6.392\ (\text{J/kg})$$

40. The internal power lost from the flow on the first passage through the rotor:

$$P_{i_{p_{r1}}} = Q \cdot Y_{i_{p_{r1}}} = 1 \cdot 6.392 = 6.392 \text{ (kW)}$$

41. The internal power available from the flow to be transformed by the rotor:

$$P_{i_{dr1}} = P_{i_4} - P_{i_{p_{r1}}} = 346.820 - 6.392 = 340.428 \text{ (kW)}$$

42. The tangential velocity of the rotor at the exit of the first flow passage, and since it is $\beta_{5*} = 90^o$, we have:

$$u_i = c_{u_{5*}} = 6.584 \text{ (m/s)}$$

43. The actual average absolute velocity at the exit of the first flow passage through the rotor:

$$c_{5*} = \sqrt{u_i^2 + w_{5*}^2} = \sqrt{6.584^2 + 14.274^2} = 15.719 \text{ (m/s)}$$

44. The angle between the direction of the tangential velocity u_i and the real absolute average velocity, c_{5*}, thus

$$\alpha_{5*} = \arcsin\frac{w_{5*}}{c_{5*}} = \arcsin\frac{14.274}{15.719} = 62.240^{\text{o}}$$

45. The internal flow power received by the rotor on its first passage, to be transformed into mechanical power on the axis:

$$P_{i_{r1}} = P_{i_{dr1}} - P_{i_{5*}} = P_{i_{dr}} - Q \cdot u_i \cdot c_{u_{5*}} = 340.428 - 1 \cdot 6.584 \cdot 6.584 = 297.079 \text{ (kW)}$$

46. The internal efficiency of the rotor in the first flow passage in relation to the internal power received:

$$\eta_{i_{r1}} = \frac{P_{i_{r1}}}{P_{i_{dr1}}} = \frac{297.079}{340.428} = 0.8727 = 87.27\ (\%)$$

47. The relative loss of available power of the flow in its first passage through the rotor with respect to P_{i_0}:

$$\eta_{i_{r1_0}} = \frac{P_{i_{r1}}}{P_{i_0}} = \frac{297.079}{392.4} = 0.7571 = 75.71\ (\%)$$

48. The internal power available in the flow at the exit of its first passage through the rotor, equal to the power at the entrance into the interior:

$$P_{i_{5*}} = P_{i_{6*}} = P_{i_{dr1}} - P_{i_{r1}} = 340.428 - 297.079 = ?$$

49. The actual absolute average velocity of the flow after its first passage through the rotor:

$$c_{6*} = \sqrt{2 \cdot 43.349} = 9.311 \text{ (m/s)}$$

50. The absolute average velocity of the flow before entering the second passage through the rotor:

$$c_{3*} = \phi_{3*} \cdot c_{6*} = 0,98 \cdot 9.311 = 9.125 \text{ (m/s)}$$

51. The specific internal work lost by the flow inside the rotor:

$$Y_{i_{p6*;3*}} = 0.5 \cdot \left(c_{6*}^2 - c_{3*}^2\right) = 0.5 \cdot \left(9.311^2 - 9.125^2\right) = 1.715 \text{ (J/kg)}$$

52. The power of the flow lost in the passage through the interior of the rotor:

$$P_{i_{p6*;3*}} = Q \cdot Y_{i_{p6*;3*}} = 1 \cdot 1.715 = 1.715 \text{ (kW)}$$

53. The internal efficiency of the flow inside the rotor in relation to its internal power in 6∗ resulting in:

$$\eta_{i_{6*;3*}} = 1 - \frac{P_{i_{p6*;3*}}}{P_{i_{6*}}} = 1 - \frac{1.715}{43.349} = 0.9604 = 96.04 \text{ (\%)}$$

54. The relative loss of the available power of the flow corresponding to its passage through the interior of the rotor in relation to P_{i_0}, resulting in:

$$\eta_{i_{6*;3*0}} = \frac{P_{i_{p6*;3*}}}{P_{i_0}} = \frac{1.715}{392.4} = 0.0044 = 0.44 \text{ (\%)}$$

55. The internal power available in the flow immediately before it enters the second passage through the rotor, equaling the power at the entrance:

$$P_{i_{3*}} = P_{i_{4*}} = P_{i_{6*}} - P_{i_{6*;3*}} = 43.349 - 1.715 = 41.634 \text{ (kW)}$$

56. The specific internal work of the flow at the entrance of its second passage through the rotor:

$$Y_{i_{4*}} = \frac{P_{i_{4*}}}{Q} = \frac{41.634}{1} = 41.634 \text{ (J/kg)}$$

57. The absolute average velocity component of the flow in the tangential direction at the entrance of the second passage through the rotor:

$$c_{u_{4*}} = \frac{Y_{i_{4*}}}{u_i} = \frac{41.634}{6.584} = 6.324\ (\text{m/s})$$

58. The average relative velocity of the flow at the entrance of the second passage through the rotor, with the condition that $c_{m_{4*}} = c_{m_{5*}} = 14{,}274\ (\text{m/s})$, resulting in:

$$w_{4*} = \sqrt{\left(u_i - c_{u_{4*}}\right)^2 + c_{m_{4*}}^2} = \sqrt{(6.584 - 6.324)^2 + 14.274^2} = 14.276\ (\text{m/s})$$

59. The internal power lost by the shock of the flow at the entrance of the second passage through the rotor, considering that the blade has an angle, $\beta_{blade} = 90^o$, resulting in:

$$P_{i_{p_{ch4*}}} = Q \cdot 0.5 \cdot \left(w_{4*}^2 - c_{m_{4*}}^2\right) = 1 \cdot 0.5 \cdot \left(14.276^2 - 14.274^2\right) = 0.029\ (\text{kW})$$

60. The absolute average velocity of the flow at the entrance of the second passage through the rotor:

$$c_{4*} = \sqrt{c_{m_{4*}}^2 + c_{u_{4*}}^2} = \sqrt{14.274^2 + 6.324^2} = 15.612\ (\text{m/s})$$

61. The angle the tangential direction of the rotor forms with the angle of the meridional average velocity at the entrance of the second flow passage through the rotor:

$$\beta_{4*} = \arcsin \frac{c_{m_{4*}}}{w_{4*}} = ?$$

62. The angle between the mean, meridional, and absolute velocity directions at the entrance of the second flow passage through the rotor:

$$\alpha_{4*} = \arcsin \frac{c_{m_{4*}}}{c_{4*}} = \arcsin \frac{14.274}{15.612} = 66.106^o$$

63. The average relative velocity of the flow at the exit of its second passage through the rotor:

$$w_5 = \psi_5 \cdot w_{4*} = 0.97 \cdot 14.276 = 13.848\ (\text{m/s})$$

64. The frictional power loss in the second flow passage through the rotor:

$$P_{i_{p4*;5}} = Q \cdot 0.5 \cdot \left(w_{4*}^2 - w_5^2\right) = 1 \cdot 0.5 \cdot \left(14.276^2 - 13.848^2\right) = 6.019\ (\text{kW})$$

65. The meridional average velocity of the flow at the exit of the second passage through the rotor, which is the average constructive angle of the blade $\beta_5 = \beta_4 = 26.501^\circ$, thus

$$c_{m_5} = w_5 \cdot \sin\beta_5 = 13.848 \cdot \sin 26.501 = 6.179\ (\text{m/s})$$

66. The absolute average velocity component of the flow in the tangential direction at the exit of the second passage through the rotor:

$$c_{u_5} = u - w_5 \cdot \cos\beta_5 = 13.168 - 13.848 \cdot \cos 26.501 = 0.775\ (\text{m/s})$$

67. The absolute average velocity of the flow at the exit of the second passage through the rotor:

$$c_5 = \sqrt{c_{m_5}^2 + c_{u_5}^2} = \sqrt{6.179^2 + 0.775^2} = 6.227\ (\text{m/s})$$

68. The angle formed by the directions of the tangential and mean absolute velocities of the flow at the exit of the second passage through the rotor:

$$\alpha_5 = \arcsin\frac{c_{m_5}}{c_5} = \arcsin\frac{6.179}{6.227} = 82.881^\circ$$

Since $\alpha_5 = 90^\circ$ was imposed and resulted in $c_{m_{5**}} = u \cdot tg\beta_5 = 13.168 \cdot tg26.501 = 6.566\ (\text{m/s})$, which is equal to the meridional average velocity of the flow at the entrance of the rotor $c_{m_4} = 6.566\ (\text{m/s})$, resulting in an internal power loss of the shock flow at the outlet (see Fig. 4.8), this will be determined in the next step, as:

$$w_{5**} = \frac{u}{\cos\beta_5} = \frac{13.168}{\cos 26.501} = 14.714\ (\text{m/s})$$

69. The loss of internal power, from shocks at the outlet of the rotor flow:

$$P_{i_{p_{ch_5}}} = Q \cdot 0.5 \cdot \left(c_{m_{5**}}^2 - c_{m_5}^2\right) = 1 \cdot 0.5 \cdot \left(6.566^2 - 6.179^2\right) = 2.489\ (\text{kW})$$

70. The internal specific work of the flow at the exit of the rotor equal to the specific work after exiting:

$$Y_{i_5} = Y_{i_6} = 0.5 \cdot c_{m_5}^2 = 0.5 \cdot 6.179^2 = 19.090\ (\text{J/kg})$$

71. The internal specific power of the flow at the exit of the rotor equal to the specific power after exiting:

$$P_{i_6} = Q \cdot Y_{i_6} = 1 \cdot 19.090 = 19.090\ (\text{kW})$$

72. The internal flow power received by the rotor in the second passage that will be transferred to its axis:

$$P_{i_{r_2}} = P_{i_{4*}} - (P_{i_{p_{ch4*}}} + P_{i_{p_{4*;5}}} + P_{i_{p_{ch_5}}} + P_{i_6} = 41.634 - (0.029 + 6.019 + 2.489 + 19.090) = 14.007\ (\text{kW})$$

73. The internal flow power received by the rotor in the two passages that will be transformed into mechanical power on the axis:

$$P_{i_r} = P_{i_{r_1}} + P_{i_{r_2}} = 297.079 + 14.007 = 311.086\ (\text{kW})$$

74. The internal efficiency of the BMHIT:

$$\eta_i = \frac{P_{i_r}}{P_{i_0}} = \frac{311.086}{392.4} = 0.7928 = 79.28\ (\%)$$

75. The portion of the BMHIT internal power received by the rotor on its first passage:

$$\eta_{i_{r_1}} = \frac{P_{i_{r_1}}}{P_{i_o}} = \frac{297.079}{392.4} = 0.7571 = 75.71\ (\%)$$

76. The portion of the *BMHIT*'s internal power received by the rotor on the second passage:

$$\eta_{i_{r_2}} = \frac{P_{i_{r_2}}}{P_{i_o}} = \frac{14.007}{392.4} = 0.0357 = 3.57\ (\%)$$

77. The sum of the internal power losses in the BMHIT, including equivalent leakage and kinetic energy losses at the rotor output:

$$P_{i_p} = P_{i_{p_{tr}}} + P_{i_{p_{1;2}}} + P_{i_{p_3}} + P_{i_{p_{Qf}}} + P_{i_{p_{r1}}} + P_{i_{p_{6*;3*}}} + P_{i_{p_{ch4*}}} + P_{i_{p_{r2}}} + P_{i_{p_{ch_5}}} + P_{i_6}$$

$$=$$

$$0.621 + 23.160 + 7.348 + 14.451 + 6.392 + 1.715 + 0.029 + 6.019 + 2.489 + 19.090$$

$$=$$

$$81.314\ (kW)$$

78. The BMHIT's internal yield, considering total losses:

$$\eta_{i_{tp}} = 1 - \frac{P_{i_p}}{P_{i_0}} = 1 - \frac{81.314}{392.4} = 0.7924 = 79.28(\%)$$

This value is practically equal to the value obtained in step 74, indicating the validity of the steps up to this point.

79. The portion of the BMHIT's internal power lost between the input and output:

$$\eta_{i_p} = \frac{P_{i_p}}{P_{i_o}} = \frac{81.314}{392.4} = 0.2072 = 20.72\ (\%)$$

80. The internal efficiency without considering kinetic energy losses at the output:

$$\eta_{i*} = 1 - \frac{P_{i_p} - P_{i_6}}{P_{i_0}} = 1 - \frac{81.314 - 19.09}{392.4} = 0.8414 = 84.14\ (\%)$$

81. The cross-sectional area of the conduit at the entrance to the BMHIT:

$$A_0 = \frac{Q}{c_o} = \frac{1}{5.603} = 0.1785\ \left(m^2\right)$$

82. The diameter of the conduit at the inlet of the BMHIT:

$$D_0 = \sqrt{\frac{4 \cdot A_0}{\pi}} = \sqrt{\frac{4 \cdot 0.1785}{\pi}} = 0.477\ (\mathrm{m})$$

83. The average flow velocity at the inlet of the injector:

$$c_1 = 2 \cdot c_0 = 2 \cdot 5.603 = 11.206\ (\mathrm{m/s})$$

84. The internal specific energy of the static pressure of the flow at the inlet of the BMHIT:

$$\frac{p_{i_0}}{\rho} = Y_{i_0} - \frac{c_0^2}{2} = 392.4 - \frac{5.603^2}{2} = 376.703\ (\mathrm{J/kg})$$

85. The internal specific energy of the static pressure of the flow at the exit of the transition piece, which is equal to the internal specific energy of the static pressure at the inlet of the injector:

$$\frac{p_{i_1}}{\rho} = Y_{i_0} - \frac{c_1^2}{2} + Y_{p_{i_{tr}}} = 392.4 - \frac{11.206^2}{2} + 0.621 = 330.234 \text{ (J/kg)}$$

86. The cross-sectional area of the transition piece at input 1:

$$A_1 = \frac{Q}{c_1} = \frac{1}{11.206} = 0.089238 \left(\text{m}^2\right)$$

87. The actual average radial velocity of the flow at the outlet of the injector:

$$c_{r_2} = c_2 \cdot \sin \alpha_2 = 27.152 \cdot \sin 14 = 6.569 \text{ (m/s)}$$

88. The BMHIT synchronous minimum and maximum rotor velocities:

$$n^*_{\text{min}} = 5 \cdot \frac{u^{15}}{\left(\eta_f \cdot Q\right)^{0.5}} = 5 \cdot \frac{13.168^{15}}{(0.96 \cdot 1)^{0.5}} = 243.845 \text{ (rpm)} \rightarrow$$

$$z^*_{\text{pp}_{\max}} = \frac{3600}{n^*_{\text{min}}} = \frac{3600}{243.845} = 14.763 \rightarrow$$

$$z_{\text{pp}_{\max}} = 14 \rightarrow$$

$$n_{\text{min}} = \frac{3600}{14} = 257.143 \text{ (rpm)}$$

and

$$n^*_{\max} = 18 \cdot \frac{u^{15}}{\left(\eta_f \cdot Q\right)^{0.5}} = 18 \cdot \frac{13.138^{15}}{(0.96 \cdot 1)^{0.5}} = 877.843 \rightarrow$$

$$z^*_{\text{pp}_{\min}} = \frac{3600}{n^*_{\max}} = \frac{3600}{877.843} = 4.101 \rightarrow$$

$$z_{\text{pp}_{\min}} = 4 \rightarrow$$

$$n_{\max} = \frac{3600}{4} = 900 \text{ (rpm)}$$

A first attempt will be made with $n = 600$ (rpm). The computer program should be able to present results stored in memory, with variations in the number of EG pole pairs, between the minimum and maximum values, so that a table with these values and corresponding diameters can be made after having the results from the next three steps. With this in hand, the design team can decide which values and calculations should continue.

89. The rotor diameters as:

$$D = \frac{60 \cdot u}{\pi \cdot n} = \frac{60 \cdot 13.168}{\pi \cdot 600} = 0.419\ (\text{m})$$

and

$$D_i = 0.5 \cdot D = 0.5 \cdot 0.419 = 0.210\ (m)$$

90. The injector widths for:

- $\delta = 60^{\circ}$
- 90°
- 120°

In this case, only one attempt will be made for $\delta = 90^{\circ}$, so:

$$B_{in_j} = \frac{360 \cdot Q}{\pi \cdot \delta_i \cdot D \cdot c_{r_2}} \rightarrow B_{\text{in}} = \frac{360 \cdot Q}{\pi \cdot \delta_{90} \cdot D \cdot c_{r_2}} = \frac{360 \cdot 1}{\pi \cdot 90 \cdot 0.419 \cdot 6.569} = 0.463\ (\text{m})$$

91. The corresponding heights of the injector:

$$b_{in_j} = \frac{A_1}{B_{in_j}} \rightarrow b_{\text{in}} = \frac{A_1}{B_{\text{in}}} = \frac{0.089238}{0.463} = 0.193\ (\text{m})$$

92. The length of the rotor:

$$B = 1.15 \cdot B_{\text{in}} = 1.15 \cdot 0.463 = 0.532\ (\text{m})$$

93. The preliminary diameter of the rotor shaft:

$$d_{\text{e}} = 118 \cdot \sqrt[3]{\frac{P_{\text{max}}}{n}} = 118 \cdot \sqrt[3]{\frac{392.4}{600}} = 102.4\ (\text{mm})$$

94. The distance $L_{\text{e}_{\text{min}}}$ of the flow inside the rotor with respect to its axis:

$$L_{\text{e}_{\text{min}}} > 0.5 \cdot 102.4 = 51.25\ (\text{mm})$$

95. The distance from the rotor axis to the internal flow limit inside:

$$L_{\text{e}} = 0.5 \cdot r_D \cdot D \cdot \cos\alpha_{5*} = 0.5 \cdot 0.5 \cdot 419 \cdot \cos 62.240 = 48.789\ (\text{mm})$$

With $L_{\text{e}} < L_{\text{e}_{\text{min}}}$ the internal part of the flow in the passage through the interior of the rotor collides with the axis (see Fig. 4.3).

96. The preliminary theoretical step at the input and output of the rotor:

$$t_4^* = 0.5 \cdot (D - D_i) = 0.5 \cdot (0.419 - 0.210) = 0.105\ (\mathrm{m})$$

97. The number of rotor blades:

$$z_\mathrm{r} = \frac{\pi \cdot D}{t_4} = \frac{\pi \cdot 0.419}{0.105} = 12.5.$$

$z_\mathrm{r} = 12$ blades will be adopted. With this, the pitch at the input and output of the rotor results in:

$$t_4 = \frac{\pi \cdot D}{z_r} = \frac{\pi \cdot 0.419}{12} = 0.110\ (\mathrm{m})$$

98. The number of blades covered by the rotor at the outlet of the injector:

$$z_\delta = \frac{\delta}{360} \cdot z_r = z_\delta = \frac{97.502}{360} \cdot 12 = 3.25 \text{ blades}$$

99. The theoretical step at the exit of the first flow passage through the rotor and at the entrance of the second:

$$t_{5*} = t_{4*} = r_D \cdot t_4 = 0.5 \cdot 0.11 = 0.055\ (\mathrm{m})$$

100. The preliminary blade thickness at the entrance and exit of the rotor in the tangential direction:

$$e_{t_4} = e_{t_5} = \left(1 - \frac{c_{m_3}}{c_{m_4}} \cdot \frac{B_\mathrm{in}}{B}\right) \cdot t_4 = 1 - \frac{6.503}{6.566} \cdot \frac{0.463}{0.532} \cdot 0.110 = 0.0159\ (\mathrm{m})$$
$$= 15.9\ (\mathrm{mm})$$

101. The constant primary rotor blade thickness:

$$e_\mathrm{blade} = e_{t_4} \cdot \sin\, \beta_4 = ?$$

This value must be verified via manual calculations.

102. The throttling factor of the blades at the inlet and outlet of the rotor:

$$f_{e_4} = f_{e_5} = 1 - \frac{e_{t_4}}{t_4} = 1 - \frac{0.0159}{0.11} = 0.855$$

103. The throttling factor of the blades at the exit of the first flow passage through the rotor and entry into the second:

$$f_{e_{5*}} = f_{e_{4*}} = 1 - \frac{e_{\text{blade}}}{t_{5*}} = 1 - \frac{0.0071}{0.055} = 0.871$$

104. The average radius of the rotor blades:

$$r_{\text{blade}} = 0.25 \cdot \frac{D^2 - D_i^2}{D \cdot \cos\beta_4 - D_i \cdot \cos\beta_{5*}} = ?$$

105. The length of the rotor blade chord, L_{blade}, and its coverage angle θ_{blade} from the following linear system of three equations:

$$\alpha_2 = 90 - (0.5 \cdot \theta_{\text{blade}} + \beta_4) = 90 - (0.5 \cdot \theta_{\text{blade}} + 26.501)$$

and

$$L_{\text{blade}} = 2 \cdot r_{\text{blade}} \cdot \sqrt{1 - \cos^2(0.5 \cdot \theta_{\text{blade}})} = 2 \cdot 88 \cdot \sqrt{1 - \cos^2(0.5 \cdot \theta_{\text{blade}})}$$

and

$$L_{\text{blade}} = L^*_{\text{blade}} = 0.5 \cdot D \cdot \cos\alpha_2 \pm 0.5 \cdot \sqrt{(D \cdot \cos\alpha_2)^2 - (D^2 - D_i^2)}$$

$$= 0.5 \cdot 419 \cdot \cos\alpha_2 - 0.5 \cdot \sqrt{(419 \cdot \cos\alpha_2)^2 - (419^2 - 210^2)}$$

$$= 209.5 \cdot cos\,\alpha_2 - 0.5 \cdot \sqrt{(419 \cdot \cos\alpha_2)^2 - 131461}$$

The solution of the system was obtained by recurrence, resulting in:

$$L_{\text{blade}} = 117 \text{ (mm)}$$

$$\theta_{\text{blade}} = 96.67^{\text{o}}$$

To program the solution you can begin with:

$$\theta_{\text{blade}} = 60^{\text{o}}$$

or use a scaled drawing of the line blade average in the arc of the circumference, and then measure angle $\theta^{\text{o}}_{\text{blade}}$

106. The radii that delimit the injector:

$$r_{\theta_j} = 0.5 \cdot D \cdot \exp\left(\frac{\pi \cdot tg\alpha_2}{180} \cdot \theta_j\right) + e_{fo} = 0.5 \cdot 0.419 \cdot \exp\left(\frac{\pi \cdot tg14}{180} \cdot \theta_j\right) + 0.002$$

$$= 0.2095 \cdot \exp\left(0.783287 \cdot \frac{\theta_j}{180}\right) + 0.002 \text{ (m)}.$$

4.7.3 *Analysis and Commentary*

4.7.3.1 Average Rotor Flow Velocity Triangles and Modeling

The results of the calculations of these average flow velocities, and the resulting angles of their directions, have been transcribed in Table 4.3, and the respective average velocities triangles have been drawn to scale in Fig. 4.15, allowing us to state the following:

- Perhaps, the biggest novelty of this algorithm is that it allows us to develop a computer program that gives initial results on the real line of average flow (RLAF), or the average real streamline (ARS), in a normal view to the axis of the BMHIT, which gives alternatives so that the real hydraulic flow path (RHFP) can be preliminarily drawn without first having to consider mechanical structural issues.
- The data were set from sub-item 4.7.1, so that the flow through the rotor, $Q_r = \eta_f \cdot Q$ (m^3/s) is constant, and consequently, $c_{m_5} = c_{m_4}$, $c_{m_{5*}} = c_{m_{4*}}$ therefore, $\alpha_5 = 90^{\circ}$ should be analyzed in two ways, first, as the flow at the entrance and exit of the rotor, $A_4 = A_5$, and so that $A_{5*} = A_{4*}$ maintains their respective geometries inside the router with respect to the cross sections i.e., they are not influenced by air resistance, which should be verified if possible at least in computational hydrodynamics studies.
- The second aspect implies that the triangle formed by the average flow velocities at the exit of the rotor has been determined in advance, since u, $\beta_5 = \beta_4$,and $\alpha_5 = 90^{\circ}$ are known. However, in most cases, similar to this study, determining the components of this triangle from the available energy at the entrance of the second flow passage through the rotor results in different values for the components, resulting in a shock component, c_{ch_5}, that is equal but in the opposite direction, canceling the specific energy at output, $u_i \cdot c_{u_5}$.
- The schematic in Fig. 4.16 models what happens at the entrance and exit of the blades, showing the regions where the so-called dead nuclei (virtually motionless water) oscillate, which arise from the simultaneous effects of rotor rotation, flow turbulence, friction, and blades thickness, which result in changes in predicted flow angle averages. Here, as was already mentioned, in the second passage of the flow through the rotor, the resulting shock is beneficial relative to the power provided by the rotor.

Table 4.3 Rotor Flow—Angles

Points	Average speeds (m/s)					Angles (degrees)	
–	u	c	w	c_u	c_m	α	β
4	13.17	27.14	14.72	26.34	6.57	14.0	26.5
5*	6.58	15.72	14.27	6.58	14.27	62.2	90.0
4*	6.58	15.61	14.18	6.32	14.27	66.1	89.0
5	13.17	6.23	13.85	0.78	6.18	82.9	26.5
5**	13.17	6.57	14.77	0.00	6.57	90.0	26.5

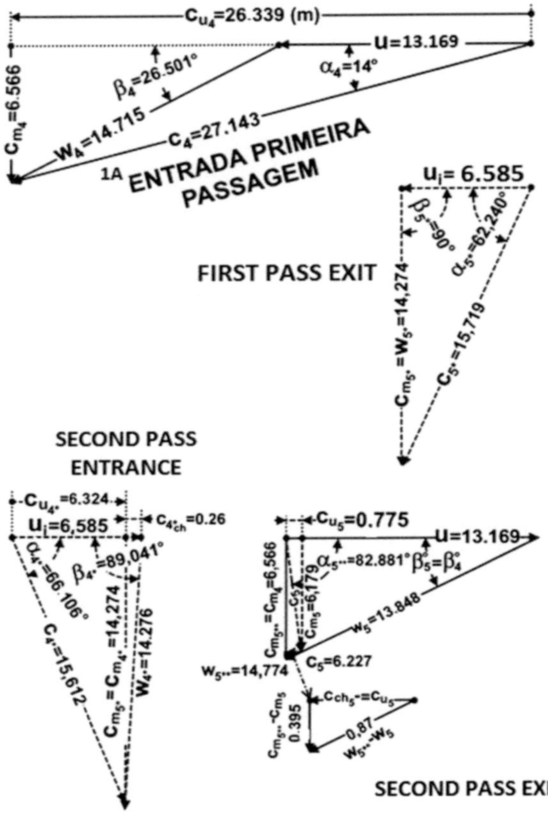

Fig. 4.15 To-scale graphic representation of the average flow velocity triangles relative to their passages through the rotor

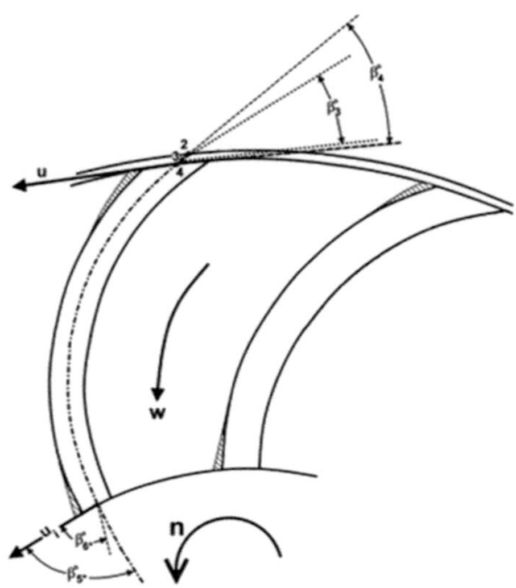

Fig. 4.16 Schematic of the first flow passage through the rotor modeling the dead spaces

4.7.3.2 Energy Flow Distribution for the BMHIT

The scaled drawing in Fig. 4.17 of the energy diagram or Sankey diagram shows the distribution of the internal powers of the flow and the respective losses in the passage through the BMHIT components studies here, allowing us to analyze and comment on:

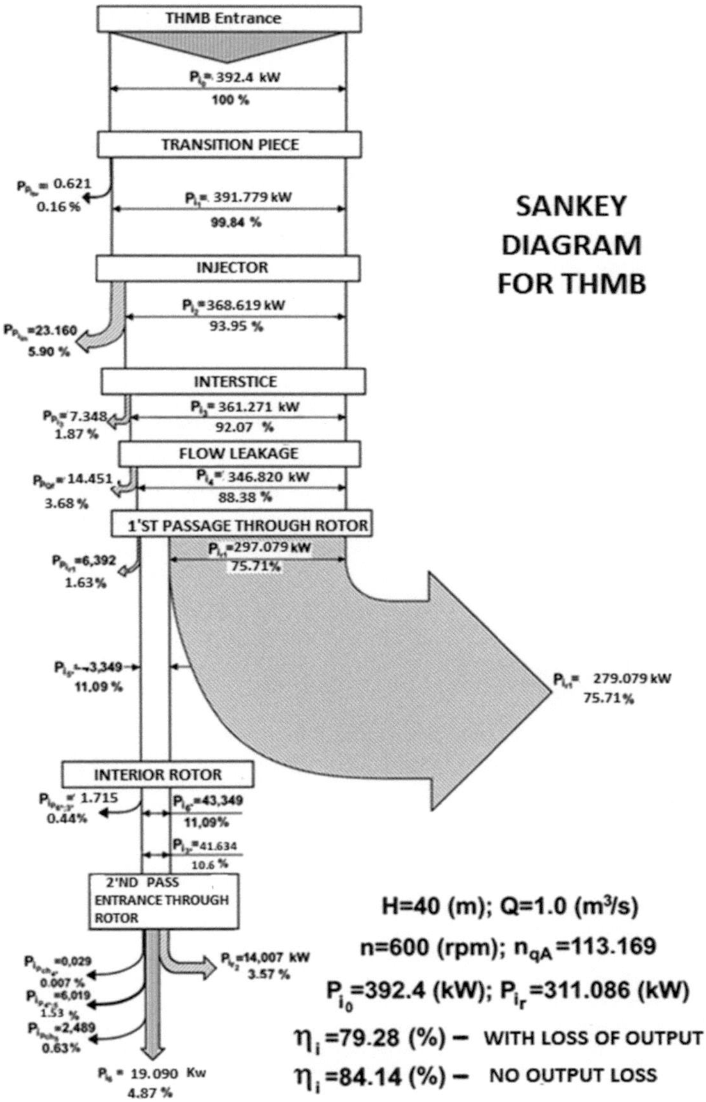

Fig. 4.17 Sankey diagram for validating the BMHIT

- The BMHIT's internal yield reached 79.28 (%), given the assumptions and set values. This will then be compared to flow values taking water as an ideal fluid.
- Most of the internal energy available in the flow at the entrance of the BMHIT, here 392.4 kW, is delivered to its rotor in the first passage, with the remaining small part at 3.57 (%) delivered in the second passage.
- Energy losses in other components, which totaled 15.85 (%), occur due to friction, shock, and water leakage when flowing. The remaining 4.87 (%) of the rotor cannot be transformed into mechanical flowing power in the form of kinetic energy at the output.
- The largest internal losses are concentrated in the injector, at 5.9 (%), followed by 4.87 (%), corresponding to the kinetic energy available at the rotor outlet and the flow leakage at 3.68 (%) in the existing interstice between the injector outlet and the rotor inlet.
- In the second passage of the flow through the rotor, little energy is delivered, but delivery occurs with many shocks, and therefore high turbulence, which is responsible for most of the vibrations, particularly under low load operation conditions.

4.7.3.3 Comparison of the Results

- To compare the results using the same basic input data, but with water flows without losses (Sect. 3.4 from Chap. 3) and with losses, we have prepared Table 4.4. The results show that the value can still be increased by $\eta_i < 14.50$ (%), provided that the limits of the characteristics used for calculating the losses can be changed and effectively obtained in the manufacturing processes.

4.7.3.4 RHFP Layout

Using the geometric results obtained in the validation transcribed in Table 4.5, we can draw a preliminary cross section of the RHFP axis in the BMHIT to scale in Fig. 4.18, observing the following steps:

1. From point 0, we start drawing in the same scale as Fig. 4.18, positioning rays r_{θ_j} every 15 degrees, with values taken from Table 4.5, and drawing section *se* that is part of the external limit of the RHFP.

Table 4.4 Flow meet without and with losses in BMHIT wiPHITo = 392.4 (kW)

Lossless			With losses		Difference
	P_r (kW)	η_i (%)	P_r (kW)	η_i (%)	$\Delta\eta_i$ (%)
P_{r1}	343.01	87.5	297.08	75.7	11.8
P_{r2}	24.66	6.3	14.01	3.6	2.7
P_r	369.45	93.8	311.09	79.3	14.5

Table 4.5 Geometric Application Results

D_0	A_0	A_1	B_{in}	b_{in}	B	L_{tr}*	δ	D	D_i	t_4	t_{5*}	
m	m^2	m^2	m	m	m	m	degrees	m	m	m	m	
0.48	0.18	0.09	0.47	0.19	0.53	0.46	90	0.42	0.21	0.11	0.06	
$e_{pá}$	$e_{t4\ t}$	f_{e4}	f_{e5}	r_θ	$z_{pá}$	z_δ	d_e	L_d	β_4	β_{5*}	$R_{pá}$	$L_{pá}$
mm	mm	–	–	Table 4.2	Num.	Num.	mm	mm	degrees	degrees	mm	mm
7.10	15.90	0.86	0.87		12.00	3.25	102.40	93.50	26,501	90	88.00	117.00

L_{tr} it was taken from Fig. 4.18

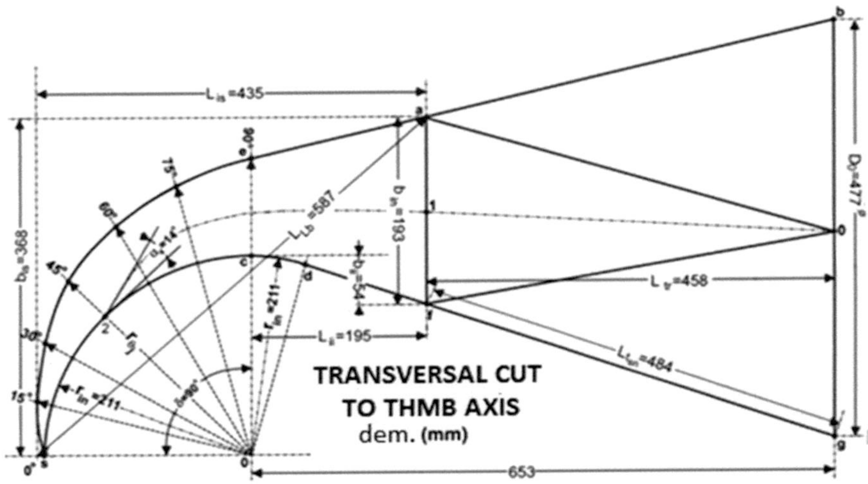

Fig. 4.18 Cross section of the BMHIT axis from the algorithm validation

Table 4.6 Values taken from Fig. 4.18

L_{is}	L_{ii}	b_{ii}	b_{is}	L_{tr}	$L_{t\ sn}$	L_{ii}
mm	mm	mm	mm	mm	mm	mm
435.00	195.00	54.00	368.00	458.00	484.00	195.00

2. From *e*, we draw a tangent line *ea*.
3. From *a*, $b_{in} = 193$ (*mm*) is marked vertically, obtaining *point f*.
4. From *s*, with a center at 0, and radius $r_{in} = 211$ (mm), an arc of circumference is drawn so that segment *df* is tangent to the circumference at *d*.
5. Next, the extensions of tangents *a* and *f* were drawn.
6. With the value at a scale of $D_0 = 477$ (mm), points *b* and *g* are determined.
7. Then, on the scale drawing, the geometric characteristics are proportionally measured, in a first approximation, transcribed in Table 4.6.

4.7.3.5 Distribution of Static Pressures and Average Speeds

Considering only the ARS cross sections of flow 0; 1; 2, in the graph in Fig. 4.19, the lines corresponding to the distribution of the average absolute real velocities and the corresponding static pressures were drawn in (mca).

These drawings show that, as a whole, and particularly with regard to the behavior of the static pressure and the average absolute velocity of the flow, the established hypotheses and criteria, when observed in studies and projects, can create satisfactory operational performance expectations for the BMHIT, given the smooth continuities.

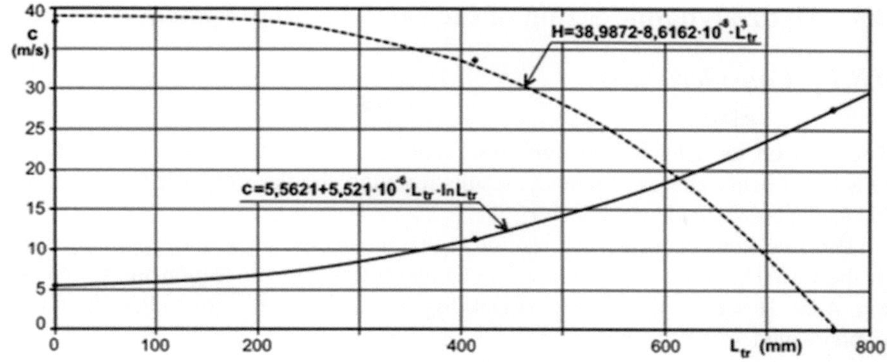

Fig. 4.19 Graphs of the mean lines of the mean absolute velocities and the static pressures in the flow from the input to the output of the BMHIT injector, obtained from the algorithm validation

4.7.3.6 Complementary Comments

- The real quantitative equation of the flow inside the rotor in the second passage, even under average conditions, occurs with a high degree of uncertainty, even when using current computational hydrodynamic techniques. This does, however, permit a more qualitative aspect for the distribution of velocities, which, when compared with measurements point at equivalent points, may provide very reliable results for BMHIT calculations and projects that are geometrically similar.
- The Sankey diagram needed to operate the BMHIT is absolutely important for analyzing behavior during its useful lifespan, and for providing data for future use in studies, projects, and developments of this type of hydraulic turbine (HT). For this, one must install the HT in places outlined in the executive project, with adequate instrumentation, and with at least a precision/accuracy of ≤ 1 (%), for local readings, at predetermined recording times, or at distances with available records, either on site, or by companies contracted to carry out the monitoring process, particularly for any occasional or preventive maintenance, as outlined in the program schedule.
- Based on this algorithm, it is possible to develop a computer program capable of calculating and generating the corresponding energy diagrams, allowing the variation of fixed characteristics such as loss coefficients or geometric coefficients. However, the flow dynamics must be preserved.
- This algorithm should be complemented with calculations and preliminary designs of the load control system, developed in the next item.

4.8 Hydrodynamic Sizing of the Load Control System

4.8.1 Considerations

The load on the BMHIT, in most installations, is controlled via flow variations, which, for better energy efficiency, results from the circular rotation of one or more blades installed inside the injector.

This load control system can take on several geometries, e.g., the types contained in the OLADE Manual—Volume I, schematized as a cross section to the axis in Fig. 4.20. However, these systems lack hypotheses and data, so studies should be implemented.

In Fig. 4.21, taken from a catalog of a German manufacturer, Ossberger, there is a solution with two blades, one covering 1/3 of the injector, and the other covering the remaining 2/3.

This solution allows one to substantially increase the operational field of the BMHIT with a small drop in the internal performance, as shown in the manufacturer's graph in Fig. 4.22. In this graph, with dimensionless coordinates, the opening range of the first flap is set between the point *A* at the beginning of the operation, and the maximum performance point *B*, at the end of the opening of the first flap.

The second flap in 2/3 up to point *F* corresponds to the maximum relative yield, for operation with the first flap closed. With the first operating at *B* and the second blade with relative efficiency in the range between points *B* and *J*, this operation reaches the maximum relative efficiency at point I, resulting in the operating range of the BMHIT with a drop in relative efficiency less than 5 (%), which is much broader than for single blade operation covering the entire flow range.

Evidently, this solution, in addition to increasing the length of the injector and rotor given the need for the divider, which is needed to house the vane trunnions, requires a dual adjustment charge regulator.

Figure 4.23 shows the possibility for using more than one blade, in the two schematics with ring sector regulation, the same system used in Francis, Helix, or Kaplan *HT* load regulation, with spiral housing. Theoretical and practical studies on this possibility are not available in literature in the public domain.

4.8.2 Injector Blade Sizing and Geometry

4.8.2.1 Blade Profile

In the OLADE Manual—Volume I, referenced in the bibliography, drawings 12/34 to 15/24 show the recommended geometric layout. Souza, Z.[2] presented a

---

[2] *Souza, Zulcy—Dimensionamento Hidrodinâmico de Injetor para THMB—VIII Seminário Nacional de Produção e Transmissão de Energia Elétrica—Grupo I/II—Brasília—DF—BR—1985.*

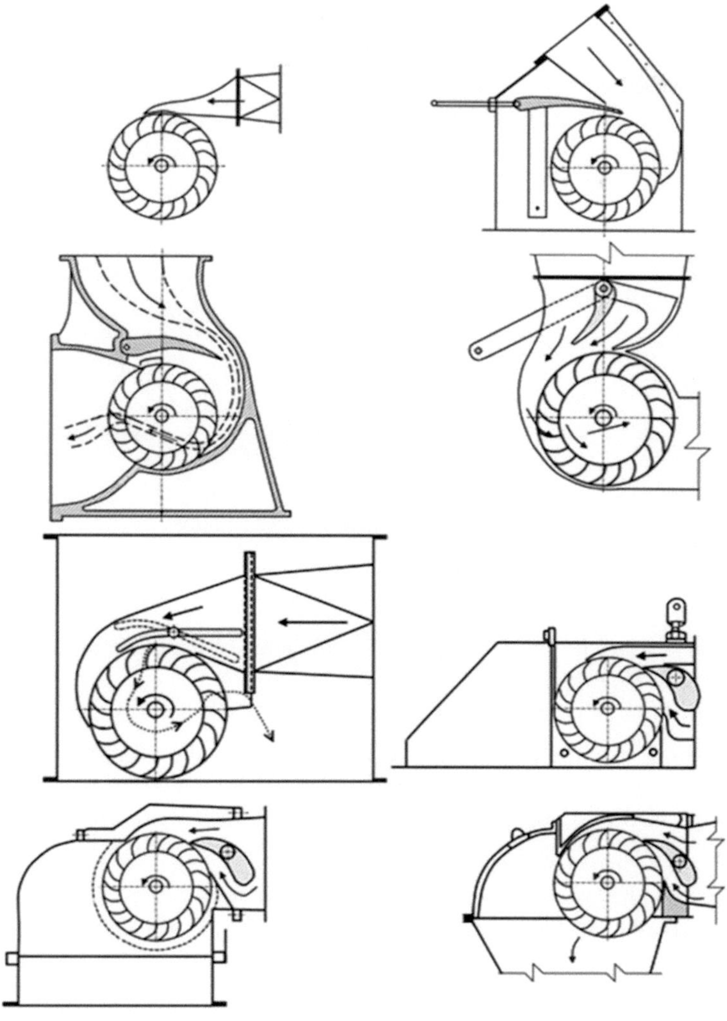

Fig. 4.20 Geometries suggested by the OLADE Manual for injectors and blade

study on the subject, which was later tested at the Hydrodynamic Laboratory for Small Hydroelectric Power Plants (LHPCH), at the Federal University of Itajuba (UNIFEI), Brazil.

In this study, we will present the possibility of using a profiled blade, which, despite eventually being more expensive to manufacturing, may result in compensatory energy compensation, particularly for flows with $N\frac{w \quad \cdot L_a{}^4}{\nu}{}_{Re}$, and roughness $\varepsilon < 0.010$ (mm). In this expression the symbols used correspond to:

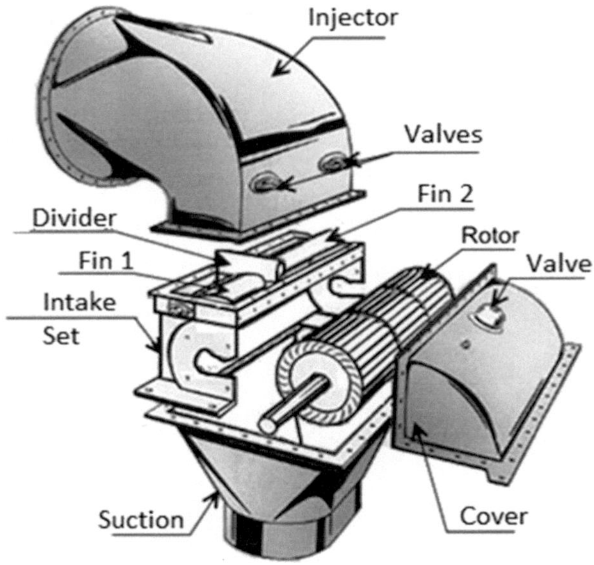

Fig. 4.21 BMHIT with two blades, with detailed named parts from the German Ossberger manufacturer's catalog

- N_{Re}–Reynolds number
- w_∞ (m/s)– water flow velocity at the leading edge of the blade profile
- L_a (m)– blade profile chord length
- ν (m^2/s)– kinematic viscosity of water $\nu \cong 10^{-6}$ (m^2/s).

Considering that the initial section of the blade in the BMHIT is practically straight, the NACA[3] profile will be used, where each number represents a characteristic, e.g., NACA 65-010, which will be used with data shown in Fig. 4.24:

- *Number 6*—characterizes the profile series.
- *Number 5*—the position of the minimum pressure for the flow at the inlet without shocks in tenths of the length of the profile, so if $L_a = 100$ (mm) the minimum pressure occurs with $x = 50$ (mm) and $y = 4.870$ (mm).
- *Number 0*—the measure of lift for an inflow without shocks, also indicating the curvature of the profile. For a symmetrical profile the curvature is 0.
- *Number 10*—the ratio of the thickness in (%) relative to the length L.

Other widely used relationships for the profile in question include:

[3] *National Advisory Committee for Aeronautics.*

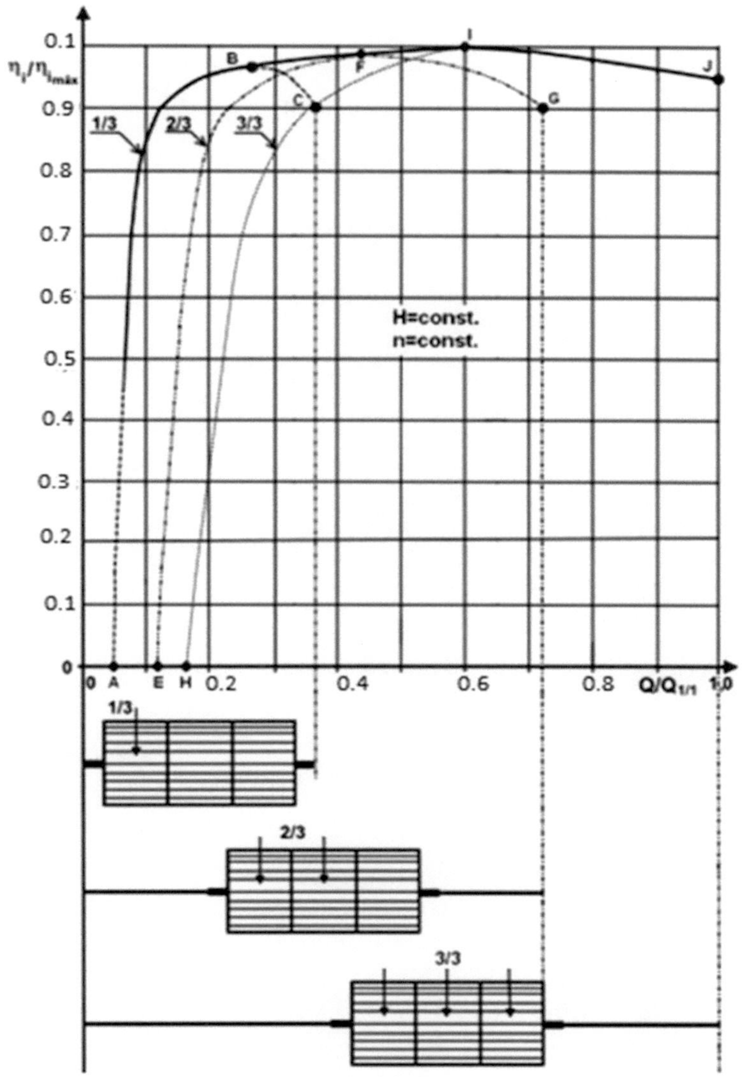

Fig. 4.22 Chart of the BMHIT operational field represented in the same perspective as Fig. 4.21

- The thickening or thinning factor:

$$f_e = \frac{\left(e_{cg}/L_a\right)_{\text{calculated}}}{\left(e_{cg}/L_p\right)_{\text{profile}}} \rightarrow f_{e_{65-010}} = \frac{\left(e_{cg}/L_a\right)_{\text{cal}}}{0.101} \tag{4.55}$$

- The profile ordinate in the center of gravity:

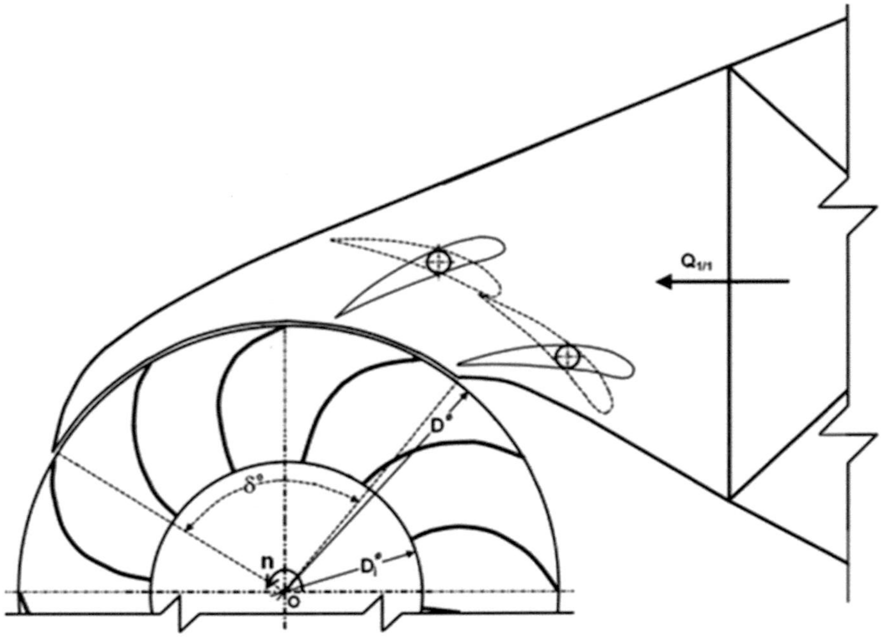

Fig. 4.23 A cross-sectional diagram with two longitudinal blades systems

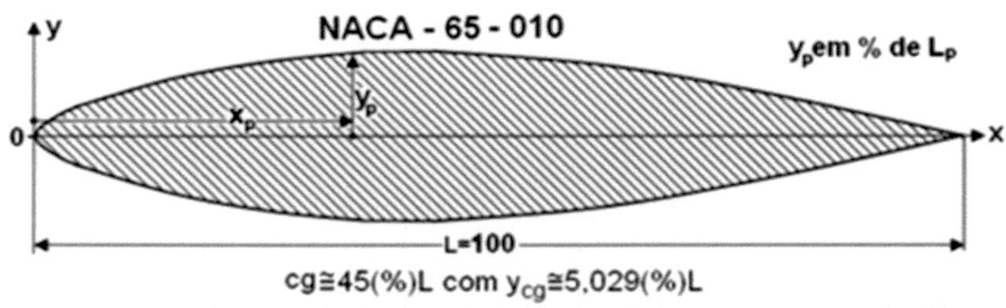

x_p	0	0.5	0.75	1.25	2.5	5.0	7.5	10
y_p	0	0.752	0.890	1.124	1.571	2.222	2.709	3.111

x_p	15	20	25	30	35	40	45	50	55
y_p	3,746	4,218	4,570	4,824	4,982	5,057	5,029	4,870	4,570

x_p	60	65	70	75	80	85	90	95	100
y_p	4,175	3,768	3,362	2,955	2,549	2,142	1,735	1,319	0

Raio na entrada do perfil: 0,666; Aresta posterior: 1,000

Fig. 4.24 Geometric characteristics of the NACA 65–10 profile

$$y_{cg} = 0.5 \cdot e_{cg} \tag{4.56}$$

- The maximum profile ordinate:

$$y_{\max} = 0.5 \cdot \left(\frac{Y_{45}}{Y_{cg}} \right) \cdot e_{cg_{cal}} + 0.028 \cdot f_e \cdot L_a \tag{4.57}$$

- The abscissas and ordinates:

$$x_a = 0.01 \cdot L_a \cdot x_p \text{ (mm)} \rightarrow L_a \text{ (mm)}$$

and

$$y_a = 0.01 \cdot f_e \cdot L_a \cdot y_p \tag{4.58}$$

4.8.2.2 Drawing the Blade Midline or Frame

Basically, the mid-line or skeletal line of the blade in profile for a BMHIT injector can be drawn using various geometries, with emphasis here on spiral, parabolic, and straight/circumferential arc shapes. To detail each of these routes, the following steps will be developed with data from the BMHIT under study here. To compare the results, the pattern will be the midline of the blade obtained from the spiral drawing.

- **Spiral Arc**

1. Start Fig. 4.25 with part of Fig. 4.18.
2. Draw the normal line r_1 passing through the middle of $D_0$and b_{in}, obtaining point 1*.
3. Copy the spiral *se* and adjust it so that it is tangential to r_1, forming angle $\alpha_2 = \alpha_4 = 14^o$ with the circle segment at radius $r_{in} = 211$ (mm) thereby determining point *B*.
4. Measure the angle formed by 0*s* and 0*A*, i.e., 30°.
5. Connect and measure, in scale with the drawing, the length of the chord axis of the fin, i.e., $L_a = 240$(mm).

- **The Arc Circumference**

1. Start Fig. 4.26 with part of Fig. 4.18.
2. Draw the side of the 40° angle from the line segment $\bar{0}s$, thus obtaining point *A*.
3. From A, mark angle $\alpha_2 = \alpha_4 = 14^o$ with a side tangent to the circle radius $r_{in} = 211$ (mm), on the scale drawing, obtaining straight line r_2.
4. With $L_a = 240$(mm), in the scale of the drawing, determine point *B* with r_1.

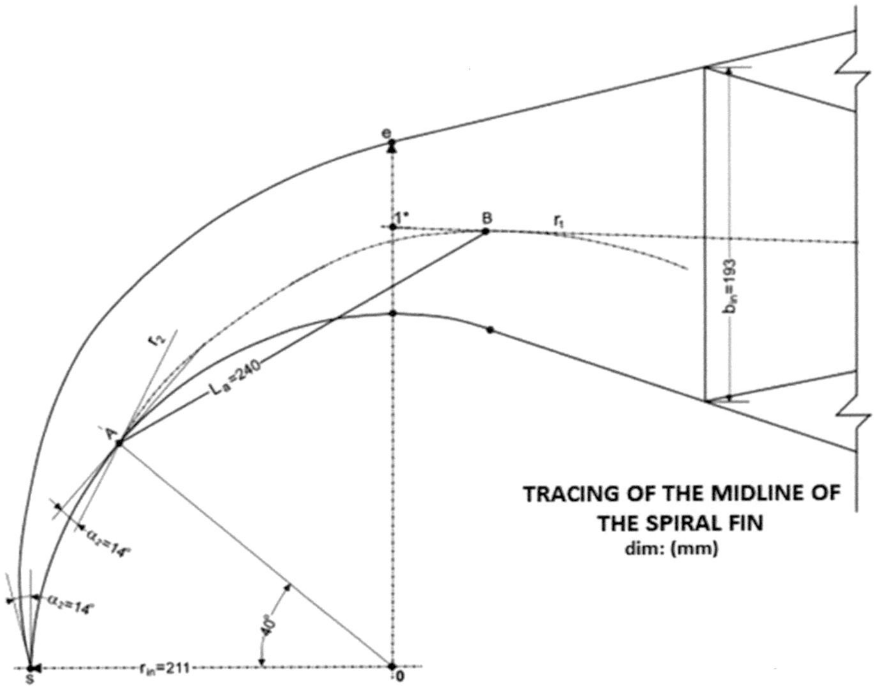

Fig. 4.25 Draw the midline of the blade in the form of a spiral arc

5. From point A and point B and from lines $r_1$and r_2, respectively, mark angles at 90°, obtaining point $0*$ at the meeting point.
6. With the center at $0*$ and radius $\bar{0A} = \bar{0B}$, draw the circumference arc AB, the midline of the injector blade.

- **Parabola Arc**

1. Start Fig. 4.27 with part of Fig. 4.18.
2. Draw the side of the 40° angle from line segment $\bar{0s}$, obtaining point A.
3. From A, mark angle $\alpha_2 = \alpha_4 = 14^o$with a tangent side to the circle at radius $r_{in} = 211$ (mm), in the scale of the drawing, obtaining line r_2 that meets line $r_1$at point C.
4. With $L_a = 240$ (mm), in the scale of the drawing, determine point B with r_1.
5. Divide segment AC into equal parts, the more the better, in this case five parts, obtaining points 1; 2; 3; 4.
6. Divide segment BC into the same number of parts as AC, obtaining points $1*$; $2*$; $3*$; $4*$.
7. Connect the points in the following order, A with $4*$; and 1 with $3*$, obtaining the point where the segments meet, then 2 with $2*$, obtaining point B where the segments meet, and so on, obtaining point C, and point D.

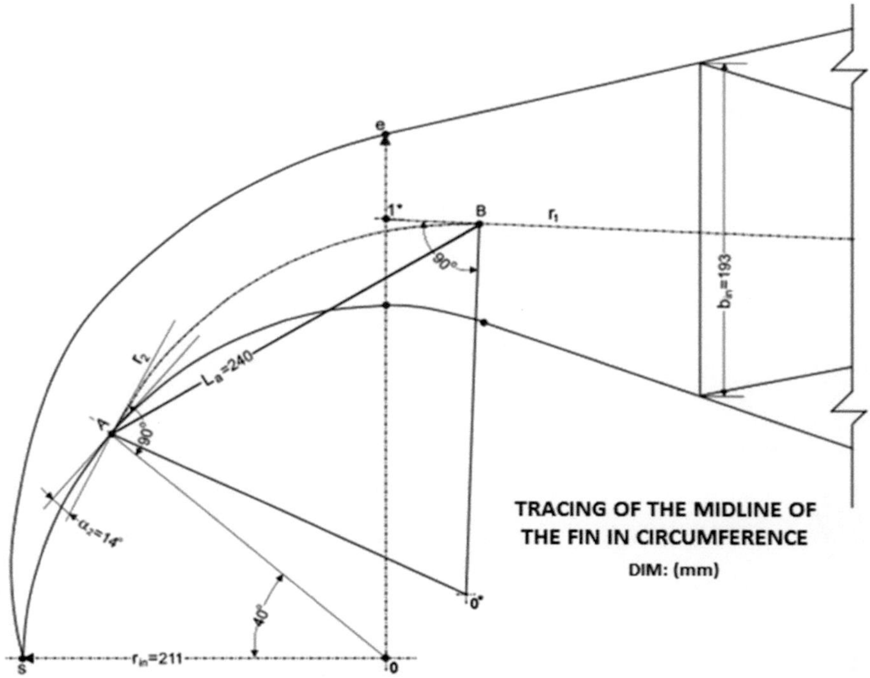

Fig. 4.26 Drawing the blade midline in the form of a circular arc

8. Connecting point A, a, B, c, d, with a curve results in B, the midline of the injector blade, in the form of a parabola arc.

4.8.2.3 Comparing the Drawings

- In Fig. 4.28, the superposition of the three arcs was drawn, which results in, geometrically, and with the same chord length $L_a = 240$(mm), the smallest arc length that is drawn in a parabolic arc, and the largest in arc of circumference. From a manufacturing standpoint, the circular arc is recommended. However, in term of flow efficiency, there are still no reliable available data.
- Regarding the length of the blade chord in profile, as a guideline, we recommended that this be within $0.5 \cdot D < L_a$ (mm) $< 1.3 \cdot D$, which happens given the value obtained in the drawing $L_a = 240$ (mm).
- The next sub-item will analyze the length of the chord linked to the movements of the blade in function of load (power) variations and operational safety, particularly in cases of sudden drops.

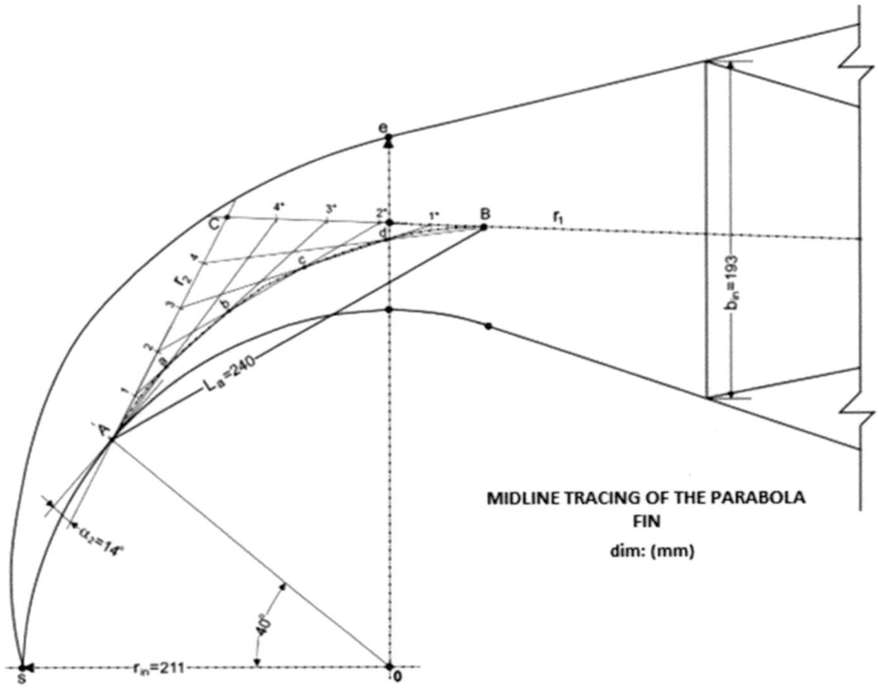

Fig. 4.27 Drawing the blade midline in the form of a parabola arc

Fig. 4.28 Comparing the drawings of the midline of the injector blade

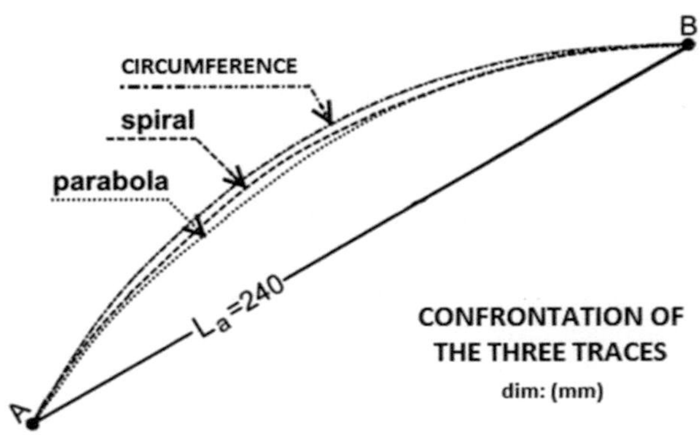

4.8.2.4 Geometric Layout of the Blade Positioned on the Injector

- **Input Data**

In this sub-item, an algorithm will be developed, and a numerical validation for the geometric layout of the blade will be given, which is already positioned in the injector, to provide a first study of the possibilities of its operational use during the load variations in the BMHIT. This study will allow us to schematize the spectrum of forces and moments that will act on the blade, which is essential for making the structural calculations.

The input data will be those from Fig. 4.25.

- **Calculations**

1. Preliminarily, fix the thickness of the blade at the center of gravity, which can be calculated by:

$$e_{cg_a} \cong 0.07 \cdot D = 0.07 \cdot 419 \cong 29 \text{ (mm)}$$

2. Calculate blade throttling factor:

$$f_{e65-010} = \frac{(e_{cg}/L_a)_{\text{cal}}}{0.101} = \frac{29/240}{0.101} = 1.196$$

3. With the values in the table from Fig. 4.24, fill in Table 4.7 as follows:

$$x_a = 0,01 \cdot L_a \cdot x_p$$

i.e.:

$$x_a = 0.01 \cdot 240 \cdot x_p = 2.4 \cdot x_p \text{ (mm)}$$

and

$$y_a = 0.01 \cdot f_{e65-010} \cdot L_a \cdot y_p$$

here

Table 4.7 Fin Coordinates—Profile NACA 65-010

%	–	0.0	0.5	0.8	1.3	2.5	5.0	7.5	10.0
x_a	mm	0.00	1.20	1.80	3.00	6.00	12.00	18.00	24.00
y_a	mm	0.00	2.16	2.55	3.23	4.51	6.38	7.77	8.93
%	–	15.0	20.0	25.0	30.0	35.0	40.0	45 CG	50.0	55.0
x_a	mm	36.00	48.00	60.00	72.00	84.00	96.00	113.40	120.00	132.00
y_a	mm	10.75	12.11	13.12	13.84	14.30	14.51	14.43	13.98	13.12
%	–	60.0	65.0	70.0	75.0	80.0	85.0	90.0	95.0	100.0
x_a	mm	144.00	156.00	168.00	180.00	192.00	214.20	216.00	228.00	240.00
y_a	mm	11.98	10.81	9.65	8.48	7.32	6.15	4.98	3.79	0.00

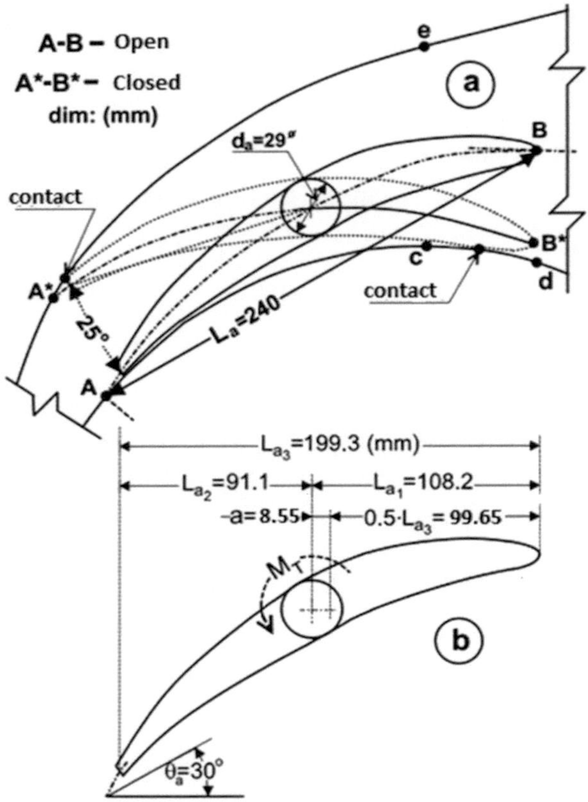

Fig. 4.29 (**a**) Scale drawing of the blade in its fully open and closed positions, and in (**b**) with its dimensions for structural calculation

$$y_a = 0.01 \cdot 1.196 \cdot 240 \cdot y_p = 2.87 \cdot y_p \text{ (mm)}$$

- **Drawing the blade in the injector**

1. Start Fig. 4.29a, scaled from Fig. 4.25, by highlighting the injector.
2. Draw the blade in its maximum opening position, AB to the same scale, with reference to the spiral midline.

$$L_a = 240 \text{ (mm)}, x_a = 113.4 \text{ (mm)}, y_a = 14.43, e_{cg_a} = 29 \text{ (mm)}.$$

3. Rotate the blade from its open position to the closed position, $A*B*$ with the center of rotation initially at cg. If the nozzle is not completely closed or if the point on the leading edge side is close to the line that inferiorly delimits the nozzle in the drawing, it must, iteratively, be sought to be close to the cg on the midline of the nozzle blade for this to occur. Once this is done, the angle that will allow

the variation of the BMHIT load from maximum values to minimum values is measured at 30°.

In this drawing, point e represents the upper limit of the spiral-shaped nozzle, point d represents its lower limit in arcs of circles, and point c represents the lower beginning of the flow in the rotor. In Fig. 4.29b, the blade was drawn to scale, with all its essential dimensions so that its structural calculation can made.

4.9 Suction Tube

4.9.1 Considerations

There are few cases in which environmental, topographical, hydrological, technical, and economic studies require fixing the floor of the engine room at a much higher level than the levels of the downstream river, with a potentially substantial reduction in the available head, and consequently in the generated power. In these few cases, the technical and economic possibility of recovering part of this loss from the head should be studied, even for the BMHIT, reducing the pressure of the discharge below local atmospheric pressure, by installing a suction tube (ST), as shown in the diagram in Fig. 4.30.

Given that, with the ST and a valve that automatically controls the pressure at the rotor outlet, p_6/γ (mca), the drag caused by the flow of water makes this pressure lower than the local barometric pressure, p_{air}/γ_{air} (mca), consequently causing an increase in the water level, h_s (mca) inside the ST.

The valve must be adjusted so that the distance between the downstream level and the lower rotor line is always, $h_r > 1.5 \cdot D$ (m). This is required for the rotor to rotate in air without water interference in the ST, depending on the housing design, and the resulting possible air volume.

Regarding the ST geometry, so that it also contributes to the increased head, it is important to reduce the kinetic energy of the water at the exit of the rotor, $\frac{c_6^2}{2 \cdot g}$ (J/kg), which can be obtained by gradually varying the cross section of the ST. Therefore, there is a reduction of the average velocity of the flow in the region of the water level in the interior, $c_{S_{TS}}$ (m/s), until the value is greater than the corresponding maximum value of the river. For this reason, it is preferable that the ST has a rectangular pyramidal shape.

4.9.2 Equation and Modeling

An equation and system modeling will be elaborated for ideal flow conditions in the rotor, in its interior, and at its exit, without considering shocks and turbulence.

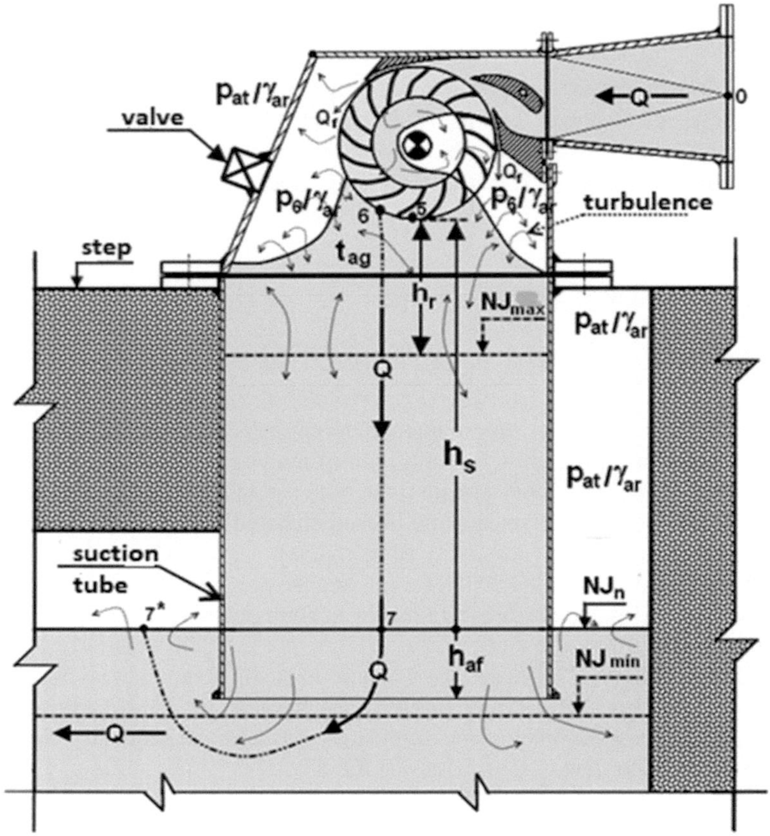

Fig. 4.30 A BMHIT with an ST

Observing these ideal conditions, and taking the ST from Fig. 4.34 as a reference, we have the specific work and the internal power of the BMHIT:

without a ST—

$$Y_{i_{sTS}} = \eta_i \cdot Y_0 - \frac{c_6^2}{2}\ (\mathrm{J/kg})$$

and

$$P_{i_{sTS}} = Q \cdot Y_{i_{sTS}}\ (\mathrm{kW}) \tag{4.59}$$

with a ST—

$$Y_{i_{cTS}} = \eta_i \cdot Y_0 + g \cdot h_s - \left(\frac{p_7}{\rho} + \frac{c_7^2}{2}\right) - Y_{p_{6;7}} (\text{J/kg})$$

and

$$P_{i_{cTS}} = Q \cdot Y_{i_{cTS}} \text{ (kW)} \tag{4.60}$$

Applying Bernoulli's theorem between sections 6 and 7 for the average values of the quantities involved, one obtains, for unit of mass of water in flow:

$$\frac{p_6}{\rho} + \frac{c_6^2}{2} + g \cdot h_s = \frac{p_7}{\rho} + \frac{c_7^2}{2} - Y_{p_{6;7}} \rightarrow g \cdot h_s = \frac{p_7}{\rho} + \frac{c_7^2}{2} - Y_{p_{6;7}} - \left(\frac{p_6}{\rho} + \frac{c_6^2}{2}\right) \tag{4.61}$$

Modeling these expressions, for lossless flows in the suction tube, $Y_{p_{TS}} = 0$, and considering $p_7 \cong p_{7*} = p_{at}$ and $c_7 = c_6$, would imply that the cross-sectional area of the flow at 6 is equal to 7, which does not physically occur, since $c_7 < c_6$, which results in a drop in kinetic energy, and an increase in the head, even though the ST is a constant section.

With this modeling, the following results for expressions (4.59) and (4.60) are:

$$Y_{i_{cTS}} = \eta_i \cdot Y_0 + g \cdot h_s - \frac{c_6^2}{2} \text{ (J/kg)}$$

and

$$P_{i_{cTS}} = Q_{1/1} \cdot Y_{i_{cTS}} \text{ (kW)} \tag{4.62}$$

$$h_s = -\frac{p_7}{\gamma} \text{ (mca)} \tag{4.63}$$

4.9.3 Application

4.9.3.1 Data

An application will be made for the BMHIT values:

- $H_0 = 40$ (m)
- $Y_0 = 392.4$ (J/kg)
- $Q = 1.0$ (m3/s)
- $P_{i_0} = 392.4$ (kW)
- $\eta_i = 0.7929$
- $\eta_{i*} = 0.8414$
- $Y_{i_6} = 0.5 \cdot 6.179^2 = 19.090$ (J/kg)

This application seeks to analyze and comment on the possible use of suction tubes in installations, assuming the following variations in suction height (vacuum in the suction of the BMHIT)

$$h_s = -p_s/\gamma = 0; 1; 2; \ldots; 7 \text{ (m)}$$

4.9.3.2 Script and Calculations

1. Internal drop/fall without a suction tube:

$$H_{i_{STS}} = \eta_{i*} H_0 - \frac{Y_{i_6}}{g} = 0.8414_* 40 - \frac{19.09}{9.81} = 31.710 \text{ (mca)}$$

2. Internal power without suction tube:

$$P_{i_{STS}} = g \cdot Q \cdot H_{i_{STS}} = 9.81 \cdot 1 \cdot 31.71 = 311.075 \text{ (kW)}$$

3. Internal fall with suction tube:

$$H_{i_{CTS}} = H_{i_{STS}} + h_s = 31.71 + h_s \text{ (mca)}$$

4. Internal power with a suction tube:

$$P_{i_{CTS}} = g \cdot Q \cdot H_{i_{CTS}} \text{ (kW)}$$

5. Internal power gains with a suction tube:

$$\Delta P_i = P_{i_{CTS}} - P_{i_{STS}} \text{ (kW)}$$

The ratio of internal power gains with a suction tube relative to the power available at the BMHIT input:

6. Table 4.8 contains the results of the respective calculations.

4.9.4 *Analysis, Comments, and Observations*

- To stop water level oscillations at the BMHIT outlet from influencing the rotor, even if it does not reach it, the automatic air pressure control valve in this location must also be sensitive to water temperature variations. In case this approaches the corresponding vaporization pressure, always observe:

Table 4.8 Results of the application

$-p_6 = h_s$	$H_{i\ c\ ts}$	$P_{i\ c\ ts}$	DP_i	DP_i/P_{io}
m	m	kW	kW	%
0.00	31.71	311.08	0.00	–
1.00	32.71	320.89	9.81	2.5
2.00	33.71	330.70	19.62	5.0
3.00	34.71	340.51	29.43	7.5
4.00	35.71	350.33	39.25	10.0
5.00	36.71	360.13	49.05	12.5
6.00	37.71	369.94	58.86	15.0
7.00	38.71	379.75	68.67	17.5

$$\frac{p_6}{\gamma_{air}} > \left(\frac{p_v}{\gamma}\right)_v = \left(0.3575 + 0.001015 \cdot t_{ag} + 0.0002702 \cdot t_{ag}^2\right)^2 \text{ (mca)}$$

For example, if

$$t_{ag} = 20\ (^{\circ}\text{C})$$

always have

$$\frac{p_6}{\gamma_{ar}} > 0.236 \text{ (mca)}$$

- The NJ_{max} must be established, considering what was mentioned in the previous item.
- An ST is recommended in cases where the engine room, due to topographical, technical, and economic conditions, must be set up a few meters above the downstream level of greater occurrence, or when the variation of this level is appreciable.
- The results obtained in the calculations that make up the last three columns of Table 4.8 are much larger than the real ones, and in most negative cases, this occurs due to the following:

  – Increased leakage flows, Q_f as the square root of the increase in pressure drops.
  – Shock and turbulence inherent to real flows, particularly in the rotor, at its exit, at the flow/air interfaces, represented in Fig. 4.30 by dashed arrows.
  – Turbulence originating from oscillations in the flow and in the air chamber can destabilize the generation system, since this can substantially vary rotation, and consequently, the frequency of the electrical system. Rendering it non-operational, compatible with the planning requirements.
  – Such practical considerations limit the use of suction tubes for most BMHITs, and in exceptional cases, due to site conditions where the reliability of the operating system load allows for the aforementioned variations.

Chapter 5
Manually Calculating a Banki-Michell Hydraulic Impulse Hydraulic Turbine (BMHIT)

In this chapter we will manually calculate the components of a BMHIT, based on the hydrodynamic equation that provided the hydrodynamic flow characteristics from the inlet to the outlet of the BMHIT, along with the real hydraulic flow path (RHFP), and the resistances of the materials that will comprise each components, in both their static and dynamic states. Therefore, initially, we will analyze the forces acting on each of them, along with their maximum and minimum values. After having established their geometries and estimated operational performance, we will determine the manufacturing materials that will be used. These conditions will allow us to establish calculation, design, and manufacturing records. Similar to previous chapters, we will develop algorithms validated them to develop a corresponding computer program.

5.1 The Rotor

5.1.1 The Forces Acting on the Blades

According to the principle of momentum, the flow of water passing through the rotor results in the following forces, F_{R1} and F_{R_2}, which are normal to the blade, respectively, and simultaneously in the first and second passage of the flow through the rotor, as shown in the graph in Fig. 5.1.

This force on a blade, considering a maximum water hammer at 30 (%), is the result of the driving force from the first passage of the flow through the rotor. This force is $F_{u_{4;5*}} = F_{u_{\text{blade}_1}} > F_{u_{\text{blade}_2}}$, or the corresponding force that is normal to it, which is meridional $F_{m_{4;5*}} = F_{m_{\text{blade}_1}} > F_{m_{\text{blade}_2}}$, the calculations of which, according to Sect. 3.1.1 in Chap. 3, results in the following expressions:

© The Author(s), under exclusive license to Springer Nature Switzerland AG 2025

Z. de Souza, G. L. Tiago Filho, *Pelton and Crossflow Impulse Hydraulic Turbines*,
https://doi.org/10.1007/978-3-031-87086-6_5

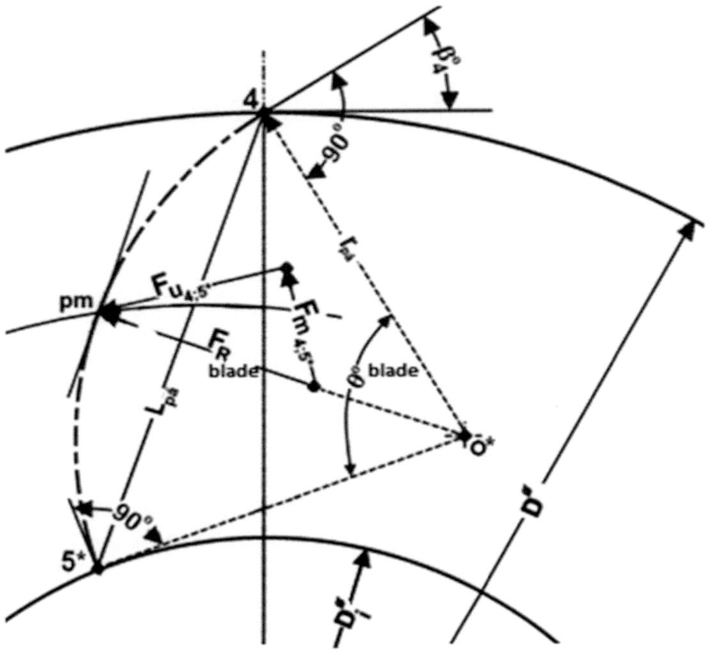

Fig. 5.1 Graphs to obtain the forces acting on the blade

$$F_{u_{4;5*}} = 1.3 \cdot \frac{\rho \cdot \eta_f \cdot Q}{z_\delta \cdot f_{e_4}} \cdot (c_{u_4} - c_{u_{5*}}) \rightarrow$$

$$F_{m_{4;5*}} = 1.3 \cdot \frac{\rho \cdot \eta_f \cdot Q}{z_\delta \cdot f_{e_4}} \cdot (c_{m_4} - c_{m_{5*}}) \rightarrow$$

$$F_{R_{\text{blade}}} = \sqrt{F_{u_{4;5*}}^2 + F_{m_{4;5*}}^2} \rightarrow$$

$$F_{R_{\text{blade}}} = 1.3 \cdot \frac{\rho \cdot \eta_f \cdot Q}{z_\delta \cdot f_{e_4}} \cdot \sqrt{(c_{u_4} - c_{u_{5*}})^2 + (c_{m_4} - c_{m_{5*}})^2} \text{ (N)} \tag{5.1}$$

The force distributed along the blade results in:

$$q_{\text{blade}} = \frac{F_{R_{\text{blade}}}}{B} \text{ (N/cm)} \tag{5.2}$$

5.1.2 Blade Thickness—Voltages—Arrows—Frequencies

The blades will be of constant thickness, and the calculations will be made for the types of supports shown in Fig. 5.2.

The maximum bending moment on the blades, for each type, results in:

Type 1:

$$M_{\max} = \frac{q_{\text{blade}} \cdot B^2}{12}\ (\text{N} \cdot \text{cm}) \rightarrow$$

Type 2:

$$M_{\max} = \frac{q_{\text{blade}} \cdot B^2}{48}\ (\text{N} \cdot \text{cm}) \tag{5.7}$$

Considering Fig. 5.3, for the scale cross-section of the blade, we can geometrically obtain:

$$h = 0.5 \cdot \sqrt{4 \cdot r_{\text{blade}}^2 - L_{\text{blade}}^2}\ (\text{cm}) \tag{5.8}$$

$$y_g = \frac{114.593}{\theta_{\text{blade}}^o} \cdot r_{\text{blade}} \cdot \sin(0.5 \cdot \theta_{\text{blade}})\ (\text{cm}) \tag{5.9}$$

$$h_1 = r_{\text{blade}} + 0.5 \cdot e_{\text{blade}} - y_g \rightarrow h_2 = y_g - r_{\text{blade}} \cdot \cos\left(0.5 \cdot \theta_{\text{blade}}^o\right) \tag{5.10}$$

For these calculations, the largest value between h_1 and h_2 must be chosen, which will be called h_3.

The central angle of the blade:

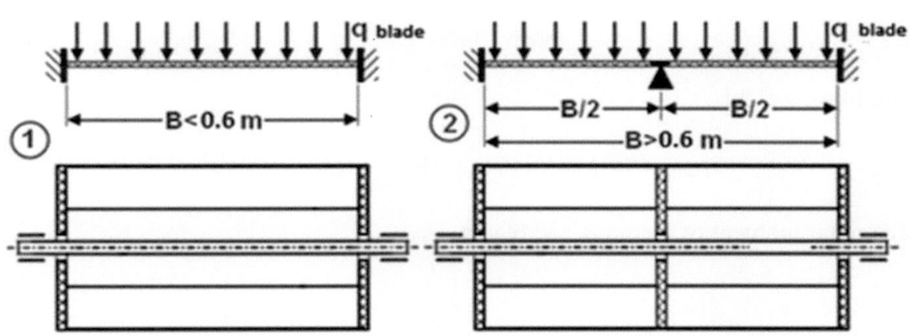

Fig. 5.2 Types of rotors: Type 1—without intermediate support; Type 2—with intermediate support

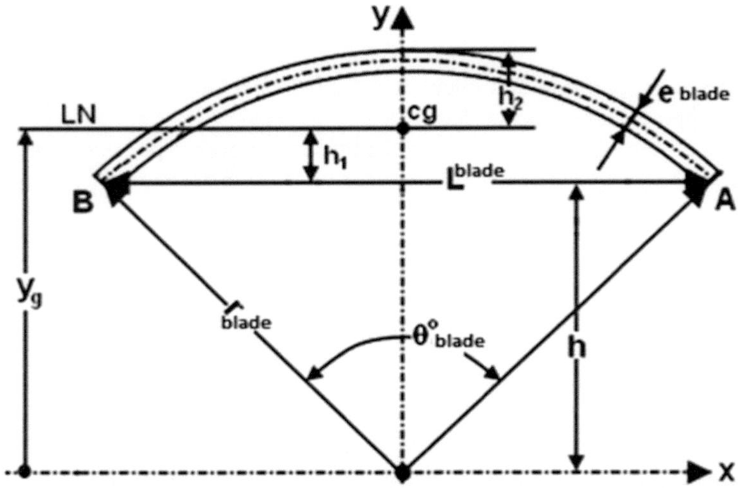

Fig. 5.3 The cross section of the blade

$$\theta = 2 \cdot \arcsin \frac{L_{\text{blade}}}{2 \cdot r_{\text{blade}}} \ (^{\circ}) \tag{5.11}$$

The arc covered by the blade:

$$\bar{AB} = \frac{\pi}{180} \cdot r_{\text{blade}} \cdot \theta^{o}_{\text{blade}} \ (\text{cm}) \tag{5.12}$$

The blade cross-sectional area:

$$A_{\text{blade}} = 0.0174 \cdot \theta^{o}_{\text{blade}} \cdot r_{\text{blade}} \cdot e_{\text{blade}} \ \left(\text{cm}^2\right) \tag{5.13}$$

The blade volume:

$$V_{\text{blade}} = A_{\text{blade}} \cdot B \ \left(\text{cm}^3\right) \tag{5.14}$$

The blade weight:

$$G_{\text{blade}} = 9.81 \cdot \gamma_{\text{steel}} \cdot V_{\text{blade}} \ (\text{N}) \rightarrow \gamma_{\text{steel}} \ \left(\text{kgf/cm}^3\right) \tag{5.15}$$

The polar moment of inertia of the blade:

$$I_{\mathrm{p}} = \left(0.0087 \cdot \theta^{\mathrm{o}}_{\mathrm{blade}} + 0.5 \cdot \sin\ \theta^{\mathrm{o}}_{\mathrm{blade}} - 229.18 \cdot \frac{\sin^2\left(0,5 \cdot \theta^{\mathrm{o}}_{\mathrm{blade}}\right)}{\theta^{\mathrm{o}}_{\mathrm{blade}}}\right) \cdot e_{\mathrm{blade}} \cdot r^3_{\mathrm{blade}}\ \left(\mathrm{cm}^4\right) \tag{5.16}$$

Given these expressions, the result for the bending stress is:

$$\sigma_f = \frac{M_{\max} \cdot h_3}{I_{\mathrm{p}}}\ \left(\mathrm{N/cm}^2\right) \tag{5.17}$$

The maximum shear force and shear stress for each type of support result are:
Type 1:

$$E_{Q_{\max}} = \frac{F_{R_{\mathrm{blade}}}}{2}\ (\mathrm{N}) \rightarrow \tau_{cis} = \frac{F_{R_{\mathrm{blade}}}}{2 \cdot A_{\mathrm{blade}}}\ \left(\mathrm{N/cm}^2\right) \tag{5.18}$$

Type 2:

$$E_{Q_{\mathrm{blade}}} = \frac{F_{R_{\mathrm{blade}}}}{4}\ (\mathrm{N}) \rightarrow \tau_{cis} = \frac{F_{R_{\mathrm{blade}}}}{4 \cdot A_{\mathrm{blade}}}\ \left(\mathrm{N/cm}^2\right) \tag{5.19}$$

Combining bending with shear, for maximum stress, we have:

$$\tau_{\max} = \sqrt{\left(\frac{\sigma_f}{2}\right)^2 + \tau^2_{cis}}\ \left(\mathrm{N/cm}^2\right) \tag{5.20}$$

The safety factor f_s is obtained by relating 5.21, where σ_{esc} is the yield stress of the blade material:

$$f_{\mathrm{s}} = \frac{0,6 \cdot \sigma_{\mathrm{esc}}}{\tau_{\max}} \rightarrow 2 < f_{\mathrm{s}} < 4 \tag{5.21}$$

The maximum arrows for each type can be determined considering the blade's own weight and considering that the weight is distributed along it, $q_{\mathrm{p}} = \frac{9.81 \cdot \gamma_{\mathrm{steel}} \cdot A_{\mathrm{blade}}}{B}$ (N/cm), resulting in the total distributed load:

$$q_{\mathrm{t}} = q_{\mathrm{p}} + q_{\mathrm{blade}}\ (\mathrm{N/cm}) \tag{5.22}$$

Thus, the arrows can be determined by the following expressions:
Type 1:

$$f_{\text{blade}} = \frac{q_t \cdot B^4}{384 \cdot I_{\text{p}} \cdot E} \quad \rightarrow .$$

Type 2:

$$f_{\text{blade}} = \frac{q_{\text{t}} \cdot B^4}{49122 \cdot I_{\text{p}} \cdot E} \text{ (cm)} \tag{5.23}$$

In this case, the ratio of the deflection, f_{blade}, per the span, which is, the length of the rotor, B, must be less than 0.9 (mm/m):

$$r_{f_{\text{blade}}} = 100 \cdot \frac{f_{\text{blade}}}{B} < 0.9 \ \ (\text{mm/m}) \tag{5.24}$$

The first five natural frequencies of the rotor blades can be determined for each type under study, as follows, for $i = 1; 2; 3; 4; 5$:

- **Type 1**

$$f_{\text{blade}_{n_i}} = \frac{[1.57 \cdot (2 \cdot i + 1)]^2}{2 \cdot \pi \cdot B^2} \cdot \sqrt{\frac{g \cdot E \cdot I_{\text{p}}}{q_{\text{p}}}} \text{ (Hz)} \tag{5.25}$$

Considering the auxiliary variable:

$$\text{VA} = \frac{1}{2 \cdot \pi \cdot B^2} \cdot \sqrt{\frac{g \cdot E \cdot I_{\text{p}}}{q_{\text{p}}}}$$

yields:

$$f_{\text{blade}_{n_1}} = 22.18 \cdot \text{VA (Hz)} \rightarrow$$

$$f_{\text{blade}_{n_2}} = 61.62 \cdot \text{VA (Hz)} \rightarrow$$

$$f_{\text{blade}_{n_3}} = 120.78 \cdot \text{VA (Hz)} \rightarrow$$

$$f_{\text{blade}_{n_4}} = 199.66 \cdot \text{VA (Hz)} \rightarrow$$

$$f_{\text{blade}_{n_5}} = 298.25 \cdot \text{VA (Hz)} \tag{5.26}$$

- **Type 2**

$$f_{\text{blade}_{n_1}} = 62.4 \cdot \text{VA (Hz)} \rightarrow$$

$$f_{\text{blade}_{n_2}} = 90.25 \cdot \text{VA (Hz)} \rightarrow$$

$$f_{\text{blade}_{n_3}} = 201.64 \cdot \text{VA (Hz)} \rightarrow$$

$$f_{\text{blade}_{n_4}} = 243.36 \cdot \text{VA (Hz)} \rightarrow$$

$$f_{\text{blade}_{n_5}} = 2530.09 \cdot \text{VA } (Hz) \tag{5.28}$$

The respective excitation frequencies on the blades can be calculated as:

$$f_{\text{blade}_{\text{exc}_i}} = i \cdot n \cdot z_{\text{r}} \text{ (Hz)} \rightarrow n \text{ (rps)} \tag{5.29}$$

where z_{r} = the number of rotor blades.

5.1.3 The Rotor Discs That Support the Blades

The rotor's lateral disks, structurally, compress the blades, with thickness e_{d} calculated from the compacting that arises from the force on each blade, which occurs in the first passage of the flow.

$$e_{\text{d}} = \frac{F_{R_{\text{blade}}}}{2 \cdot L_{\text{blade}} \cdot \sigma_{c_{\text{ad}}}} \text{ (cm)} \tag{5.30}$$

Since, in most cases, the blades are welded to the discs, there is a risk of deformation from the welding process if the thicknesses are small. To avoid this problem, a minimum thickness should be standardized, as suggested in 5.31:

$$e_{\text{d}_{\min}} = 1.0 \text{ (cm)} \geq 2 \cdot e_{\text{blade}} \tag{5.31}$$

The maximum disk closure can be obtained by:

$$f_{\text{d}_{\max}} = 0.097 \cdot \frac{D \cdot q_{\text{t}} \cdot B^2 \cdot (1 - \mu^2)}{E \cdot e_{\text{d}}^3 \cdot (17 + 15 \cdot \mu)} \text{ (cm)} \rightarrow \mu_{\text{steel}} = 0.3 - \text{Poisson } Modulus \tag{5.32}$$

The disk weight can be calculated by:

$$G_{\text{d}} = \frac{\pi \cdot D^2}{4} \cdot e_{\text{d}} \cdot 9.81 \cdot \gamma_{\text{steel}} \text{ (N)} \rightarrow \left(\text{kgf/cm}^3\right) \tag{5.33}$$

5.1.4 The Spindle or Rotor Shaft

5.1.4.1 Forces and Momentum

Figure 5.4 schematically shows the action system and reaction forces for the Type 1 shaft used in the BMHIT, which will be used in this manual equation.

The acting forces coming from the flow in the rotor, e_{r_r}, and the forces from the rotor weight, G_r,are therefore, the corresponding result, $F_R = F_{r_\delta} + G_r$. The reaction forces on the bearings, from geometric symmetry, are equal to $R_m = 0.5 \cdot F_R$.

The resulting action of the flow e_{r_r} is equal to the motor, $F_{u_{4;5}}$since the normal value is null, $c_{m_4} = c_{m_5}$, therefore,

$$F_{u_{4;5}} = F_{r_\delta} = 1.3 \cdot \rho \cdot \eta_f \cdot Q \cdot (c_{u_4} - c_{u_5}) \rightarrow q_r = \frac{F_{R_r}}{B} \tag{5.34}$$

The operating G_r of the rotor weight is equal to the sum of the weights of the blades, $z_r \cdot G_{\text{blade}}$, of the two discs, $2 \cdot G_d G_d$ and of the shaft G_e, so:

$$G_r = z_r \cdot G_{\text{blade}} + 2 \cdot G_d + G_e \text{ (N)} \rightarrow \tag{5.35}$$

The axle weight is calculated as:

$$G_e = \frac{\pi \cdot d_e^2}{4} \cdot (B + 2 \cdot x_1) \cdot 9.81 \cdot \gamma_{\text{steel}} \text{ (N)} \rightarrow \gamma_{\text{steel}} \left(\text{kgf/cm}^3\right) \tag{5.36}$$

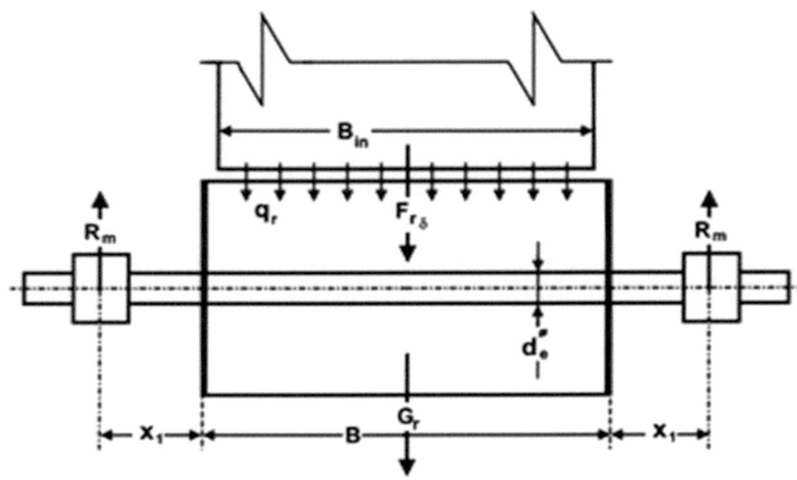

Fig. 5.4 Schematic of the forces on the BMHIT Type 1 shaft

To first calculate G_e, the minimum shaft $d_{e_{min}}$ diameter can be constant in the data, imposing the preliminary condition of $d_{e_{min}} = d_e = x_1$. The momentum from the flow is determined by:

$$M = 0.25 \cdot F_{u_\delta} \cdot D\ (\text{N} \cdot \text{m}) \tag{5.37}$$

The reaction in the bearings is:

$$R_m = 0.5 \cdot (F_{n_\delta} + z_r \cdot G_{\text{blade}} + 2 \cdot G_d + G_e)\ (\text{N}) \tag{5.38}$$

Thus, for type I, the bending moment can be calculated as:

$$M_{f_{max}} = R_m \cdot (x_1 + 0.5 \cdot B)\ (\text{N} \cdot \text{cm}) \tag{5.39}$$

Considering constant torsion, and alternating bending, the shaft diameter can be determined by the following expression from ANSI/ASME—1985:

$$d_e = \sqrt[3]{\frac{32 \cdot f_s}{\pi} \cdot \sqrt{\left(f_{ci} \cdot \frac{M_{fe}}{\sigma_{fe_{adm}}}\right)^2 + 0.75 \cdot \left(\frac{M}{\sigma_{es}}\right)}}\ (\text{cm}) \tag{5.40}$$

In this expression we have:

- The safety factor: $f_s = 3.5$,
- The stress concentration factor: $f_{ct} = 1.3$, and
- The permissible fatigue stress: $\sigma_{fd_{adm}} = 0.5 \cdot \sigma_{esc}$.

Substituting these values into expression (5.40) gives:

$$d_e = 3.138 \cdot \sqrt[6]{\frac{M^2 + 9{,}013 \cdot M_{f_{max}}^2}{\sigma_{es}^2}}\ (\text{cm}) \tag{5.41}$$

The maximum arrow is determined by:

$$f_{e_{max}} = \frac{R_A \cdot (B + 2 \cdot x_1)^2}{24 \cdot E \cdot I_{p_e}} \cdot \left[\frac{3 \cdot x_1}{B + 2 \cdot x_1} - 4 \cdot \left(\frac{x_1}{B + 2 \cdot x_1}\right)^3\right] \tag{5.42}$$

Considering $I_{p_e} = \frac{\pi \cdot d_e^4}{64}\ (\text{cm}^4)$ and substituting in (5.42) we have:

$$f_{e_{max}} = \frac{3.4 \cdot R_m \cdot x_1}{E \cdot d_e^4} \cdot \left[\frac{0.75 \cdot (B + 2 \cdot x_1)^2 - x_1^2}{B + 2 \cdot x_1}\right]\ (\text{mm}) \tag{5.43}$$

The maximum arrow ratio and its limit are given by:

$$R_{f_{e_{max}}} = 1000 \cdot \frac{f_{e_{max}}}{B} \leq 0.9\ (\text{mm/m}) \tag{5.44}$$

Once the axis has been sized, and if the corresponding relationship is not met, it must be varied from d_e and recalculated until it is met.

The critical shaft speed can be determined as follows.

Once this has been achieved, the auxiliary deflection $f_{e_{aux}}$ is first calculated, and then the critical rotation of the shaft $n_{e_{crit}}$ is determined as follows:

$$f_{e_{aux}} = \frac{R_m - 0.5 \cdot F_{n_\delta}}{R_m} \cdot f_{e_{max}}\ (\text{cm}) \rightarrow n_{e_{crit}} = \frac{1}{2 \cdot \pi} \cdot \sqrt{\frac{g}{f_{e_{aux}}}}\ (\text{rps}) \tag{5.45}$$

5.2 Injector: Blade

5.2.1 *Injector*

The structural design of the injector must be carried out considering eventual sudden, practically instantaneous load losses, which will result in additional forces to the walls that comprise it, along with the fin, called the water hammer. At the Basic Project level and even in the Executive Project level, we recommended using pressure at 30 (%) over the available design head, *H*(*m*) or preferably, if one can, over the maximum $H_{max}(m)$, resulting in:

$$p_{max} = 1.3 \cdot \gamma \cdot g \cdot H_{max} \left(\frac{\text{N}}{\text{m}^2}\right) = 1.28 \cdot H_{max} \left(\frac{\text{N}}{\text{cm}^2}\right) = 0.13 \cdot H_{max}\ \left(\text{kgf/cm}^2\right) \tag{5.46}$$

This pressure, which is taken as being constant, gives rise to normal forces upon the surfaces of the fixed components of the injector, schematized and named in Fig. 5.5.

These resulting normal forces can be generally determined by the following expression:

$$F_j = p_{max} \cdot A_j = 1.28 \cdot H_{max} \cdot A_j \tag{5.47}$$

A_j (cm^2) is the area of the normal projection of the corresponding wall, or the inlet pipe, $A_0 = D_0 \cdot L_0$ (cm^2), for example, which yields the following results for the wall thickness:

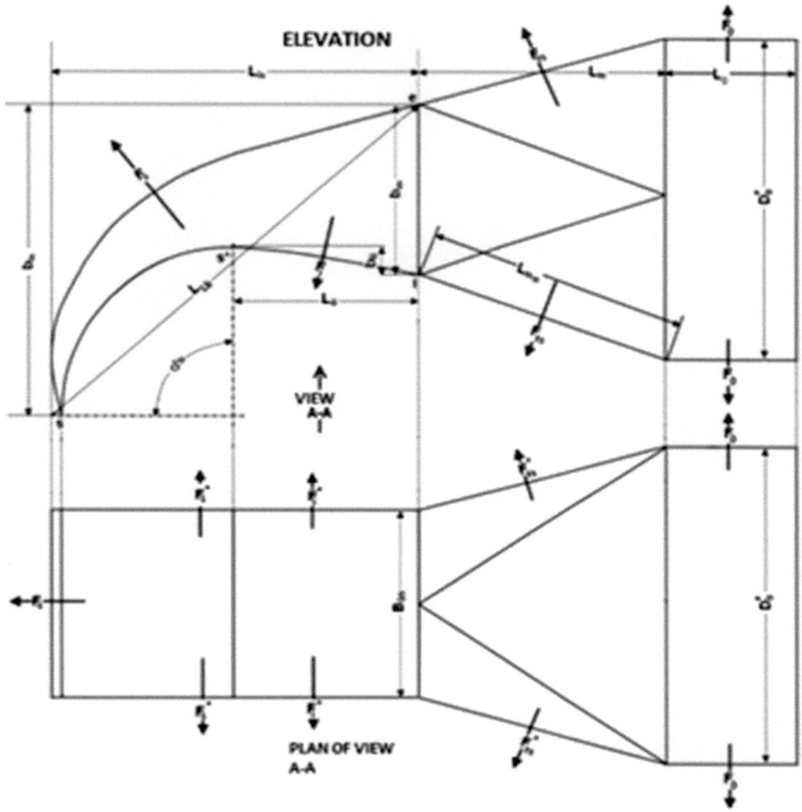

Fig. 5.5 Schematic of the BMHIT injector at the plant and at elevation, characterizing the various forces that normally act on the walls of the components

$$e_0 = \frac{p0_{\max}}{2 \cdot \sigma_{t_{adm}} \cdot k_{f\,su} \frac{1.3 \cdot 10^3 \cdot 9.81 \cdot H_{\max} \cdot D_0}{10^4 \cdot 2 \cdot \sigma_{t_{adm}} \cdot k_f} \frac{0.64 \cdot H_{\max} \cdot D_0}{\sigma_{t_{adm}} \cdot k_f} \quad su0_{\min}} \qquad (5.48)$$

With:

- $H_{\max}$ (m)
- D_0 (cm)
- $\sigma_{t_{adm}}$ $(\mathrm{N/cm^2}) \rightarrow \mathrm{steel} \rightarrow 8829 \leq \sigma_{t_{adm}} (\mathrm{N/cm^2}) \leq 13734$
- e_{su} (cm)—the supplementary thickness
- k_f (1)—the coefficient that considers manufacturing defects in the tube, we have the following values for steel tubes:
  - $k_f = 1.0$—a seamless tube or welded steel plate, X-rayed and found to be without flaws, and with stress relief;

Fig. 5.6 A BMHIT manufactured using the casting process, from *Betta Hidroturbinas*. (Photo from the manufacturer)

- $k_f = 0.9$—a seamless tube or welded steel plate, X-rayed and found to be without flaws, without stress relief;
- $k_f = 0.8$—a seamless tube or welded steel sheet, not X-rayed, without apparent flaws, and without stress relief.

We recommend that the minimum thickness of the steel tube maintains its shape when empty:

$$e_{0_{\min}} = \frac{D_0 + 51}{400} \geq 0.7 \text{ (cm)} \rightarrow D_0 \text{ (cm)}$$

Determining the thickness of the transition part and the injector depends not only on the manufacture material but also on the manufacturing process. Thus, Fig. 5.6 shows a photograph of a BMHIT with steel castings from *Betta Hidroturbinas—Franca—São Paulo*—Brazil. Figure 5.7 shows a turbine from the same manufacturer, as a schematic showing most of the components of the welded steel sheets.

To first evaluate the thicknesses of the components, we can consider the transition part as an equivalent tube, and use expression (5.48), with:

$$D_{tr_m} = 4 \cdot \frac{A_m}{U_m} \rightarrow A_m = \frac{\pi \cdot D_0^2 + 4 \cdot B_{\text{in}} \cdot b_{\text{in}}}{8} \rightarrow U_m = \frac{\pi \cdot D_0 + 2 \cdot (B_{\text{in}} + b_{\text{in}})}{2}$$

Thus:

$$D_{tr_m} = \frac{4.A_m}{U_m}$$

or

$$D_{tr} = \frac{\pi \cdot D_0^2 + 4 \cdot B_{\text{in}} \cdot b_{\text{in}}}{\pi \cdot D_0 + 2 \cdot (B_{\text{in}} + b_{\text{in}})} \text{ (cm)} \rightarrow e_{tr} = \frac{0,64 \cdot H_{\max} \cdot D_{tr}}{\sigma_{t_{\text{adm}}} \cdot k_f} + e_{\text{su}} \text{ (cm)} \quad (5.49)$$

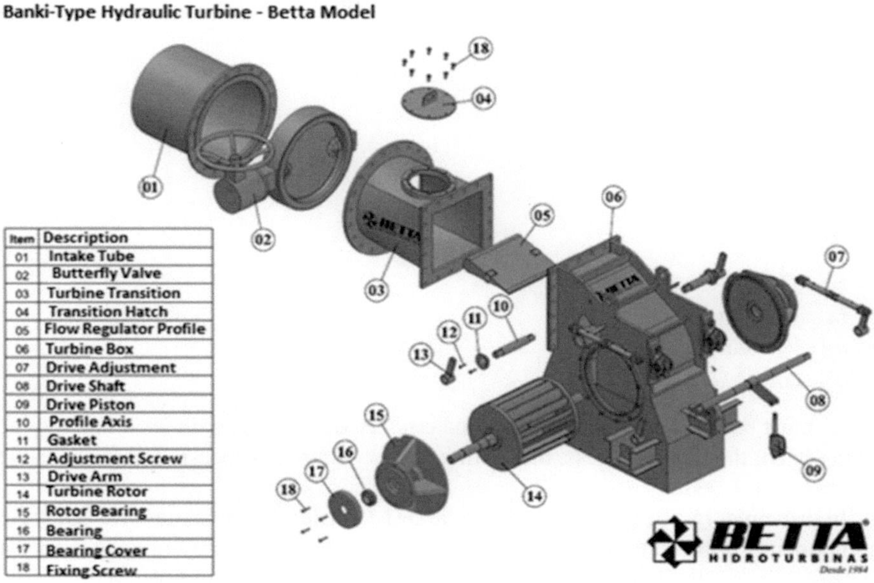

Fig. 5.7 Exploded perspective of a BMHIT manufactured as a welded sheet from *Betta Hidroturbinas*. (Figure from the manufacturer)

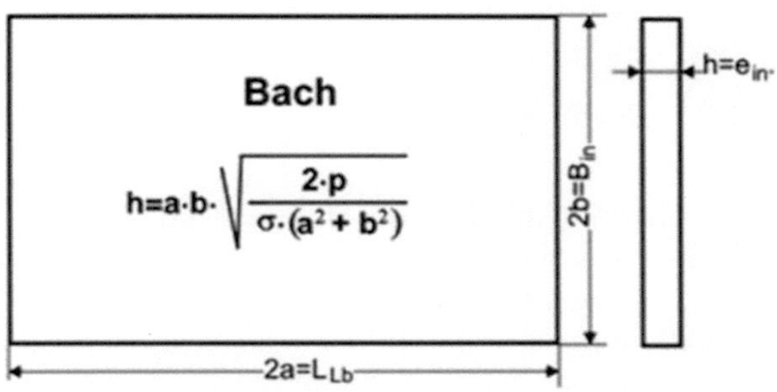

Fig. 5.8 Bach theory for bi-beam rectangular plates

For other cases, preliminarily, manufactured stamped and welded steel sheets can be considered as crimped plates subjected to bending.

Considering the largest injector plate, $B_{in}xL_{Lb}$, from Fig. 5.5, and the Bach theory for rectangular plates in Fig. 5.8, applied to this plate, we have:

$$e_{\text{in}} = \frac{L_{\text{Lb}} \cdot B_{\text{in}}}{4} \cdot \sqrt{\frac{2 \cdot 1.28 \cdot H_{\max}}{\sigma_{f_{\text{adm}}} \cdot \left[\left(\frac{L_{\text{Lb}}}{4}\right)^2 + \left(\frac{B_{\text{in}}}{4}\right)^2\right]}} = 0.8 \cdot L_{\text{Lb}} \cdot B_{\text{in}}$$

$$\cdot \sqrt{\frac{H_{\max}}{\sigma_{f_{\text{adm}}} \cdot \left(L_{\text{Lb}}^2 + B_{\text{in}}^2\right)}} + e^*_{\text{su}} \text{ (cm)} \qquad (5.50)$$

Additional thickness:

- $H_{\max}$ (m)
- L_{Lb}, B_{in} (cm)
- $\sigma_{f_{\text{adm}}}$ (N/cm^2)
- e^*_{su} (cm)

… must be fixed depending on the material used, any uncertainties related to the fixing itself, the flexural stress, the casting and/or stamping process, and the possible machining processes.

In principle, (5.50) is applied for the largest L_j, using this thickness for the other components. If the component is cast, the calculated thickness must only be used if it is greater than what is permitted by the material used to cast and finish it.

The sizing of the screws that interconnect the components and guarantee the respective seals, or in practice, their traction, is calculated referencing the normal forces to their axes.

5.2.2 *Blade Trunnion*

The geometry of the blade was already hydrodynamically established in Sect. 4.8.2.4 in Chap. 4. Considering the transverse axis passing through the CG located at 45 (%) of the leading profile edge, the diameter, $d_{\text{cg}} \cong 0.07 \cdot D$, in the example, is $d_{\text{cg}} \cong 29$ (mm).

This diameter, which is considered the maximum thickness, can be compared with the e_{al} thickness of a flat beam supported on the bearing and trunnion, as schematized in Fig. 5.9.

This allows e_{al} to be obtained as follows:

$$M_f = \frac{F \cdot B_{\text{in}}}{4} = \frac{1.28 \cdot H_{\max} \cdot L_{\text{a3}} \cdot B_{\text{in}}^2}{4} \rightarrow W = \frac{B_{\text{in}} \cdot e_{\text{al}}^2}{6} \rightarrow \sigma_{f_{\text{adm}}} = 1.92 \cdot \frac{H \cdot L_{\text{a3}} \cdot B_{\text{in}}}{e_{\text{al}}^2}$$

resulting in:

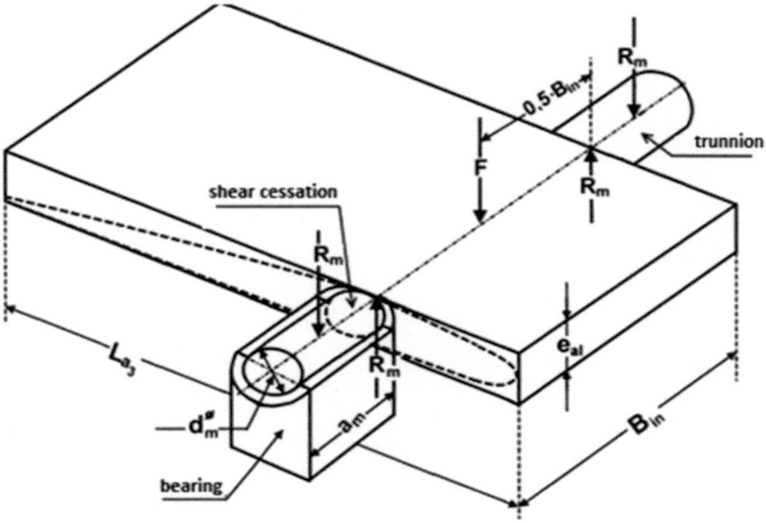

Fig. 5.9 Modeling the fin

$$e_{\mathrm{al}} = 1.4 \cdot \sqrt{\frac{H_{\max} \cdot L_{a3} \cdot B_{\mathrm{in}}}{\sigma_{f_{\mathrm{adm}}}}}\ (\mathrm{cm}) \tag{5.51}$$

$H_{\max}$ (mca)
L_{a3} (cm)
B_{in} (cm)
$\sigma_{f_{\mathrm{adm}}}$ ($\mathrm{N/cm^2}$)

The shaft diameter d_m (cm) of the support on the bearing and on the trunnion fixed in the respective CG of the fin, in Fig. 5.9, can be calculated considering simple shear:

$$\tau_{cis_{\mathrm{adm}}} = \frac{4 \cdot R_m}{\pi \cdot d_m^2}$$

or:

$$\tau_{cis_{\mathrm{adm}}} = \frac{2 \cdot F}{\pi \cdot d_m^2}$$

thus:

$$d_m = \sqrt{\frac{2 \cdot 1.28 \cdot H_{\max} \cdot L_{a3} \cdot B_{in}}{\pi \cdot \tau_{cis_{ad}}}} = 0.9 \cdot \sqrt{\frac{H_{\max} \cdot L_{a3} \cdot B_{in}}{\tau_{cis_{ad}}}} \text{ (cm)} \tag{5.52}$$

where:

- $H_{\max}$ (mca)
- L_{a3} (cm)
- B_{in} (cm)
- $\tau_{cis_{adm}}$ (N/cm^2)

5.2.3 Load Control

In BMHITs, similar to most other hydraulic turbines (HTs), the load control is obtained by varying the flow rate Q, therefore via the positioning of the blade in the injector. This positioning can be done manually, through the devices shown in Fig. 5.10, for example, one with a connecting rod which serves the engine, and another with a connecting rod and crank, or automatically, with mechanisms activated by speed regulators, as shown in Fig. 5.11.

What characterizes a velocity regulator (VR) is its working capacity, T_{cap}, which is the total work required to turn the BMHIT inlet vane from fully open to fully

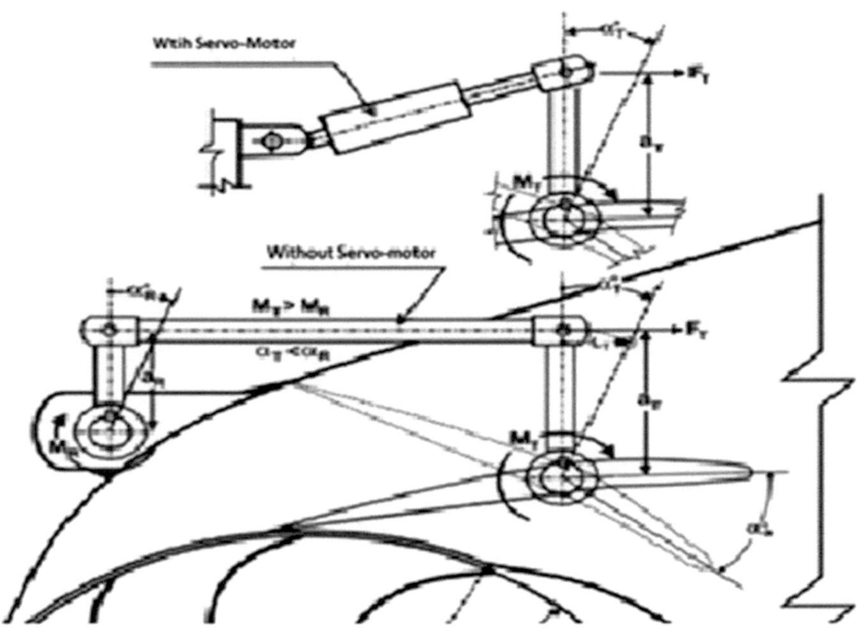

Fig. 5.10 Mechanisms for load control

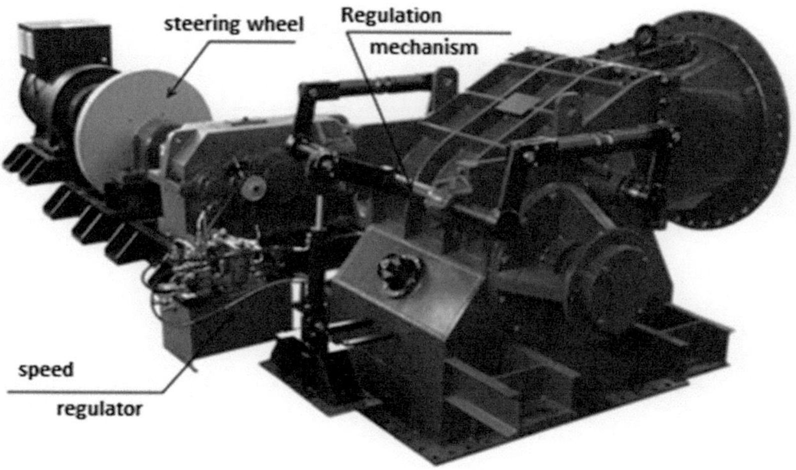

Fig. 5.11 Photograph of a cast manufactured BMHIT GG from *Betta Turbinas*, showing the automatic load regulation system

closed. T_{cap} is the product of the force F_{T} from the linear displacement L_{T} or maximum moment M_{T}, from the maximum angular displacement αT:

$$T_{\text{cap}} = F_{\text{T}} \cdot L_{\text{T}}$$

or

$$T_{\text{cap}} = M_{\text{T}} \cdot \alpha T \tag{5.53}$$

Through the mechanisms shown in Figs. 5.10 and 5.11, the same VR can be adjusted for several HTs, provided that, $T_{\text{cap}_{\text{TH}}} \leq T_{\text{cap}_{\text{RV}}}$.

The motor or driving moment M_{T} (N · cm) is variable with the position of the injector vane α_a, in Fig. 5.9. The maximum value occurs at a position very close to the closed position, when the value is given by the sum of the following moments, in Fig. 5.12.

The hydraulic pressure moment on the blade is:

$$M_{\text{h}} = F_{\text{max}} \cdot a = 0.128 \cdot H_{\text{max}} \cdot B_{\text{in}} \cdot L_{a_3} \cdot a \ (\text{N} \cdot \text{cm}) \tag{5.54}$$

The sum of moments coming from friction forces in the bearing/fin of the blade and pins of the control mechanism, in Fig. 5.10, is:

$$M_a = 0.5 \cdot \mu \cdot \sum (F_n \cdot d)_j$$

For steel, it is:

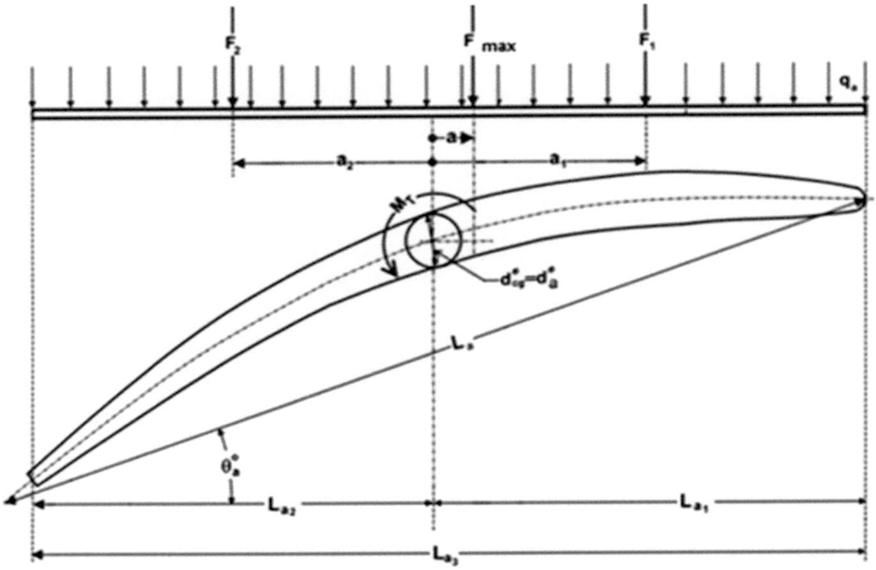

Fig. 5.12 Blade with maximum hydraulic force and driving moment, M_T

$$\mu = 0.2 \rightarrow M_a = 0.1 \cdot \sum (F_n \cdot d)_j (\mathrm{N \cdot cm}) \tag{5.55}$$

This results in the maximum M_T needed to close the injector flap:

$$M_T = M_h + M_a = 0.128 \cdot H_{max} \cdot B_{in} \cdot L_{a_3} \cdot a + 0.1 \cdot \sum (F_n \cdot d)_j \ (\mathrm{N \cdot cm}) \tag{5.56}$$

The T_{cap} for the BMHIT can first be estimated using the following expression:

$$T_{cap} = k_r \cdot \frac{P_e}{\sqrt{H}} \rightarrow 1.5 \leq \ k_r \leq 2.5 \tag{5.57}$$

T_{cap} (kgf · m)
$P_e = \frac{\gamma \cdot H \cdot Q \cdot \eta_t}{75}$ (cv)
$\gamma = 1000$ (kgf/m^3)
$H(m)$; Q (m^3/s)

With the T_{cap} (kgf · m), and the blade closing time, $t_d(s)$, in hand, the servo-motor and the VR oil supply system can be calculated, as long as the mechanism that will activate the blade is known, via expression:

$$p_o \cdot Q_o \cdot \eta_o = \frac{T_{cap}}{t_d} \tag{5.58}$$

where:

- p_o (kgf · m^2)—fixed oil pressure
- Q_o (m^3/s)—required oil flow
- T_{cap} (kgf · m)
- t_d (s)
- η_o (1)—oil supply system performance

5.3 Algorithm for Manually Calculating the BMHIT Components

5.3.1 *Input Data*

1. Preliminary fixed characteristics from Sect. 4.7.1.

   - $\phi_0 = 0.2$
   - $c_1 = 2 \cdot c_o$
   - $\phi_{tr} = 0.98$
   - $\varepsilon_{tr} = 0.03$ (mm) $\phi_{in} = \phi_2 = 0.97$
   - $\phi_3 = 0.99$
   - $\eta_f = 0.96$
   - $\alpha_2 = \alpha_4 = 14^o$
   - $\psi_5 = 0.97$
   - $e_{fo} = 2$ (mm)
   - $f_{e_{5*}} = f_{e_{4*}} = 0.80$
   - $\phi_{3*} = 0.98$
   - $r_D = 0.5$
   - $\beta_{5*} = \beta_{5*_{blade}} = \beta_{4*_{blade}} = 90^o$
   - $\rho = 1000$ (kg/m^3)
   - $g = 9.81$ (m/s^2)
     Considering that the flow through the rotor $Q_r = \eta_f \cdot Q = 0.96 \cdot 1 = 0.96$ (m^3/s) is constant, $c_{m_5} = c_{m_4}$ and $c_{m_{5*}} = c_{m_{4*}}$ will be established, which implies $\alpha_5 = 90^o$.

     Make the computer program give possible results for:

$\delta = 60^o$
90°
120°

Table 5.1 Spiral radius

q_j	r_{qj}	q_j	r_{qj}
degrees	m	degrees	m
0	0.21	75	0.29
15	0.23	90	0.31
30	0.24	97.5	0.32
45	0.26	105	0.33
60	0.27	120	0.36

Table 5.2 Average flow speeds in the rotor—angles

	Average speeds (m/s)					Angles (degrees)	
Points	u	c	w	c_u	c_m	α	β
4	13.00	27.14	14.72	26.34	6.57	14.00	26.50
5*	7.00	15.72	14.27	6.58	14.27	62.24	90.00
4*	7.00	15.61	14.18	6.32	14.27	66.11	89.04
5	13.00	6.23	13.85	0.78	6.18	82.88	26.50
5**	13.17	6.57	14.77	0.00	6.57	90.00	26.50

2. Characteristics obtained from the validation, Tables 5.1, 5.2, 5.3, and 5.4.

3. **Structural Features**: ASTM steel and SAE/ABNT 1020 steel with:

- $\sigma_{f_{\text{ad}}} = 14469.75\ (\text{N/cm}^2)$
- $\sigma_{c_{\text{ad}}} = 19620\ (\text{N/cm}^2)$
- $\sigma_{\text{t}_{\text{adm}}} = 12000\ (\text{N/cm}^2)$
- $\sigma_{f_{\text{adm}}} = 14470\ (N/\text{cm}^2)$
- $\sigma_{\text{e}} = 24623\ (\text{N/cm}^2)$
- $\tau_{cis_{\text{adm}}} = 9565\ (\text{N/cm}^2)$
- $\gamma_{\text{steel}} = 0.00785\ (\text{kgf/cm}^3)$
- $e_{\text{su}} = 0.2\ (\text{cm})$
- $e^*_{\text{su}} = 0.1\ (\text{cm})$
- $k_f = 0.9$. esu $= 0.2\ (\text{cm})$
- $e^*_{\text{su}} = 0.1\ (\text{cm})$
- $k_f = 0.9$

Table 5.3 Geometric results of the application

D_0	A_0	A_1	B_{in}	b_{in}	B	δ	D	D_i	t_4	t_{5*}
m	m^2	m^2	m	m	m	degrees	m	m	m	m
0.48	0.18	0.09	0.47	0.19	0.53	90	0.42	0.21	0.11	0.06

epá	e_t	f_{e4}	f_{e5*}	r_θ	*zpá*	z_δ	d_e	L_d	β_4	β_{5*}	r
mm	mm	–	–		number	number	mm	mm	degrees	degrees	m
7.10	15.90	0.86	0.87	Table 4.2	12.00	3.25	102.40	93.50	26,501	90	0.09

L_{is}	L_{ii}	b_{ii}	b_{is}	L_{tr}	$L_{t\ sr}$	L_{ii}	L_a	L_{a1}	L_{a2}	L_{a3}	a	d_a
mm												
435.00	195.00	54.00	368.00	458.00	484.00	195.00	240.00	108.20	91.10	199.30	8.55	29.00

Table 5.4 Fin Coordinates—Profile NACA 65-010

%	–	0.0	0.5	0.8	1.3	2.5	5.0	7.5	10.0
x_a	mm	0.00	1.20	1.80	3.00	6.00	12.00	18.00	24.00
y_a	mm	0.00	2.16	2.55	3.23	4.51	6.38	7.77	8.93
%	–	15.0	20.0	25.0	30.0	35.0	40.0	45 CG	50.0	55.0
x_a	mm	36.00	48.00	60.00	72.00	84.00	96.00	113.40	120.00	132.00
y_a	mm	10.75	12.11	13.12	13.84	14.30	14.51	14.43	13.98	13.12
%	–	60.0	65.0	70.0	75.0	80.0	85.0	90.0	95.0	100.0
x_a	mm	144.00	156.00	168.00	180.00	192.00	214.20	216.00	228.00	240.00
y_a	mm	11.98	10.81	9.65	8.48	7.32	6.15	4.98	3.79	0.00

5.3.2 *Roadmap and Validation*

1. The maximum resultant force:

$$F_{R_{\text{blade}}} = 1.3 \cdot \frac{\rho \cdot \eta_f \cdot Q}{z_\delta \cdot f_{e_4}} \cdot \sqrt{(c_{u_4} - c_{u_{5*}})^2 + (c_{m_4} - c_{m_{5*}})^2} = 1.3 \cdot \frac{1000 \cdot 0.96 \cdot 1}{3.5 \cdot 0.855}$$

$$\cdot \sqrt{(26.337 - 6.584)^2 + (6.566 - 14.274)^2} = 8842.821 \text{ (N)}$$

2. The load evenly distributed along the blade:

$$q_{\text{blade}} = \frac{F_{R_{\text{blade}}}}{B} = \frac{8842.821}{0.532} = 16907.880 \text{ (N/m)} = 169.079 \text{ (N/cm)}$$

3. The maximum bending moment on the blade:

$$M_{\max} = \frac{q_{\text{blade}} \cdot B^2}{12} = \frac{169.079 \cdot 53.2^2}{12} = 39877.846 \ \text{(N} \cdot \text{cm)}$$

4. The distance from the shaft to the blade chord, Fig. 5.3:

$$h = 0.5 \cdot \sqrt{4 \cdot r_{\text{blade}}^2 - L_{\text{blade}}^2} = 0.5 \cdot \sqrt{4 \cdot 8.8^2 - 11.7^2} = 6.574 \text{ (cm)}$$

5. The central angle of the blade:

$$\theta = 2 \cdot \arcsin \frac{L_{\text{blade}}}{2 \cdot r_{\text{blade}}} = 2 \cdot \arcsin \frac{11.7}{2 \cdot 8.8} = 83.330 \ (^\circ)$$

6. The ordinate of the center of gravity:

$$y_g = \frac{114.593}{\theta^o_{\text{blade}}} \cdot r_{\text{blade}} \cdot \sin\left(0.5 \cdot \theta^o_{\text{blade}}\right) = \frac{114.593}{83.33} \cdot 8.8 \cdot \sin(0.5 \cdot 83.33)$$
$$= 8.05 \text{ (cm)}$$

7. Distances from the center of gravity:

$$h_1 = r_{\text{blade}} + 0.5 \cdot e_{\text{blade}} - y_g = 8.8 + 0.5 \cdot 0.71 - 8.05 = 1.11 \text{ (cm)}$$

and

$$h_2 = y_g - r_{\text{blade}} \cdot \cos(0.5 \cdot \theta_{\text{blade}}) = 8.05 - 8.8 \cdot \cos(0.5 \cdot 83.33) = 1.48$$

Continue with …

$$h_3 = h_2 = 1.48 \text{ (cm)}$$

8. The arc covered by the blade:

$$\bar{AB} = \frac{\pi}{180} \cdot r_{\text{blade}} \cdot \theta^o_{\text{blade}} = \frac{\pi}{180} \cdot 8.8 \cdot 83.33 = 12.80 \text{ (cm)}$$

9. The cross-sectional area of the blade:

$$A_{\text{blade}} = 0.0174 \cdot \theta^o_{\text{blade}} \cdot r_{\text{blade}} \cdot e_{\text{blade}} = 0.0174 \cdot 83.33 \cdot 8.8 \cdot 0.71 = 9.059 \left(\text{cm}^2\right)$$

10. The volume occupied by the blade:

$$V_{\text{blade}} = A_{\text{blade}} \cdot B = 9.059 \cdot 53.2 = 481.951 \left(\text{cm}^3\right)$$

11. The weight of the blade:

$$G_{\text{blade}} = 9.81 \cdot \gamma_{\text{steel}} \cdot V_{\text{blade}} = 9.81 \cdot 0.00785 \cdot 481.951 = 37.076 \text{ (N)}$$

12. The polar moment of inertia:

$$I_p = \left(0.0087 \cdot \theta^o_{\text{blade}} + 0.5 \cdot \sin\, \theta^o_{\text{blade}} - 229.18 \cdot \frac{\sin^2\left(0.5 \cdot \theta^o_{\text{blade}}\right)}{\theta^o_{\text{blade}}}\right) \cdot e_{\text{blade}} \cdot r^3_{\text{blade}}$$
$$= \left(0.0087 \cdot 83.33 + 0.5 \cdot \sin\, 83.33 - 229.18 \cdot \frac{\sin^2(0.5 \cdot 83.33)}{83.33}\right) \cdot 0.71 \cdot 8.8^3$$
$$= 2.988 \left(\text{cm}^4\right)$$

13. The bending stress:

$$\sigma_f = \frac{M_{\max} \cdot h_3}{I_p} = \frac{39877.846 \cdot 1.48}{2.988} = 19752.079\ \left(\mathrm{N/cm^2}\right)$$

14. The maximum shear force:

$$E_{Q_{\max}} = \frac{F_{R_{\text{blade}}}}{2} = \frac{8842.821}{2} = 4421.411\ (\mathrm{N})$$

15. The shear stress:

$$\tau_{\text{cis}} = \frac{F_{R_{\text{blade}}}}{4 \cdot A_{\text{blade}}} = \frac{8842.821}{4 \cdot 9.059} = 244.034\ \left(\mathrm{N/cm^2}\right)$$

16. The maximum combined bending and shearing stress:

$$\tau_{\max} = \sqrt{\left(\frac{\sigma_f}{2}\right)^2 + \tau_{\text{cis}}^2} = \sqrt{\left(\frac{19752.079}{2}\right)^2 + 244.034^2} = 9879.054\ \left(\mathrm{N/cm^2}\right)$$

17. The safety factor:

$$f_s = \frac{0.7 \cdot \sigma_{\text{esc}}}{\tau_{\max}} = \frac{0.7 \cdot 24623}{9879.054} = 1.7$$

Since …

$$2 < f_s < 4$$

… the result is not satisfactory, and the blade thickness must be increased, and the calculations must be re-run from step 9. After these attempts, $e_{\text{blade}} = 12$ (mm) was fixed, which allows us to recalculate steps:

9.

$$h_1 = 8.8 + 0.6 - 8.05 = 1.35\ (\mathrm{cm})$$

$$h_2 = 1.48$$

Continue with …

$$h_3 = h_2 = 1.48\ (\mathrm{cm})$$

10. ?

11.

$$A_{\text{blade}} = 0.0174 \cdot 83.33 \cdot 8.8 \cdot 1.2 = 15.311 \ \left(\text{cm}^2\right)$$

12.

$$V_{\text{blade}} = 15.311 \cdot 53.2 = 814.566 \ \left(\text{cm}^3\right)$$

13.

$$G_{\text{blade}} = 9.81 \cdot 0.00785 \cdot 814.566 = 62.729 \ (\text{N})$$

14.

$$I_{\text{p}} = \frac{2.988 \cdot 2}{0.71} = 5.050 \ (\text{cm})$$

15.

$$\sigma_f = \frac{M_{\max} \cdot h_3}{I_{\text{p}}} = \frac{39877.846 \cdot 1.48}{5.05} = 11686.973 \ \left(\text{N/cm}^2\right)$$

17.

$$\tau_{\text{cis}} = \frac{F_{R_{\text{blade}}}}{4 \cdot A_{\text{blade}}} = \frac{8842.821}{4 \cdot 15.311} = 144.387 \ \left(\text{N/cm}^2\right)$$

$$\tau_{\max} = \sqrt{\left(\frac{11686.972}{2}\right)^2 + 144.387^2} = 5845.270 \ \left(\text{N/cm}^2\right)$$

$$f_{\text{s}} = \frac{0.7 \cdot \sigma_{\text{esc}}}{\tau_{\max}} = \frac{0.7 \cdot 24623}{5845.270} = 2.9.$$

Since $2 < f_{\text{s}} < 4$ is a satisfactory result, continue with the recalculated values in this item:

- $F_{R_{\text{blade}}} = 8842.821$ (N)
- $e_{\text{blade}} = 1.2$ (cm)
- $I_{\text{p}} = 5.050$ (cm^4)
- $A_{\text{blade}} = 15.311$ (cm^2)
- $V_{\text{blade}} = 814.566$ (cm^3)
- $G_{\text{blade}} = 62.729$ (N)

18. The load distributed on the blade from its own weight:

$$q_{\mathrm{p}} = \frac{9.81 \cdot \gamma_{\text{steel}} \cdot A_{\text{blade}}}{B} = \frac{9.81 \cdot 0.00785 \cdot 15.311}{53.2} = 0.022 \text{ (N/cm)}$$

19. The total load distributed on the blade:

$$q_{\mathrm{t}} = q_{\mathrm{p}} + q_{\text{blade}} = 0.022 + 169.079 = 169.101 \text{ (N/cm)}$$

20. The arrow of blades:

$$f_{\text{blade}} = \frac{q_t \cdot B^4}{384 \cdot I_{\mathrm{p}} \cdot E} = \frac{169.101 \cdot 53.2^4}{384 \cdot 5.05 \cdot 20601000} = 0.034 \text{ (cm)}$$
$$= 0.34 \text{ (mm)}, 0.19 \text{ (mm)}$$

21. The ratio of the arrow of the blades per meter of span:

$$r_{f_{\text{blade}}} = 100 \cdot \frac{f_{\text{blade}}}{B} = 100 \cdot \frac{0.34}{53.2} = 0.64 < 0.9 \text{ (mm/m)}.$$

… which is a satisfactory value.

22. The auxiliary variable:

$$\mathrm{VA} = \frac{1}{2 \cdot \pi \cdot B^2} \cdot \sqrt{\frac{g \cdot E \cdot I_p}{q_{\mathrm{p}}}} = \frac{1}{2 \cdot \pi \cdot 53.4^2} \cdot \sqrt{\frac{981 \cdot 20601000 \cdot 5.05}{0.022}} = 120.213 \text{ (Hz)}$$

23. The natural frequencies:

- $f_{\text{blade}_{n_1}} = 22.18 \cdot \mathrm{VA} = 22.18 \cdot 120.213 = 2666.324 \text{ (Hz)} \rightarrow$
- $f_{\text{blade}_{n_2}} = 61.62 \cdot \mathrm{VA} = 61.62 \cdot 120.213 = 7407.525 \text{ (Hz)} \rightarrow$
- $f_{\text{blade}_{n_3}} = 120.78 \cdot \mathrm{VA} = 120.78 \cdot 120.213 = 14519.326 \text{ (Hz)} \rightarrow$
- $f_{\text{blade}_{n_4}} = 199.66 \cdot \mathrm{VA} = 199.66 \cdot 120.213 = 24001.728 \text{ (Hz)} \rightarrow$
- $f_{\text{blade}_{n_5}} = 298.25 \cdot \mathrm{VA} = 298.25 \cdot 120.213 = 35853.527 \text{ (Hz)}.$

24. The excitation frequencies:

- $f_{\text{blade}_{\text{exc}_1}} = 1 \cdot n \cdot z_{\mathrm{r}} = 1 \cdot 10 \cdot 12 = 120 \text{ (Hz)} \rightarrow$
- $f_{\text{blade}_{\text{exc}_2}} = 2 \cdot n \cdot z_{\mathrm{r}} = 2 \cdot 10 \cdot 12 = 240 \text{ (Hz)} \rightarrow$
- $f_{\text{blade}_{\text{exc}_3}} = 3 \cdot n \cdot z_{\mathrm{r}} = 3 \cdot 10 \cdot 12 = 360 \text{ (Hz)} \rightarrow$
- $f_{\text{blade}_{\text{exc}_4}} = 4 \cdot n \cdot z_{\mathrm{r}} = 4 \cdot 10 \cdot 12 = 480 \text{ (Hz)} \rightarrow$
- $f_{\text{blade}_{\text{exc}_5}} = 5 \cdot n \cdot z_r = 5 \cdot 10 \cdot 12 = 600 \text{ (Hz)}.$

Since $f_{\text{blade}_{n_i}} > f_{\text{blade}_{\text{exc}_i}}$ in this example, the operational reliability of the rotor, under the vibrating aspect of the blades, is safe.

25. The thickness of the discs:

$$e_{\mathrm{d}} = \frac{F_{R_{\mathrm{blade}}}}{2 \cdot L_{\mathrm{blade}} \cdot \sigma_{c_{\mathrm{ad}}}} = \frac{8842.821}{2 \cdot 11.7 \cdot 19620} = 0.0193\ (\mathrm{cm})$$

The minimum thickness of the disks:

$$e_{\mathrm{d_{min}}} = 1.0\ (\mathrm{cm}) \geq 2 \cdot e_{\mathrm{blade}} = 2 \cdot 1.2 = 2.4\ (\mathrm{cm})$$

Since the calculated value was smaller, the calculation will continue with:

$$e_{\mathrm{d}} = e_{\mathrm{d_{min}}} = 2.4$$

To proceed with the calculations, the value of B must be verified, since:

$$B > B_{\mathrm{in}} + 2 \cdot e_{\mathrm{d}} = 46.3 + 2 \cdot 2.4 = 51.1\ (\mathrm{cm})$$

… indicating, in this case, that we should continue with:

$$B = 53.2\ (\mathrm{cm})$$

… established in the hydrodynamic calculation.

26. The maximum arrow on the disks:

$$f_{\mathrm{d_{max}}} = 0.097 \cdot \frac{D \cdot q_t \cdot B^2 \cdot (1 - \mu^2)}{E \cdot e_{\mathrm{d}}^3 \cdot (17 + 15 \cdot \mu)}\ (\mathrm{cm})$$

$$f_{\mathrm{d_{max}}} = 0.097 \cdot \frac{41.9 \cdot 169.101 \cdot 53.2^2 \cdot (1 - 0.3^2)}{20601000 \cdot 2.4^3 \cdot (17 + 15 \cdot 0.3)} = 0.0002891\ (\mathrm{cm})$$

27. The weight of a disk:

$$G_{\mathrm{d}} = \frac{\pi \cdot D^2}{4} \cdot e_{\mathrm{d}} \cdot 9.81 \cdot \gamma_{\mathrm{steel}} = \frac{\pi \cdot 41.9^2}{4} \cdot 2.4 \cdot 9.81 \cdot 0.00785 = 254.840\ (\mathrm{N})$$

28. The tangential force received from the flow by the rotor:

$$F_{r_\delta} = 1,3 \cdot \rho \cdot \eta_f \cdot Q \cdot (c_{u_4} - c_{u_5}) = 1.3 \cdot 1000 \cdot 0.96 \cdot 1 \cdot (26.337 - 0.775)$$
$$= 31901.376\ (\mathrm{N})$$

29. The moment on the shaft from the flow:

$$M = 0.25 \cdot F_{u_\delta} \cdot D = 0.25 \cdot 31901.376 \cdot 41.9 = 334166.914\ (\mathrm{N \cdot cm})$$

30. The weight of the shaft in the length of the rotor:

$$G_\mathrm{e} = \frac{\pi \cdot d_{\mathrm{e_{min}}}^2}{4} \cdot (B + 2 \cdot x_1) \cdot 9.81 \cdot \gamma_\mathrm{steel} = \frac{\pi \cdot 10.24^2}{4} \cdot (53.2 + 2 \cdot 10.24) \cdot 9.81 \cdot 0.00785 = 467.281\ (\mathrm{N})$$

31. The reaction at the supports:

$$R_m = 0.5 \cdot (F_{r_\delta} + z_\mathrm{r} \cdot G_\mathrm{blade} + 2 \cdot G_\mathrm{d} + G_\mathrm{e}) = 0.5 \cdot (31901.376 + 12 \cdot 37.076 + 2 \cdot 254.84 + 467.281) = 16661.625\ (\mathrm{N})$$

32. The distance between the disc and the bearing:

$$x_1 = d_{\mathrm{e_{min}}} = 10.24\ (\mathrm{cm})$$

33. The bending moment:

$$M_{f_\mathrm{max}} = R_m \cdot (x_1 + 0.5 \cdot B) = 16661.625 \cdot (10.24 + 0.5 \cdot 53.2) = 613814.265\ (\mathrm{N \cdot cm})$$

34. The diameter considering constant torsion and alternating bending:

$$d_\mathrm{e} = 3.138 \cdot \sqrt[6]{\frac{M^2 + 9.013 \cdot M_{f_\mathrm{max}}^2}{\sigma_\mathrm{es}^2}} = 3.138 \cdot \sqrt[6]{\frac{334166.914^2 + 9.013 \cdot 613814.265^2}{24525^2}} = 13.3\ (\mathrm{cm})$$

Recalculating we have …

$$G_\mathrm{e} = \frac{\pi \cdot d_\mathrm{e}^2}{4} \cdot (B + 2 \cdot x_1) \cdot \gamma_\mathrm{steel} = \frac{\pi \cdot 13.3^2}{4} \cdot (53.2 + 2 \cdot 10.24) \cdot 9.81 \cdot 0.00785 = 788.282\ (\mathrm{N})$$

$$R_m = 0.5 \cdot (F_{n_\delta} + z_\mathrm{r} \cdot G_\mathrm{blade} + 2 \cdot G_\mathrm{d} + G_\mathrm{e}) = 0.5 \cdot (31901.376 + 12 \cdot 37.076 + 2 \cdot 254.84 + 788.282) = 16822.125$$

$$M_{f_\mathrm{max}} = R_m \cdot (x_1 + 0.5 \cdot B) = 16822.125 \cdot (10.24 + 0.5 \cdot 53.2) = 619727.085\ (\mathrm{N \cdot cm})$$

$$d_{\text{e}} = 3.138 \cdot \sqrt[6]{\frac{M^2 + 9.013 \cdot M_{f_{\max}}^2}{\sigma_{\text{es}}^2}} = 3.138$$

$$\cdot \sqrt[6]{\frac{334166.914^2 + 9.013 \cdot 619727.085^2}{24525^2}} = 13.35\ (\text{cm})$$

… and …

$$d_{\text{e}} = 134\ (\text{mm})$$

… will be adopted in the project.

35. The maximum arrow:

$$f_{\text{e}_{\max}} = \frac{3.4 \cdot R_m \cdot x_1}{E \cdot d_{\text{e}}^4} \cdot \left[\frac{0.75 \cdot (B + 2 \cdot x_1)^2 - x_1^2}{B + 2 \cdot x_1}\right] \rightarrow$$

$$= \frac{3.4 \cdot 16822.125 \cdot 10.24}{20601000 \cdot 13.4^4} \cdot \left[\frac{0.75 \cdot (53.2 + 2 \cdot 10.24)^2 - 10.24^2}{53.2 + 2 \cdot 10.24}\right]$$
$$= 0.0000475\ (\text{cm})$$

36. The maximum arrow ratio:

$$R_{f_{\text{emax}}} = 1000 \cdot \frac{f_{\text{e}_{\max}}}{B} = 1000 \cdot \frac{0.00000475}{0.532} = 0.09\ (\text{mm/m}) < 0.9$$

37. The auxiliary arrow:

$$f_{\text{e}_{\text{aux}}} = \frac{R_m - 0.5 \cdot F_{n_\delta}}{R_m} \cdot f_{\text{e}_{\max}} = \frac{16822.125 - 0.5 \cdot 31901.376}{16822.125} \cdot 0.0000475$$
$$= 0.00000246$$

38. Critical shaft rotation:

$$n_{\text{e}_{\text{crit}}} = \frac{1}{2 \cdot \pi} \cdot \sqrt{\frac{g}{f_{e_{\text{aux}}}}} = \frac{1}{2 \cdot \pi} \cdot \sqrt{\frac{981}{0.00000246}} = 3178\ (\text{rps}).$$

The axis is highly subcritical.

39. The maximum pressure on the blade:

$$p_{\max} = 1.28 \cdot H_{\max} = 1.28 \cdot 40 = 51.2 \ \left(\mathrm{N/cm^2}\right)$$

40. The hydraulic diameter of the transition piece:

$$D_{\mathrm{tr}} = \frac{\pi \cdot D_0^2 + 2 \cdot B_{\mathrm{in4}} \cdot b_{\mathrm{in}}}{\pi \cdot D_0 + 2 \cdot (B_{\mathrm{in}} + b_{\mathrm{in}})} = \frac{\pi \cdot 47.7^2 + 4 \cdot 46.3 \cdot 19.3}{\pi \cdot 47.7 + 2 \cdot (46.3 + 19.3)} = 38.15 \ (\mathrm{cm})$$

41. The thickness of the transition piece:

$$e_{\mathrm{tr}} = \frac{0.64 \cdot H_{\max} \cdot D_{\mathrm{tr}}}{\sigma_{\mathrm{t_{adm}}} \cdot k_f} + e_{\mathrm{su}} = \frac{0.64 \cdot 40 \cdot 38.15}{12000 \cdot 0.9} + 0.2 = 0.09 + 0.2 = 0.29 \ (\mathrm{cm})$$

since …

$$e_{\mathrm{tr_{min}}} = \frac{D_{tr} + 51}{400} = \frac{38.15 + 51}{400} = 0.22 \ (\mathrm{cm}) \geq 0.7 \ (\mathrm{cm})$$

then …

$$e_{\mathrm{tr}} = 0.7 \ (\mathrm{cm})$$

42. Thickness of the largest injector part L_j in this case $L_{is} = 43.5$ (*cm*):

$$e_{\mathrm{in}} = 0.8 \cdot L_{\mathrm{Lb}} \cdot B_{\mathrm{in}} \cdot \sqrt{\frac{H_{\max}}{\sigma_{f_{\mathrm{adm}}} \cdot \left(L_{\mathrm{Lb}}^2 + B_{\mathrm{in}}^2\right)}} + e_{\mathrm{su}}^* = 0.8 \cdot 58.7 \cdot 46.3$$

$$\cdot \sqrt{\frac{40}{14470 \cdot \left(58.7^2 + 46.3^2\right)}} + 0.1 = 1.53 + 0.1 = 1.63 \ (\mathrm{cm})$$

43. The blade thickness in the *cg*:

$$e_{al} = 1.4 \cdot \sqrt{\frac{H_{\max} \cdot L_{a3} \cdot B_{in}}{\sigma_{f_{\mathrm{adm}}}}} = 1.4 \cdot \sqrt{\frac{40 \cdot 19.93 \cdot 46.3}{14470}} = 2.24 \ (\mathrm{cm})$$

Since the preliminary value calculated was $d_a = 2.9$ (cm) > 2.24 (cm), we can continue with 2.9 (cm), or adjust the blade to result in $d_a = 2.5$ (cm) in the final project.

44. Trunnion diameter:

$$d_m = 0.9 \cdot \sqrt{\frac{H_{\max} \cdot L_{a3} \cdot B_{in}}{\tau_{\mathrm{cis_{adm}}}}} = 0.9 \cdot \sqrt{\frac{40 \cdot 23.1 \cdot 46.3}{9565}} = 1.9 \ (\mathrm{cm})$$

Since the result was less than $e_{al} = 2.0$ (cm), $d_m = 2.0$ (cm) will be adopted in the project.

5.3.3 *Reviews, Comments, and Recommendations*

- This script was designed to minimize the number of return steps, when needing to change the value of the calculated characteristic, as occurred in step 17, for example, and eventually in other steps.
- When some predetermined characteristics need to be removed from the drawing, we recommend that, whenever possible, this be done using a precision graphical software program. To facilitate the elaboration of the computational program for calculating and designing a BMHIT, the algorithms thus far established will be unified.

Chapter 6
An Algorithm for Calculating and Designing a Banki-Michell Hydraulic Impulse Turbine (*BMHIT*)

This chapter will synthesize the developments, applications, and validations that have been made on the Banki-Michell Hydraulic Turbine (*BMHIT*) in the previous chapters, particularly in Chaps. 4 and 5, into a single algorithm that give designers an initial physical view (calculations and basic drawings) of basic *BMHIT* projects by observing the characteristics we have already established. When applicable, the algorithm will detail the alternatives that must be established, to continue to give results with the best possible alternatives within the confines of the hypotheses.

In principle, the algorithm validations used and developed in Chaps. 4 and 5 will be taken here as input data.

6.1 Considerations and Hypotheses

- In general, in each sub-item, after its characterization, observations can be made that must be considered by the calculator and/or programmer, and must be included in the corresponding software user manual.
- The input data will be subdivided into two groups:
  - *Determinants* that cannot be changed, since they are included in the contract clauses between the parties (contractor/contractor).
  - *Fixed data* set by the Designer, which can be changed within limitations of the *BMHIT* bibliography at the end of the book, and/or that fall outside the designer's responsibility, which are analyzed and detailed together with the programmer.
- We recommended adopting indirect coupling, generally $n_{GE} > n_T$, for $P_e \leq 1000$ (kW) and $n_T < 600$ (rpm), with a transmission ratio, and Electric Generator/Banki-Michell Hydraulic Impulse Turbine (*EG/BMHIT*) up to 5, and adopting direct coupling $n_{GE} = n_T$, for $P_e > 1000$ (kW) and $n_T \geq 600$ (rpm).

© The Author(s), under exclusive license to Springer Nature Switzerland AG 2025

Z. de Souza, G. L. Tiago Filho, *Pelton and Crossflow Impulse Hydraulic Turbines*,
https://doi.org/10.1007/978-3-031-87086-6_6

Table 6.1 Injector vane profile coordinate dimensions

X_p (%)	0.0	0.5	0.8	1.3	2.5	5.0	7.5	10.0
Y_p (1)	0.00	0.75	0.89	1.12	1.57	2.22	2.71	3.11
X_p (%)	15.0	20.0	25.0	30.0	35.0	40.0	45CG	50.0	55.0
Y_p (1)	3.75	4.22	4.57	4.82	4.98	5.06	5.03	4.87	4.57
X_p (%)	60.0	65.0	70.0	75.0	80.0	85.0	90.0	95.0	100.0
Y_p (1)	4.18	3.77	3.36	2.96	2.55	2.14	1.74	1.32	0.00

- In steps where scale drawings are needed with the removal of characteristic values to proceed with calculations in the following steps, this can be done with a computational routine in a graphical software program, for example, AutoCAD, and these results must be stored in usable format for sequential steps in the computer program.
- Symbology, data, and turbine characteristics, that are determined as final values, must be stored for future analysis, for the continuity of calculations, and for output into tables and graphs at the discretion of the user.
- The computer program (software) must initially be able to, at the end of the hydrodynamic and geometric steps, perform recalculations from step 1, modifying the input data set by the designer, tabulating the resulting alternatives, so that these alternatives can be chosen to continue with the calculations and decision-making processes from the specified steps with new data, for example, defining angle δ°.
- The algorithm presented here is for a simple rotor, called a Type 1 rotor (only one blade and a rotor with a single sector).
- The magnitude conventions are characterized in the previous chapters and synthetically named at the beginning of this publication.
- Hydrodynamically, the transition part and the injector must be designed for $Q_{\max}$ (m^3/s), $H_{\max}$ (m), the blade for Q (m^3/s), H (m), and the rotor for $Q \cdot \eta_f$ (m^3/s), H (m).
- Mechanically, all *BMHIT* components must be designed, for operational safety, for $Q_{\max}$ (m^3/s), $H_{\max}$ (m), and the vibrating part must also be verified to meet the full range of hydrodynamic values.
- The transition piece will be convergent, observing the following increase in the average flow velocity, $c_1 = 2 \cdot c_0$ (m/s).
- Following the transition part, the injector will have its outer surface be flat and spiral-shaped, $r_{\theta_j} = 0.5 \cdot D \cdot \exp\left(\frac{\pi \cdot tg\alpha_2}{180} \cdot \theta_j\right) + e_{fo}$, until exiting before the rotor inlet.
- The injector blade is calculated and projected in the form of a profile from the National Advisory Committe for Aeronautics (NACA) profile—65 – 10 from fundamental coordinates shown in the following table (Table 6.1).
- Steps for calculated characteristics requiring the injector outlet angle must be calculated for at least $\delta_j = 60^{\circ}; 90^{\circ}; 120^{\circ}$, for further analysis and deliberation.

- The specific energy loss from leakage, equivalent to specific work on the rotor, must follow the expression, $Y_{i_{PQf}} = \left(1 - \eta_f\right) \cdot Y_{i_3}$ (J/kg).
- The real absolute average velocity of the flow at the entrance of the rotor, and the consequent tangential velocity of the rotor at the entrance, will be determined from the condition of maximum theoretical internal efficiency of the *BMHIT*, analytically determined from the following expressions, $c_4 = \frac{\sqrt{2 \cdot Y_{i_4}}}{\cos \alpha_4}$ *(m/s)* and $u = 0.5 \cdot c_4 \cdot \ cos\ \alpha_4$ (m/s).
- The meridional velocities at the entrance and exit of the rotor will be equal, $c_{m_4} = c_{m_5}$, along with those at the exit of the first passage through the rotor, and entry into the second passage, $c_{m_{5*}} = c_{m_{4*}}$ which means $\alpha_5 = 90°$.
- The axis of the blade will be in the form of a circular radius arc, r_{balde} and exit angle $\beta_{5*} = 90^{\circ}$.

6.2 The Input Data

- $Q_{\max}$
- Q
- $Q_{\min}$ (m^3/s)
- $H_{\max}$
- H
- $H_{\min}$ (mca)
- Network frequency 50 or 60 (Hz)
- *NACA* profile coordinates—65 – 10

6.3 Input Data Set from the Designer

- ϕ_0
- ϕ_{tr}
- ε_{tr}
- ϕ_2
- ϕ_3
- η_f
- $\alpha_2 = \alpha_4^o$
- ψ_5
- ϕ_{3*}
- k_f
- e_{fo}
- e_{su}
- $e_{su}^*(mm)$

- δ^o
- r_D
- $\eta_m = 0.95$
- η_{el}
- $g = 9.81$ (m/s^2)
- $p = 1000$ (kg/m^3)
- *NACA* profile coordinates—65—10, and
- the material characteristics that will be used to design and make the calculations for each *BMHIT* component.

6.4 Validation

The step by step validation will be done using the following input data:

- **Nominal characteristics established**:
  - $H_0 = H = 40$ (m)
  - $Q = Q_{1/1} = 1.0$ (m^3/s)
  - $60 =$ (Hz)
- **Preliminary fixed characteristics**:
  - $\phi_0 = 0.2$
  - $\phi_{tr} = 0.98$
  - $\varepsilon_{tr} = 0.03$ (mm)
  - $\phi_2 = 0.97$
  - $\phi_3 = 0.99$
  - $\eta_f = 0.96$
  - $\alpha_2 = \alpha_4^o = 14^o$
  - $\psi_5 = 0.97$
  - $\phi_{3*} = 0.98$
  - $k_f = 0.9$
  - $e_{fo} = 2$ (mm)
  - $e_{su} = 2$ (mm)
  - $e_{su}^* = 1$ (mm)
  - $\delta^o = 90^\circ$
  - $r_D = 0.5$
  - $\eta_m = 0.95$
  - $\eta_{el} = 0.98$
  - $g = 9.81$ (m/s^2)
  - $p = 1000$ (kg/m^3)
  - *NACA* profile coordinates—65—10
  - American Society for Testing and Materials (*ASTM*) steels

– The Society of Automotive Engineers (*SAE*)
– Brazilian Association of Norms and Standards (*ABNT*) 1020 with:

$$\sigma_{f_{ad}} = 14469.75\ (\mathrm{N/cm^2})$$
$$\sigma_{c_{ad}} = 19620\ (\mathrm{N/cm^2})$$
$$\sigma_{t_{adm}} = 12000\ (\mathrm{N/cm^2})$$
$$\sigma_{f_{adm}} = 14470\ (\mathrm{N/cm^2})$$
$$\sigma_e = 24623\ (\mathrm{N/cm^2})$$
$$\tau_{cis_{adm}} 9565\ (\mathrm{N/cm^2})$$

6.5 Roadmap and Decisions

Calculate the following steps (1–156), making approximations to the third decimal place:

1. The specific internal work available in the flow at the entrance of the *BMHIT*:

$$Y_{i_0} = g \cdot H = 9.81 \cdot 40 = 92.4\ (\mathrm{j/kg})$$

2. Internal power available in the flow at the *BMHIT* inlet:

$$P_{i_0} = Q \cdot Y_{i_0} = P_{i_0} = 1 \cdot 392.4 = 392.4\ (\mathrm{kW})$$

3. The absolute average velocity of the flow at the inlet of the *BMHIT*:

$$c_0 = \phi_0 \cdot \sqrt{2 \cdot Y_0} = 0.2 \cdot \sqrt{2 \cdot 392.4} = 5.603\ (\mathrm{m/s})$$

4. The loss of available internal work in the flow in the passage through the transition:

$$Y_{i_{Ptr}} = 0.5 \cdot \left(1 - \phi_{tr}^2\right) \cdot c_0^2 = 0.5 \cdot \left(1 - 0.98^2\right) \cdot 5.603^2 = 0.621\ (\mathrm{J/kg})$$

5. The loss of available internal power of the flow and its passage through the transition:

$$P_{p_{i_{tr}}} = Q \cdot Y_{p_{i_{tr}}} = 1 \cdot 0.621 = 0.621 \text{ (kW)}$$

6. The transmission's internal performance:

$$\eta_{i_{tr}} = 1 - \frac{P_{p_{i_{tr}}}}{P_0} = 1 - \frac{0.621}{392.4} = 0.9984 = 99.84 \ (\%)$$

7. The specific internal work available in the flow at the inlet of the injector:

$$Y_{i_{in}} = Y_{i_1} = Y_{i_0} - Y_{i_{Ptr}} =$$

8. The internal flow power available at the inlet of the injector:

$$P_{i_1} = Q \cdot Y_{i_1} = 1 \cdot 391.779 = 391.779 \text{ (kW)}$$

9. The theoretical average absolute velocity of the flow at the outlet of the injector:

$$c_{2_t} = \sqrt{2 \cdot Y_{i_1}} = \sqrt{2 \cdot 391.779} = 27.992 \text{ (m/s)}$$

10. The actual average absolute velocity of the flow at the outlet of the injector:

$$c_2 = \phi_2 \cdot c_{2_t} = 0.97 \cdot 27.992 = 27.152 \text{ (m/s)}$$

11. The specific internal work of the flow lost in the injector:

$$Y_{i_{Pin}} = 0.5 \cdot \left(c_{2_t}^2 - c_2^2 \right) = 0.5 \cdot \left(27.992^2 - 27.152^2 \right) = 23.160 \text{ (J/kg)}$$

12. The internal power lost from the flow as it passes through the injector:

$$P_{i_{Pin}} = Q \cdot Y_{i_{Pin}} = 1 \cdot 23.160 = 23.160 \text{ (kW)}$$

13. The internal efficiency of the injector:

$$\eta_{i_{in}} = 1 - \frac{P_{i_{p_{in}}}}{P_{i_1}} = 1 - \frac{23.160}{391.779} = 0.9409 = 94.09\ (\%)$$

14. The relative loss of available power of the flow in the injector with respect to P_{i_0}:

$$\eta_{i_{in_0}} = \frac{P_{i_{p_{in}}}}{P_{i_0}} = \frac{23.160}{392.4} = 0.0590 = 5.90\ (\%)$$

15. The available internal flow specific work at the outlet of the injector:

$$Y_{i_2} = Y_{i_1} - Y_{i_{p_i}} = 391.779 - 23.160 = 368.619\ (\text{J/kg})$$

16. The internal power available in the flow at the outlet of the injector:

$$P_{i_2} = Q \cdot Y_{i_2} = 1 \cdot 368.619 = 368.619\ (\text{kW})$$

17. The actual average absolute velocity of the flow in the interstice, injector/rotor:

$$c_3 = \phi_3 \cdot c_2 = 0.99 \cdot 27.152 = 26.880\ (\text{m/s})$$

18. The true average meridional velocity of the flow in the interstice, injector/rotor:

$$c_{m_3} = c_3 \cdot \sin \alpha_2 = 26.880 \cdot \sin 14 = 6.503\ \ (\text{m/s})$$

19. The specific internal work of the flow lost in the injector/rotor interstice:

$$Y_{i_{p_3}} = 0.5 \cdot \left(c_2^2 - c_3^2\right) = 0.5 \cdot \left(27.152^2 - 26.880^2\right) = 7.348\ (\text{J/kg})$$

20. The internal power lost from the flow as it passes through the injector/rotor interstice:

$$P_{i_{p_3}} = Q \cdot Y_{i_{p_3}} = 1 \cdot 7.348 = 7.348\ (\text{kW})$$

21. The internal efficiency of the injector/rotor interstice:

$$\eta_{i_3} = 1 - \frac{P_{i_{p3}}}{P_{i_2}} = 1 - \frac{7.348}{368.619} = 0.9801 = 98.01\ (\%)$$

22. The relative loss of available flow power in the injector/rotor interstice with respect to P_{i_0}:

$$\eta_{i_{3_0}} = \frac{P_{p_3}}{P_{i_0}} = \frac{7.348}{392.4} = 0.0187 = 1.87\ (\%)$$

23. The specific internal flow work available in the injector/rotor interstice:

$$Y_{i_3} = Y_{i_2} - Y_{i_{p_3}} = 368.619 - 7.348 = 361.271\ (\mathrm{J/kg})$$

24. The internal power available in the flow in the injector/rotor interstice:

$$P_{i_3} = Q \cdot Y_{i_3} = 1 \cdot 361.271 = 361.271\ (\mathrm{kW})$$

25. The loss of specific energy from leakage equivalent to specific work on the rotor:

$$Y_{i_{pQf}} = \left(1 - \eta_f\right) \cdot Y_{i_3} = (1 - 0.96) \cdot 361.271 = 14.451\ (\mathrm{J/kg})$$

26. The internal power of the flow lost to leakage:

$$P_{i_{pQf}} = Q \cdot Y_{i_{pQf}} = 1 \cdot 14.451 = 4.451\ (\mathrm{kW})$$

27. The internal efficiency corresponding to the power lost to leakage:

$$\eta_{i_{Qf}} = 1 - \frac{P_{i_{pQf}}}{P_{i_3}} = 1 - \frac{14.451}{361.271} = 0.9600 = 96.00\ (\%)$$

This value confirms the calculations up to this step since $\eta_f = \eta_{i_{Qf}} = 0.96$, and should therefore be observed in studies for fixing the sizes of the corresponding labyrinths.

28. The relative loss of available flow power corresponding to the leakage relative to P_{i_0}:

$$\eta_{i_{Qf_0}} = \frac{P_{i_{pQf}}}{P_{i_0}} = \frac{14.451}{392.4} = 0.0368 = 3.68\ (\%)$$

29. The specific internal work available from the flow at the impeller inlet:

$$Y_{i_4} = Y_{i_3} - Y_{i_{pQf}} = 361.271 - 14.451 = 346.820\ (\text{kJ/kg})$$

30. The internal power available in the flow at the rotor inlet:

$$P_{i_4} = Q \cdot Y_{i_4} = 1 \cdot 346.820 = 346.820\ (\text{kW})$$

31. The actual absolute average speed at the rotor inlet:

$$c_4 = \frac{\sqrt{2 \cdot Y_{i_4}}}{\cos \alpha_4} = \frac{\sqrt{2 \cdot 346.820}}{\cos 14} = 27.143\ (\text{m/s})$$

32. The tangential speed at the rotor inlet:

$$u = 0.5 \cdot c_4 \cdot \cos \alpha_4 = 0.5 \cdot 27.143 \cdot \cos 14 = 13.168\ (\text{m/s})$$

33. The component of the actual average absolute velocity at the rotor input in the tangential direction:

$$c_{u_4} = c_4 \cdot \cos \alpha_4 = 27.143 \cdot \cos 14 = 26.337\ (\text{m/s})$$

The product $u \cdot c_{u_4} = 13.168 \cdot 26.337 = 346.806\ (\text{J/kg})$ differs from $\pm 0.004\ (\%)$ from what was determined in step 30, $Y_{i_4} = 346.820\ (\text{J/kg})$. This difference comes from the approximation adopted to only three decimal places, indicating that the calculations can continue when this difference is less than 1 (%). If it is greater, stop the calculations and print the results for analysis, and once correcting what needs to be corrected, restart the program at appropriate step.

34. The tangential velocity at the inside diameter of the rotor:

$$u_i = r_D \cdot u = 0.5 \cdot 13.168 = 6.584\ (\text{m/s})$$

35. The component of the absolute mean velocity of the flow at the rotor inlet in the meridional direction:

$$c_{m_4} = c_4 \cdot \sin\alpha_4 = 27.143 \cdot \sin\ 14 = 6.566\ (\mathrm{m/s})$$

36. The actual average relative velocity of the flow at the impeller inlet:

$$w_4 = \sqrt{c_{m_4}^2 + (c_{u_4} - u)^2} = \sqrt{6.566^2 + (26.337 - 13.168)^2} = 14.715\ (\mathrm{m/s})$$

37. The average actual angle of the rotor blade in the tangential direction with the direction of the true average relative velocity of the flow at the entrance and exit of the rotor:

$$\beta_4 = \beta_5 = \beta_{blade} = \arcsin\frac{c_{m_4}}{w_4} = \arcsin\frac{6.566}{14.715} = 26.501^\circ$$

38. The true average relative velocity of the flow at the exit of the first passage through the rotor, equal to the corresponding average meridional, for the true angle of the blade $\beta_{5*} = 90^\circ$, thus equal to $c_{m_{5*}}$:

$$w_{5*} = c_{m_{5*}} = \psi_5 \cdot w_4 = 0.97 \cdot 14.715 = 14.274\ (\mathrm{m/s})$$

39. The specific internal work of the flow lost in the first passage through the rotor:

$$Y_{i_{Pr1}} = 0.5 \cdot \left(w_4^2 - w_{5*}^2\right) = 0.5 \cdot \left(14.715^2 - 14.274^2\right) = 6.392\ (\mathrm{J/kg})$$

40. The internal power lost from the flow on the first passage through the rotor:

$$P_{i_{Pr1}} = Q \cdot Y_{i_{Pr1}} = 1 \cdot 6.392 = 6.392\ (\mathrm{kW})$$

41. The internal power available from the flow that will be transformed by the rotor:

$$P_{i_{dr1}} = P_{i_4} - P_{i_{Pr1}} = 346.820 - 6.392 = 340.428\ (\mathrm{kW})$$

42. The tangential velocity of the rotor at the exit of the first flow passage, with $\beta_{5*} = 90^\circ$, resulting in:

$$u_i = c_{u_{5*}} = 6.584\ (\mathrm{m/s})$$

43. The actual average absolute velocity at the exit of the first flow passage through the impeller:

$$c_{5*} = \sqrt{u_i^2 + w_{5*}^2} = \sqrt{6.584^2 + 14.274^2} = 15.719\ (\mathrm{m/s})$$

44. The angle between the direction of the tangential velocity u_i and the real absolute average velocity c_{5*}:

$$\alpha_{5*} = \arcsin\frac{w_{5*}}{c_{5*}} = \arcsin\frac{14.274}{15.719} = 62.240^\circ$$

45. The internal flow power received by the rotor in the first passage, which will be transformed into mechanical power on its axis:

$$P_{i_{r1}} = P_{i_{dr1}} - P_{i_{5*}} = P_{i_{dr}} - Q \cdot u_i \cdot c_{u_{5*}} = 340.428 - 1 \cdot 6.584 \cdot 6.584 = 297.079\ (\mathrm{kW})$$

46. The internal efficiency of the rotor in the first flow passage in relation to the internal power received:

$$\eta_{i_{r1}} = \frac{P_{i_{r1}}}{P_{i_{dr1}}} = \frac{297.079}{340.428} = 0.8727 = 87.27\ (\%)$$

47. The relative loss of available power of the flow in its first passage through the rotor with respect to P_{i_0}:

$$\eta_{i_{r1_0}} = \frac{P_{i_{p_{r1}}}}{P_{i_0}} = \frac{6.392}{392.4} = 0.0163 = 1.63\ (\%)$$

48. The internal power available in the flow at the exit of its first passage through the rotor, equal to that at the entrance into its interior:

$$P_{i_{5*}} = P_{i_{6*}} = P_{i_{dr1}} - P_{i_{r1}} = 340.428 - 297.079 = 43.349\ (\mathrm{kW})$$

49. The true absolute average velocity of the flow after its first passage through the rotor:

$$c_{6*} = \sqrt{2 \cdot 43.349} = 9.311 \text{ (m/s)}$$

50. The absolute average velocity of the flow before entering the second passage through the rotor:

$$c_{3*} = \phi_{3*} \cdot c_{6*} = 0.98 \cdot 9.311 = 9.125 \text{ (m/s)}$$

51. The specific internal work lost by the flow inside the rotor:

$$Y_{i_{P6*;3*}} = 0.5 \cdot \left(c_{6*}^2 - c_{3*}^2\right) = 0.5 \cdot \left(9.311^2 - 9.125^2\right) = 1.715 \text{ (J/kg)}$$

52. The power of the flow lost in its passage through the interior of the rotor:

$$P_{i_{P6*;3*}} = Q \cdot Y_{i_{P6*;3*}} = 1 \cdot 1715 = 1.715 \text{ (kW)}$$

53. The internal efficiency of the flow inside the rotor in relation to its internal power in 6* results in:

$$\eta_{i_{6*;3*}} = 1 - \frac{P_{i_{P6*;3*}}}{P_{i_{6*}}} = 1 - \frac{1.715}{43.349} = 0.9604 = 96.04 \text{ (\%)}$$

54. The relative loss of the available power of the flow corresponding to its passage through the interior of the rotor in relation to P_{i_0}, results in:

$$\eta_{i_{6*;3*0}} = \frac{P_{i_{P6*;3*}}}{P_{i_0}} = \frac{1.715}{392.4} = 0.0044 = 0.44 \text{ (\%)}$$

55. The internal power available in the flow immediately before it enters the second passage through the rotor equals the power at its entrance:

$$P_{i_{3*}} = P_{i_{4*}} = P_{i_{6*}} - P_{i_{6*;3*}} = 43.349 - 1.715 = 41.634 \text{ (kW)}$$

56. The specific internal work of the flow at the entrance of its second passage through the rotor:

$$Y_{i_{4*}} = \frac{P_{i_{4*}}}{Q} = \frac{41.634}{1} = 41.634 \text{ (J/kg)}$$

57. The absolute average velocity component of the flow in the tangential direction at the entrance of its second passage through the rotor:

$$c_{u_{4*}} = \frac{Y_{i_{4*}}}{u_i} = \frac{41.634}{6.584} = 6.324 \text{ (m/s)}$$

58. The average relative velocity of the flow at the entrance of its second passage through the rotor, with the condition that $c_{m_{4*}} = c_{m_{5*}} = 14.274 \text{ (m/s)}$, resulting in:

$$w_{4*} = \sqrt{\left(u_i - c_{u_{4*}}\right)^2 + c_{m_{4*}}^2} = \sqrt{(6.584 - 6.324)^2 + 14.274^2} = 14.276 \text{ (m/s)}$$

59. The internal power lost by the flow shock at the entrance of the second passage through the rotor, considering that the blade has an angle, $\beta_{blade} = 90^o$, resulting in:

$$P_{i_{P_{ch4*}}} = Q \cdot 0.5 \cdot \left(w_{4*}^2 - c_{m_{4*}}^2\right) = 1 \cdot 0.5 \cdot \left(14.276^2 - 14.274^2\right) = 0.029 \text{ (kW)}$$

60. The absolute average velocity of the flow at the entrance of the second passage through the rotor:

$$c_{4*} = \sqrt{c_{m_{4*}}^2 + c_{u_{4*}}^2} = \sqrt{14.274^2 + 6.324^2} = 15.612 \text{ (m/s)}$$

61. The angle that the tangential direction of the rotor forms with the direction of the meridional average velocity at the entrance of the second flow passage through the rotor:

$$\beta_{4*} = \arcsin \frac{c_{m_{4*}}}{w_{4*}} = \arcsin \frac{14.274}{14.276} = 89.041^o$$

62. The angle between the directions of the mean, meridional, and absolute velocities at the entrance of the second flow passage through the rotor:

$$\alpha_{4*} = \arcsin\frac{c_{m_{4*}}}{c_{4*}} = \arcsin\frac{14.274}{15.612} = 66.106^{\circ}$$

63. The average relative velocity of the flow at the exit of the second passage through the rotor:

$$w_5 = \psi_5 \cdot w_{4*} = 0.97 \cdot 14.276 = 13.848\ (\mathrm{m/s})$$

64. The frictional power loss in the second flow passage through the rotor:

$$P_{i_{p_{4*;5}}} = Q \cdot 0.5 \cdot \left(w_{4*}^2 - w_5^2\right) = 1 \cdot 0.5 \cdot \left(14.276^2 - 13.848^2\right) = 6.019\ (\mathrm{kW})$$

65. The average meridional velocity of the flow at the exit of the second passage through the rotor, with the average constructive angle of the blade calculated in step 37, $\beta_5 = \beta_4 = 26.501^{\circ}$, thus:

$$c_{m_5} = w_5 \cdot \sin\ \beta_5 = 13.848 \cdot \sin\ 26.501 = 6.179\ (\mathrm{m/s})$$

66. The absolute average velocity component of the flow in the tangential direction at the exit of the second passage through the rotor:

$$c_{u_5} = u - w_5 \cdot \cos\beta_5 = 13.168 - 13.848 \cdot \cos 26.501 = 0.775\ (\mathrm{m/s})$$

67. The absolute average velocity of the flow at the exit of the second passage through the rotor:

$$c_5 = \sqrt{c_{m_5}^2 + c_{u_5}^2} = \sqrt{6.179^2 + 0.775^2} = 6.227\ (\mathrm{m/s})$$

68. The angle formed by the directions of the tangential and mean absolute velocities of the flow at the exit of its second passage through the rotor.
69. $\alpha_5 = \arcsin\frac{c_{m_5}}{c_5} = \arcsin\frac{6.179}{6.227} = 82.881^o$
    Since $\alpha_5 = 90^{\circ}$, we have:

$$c_{m_{5**}} = u \cdot tg\beta_5 = 13.168 \cdot tg26.501 = 6.566\ (\mathrm{m/s})$$

… equal to the meridional average velocity of the flow at the entrance of the rotor, $c_{m_4} = 6.566\ (\mathrm{m/s})$, resulting in an internal power loss of the flow from shocks at the exit, which will be determined in the next step. For current validation:

$$w_{5**} = \frac{u}{\cos \beta_5} = \frac{13.168}{\cos 26.501} = 14.714\ (\mathrm{m/s})$$

70. The internal power loss, from shocks, at the rotor flow output:

$$P_{i_{i_{Pch_5}}} = Q \cdot 0.5 \cdot \left(c_{m_{5**}}^2 - c_{m_5}^2\right) = 1 \cdot 0.5 \cdot \left(6.566^2 - 6.179^2\right) = 2.489\ (\mathrm{kW})$$

71. The internal specific work of the flow at the blade outlet equals the internal specific work at the rotor outlet:

$$Y_{i_5} = Y_{i_6} = 0.5 \cdot c_{m_5}^2 = 0.5 \cdot 6.179^2 = 19.090\ (\mathrm{J/kg})$$

72. The internal specific power of the flow at the exit of the rotor equals the internal specific power after exiting:

$$P_{i_6} = Q \cdot Y_{i_6} = 1 \cdot 19.090 = 19.090\ (\mathrm{kW})$$

73. The internal flow power received by the rotor in the second passage that will be transformed to the exit:

$$P_{i_{r_2}} = P_{i_{4*}} - \left(P_{i_{i_{Pch_{4*}}}} + P_{i_{i_{P4*;5}}} + P_{i_{i_{Pch_5}}} + P_{i_6}\right) = 41.634 - (0.029 + 6.019 + 2.489 + 19.090) = 14.007\ (\mathrm{kW})$$

74. The internal flow power received by the rotor in the two passages that will be transformed into mechanical power on the axis:

$$P_{i_r} = P_{i_{r_1}} + P_{i_{r_2}} = 297.079 + 14.007 = 311.086\ (\mathrm{kW})$$

75. The *BMHIT*'s internal yield:

$$\eta_i = \frac{P_{i_r}}{P_{i_0}} = \frac{311.086}{392.4} = 0.7928 = 79.28\ (\%)$$

76. The portion of the *BMHIT*'s internal power received by the rotor on the first passage:

$$\eta_{i_{r_{1;1}}} = \frac{P_{i_{r_1}}}{P_{i_o}} = \frac{297.079}{392.4} = 0.7571 = 75.71\ (\%)$$

77. The portion of the *BMHIT*'s internal power received by the rotor on the second passage:

$$\eta_{i_{r_2}} = \frac{P_{i_{r_2}}}{P_{i_o}} = \frac{14.007}{392.4} = 0.0357 = 3.57\ (\%)$$

78. The sum of the internal power losses in the *BMHIT*, including equivalent leakage and kinetic energy losses at the rotor outputL

$$P_{i_p} = P_{i_{P_{tr}}} + P_{i_{P_{1;2}}} + P_{i_{P_3}} + P_{i_{P_{Qf}}} + P_{i_{P_{r1}}} + P_{i_{P_{6*;3*}}} + P_{i_{P_{ch4*}}} + P_{i_{P_{r2}}} + P_{i_{P_{ch_5}}} + P_{i_6}$$

$$=$$

$$0.621 + 23.160 + 7.348 + 14.451 + 6.392 + 1.715 + 0.029 + 6.019 + 2.489 + 19.090$$

$$=$$

$$81.314\ (\text{kW})$$

79. The *BMHIT*'s internal yield considering total losses:

$$\eta_{i_{tp}} = 1 - \frac{P_{i_p}}{P_{i_0}} = 1 - \frac{81.314}{392.4} = 0.7924 = 79.28\ (\%)$$

This value is practically equal to the value obtained in step 74, indicating the validity of what we have done up to this step.

80. The portion of the *BMHIT*'s internal power lost between the input and output:

$$\eta_{i_p} = \frac{P_{i_p}}{P_{i_o}} = \frac{81.314}{392.4} = 0.2072 = 20.72\ (\%)$$

81. The internal efficiency without considering kinetic energy losses at the output:

$$\eta_{i*} = 1 - \frac{P_{i_p} - P_{i_6}}{P_{i_0}} = 1 - \frac{81.314 - 19.09}{392.4} = 0.8414 = 84.14\ (\%)$$

82. The cross-sectional area of the conduit at the entrance:

$$A_0 = \frac{Q}{c_o} = \frac{1}{5.603} = 0.1785\ (\mathrm{m}^2)$$

83. The diameter of the conduit at the entrance to the *BMHIT*:

$$D_0 = \sqrt{\frac{4 \cdot A_0}{\pi}} = \sqrt{\frac{4 \cdot 0.1785}{\pi}} = 0.477\ (\mathrm{m})$$

84. The average flow velocity at the injector inlet:

$$c_1 = 2 \cdot c_0 = 2 \cdot 5.603 = 11.206\ (\mathrm{m/s})$$

85. The internal specific energy of the static pressure of the flow at the inlet of the *BMHIT*:

$$\frac{p_{i_0}}{\rho} = Y_{i_0} - \frac{c_0^2}{2} = 392.4 - \frac{5.603^2}{2} = 376.703\ (\mathrm{J/kg})$$

86. The internal specific energy of the static pressure of the flow at the exit of the transition piece, which is equal to the internal specific energy of the static pressure of the flow at the inlet of the injector:

$$\frac{p_{i_1}}{\rho} = Y_{i_0} - \frac{c_1^2}{2} + Y_{p_{i_{tr}}} = 392.4 - \frac{11.206^2}{2} + 0.621 = 330.234\ (\mathrm{J/kg})$$

87. The cross-sectional area of the transition piece at input 1:

$$A_1 = \frac{Q}{c_1} = \frac{1}{11.206} = 0.089238\ (\mathrm{m}^2)$$

88. The actual average radial velocity of the flow at the outlet of the injector:

$$c_{r_2} = c_2 \cdot \sin\ \alpha_2 = 27.152 \cdot \sin\ 14 = 6.569\ (\mathrm{m/s})$$

The synchronous minimum and maximum rotor speed of the *BMHIT*,

$$n^*_{\min} = 5 \cdot \frac{u^{15}}{(\eta_f \cdot Q)^{0.5}} = 5 \cdot \frac{13.168^{15}}{(0.96 \cdot 1)^{0.5}} = 243.845\ (\mathrm{rpm}) \rightarrow$$

$$z^*_{pp_{\max}} = \frac{3600}{n^*_{\min}} = \frac{3600}{243.845} = 14.763 \rightarrow$$

$$z_{pp_{\max}} = 14 \rightarrow$$

$$n_{\min} = \frac{3600}{14} = 257.143\ (\mathrm{rpm})$$

and

$$n^*_{\max} = 18 \cdot \frac{u^{15}}{(\eta_f \cdot Q)^{0.5}} = 18 \cdot \frac{13.138^{15}}{(0.96 \cdot 1)^{0.5}} = 877.843 \rightarrow$$

$$z^*_{pp_{\min}} = \frac{3600}{n^*_{\max}} = \frac{3600}{877.843} = 4.101 \rightarrow$$

$$z_{pp_{\min}} = 4 \rightarrow$$

$$n_{\max} = \frac{3600}{4} = 900\ (\mathrm{rpm}).$$

Only one attempt will be made at this validation with $n = 600$ (rpm). The computer program should be able to present results stored in memory, with variations in the number of pairs of *EG* poles, between the minimum and maximum values, so that after having the results of the next three steps, a table with these values can be made available along with the corresponding diameters, so that the project team can decide which values the calculations should continue with and store.

89. The rotor diameters, here:

$$D = \frac{60 \cdot u}{\pi \cdot n} = \frac{60 \cdot 13.168}{\pi \cdot 600} = 0.419\ (\mathrm{m}) = 419\ (\mathrm{mm})$$

and

$$D_i = r_D \cdot D = 0.5 \cdot 0.419 = 0.210\ (\mathrm{m}) = 210\ (\mathrm{mm})$$

90. The radius at the injector outlet:

$$D_{in} = D + 3 = 419 + 3 = 422 \text{ (mm)}$$

91. The injector's widths and coverage arcs to allow for making at least $\delta = 60^{\circ}$, 90°, and 120°. In this case, only one attempt will be made for $\delta = 90^{\circ}$, so:

$$B_{in_j} = \frac{360 \cdot Q}{\pi \cdot \delta \cdot D_{in} \cdot c_{r_2}} \rightarrow A\bar{C}_{\delta_{in}} = \frac{\delta^o \cdot \pi \cdot D_{in}}{360}$$

In this case …

$$B_{in} = \frac{360 \cdot 1}{\pi \cdot 90 \cdot 0.422 \cdot 6.569} = \rightarrow = 0.459 \text{ (m)} = 459 \text{ (mm)} \rightarrow$$

$$\overline{AC_{\delta_{in}}} = \frac{90 \cdot \pi \cdot 0.422}{360} = 0.331 \text{ (m)} = 331 \text{ (mm)}$$

92. The corresponding injector heights:

$$b_{in_j} = \frac{A_1}{B_{in_j}} \rightarrow b_{in} = \frac{A_1}{B_{in}} = \frac{0.089238}{0.459} = 0.194 \text{ (m)}$$

93. The minimum preliminary diameter of the rotor shaft:

$$d_{e_{\min}} = 118 \cdot \sqrt[3]{\frac{P_{\max}}{n}} = 188 \cdot \sqrt[3]{\frac{392.4}{600}} = 102.4 \text{ (mm)}$$

This value must be verified by the structural calculations.

94. At distance $L_{e_{\min}} =$ of the flow inside the rotor relative to the axis:

$$L_{e_{\min}} > 0.5 \cdot 102.4 = 51.25 \text{ (mm)}$$

95. The distance from the rotor axis to the internal flow limit inside it:

$$L_e = 0.5 \cdot r_D \cdot D \cdot \cos \alpha_{5*} = 0.5 \cdot 0.5 \cdot 419 \cdot \cos 62.240 = 48.789 \text{ (mm)}$$

With $L_e < L_{e_{\min}}$, the internal part of the flow in the passage through the interior of the rotor collides with its axis.

96. The preliminary theoretical step at the input and output of the rotor:

$$t_4^* = 0.5 \cdot (D - D_i) = 0.5 \cdot (0.419 - 0.210) = 0.105 \text{ (m)}$$

97. The number of rotor blades:

$$z_r = \frac{\pi \cdot D}{t_4} = \frac{\pi \cdot 0.419}{0.105} = 12.5$$

$z_r = 12$ blades will be adopted.
With this, the theoretical step at the input and output of the rotor results in:

$$t_4 = t_5 = \frac{\pi \cdot D}{z_r} = \frac{\pi \cdot 0.419}{12} = 0.110 \text{ (m)}$$

This value must be stored for later use.

98. The number of blades covered by the rotor at the outlet of the injector:

$$z_\delta = \frac{\delta}{360} \cdot z_r = z_\delta = \frac{90}{360} \cdot 12 = 3 \text{ blades}$$

This integer or fractional value must be stored to be used in the following steps.

99. The theoretical step at the exit of the first flow passage through the rotor and at the entrance of the second:

$$t_{5*} = t_{4*} = r_D \cdot t_4 = 0.5 \cdot 0.11 = 0.055 \text{ (m)}$$

100. The average radius of the rotor blades:

$$r_{blade} = 0.25 \cdot \frac{D^2 - D_i^2}{D \cdot \cos\beta_4 - D_i \cdot \cos\beta_{5*}} = 0.25 \cdot \frac{0.419^2 - 0.21^2}{0.419 \cdot \cos 26.501 - 0.21 \cdot \cos 90}$$
$$= 0.088 \text{ (m)}$$

101. The rotor blade chord length L_{blade}, its coverage angle θ_{blade}, and respective arc through the following linear system of three equations:

$$\alpha_2 = 90 - (0.5 \cdot \theta_{blade} + \beta_4) = 90 - (0.5 \cdot \theta_{blade} + 26.501)$$

$$L_{blade} = 2 \cdot r_{blade} \cdot \sqrt{1 - \cos^2(0.5 \cdot \theta_{blade})} = 2 \cdot 88 \cdot \sqrt{1 - \cos^2(0.5 \cdot \theta_{blade})}$$

$$L_{blade} = L^*_{blade} = 0.5 \cdot D \cdot \cos \alpha_2 \pm 0.5 \cdot \sqrt{(D \cdot \cos \alpha_2)^2 - (D^2 - D_i^2)} = = 0.5 \cdot 419$$
$$\cdot \cos \alpha_2 - 0.5 \cdot \sqrt{(419 \cdot \cos \alpha_2)^2 - (419^2 - 210^2)} = = 209.5 \cdot \cos \alpha_2 - 0.5$$
$$\cdot \sqrt{(419 \cdot \cos \alpha_2)^2 - 131461}$$

The solution to the system was obtained

To program the solution, one can start a scale drawing of the average line of the blade in a circular arc, and then measure the angle.

102. The radii delimiting the injector:

$$r_{\theta_j} = 0.5 \cdot D \cdot \exp\left(\frac{\pi \cdot tg\alpha_2}{180} \cdot \theta_j\right) + e_{fo} = 0.5 \cdot 0.419 \cdot \exp\left(\frac{\pi \cdot tg14}{180} \cdot \theta_j\right) + 0.002$$
$$= 0.2095 \cdot exp\left(0.783287 \cdot \frac{\theta_j}{180}\right) + 0.002\ (m)$$

Values must be stored for later use.

103. Preliminarily fix the thickness of the blade at its center of gravity by:

$$e_{cg_a} \cong 0.07 \cdot D = 0.07 \cdot 419 \cong 29\ (\text{mm})$$

104. Using a graphical software program, draw the middle or skeleton line of the blade in the form of a spiral, to scale, observing the following steps:

1. Start Fig. 6.1, establishing a scale and drawing a system of orthogonal axes from point 0.
2. From the horizontal axis with the r_{θ_j}(mm) values and angles 0°, 15°, and 90° obtain the corresponding points of the rays from e to s.
3. Connect the ends of the spokes, obtaining the spiral part of the injector.
4. Draw a line r_t from point e, tangent to the spiral.
5. Draw the $\overline{si}$ arc.
6. Vertically, using r_t as a reference, mark on the scale drawing $b_{in} = 194$ (mm), obtaining points a; b, and the midpoint of this segment, c.
7. Connect b with i with line and circle segments with center 0, obtaining point d.
8. Connect the midpoint $1*$ of segment ei with c, obtaining line r_1.
9. Copy spiral se, adjusting it so that it is tangent to r_1 at point B and on the circumference segment at A.
10. Measure the length of the chord of the spiral blade profile on the scaled drawing, in this case, $L_a = 240$ (mm), and the angle 40° that makes a trailing edge A with the horizontal axis.

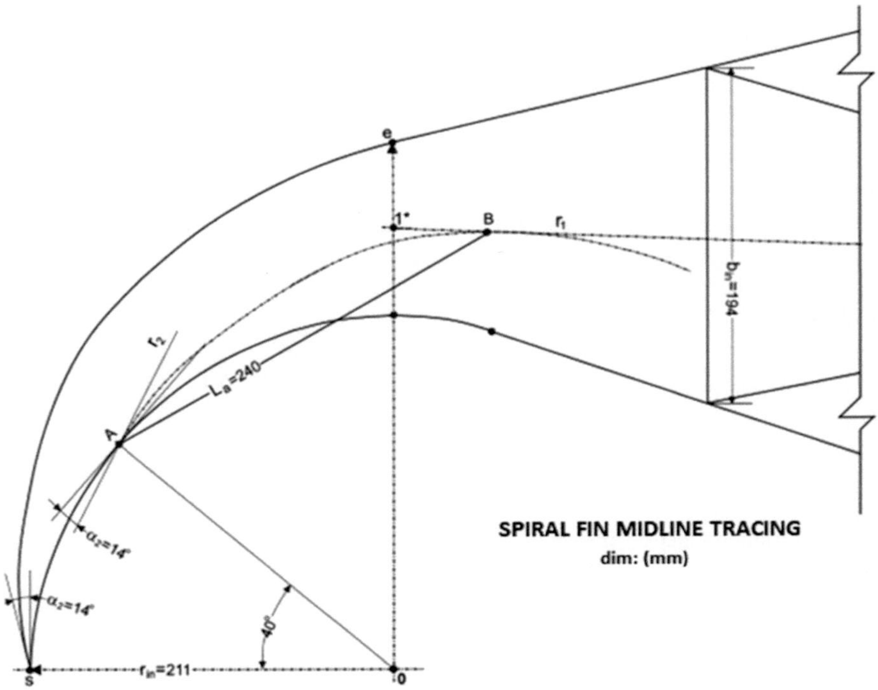

Fig. 6.1 Cross section to the axis in the *BMHIT* to obtain the length of the blade chord

105. Calculate the blade throttling factor:

$$f_{e_{65-010}} = \frac{\left(e_{cg}/L_{blade}\right)_{cal}}{0.101} = \frac{29/240}{0.101} = 1.196$$

106. With the values stored in the NACA profile data Table—65—10, calculate the blade coordinates and store them from the following expressions:

$$x_a = 0.01 \cdot L_a \cdot x_p$$

in this case

$$x_a = 0.01 \cdot 240 \cdot x_p = 2.4 \cdot x_p \text{ (mm)}$$

and

$$y_a = 0.01 \cdot f_{e_{65-010}} \cdot L_a \cdot y_p$$

in this case

$$y_a = 0.01 \cdot 1.196 \cdot 240 \cdot y_p = 2.87 \cdot y_p \text{ (mm)}$$

$$r_{a_{ba}} = 2.87 \cdot r_{a_{ba}} = 2.87 \cdot 0.666 = 1.911 \text{ (mm)}$$

$$y_{a_{bf}} = 2.87 \cdot 1 = 2.87 \text{ (mm)}$$

107. The maximum resulting force on a rotor blade is:

$$F_{R_{blade}} = 1.3 \cdot \frac{\rho \cdot \eta_f \cdot Q}{z_\delta \cdot f_{e_4}} \cdot \sqrt{(c_{u_4} - c_{u_{5*}})^2 + (c_{m_4} - c_{m_{5*}})^2} = 1.3 \cdot \frac{1000 \cdot 0.96 \cdot 1}{3.5 \cdot 0.855} \cdot \sqrt{(26.337 - 6.584)^2 + (6.566 - 14.274)^2} = 8842.821 \text{ (N)}$$

108. The distance from the shaft to the blade chord is:

$$h = 0.5 \cdot \sqrt{4 \cdot r_{blade}^2 - L_{blade}^2} = 0.5 \cdot \sqrt{4 \cdot 8.8^2 - 11.7^2} = 6.574 \text{ (cm)}$$

109. The ordinate of the center of gravity is:

$$y_g = \frac{114.593}{\theta_{blade}^o} \cdot r_{blade} \cdot \sin\left(0.5 \cdot \theta_{blade}^o\right) = \frac{114.593}{83.33} \cdot 8.8 \cdot \sin(0.5 \cdot 83.33) = 8.05 \text{ (cm)}$$

110. The arc covered by the blade is:

$$\overline{AB} = \frac{\pi}{180} \cdot r_{blade} \cdot \theta_{blade}^o = \frac{\pi}{180} \cdot 8.8 \cdot 83.33 = 12.80 \text{ (cm)}$$

111. In the first attempt, adopt the useful length of the rotor, initially considering the same order as:

$$c_{m_4} / c_{m_3} = 6.566/5.503 = 1.19$$

therefore

$$B = 1.2 \cdot B_{in} = 1.2 \cdot 463 = 1.2 \cdot 463 = 556\ (\mathrm{mm}) = 0.556\ (\mathrm{m})$$

112. The load evenly distributed along the blade is:

$$q_{blade} = \frac{F_{R_{blade}}}{B} = \frac{8842.821}{0.556} = 15904.354\ (\mathrm{N/m}) = 159.044\ (\mathrm{N/cm})$$

113. The maximum bending moment on the blade is:

$$M_{f_{\max}} = \frac{q_{blade} \cdot B^2}{12} = \frac{159.044 \cdot 55.6^2}{12} = 40971.855\ (\mathrm{N \cdot cm})$$

114. The arc of coverage of the flow at its entrance to the rotor is:

$$\overline{AC_{\delta_r}} = \frac{\delta \cdot \pi \cdot D}{360} = \frac{90 \cdot \pi \cdot 0.419}{360} = 0.329\ (\mathrm{m}) = 329\ (\mathrm{mm})$$

115. The preliminary tangential thickness of the blade at the inlet and outlet of the rotor is:

$$e_{blade_t} = \frac{A\bar{C}_{\delta_r}}{z_\delta} - \frac{A\bar{C}_{\delta_{in}}}{z_\delta} \cdot \frac{B_{in}}{B} \cdot \frac{c_{m_3}}{c_{m_4}} = \frac{329}{3} - \frac{331}{3} \cdot \frac{463}{556} \cdot \frac{6.503}{6.566} = 18.7 \cong 19\ (\mathrm{mm})$$

116. The constant primary rotor blade thickness is:

$$e_{blade} = e_{t_4} \cdot \sin\ \beta_4 = 19 \cdot \sin\ 26.501 = 0.0071\ (\mathrm{m}) = 8.5\ (\mathrm{mm})$$

117. The distances from the center of gravity are:

$$h_1 = r_{blade} + 0.5 \cdot e_{blade} - y_g = 8.8 + 0.5 \cdot 0.85 - 8.05 = 1.18\ (\mathrm{cm})$$

and

$$h_2 = y_g - r_{blade} \cdot cos\left(0.5 \cdot \theta^o_{blade}\right) = 8.05 - 8.8 \cdot cos(0.5 \cdot 83.33) = 1.48$$

Continue with

$$h_3 = h_2 = 1.48 \text{ (cm)}$$

118. The cross-sectional area of the blade is:

$$A_{blade} = 0.0174 \cdot \theta^o_{blade} \cdot r_{blade} \cdot e_{blade} = 0.0174 \cdot 83.33 \cdot 8.8 \cdot 0.85 = 10.846 \left(\text{cm}^2\right)$$

119. The volume occupied by the blade is:

$$V_{blade} = A_{blade} \cdot B = 10.846 \cdot 55.6 = 603.038 \left(\text{cm}^3\right)$$

120. The polar moment of inertia is:

$$\begin{aligned} I_p &= \left(0.0087 \cdot \theta^o_{blade} + 0.5 \cdot \sin \theta^o_{blade} - 229.18 \cdot \frac{\sin^2\left(0.5 \cdot \theta^o_{blade}\right)}{\theta^o_{blade}}\right) \cdot e_{blade} \cdot r^3_{blade} \\ &= \left(0.0087 \cdot 83.33 + 0.5 \cdot \sin 83.33 - 229.18 \cdot \frac{\sin^2(0.5 \cdot 83.33)}{83.33}\right) \cdot 0.85 \cdot 8.8^3 \\ &= 3.577 \left(\text{cm}^4\right) \end{aligned}$$

121. The weight of the blade is:

$$G_{blade} = 9.81 \cdot \gamma_{steel} \cdot V_{blade} = 9.81 \cdot 0.00785 \cdot 603.038 = 46.439 \text{ (N)}$$

122. The bending stress is:

$$\sigma_f = \frac{M_{\max} \cdot h_3}{I_p} = \frac{40971.855 \cdot 1.48}{3.577} = 16952.291 \left(\text{N/cm}^2\right)$$

123. The maximum shear force is:

$$E_{Q_{\max}} = \frac{F_{R_{blade}}}{2} = \frac{8842.821}{2} = 4421.411 \text{ (N)}$$

124. The shear stress is:

$$\tau_{cis} = \frac{F_{R_{blade}}}{4 \cdot A_{blade}} = \frac{8842.821}{4 \cdot 10.846} = 203.827 \ (\mathrm{N/cm^2})$$

125. The maximum combined bending and shearing stress is:

$$\tau_{\max} = \sqrt{\left(\frac{\sigma_f}{2}\right)^2 + \tau_{cis}^2} = \sqrt{\left(\frac{16952.291}{2}\right)^2 + 203.827^2} = 8478.596 \ (\mathrm{N/cm^2})$$

126. The safety factor is:

$$f_s = \frac{0.6 \cdot \sigma_{esc}}{\tau_{\max}} = \frac{0.6 \cdot 24623}{8478.596} = 2.53$$

Since $2 < f_s < 4$, the result is satisfactory. If it were out of range, return to step 113 and interactively change the B/B_{in} relationship, recalculate it until the value obtained from f_s falls within the range, continuing with the last calculated values:

127. The load distributed on the blade from its own weight is:

$$q_p = \frac{9.81 \cdot \gamma_{steel} \cdot A_{blade}}{B} = \frac{9.81 \cdot 0.00785 \cdot 10.846}{55.6} = 0.015 \ (\mathrm{N/cm})$$

128. The total load distributed on the blade is:

$$q_t = q_p + q_{blade} = 0.015 + 159.044 = 159.059 \ (\mathrm{N/cm})$$

129. The arrow in the blades is:

$$f_{blade} = \frac{q_t \cdot B^4}{384 \cdot I_p \cdot E} = \frac{159.059 \cdot 55.6^4}{384 \cdot 3.557 \cdot 20601000} = 0.0540 \ (\mathrm{cm}) = 0.540 \ (\mathrm{mm})$$

130. The ratio of the arrow on the blades per meter of span is:

$$r_{f_{blade}} = 100 \cdot \frac{f_{blade}}{B} = 100 \cdot \frac{0.540}{55.6} = 0.97 < 0.9 \ \ (\mathrm{mm/m})$$

This value is satisfactory, recalculated from step 113 with $B = 1.25 \cdot B_{in} = 579$ (mm), resulting in the following values for the following steps:

- 114 – $q_{blade} = 153.390$ (N/cm)
- 115 – $M_{f_{max}} = 39515.309$ (N · cm)
- 116 – $\overline{AC_{\delta_r}} = 329$ (mm)
- 117 – $e_{blade_t} = 22.3$ (mm)
- 118 – $e_{blade} = 10$ (mm)
- 119 – $h_1 = 1.25$ (cm)
- 120 – $A_{blade} = 12.759$ (cm^2)
- 121 – $V_{blade} = 738.746$ (cm^3)
- 122 – $I_p = 4.208$ (cm^4)
- 123 – $G_{blade} = 56.890$ (N)
- 124 – $\sigma_f = 14410.253$ (N/cm^2)
- 125 – $E_{Q_{max}} = 4421.411$ (N)
- 126 – $\tau_{cis} = 173.266$ (N/cm^2)
- 127 – $\tau_{max} = 7207.210$ (N/cm^2)
- 128 – $f_s = 2.39$
- 129 – $q_p = 0.017$ (N/cm)
- 130 – $q_t = 153.497$ (N/cm)
- 131 – $f_{blade} = 0.518$ (mm)
- 132 – $r_{f_{blade}} = 0.89 < 0.9$.

131. The auxiliary variable is:

$$VA = \frac{1}{2 \cdot \pi \cdot B^2} \cdot \sqrt{\frac{g \cdot E \cdot I_p}{q_p}} = = \frac{1}{2 \cdot \pi \cdot 57.9^2} \cdot \sqrt{\frac{981 \cdot 20601000 \cdot 4.208}{0.017}}$$
$$= 106.183 \text{ (Hz)}$$

132. The natural frequencies are:

$$f_{blade_{n_1}} = 22.18 \cdot VA = 22.18 \cdot 106.183 = 2355.139\ (Hz) \rightarrow$$
$$f_{blade_{n_2}} = 61.62 \cdot VA = 61.62 \cdot 106.183 = 6542.996\ (Hz) \rightarrow$$
$$f_{blade_{n_3}} = 120.78 \cdot VA = 120.78 \cdot 106.183 = 12824.783\ (Hz) \rightarrow$$
$$f_{blade_{n_4}} = 199.66 \cdot VA = 199.66 \cdot 106.183 = 21200.498\ (Hz) \rightarrow$$
$$f_{blade_{n_5}} = 298.25 \cdot VA = 298.25 \cdot 106.183 = 31669.080\ (Hz)$$

133. The excitation frequencies are:

$$f_{blade_{exc_1}} = 1 \cdot n \cdot z_r = 1 \cdot 10 \cdot 12 = 120\ (\text{Hz}) \rightarrow$$

$$f_{blade_{exc_2}} = 2 \cdot n \cdot z_r = 2 \cdot 10 \cdot 12 = 240\ (\text{Hz}) \rightarrow$$

$$f_{blade_{exc_3}} = 3 \cdot n \cdot z_r = 3 \cdot 10 \cdot 12 = 360\ (\text{Hz}) \rightarrow$$

$$f_{blade_{exc_4}} = 4 \cdot n \cdot z_r = 4 \cdot 10 \cdot 12 = 480\ (\text{Hz}) \rightarrow$$

$$f_{blade_{exc_5}} = 5 \cdot n \cdot z_r = 5 \cdot 10 \cdot 12 = 600\ (\text{Hz})$$

Since all $f_{blade_{n_i}} > f_{blade_{exc_i}}$ in this example, the operational reliability of the rotor is safe from a blade vibrational standpoint.

134. The thickness of the discs is:

$$e_d = \frac{F_{R_{blade}}}{2 \cdot L_{blade} \cdot \sigma_{c_{ad}}} = \frac{8842.821}{2 \cdot 11.7 \cdot 19620} = 0.0193\ (\text{cm})$$

135. The minimum thickness of the discs is:

$$e_{d_{min}} = 1.0\ (\text{cm}) \geq 2 \cdot e_{blade} = 2 \cdot 1.0 = 2.0\ (\text{cm})$$

Since the calculated value was smaller, the calculation will continue with

$$e_d = e_{d_{\min}} = 2.0\ (\text{cm}).$$

To proceed with the calculations, the adopted B value must be verified, since $B > B_{in} + 2 \cdot e_d = 46.3 + 2 \cdot 2.0 = 50.3$ (cm) which indicates, in this case, that we must continue with $B = 57.6$ (cm), the calculated value.

136. The maximum arrow on the discs is:

$$f_{d_{\max}} = 0.097 \cdot \frac{D \cdot q_t \cdot B^2 \cdot (1 - \mu^2)}{E \cdot e_d^3 \cdot (17 + 15 \cdot \mu)}\ (\text{cm})$$

$$f_{d_{\max}} = 0.097 \cdot \frac{41.9 \cdot 153.407 \cdot 57.9^2 \cdot \left(1 - 0.3^2\right)}{20601000 \cdot 2^3 \cdot (17 + 15 \cdot 0.3)} = 0.0005368\ (\text{cm})$$

137. The weight of a disk is:

$$G_d = \frac{\pi \cdot D^2}{4} \cdot e_d \cdot 9.81 \cdot \gamma_{steel} = \frac{\pi \cdot 41.9^2}{4} \cdot 2 \cdot 9.81 \cdot 0.00785 = 212.367\ (\text{N})$$

138. The tangential force received from the flow by the rotor is:

$$F_{r_\delta} = 1.3 \cdot \rho \cdot \eta_f \cdot Q \cdot (c_{u_4} - c_{u_5}) = 1.3 \cdot 1000 \cdot 0.96 \cdot 1 \cdot (26.337 - 0.775)$$
$$= 31901.376 \text{ (N)}$$

139. The moment on the shaft from the flow is:

$$M = 0.25 \cdot F_{u_\delta} \cdot D = 0.25 \cdot 31901.376 \cdot 41.9 = 334166.914 \text{ (N} \cdot \text{cm)}$$

140. The weight of the shaft in the length of the rotor is:

$$G_e = \frac{\pi \cdot d_{e_{\min}}^2}{4} \cdot (B + 2 \cdot x_1) \cdot 9.81 \cdot \gamma_{steel} = \frac{\pi \cdot 10.24^2}{4} \cdot (57.9 + 2 \cdot 10.24) \cdot 9.81$$
$$\cdot 0.00785 = 497.089 \text{ (N)}$$

141. The reaction at the supports is:

$$R_m = 0.5 \cdot (F_{r_\delta} + z_r \cdot G_{blade} + 2 \cdot G_d + G_e) = 0.5$$
$$\cdot (31901.376 + 12 \cdot 56.89 + 2 \cdot 212.367 + 497.089) = 16752.940 \text{ (N)}$$

142. The distance between the disc and the bearing is:

$$x_1 = d_{e_{\min}} = 10.24 \text{ (cm)}$$

143. The bending moment is:

$$M_{f_{\max}} = R_m \cdot (x_1 + 0.5 \cdot B) = 16752.94 \cdot (10.24 + 0.5 \cdot 57.9) = 656547.719 \text{ (N} \cdot \text{cm)}$$

144. The diameter considering constant torsion and alternating bending is:

$$d_e = 3.138 \cdot \sqrt[6]{\frac{M^2 + 9.013 \cdot M_{f_{\max}}^2}{\sigma_{es}^2}} = 3.138 \cdot \sqrt[6]{\frac{334166.914^2 + 9.013 \cdot 656547.719^2}{24525^2}} = 13.6\ (\mathrm{cm})$$

Recalculating:

$$G_e = \frac{\pi \cdot d_e^2}{4} \cdot (B + 2 \cdot x_1) \cdot \gamma_{steel} = \frac{\pi \cdot 13.6^2}{4} \cdot (57.9 + 2 \cdot 10.24) \cdot 9.81 \cdot 0.00785 = 886.822\ (\mathrm{N})$$

$$R_m = 0.5 \cdot (F_{n_\delta} + z_r \cdot G_{blade} + 2 \cdot G_d + G_e) = 0.5 \cdot (31901.376 + 12 \cdot 56.89 + 2 \cdot 212.367 + 886.822) = 16947.806$$

$$M_{f_{max}} = R_m \cdot (x_1 + 0.5 \cdot B) = 16752.94 \cdot (10.24 + 0.5 \cdot 57.9) = 656547.719\ (\mathrm{N} \cdot \mathrm{cm})$$

$$d_e = 3.138 \cdot \sqrt[6]{\frac{M^2 + 9.013 \cdot M_{f_{max}}^2}{\sigma_{es}^2}} = 3.138 \cdot \sqrt[6]{\frac{334166.914^2 + 9.013 \cdot 656547.719^2}{24525^2}} = 13.6\ (\mathrm{cm})$$

We will adopt:

$$d_e = 13.6\ (\mathrm{cm})$$

$$f_{e_{\max}} = \frac{3.4 \cdot R_m \cdot x_1}{E \cdot d_e^4} \cdot \left[\frac{0.75 \cdot (B + 2 \cdot x_1)^2 - x_1^2}{B + 2 \cdot x_1}\right] \rightarrow$$

$$= \frac{3.4 \cdot 16752.94 \cdot 10.24}{20601000 \cdot 13.6^4} \cdot \left[\frac{0.75 \cdot (57.9 + 2 \cdot 10.24)^2 - 10.24^2}{57.9 + 2 \cdot 10.24}\right] = 0.0000474\ (\mathrm{cm})$$

145. The maximum arrow ratio is:

$$R_{f_{e_{\max}}} = 1000 \cdot \frac{f_{e_{\max}}}{B} = 1000 \cdot \frac{0.00000474}{0.579} = 0.082\ (\text{mm/m}) < 0.9$$

146. The auxiliary arrow is:

$$f_{e_{aux}} = \frac{R_m - 0.5 \cdot F_{n_\delta}}{R_m} \cdot f_{e_{\max}} = \frac{16752.94 - 0.5 \cdot 31901.376}{16752.94} \cdot 0.0000474$$
$$= 0.0000023\ (\text{cm})$$

147. Critical axis rotation is:

$$n_{e_{crit}} = \frac{1}{2 \cdot \pi} \cdot \sqrt{\frac{g}{f_{e_{aux}}}} = \frac{1}{2 \cdot \pi} \cdot \sqrt{\frac{981}{0.0000023}} = 3286.932\ (\text{rps})$$

which is a highly subcritical axis.

148. The maximum pressure on the blade is:

$$p_{\max} = 1.28 \cdot H_{\max} = 1.28 \cdot 40 = 51.2\ \left(\text{N/cm}^2\right)$$

149. The hydraulic diameter of the transition piece is:

$$D_{tr} = \frac{\pi \cdot D_0^2 + 2 \cdot B_{in4} \cdot b_{in}}{\pi \cdot D_0 + 2 \cdot (B_{in} + b_{in})} = = \frac{\pi \cdot 47.7^2 + 4 \cdot 46.3 \cdot 19.4}{\pi \cdot 47.7 + 2 \cdot (46.3 + 19.4)} = 38.15\ (\text{cm})$$

150. The thickness of the transition piece is:

$$e_{tr} = \frac{0.64 \cdot H_{\max} \cdot D_{tr}}{\sigma_{t_{adm}} \cdot k_f} + e_{su} = \frac{0.64 \cdot 40 \cdot 38.15}{12000 \cdot 0.9} + 0.2 = 0.09 + 0.2 = 0.29\ (\text{cm})$$

Since:

$$e_{tr_{\min}} = \frac{D_{tr} + 51}{400} = \frac{38.15 + 51}{400} = 0.22\ (\text{cm}) \geq 0.7\ (\text{cm})$$

we will use:

$$e_{tr} = 0.7\ (\text{cm})$$

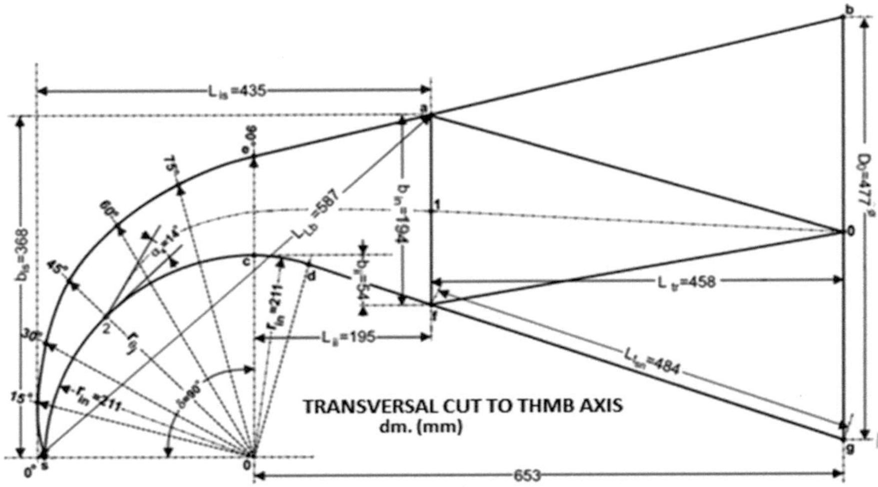

Fig. 6.2 Cross section of the *RHFP* axis in the *BMHIT* between sections 0 and 1

151. Using a graphical software program draw a cross-section of the Real Hydraulic Flow Path (*RHFP*) axis to scale, Fig. 6.2, from the *BMHIT* inlet section 0 to the injector outlet section 1, using the following steps:

1. From point 0, draw a horizontal line starting with Fig. 6.1.
2. From this straight line, mark successive angles from 15° to 15°, starting at 0° and ending at 90°.
3. Choose a scale for the drawing, then mark the values of the stored θ_j rays on the sides of the angles, in the chosen scale, indicating their limits with arrows.
4. Connect the limits of the rays by a curve which will determine the spiral curve between *s* and *a*.
5. From point *e*, draw a tangent to the spiral.
6. From point *c*, draw an arc of radius, $r_{in} = 211$ (mm).
7. Vertically, downwards, from the tangent *a*, detach the $b_{in} = 193$ (mm) value to scale in such a way to lower point *f*, drawing a line segment *fd* that is tangent to the arc drawn from $r_{in} = 211$ (mm).
8. Then draw the extensions of the tangents in *e* and in *f*.
9. With this value, in the scale of $D_0 = 477$ (mm), vertically determine points *b* and *g*.
10. Draw the part of the *ARS* connecting the midpoints of cross section 0 and cross section 1.
11. Then, on the drawing scale, measure the geometric features shown in Fig. 6.1, storing them and transcribing them in Table 6.1.

152. Calculate the thickness of the curved part of the longitudinal injector:

$$e_{in} = 0.8 \cdot L_{Lb} \cdot B_{in} \cdot \sqrt{\frac{H_{\max}}{\sigma_{f_{adm}} \cdot \left(L_{Lb}^2 + B_{in}^2\right)}} + e_{su}^* = 0.8 \cdot 58.7 \cdot 46.3$$

$$\cdot \sqrt{\frac{40}{14470 \cdot \left(58.7^2 + 46.3^2\right)}} + 0.1 = 1.53 + 0.1 = 1.63 \text{ (cm)}$$

153. Using a graphical software program, draw a blade with a spiral-shaped axis to scale, referencing values from the y_a/x_a coordinates obtained and stored in the validation step 108 transcribed in Table 6.2, using the following steps:

1. Transport, preferably to a 1:1 scale, the midline of the chosen fin, for the spiral from Figs. 6.2 and 6.3, in an orthogonal coordinate system, $x_a(\%)$, y (mm).
2. From leading edge 0, divide length $L_a = 240$ (mm) into 20 equal parts, each corresponding to 10%, then mark the percentages for the first 10% as determined in Table 4.7. Only points 2.5% and 5% were marked in the drawing.
3. Draw the respective verticals from each point marked on the horizontal axis until they exceed the midline of the fin.
4. Draw a normal spiral through the points common to the verticals and the midline of the fin.
5. Mark the y_a values taken from Table 6.2 on the respective normal spiral as shown for point a, thus obtaining the points corresponding to the lines that delimited the fin.
6. Connect these points to obtain a true size drawing of the fin, in Fig. 6.4, and to obtain details on the leading and trailing edges, establishing agreement with a circumference of $r_{a_{ba}} = 1.91$ (mm) in the leading edge. For reasons related to wear set to arc of circumference $y_{a_{bf}} = 2.87$ (mm) in the trailing edge.

154. Using a graphical software program, draw a blade in its open and closed position, to scale, observing the steps below:

1. Start Fig. 6.5a, scaled from Fig. 6.2, by highlighting the injector and the midline of the spiral blade in its open position, *AB*.
2. Referring to Fig. 6.3, draw the blade in its maximum opening position, *AB*, to the same scale, with reference to the spiral midline, $L_a = 240$ (mm), $x_a = 113.4$ (mm), and $y_a = 14.43$, $e_{cg_a} = 29$ (mm).

Table 6.2 Rotor dimensions (part I)

D_0	B_{in}	b_{in}	B	L_{tr}	δ	D	D_i	t_4	t_{5*}	$e_{pá}$	$e_{pát}$	$z_{pá}$	z_δ
mm					degrees	mm						number	
477.00	463.00	194.00	579.00	458.00	90	419.00	210.00	110.00	55.00	10.00	22.30	12.00	3.25

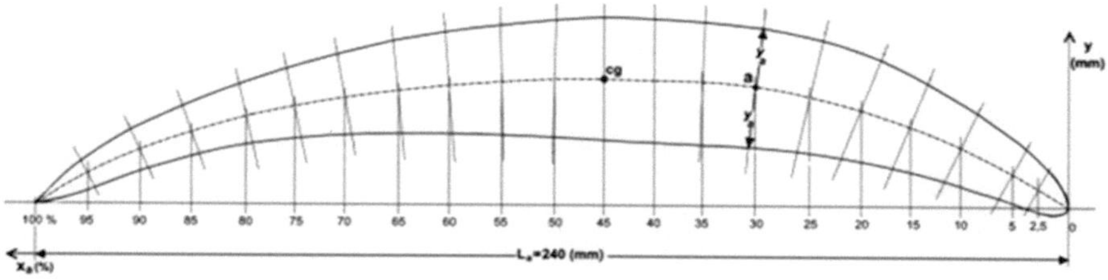

Fig. 6.3 Real size drawing of the fin

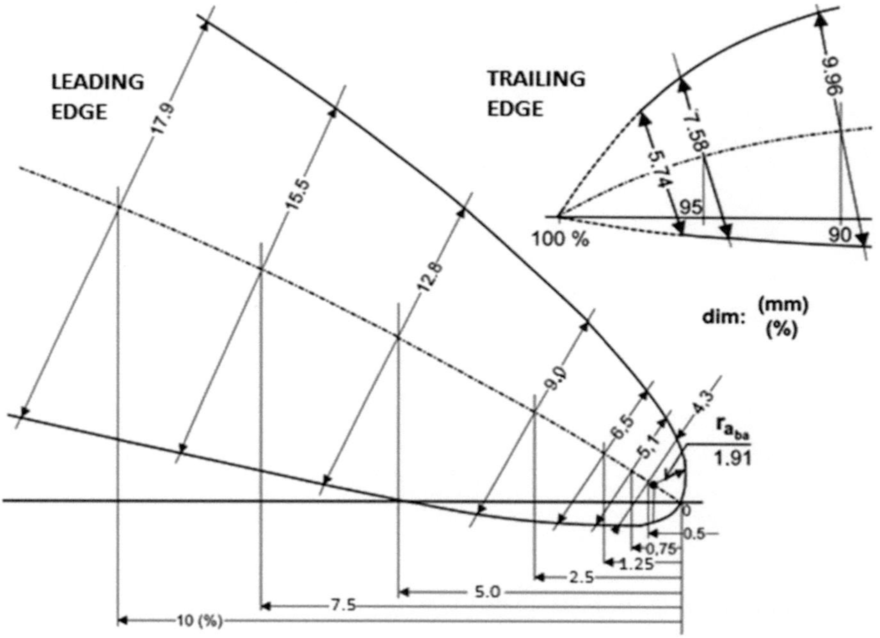

Fig. 6.4 Details of the leading and trailing edges of the fin

3. Rotate the flap from its open position to the closed position, $A*B*$ initially setting the center of rotation *cg*. If the injector does not completely close, or, if at the very least on the drawing, the point on the border of the attack is close to the line that inferiorly delimits the injector, you must iteratively seek out a point close to the *CG* on the midline of the blade so that the injector can totally close. Once this is done, the angle that will allow for *BMHIT* load variations from maximum to minimum can be measured. Here, the angle was measured at 30°.
4. In Fig. 6.5b, to scale, obtain the geometric characteristic values, and store and transcribing them in Table 6.3.

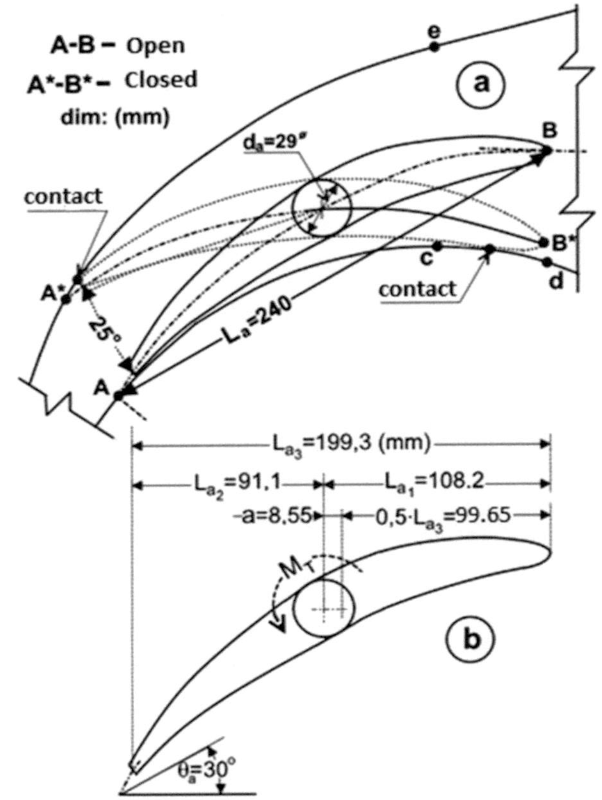

Fig. 6.5: (**a**) Scale drawing of a blade in its fully open and closed position. (**b**) Dimensions for a structural design

155. The thickness of the blade in its *CG* is:

$$e_{al} = 1.4 \cdot \sqrt{\frac{H_{\max} \cdot L_{a3} \cdot B_{in}}{\sigma_{f_{adm}}}} = 1.4 \cdot \sqrt{\frac{40 \cdot 19.93 \cdot 46.3}{14470}} = 2.24 \text{ (cm)}$$

Since the preliminary calculated value is $d_a = 2.9$ (cm) > 2.24 (cm), we can continue with 2.9 (cm), or fine tune the blade to $d_a = 2.5$ (cm), for the final design, for example.

156. The trunnion diameter is:

$$d_m = 0.9 \cdot \sqrt{\frac{H_{max} \cdot L_{a3} \cdot B_{in}}{\tau_{cis_{adm}}}} = 0.9 \cdot \sqrt{\frac{40 \cdot 23.1 \cdot 46.3}{9565}} = 1.9 \text{ (cm)}$$

A result smaller than $e_{al} = 2.0$ (cm) means that $d_m = 2.0$ (cm) will be adopted in the project.

Table 6.3 Rotor dimensions (part II)

d_e	L_d	e_d	x_1	e_{tr}	$r_{pá}$	$L_{pá}$	e_{in}	e_{cga}	d_m	L_a	β_4	β_{5*}	$\alpha_2 = \alpha_4$	$\theta pá$
mm											degrees			
136.00	93.50	20.00	102.40	7.00	88.00	117.00	16.30	29.00	20.00	240.00	26501.00	90.00	14.00	96.7

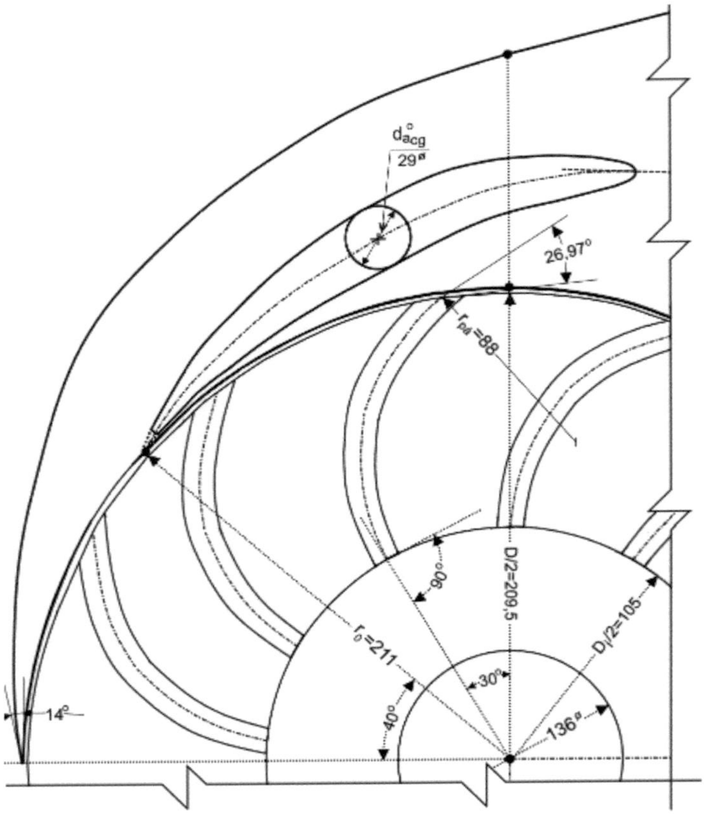

Fig. 6.6 Cross section of the *BMHIT* axis with details on the load control system

157. Using a graphical software program, draw the rotor with its crown of blades and axis from the axis of the *BMHIT* and from the injector and vane, to scale and in cross section, using the following steps:

 1. Start Fig. 6.6, to scale, by drawing two concentric circles with their respective radii, one with $D/2 = 209{,}5$ (*mm*), $D_i/2 = 105$ (*mm*), corresponding to the outer radius of the rotor, and the other with $D_e = 137$ (*mm*), corresponding to the rotor axis of the *BMHIT*.
 2. Calculate the rotor blade radius as:

$$r_{blade} = 0.25 . \frac{D^2 - D_i^2}{D . \cos\,\beta_5 - D_i . \cos\,\beta_5} = 0.25 . \frac{419^2 - 210^2}{419 . \cos\,\beta_5 - 210 . \cos 90} = 88\ (\text{mm})$$

 3. Using point 4 from Fig. 6.6, draw a line tangent to circle diameter *D*, and mark angle $\beta_4 = 26,\,501^\circ$ from it.

4. Starting from 4, normally next to angle β_4, mark $r_{blade} = 88$ (mm) in scale, obtaining point $O*$.
5. With $O*$ center, and ray $O*$, draw the arc of circle 4; $5*$ that represents the axis of the blade in the projection of the figure.
6. Mark the thickness of the blade, $e_{blade} = 12$ (mm), in $5*$, then draw the blade with the center in $O*$.
7. Copy the blade, distributing it every 15 degrees, thus obtaining the rotor cutaway transversely to the *BMHIT* axis to scale.

6.6 Output: Tables

Table the geometric values obtained, which are indispensable for carrying out a preliminary basic project for a *BMHIT*, i.e., the values resulting from the validation (Tables 6.4, 6.5 and 6.6).

Table the characteristics and plot graphs with the data and calculation results for analysis and deliberation among the technical team, using the validation information listed below (Tables 6.7, 6.8 and 6.9):

A diagram of the power distribution can be made with the tabulated results, for a more visual study, as was presented at the end of Chap. 4.

Finally, in Fig. 6.7, a preliminary study was made, in cross section to the dimensioned *BMHIT* axis, already projecting some dimensions, and forecasting how some of the components will be manufactured, for example: the transition piece and the injector fused in steel with external roughing and internal machining, both with flanges, with the injector in its upper cover, with at least one longitudinal flap and two transversal flaps. The rotor will have blades stamped in steel, machined and fitted to the discs in their slots, milled and welded externally.

Table 6.4 Injector dimensions

Lis	*Lii*	*bii*	*bis*	*Ltr*	*Lt sr*	*Lii*	*La*	*La*1	*La*2	*La*3	*a*	*θa*
mm												degrees
435.00	195.00	54.00	368.00	458.00	587.00	484.00	195.00	108.20	91.10	199.30	−8.55	30

Table 6.5 Rays of spiral

Spiral Rays			
q_j	r_{qj}	q_j	r_{qj}
degrees	mm	degrees	mm
0	211.00	75	292.00
15	226.00	90	312.00
30	241.00	97,502	322.00
45	257.00	105	333.00
60	274.00	120	355.00

Table 6.6 Injector vane coordinates

Fin coordinates—NACA Profile 65–010

%	–	0.0	0.5	0.8	1.3	2.5	5.0	7.5	10.0	15.0
x_a	mm	0.00	1.20	1.80	3.00	6.00	12.00	18.00	24.00	36.00
y_a	mm	0.00	2.16	2.55	3.23	4.51	6.38	7.77	8.93	10.75

20.00	25.00	30.00	35.00	40.00	45 CG	50.00	55.00
48.00	60.00	72.00	84.00	96.00	113.40	120.00	132.00
12.11	13.12	13.84	14.30	14.51	14.43	13.98	13.12

60.00	**65.00**	**70.00**	**75.00**	**80.00**	**85.00**	**90.00**	**95.00**	**100.00**
144.00	156.00	168.00	180.00	192.00	204.20	216.00	228.00	240.00
11.98	10.81	9.65	8.48	7.32	6.15	4.98	3.79	0.00

$r_{a\ ba}$ = 1,91 (mm); $y_{a\ \mathrm{bf}}$ = 2,87 (mm)

Table 6.7 Characteristics of the rotor and the injector vane

Determining characteristics				
H	Q	Freq.	Fin	Blade
m	m^3/s	Hz	Perfil NACA	Medium line segment of circumference.
40	1	60	65–10	$\beta_5{}^* = \beta_{blade} = 90o$

Table 6.8 Characteristics fuixed by the designer

Characteristics fixed by the designer									
φ_0	φ_{tr}	φ_2	φ_3	φ_{3*}	ψ_5	k_f	r_D	η_f	$\alpha_2 = \alpha_4$
–									degrees
0.20	0.98	0.97	0.99	0.98	0.97	0.90	0.50	0.96	14

Table 6.9 Characteristics fixed by designer

ε_{tr}	e_{fo}	e_{su}	$e^*{}_{su}$	g	ρ	Materials
mm				m/s^2	kg/m^3	ASTM Steel
0.03	2.00	2.00	1.00	9.81	1000.00	SAE/ABNT

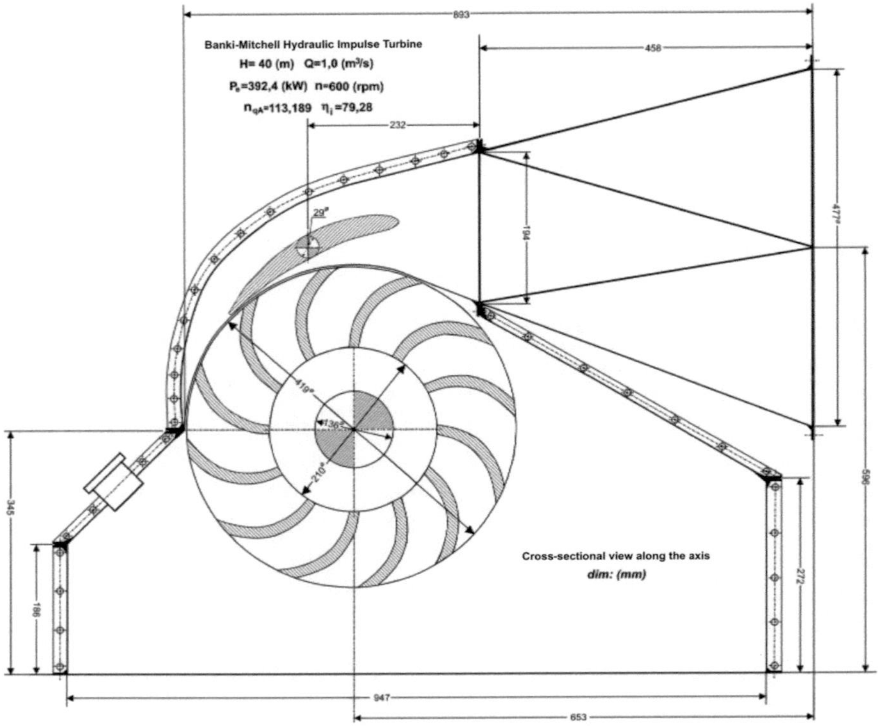

Fig. 6.7 Cross-section of the axis for initial studies on the calculated *BMHIT* for an executive project

Part III
Pelton Hydraulic Impulse Turbines (PHITs)

Chapter 7
Pelton Hydraulic Impulse Turbines (*PHIT*s), A Theoretical Hydrodynamic Analysis

In principle, everything that appeared in the introduction of Chap. 3 is valid for this chapter. The only difference is that the acronym *BMHIT* should be substituted as *PHIT*. However, as we will show, the complexity of a *PHIT* is much greater in almost every regard. It is well established among researchers and engineers who study, systematize, calculate, and design Pelton-type impulse turbines (PHITs) that the technology was invented and patented in 1859 by the American engineer Lester Allan Pelton (08/05/1829–03/15/1908). In 1932, André Tenot stated the following:

> *L'étude de turbine Pelton, d'apparence simple et plus simple que celle dês turbines Francis et roues-hélices, est, au contaire, três complexe, par suíte de la diversité dês triangles de vitesses suivant lês points du jet e suivant la position du Jet dans son arc d'activité. L'étude théorique exacte est impossible, et l'expériense seule a permis de déterminer les meilleures formes et proportion à donner aux augets*[1]

> The Pelton turbine simulations are, without doubt, the most complex and difficult of all hydraulic turbomachinery simulations. Pelton turbines involve a number of special flow characteristics that are extremely difficult to simulate. The interaction between the jet and the shell is completely transient, and depends on the motion geometry of the shells. An even greater challenge is the multiphase air-water system that regulates the formation of flows and free jets along the shells. In the past, developing flow simulations to allow for realistic analyses of these phenomena seemed an unsolvable task. Voith's numerical development group was finally able to overcome these problems by simulating a typical flow along the shell of a Pelton turbine. The simulation results showed remarkable convergence to the experimental data. Designers now have new knowledge on the complex shell flow phenomenon, which serves as a solid base for the development of new shell profiles to improve Pelton turbine designs.[2]

Figure 7.1 shows a copy of Fig. 4 from Brazilian Standard NB 11374, which is called the Hydraulic Turbine Field test. It shows the horizontal axis of a 1- and 2-jet *PHIT*, the conventions, symbols, and expressions for determining the respective drop, or

[1] André Tenot—Turbines Hydrauliques—Livre II—pg. 446—Paris—1932.
[2] Voith—Turbinas Pelton—Internet—2017.

© The Author(s), under exclusive license to Springer Nature Switzerland AG 2025

Z. de Souza, G. L. Tiago Filho, *Pelton and Crossflow Impulse Hydraulic Turbines*,
https://doi.org/10.1007/978-3-031-87086-6_7

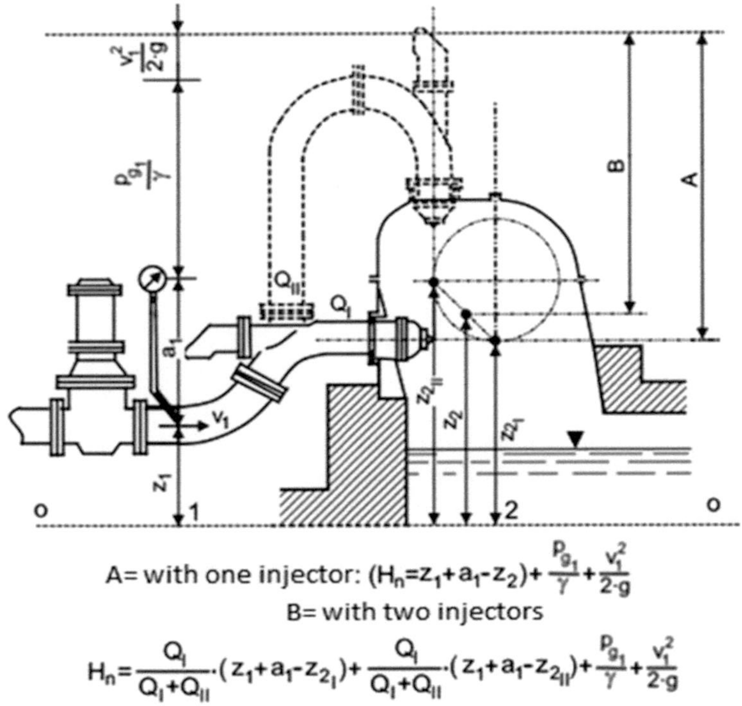

Fig. 7.1. A copy of Fig. 7.4 from *NBR* 11374

nominal specific energy units, H_n (meters of the water column) = (*mca*), in the Technical System Unit (*TSU*), which will be equal to the corresponding symbols: $H_n = H = H_0$ (*mca*) for the entire study that will be carried out in sequence. Regarding the Average Absolute Flow Velocity, v (m/s) will be replaced by c (m/s). For the vertical *PHIT* axis, this head, or the nominal specific energy, is always given with respect to the horizontal axis that contains the vertical projection of the circumference that contains the dividing edge of the shells, and preferably in the *TSU*, $Y = g \cdot H$ (J/kg), it is called the specific work.

7.1 Modeling the Hydraulic Flow Path

The study in this chapter considers average water flow conditions from entry into the *PHIT* in the A_o cross section, up to exit at the A_5 cross-section shells.

A photograph of a Pelton shell was taken and shown in Fig. 7.2 to give a first sense of the flow in the *PHIT* rotor shells. Then, we schematized the theoretical three-dimensional water flow in a drawing, from the cylindrical part after the outlet of the injector, flow Q, to the entrance at the dividing edge of the shell, where the

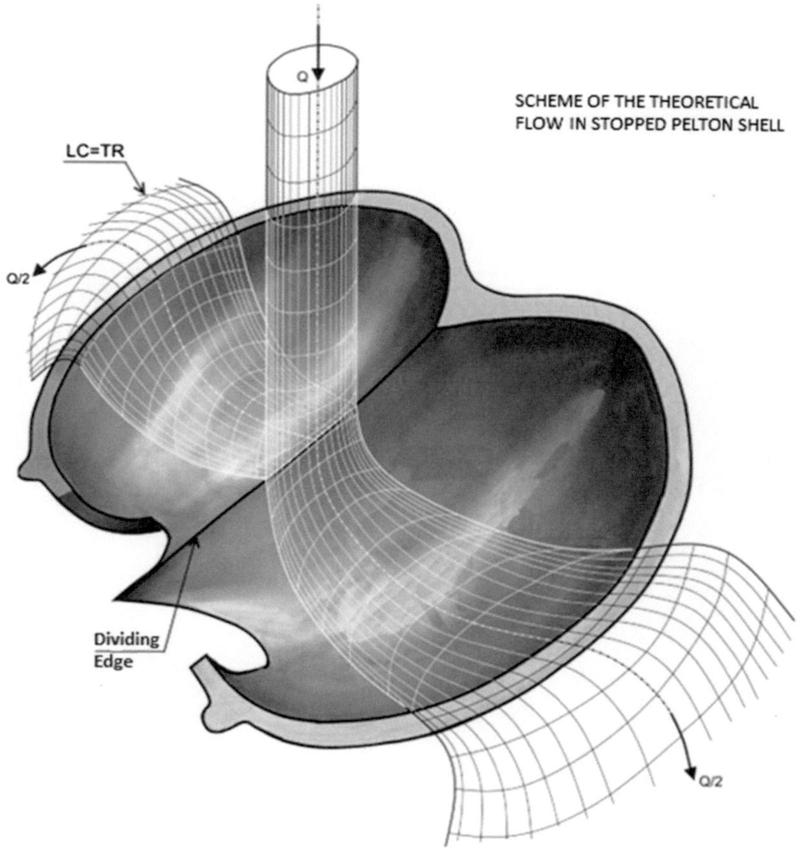

Fig. 7.2 Water flow diagram of a nonmoving Pelton shell

flow of the circular cross section is divided into two flows $Q/2$, which is initially semicircular. These flows go through the respective internal surfaces of the shell, modifying the normal sections, due to the effect of gravity on the external surface with contact with air, and internally with the surface of the shell, which of course, must be as close as possible to being hydraulically smooth.

In this configuration, since the shell is stationary, streamlines *LC* coincide with the respective relative trajectories (*TR*). Since the shells are parts of the rotor, if the rotor is rotating with n (rpm), the diameter D shells move with tangential speed u(m/s), and the *TR* differs from the LC, as will be shown in Chap. 9.

We will use a very simplified configuration to model and develop the studies of the first approximation of the Average Theoretical Streamline (*ATS*) considered in the first paragraph of this item, as shown in Fig. 7.3a. Figure 7.3 shows a cross section of the horizontal axis of a $1 - jet\ PHIT$, and the trajectory of the water flow,

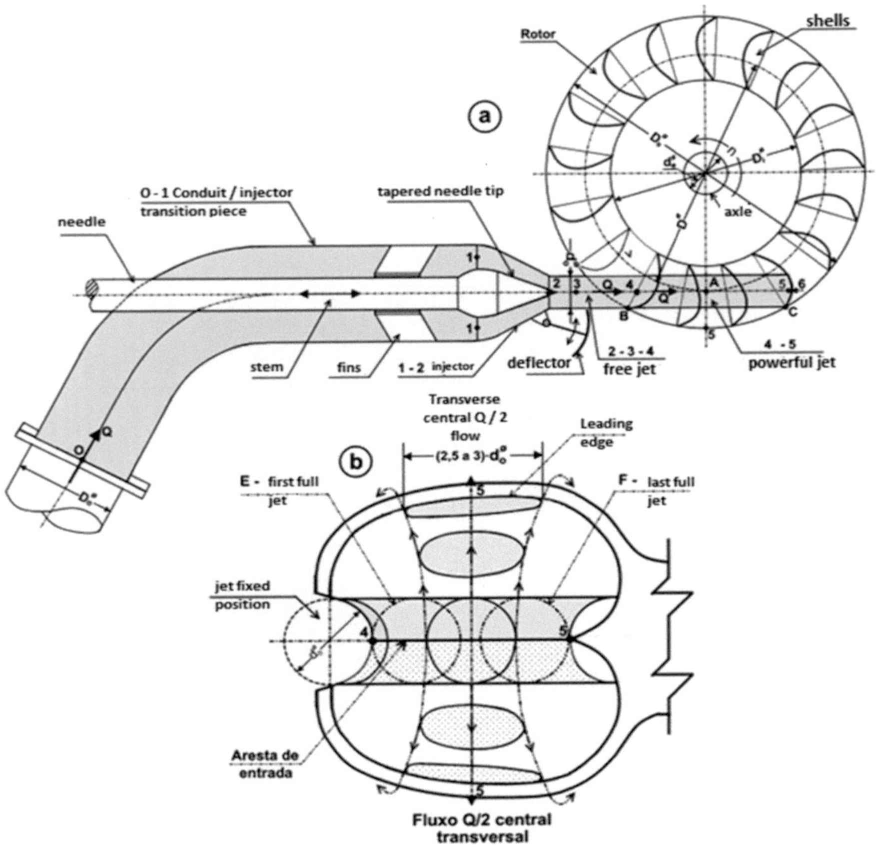

Fig. 7.3 A model of the *HFP* in a 1-jet *PHIT* and horizontal axis

i.e., the Hydraulic Flow Path (*HFP*), from inlet section 0, to the outlet shells at 5, characterizing and naming the main components.

As one can see, the flow between section 0 and section 2, which we will call the Inlet System (*IS*), occurs with decreasing absolute static pressure, with maximum at 0, and local atmospheric pressure at 2.

From this section, the flow travels through the sections 2; 3; 4 as a free jet in a two-phase flow state (water/air in stationary movement), with energy losses arising from absorbing the surrounding air, causing reductions in the specific mass of water, from leakage at the water/air interfaces. This also occurs in the paths between the shells.

Between points 4 and 5, the jet travels through the bipartite shells (always in 4 rotor shells in the schematic), with cross sections, and some parts come into contact with the material surfaces, while the stream travels through other parts interacting with the surrounding air.

In this last section, the flow transfers its available kinetic energy to the rotor shells to be transformed into mechanical energy, which is transferred to the rotor shaft, which rotates at n (rmp)

From an energy standpoint, the internal energy per unit of time for the flow, i.e., the available power at the entrance of section 0 in the *PHIT*, can be expressed as:

$$P_{i_0} = \rho \cdot Q \cdot g \cdot \left(\frac{p_0}{\gamma} + \frac{c_0^2}{2 \cdot g} - y_0 \right) = \rho \cdot Q \cdot g \cdot H_0 = \rho \cdot Q \cdot Y_{i_0} \ (\mathrm{kW}) \rightarrow$$

$$\rho \ (\mathrm{kg/m^3}) \rightarrow Q \ (m^3/s) \rightarrow g \ (\mathrm{m/s^2}) \rightarrow$$

$$\frac{p}{\gamma} \ (\mathrm{mca}) \rightarrow$$

$$\frac{c_0^2}{2 \cdot g} \ (\mathrm{mca}) \rightarrow$$

$$y_0 \ (\mathrm{mca}) \rightarrow$$

$$H_0 \ (\mathrm{mca}) \rightarrow$$

$$Y_{i_0} \ (\mathrm{J/kg})$$

The power in the Inlet System, which comprises the flow section between cross section 0 and section 2, and the injector outlet, must be transformed with minimum losses to meet the energy needs of the consumer at the Electric Generator (*EG*) outlet.

This transformation mostly occurs in the Injector via the translational movement of a needle that varies flow rate Q (Fig. 7.3a).

For the flow to reach inlet 1 of the Injector with minimum vorticity, the blades or flow drivers are fixed inside the transition piece.

As shown in Fig. 7.3a, the free part jet in 2; 3; 4 with respect to the Rotor (*RO*) should be positioned so that the centerline is normally tangent to diameter D of the *RO* at A, and its bottom edge line to the lower end of the dividing edge of the shell at the diameter D_e of the RO at B. This will ensure full attack on the shell upon entry. To ensure that all the specific kinetic energy available in the water stream from point B, given by $Y_4 = 0.5 \cdot c_4^2$ (J/kg), is transmitted to the *RO* minus the losses, including the output, $Y_5 = 0.5 \cdot c_5^2$ (J/kg), the number of shells z_c must be set so that the bottom edge line finds the shell at C at the output.

Here, 4 shells, at $z_c = 17$, of the *RO*, receive the kinetic hydraulic energy available from the jet to be transformed into mechanical energy on the axis.

Figure 7.3b, gives a horizontal projection of a shell and seeks to schematize the displacement of the fixed jet in the shell in function of the rotation of the *RO*, which occurs counterclockwise (see Fig. 7.3a). The jet reaches the shell at inlet dividing edge 4, which divides it in half, until reaching the full value at position *E*, and is maintained until position *F*.

Referencing the central full jet and the stationary shell in Fig. 7.3b, we can schematize the flows of flow rates *Q*/2 (m^3/s) in the respective axial directions of the shell, with their cross sections altered as a function of the geometry of the inner three-dimensional surface structure of the shell, until exiting through the corresponding edge.

Figure 7.3a also shows the deflector (DF), which operates rapidly in response to load variations required by the consumer. In the PHIT, this is achieved by sequentially adjusting the flow rate Q, initially through the action of the deflector and subsequently through the action of the needle.

7.2 Energy Transformations

7.2.1 Considerations

Similar to *HIT*s in general, and in the *PHIT* in particular, the *IS* of the flow is, in theory, independent of the *RO*, without direct dependence between the flow and rotation n, since both the H_0 and the needle position, α_{ag} are fixed. Thus, function $f = (Q \times n)$ are the horizontal lines, as shown in Fig. 7.4a.

The two-phase flow in 2; 3; 4, with minimum diameter d_0, is influenced in the interior by the boundary layers at the conical tip of the needle, and is smaller at the entrance edge of the shell, which alters the two-phase cylindrical periphery, thus modifying the speed paraboloids, schematized here without considering the effects of turbulence, as shown in the flat section of Fig. 7.4b.

In this same figure, we have the flow in the shell schematized, with the respective *AS*s, while in Fig. 7.4c, we have drawings of the corresponding average velocity triangles for the flow entry and exit in the shells.

Using these figures as models, we can apply Euler's Equation, and can analyze the behavior of the delivery of hydraulic energy from the available flow to the *RO* of the *PHIT* so it can be transformed into mechanical energy on the axis.

Studies will be carried out for the following cases:

- **Flows without internal and surface energy losses**: to analyze only losses inherent to the *RO* shell geometry.

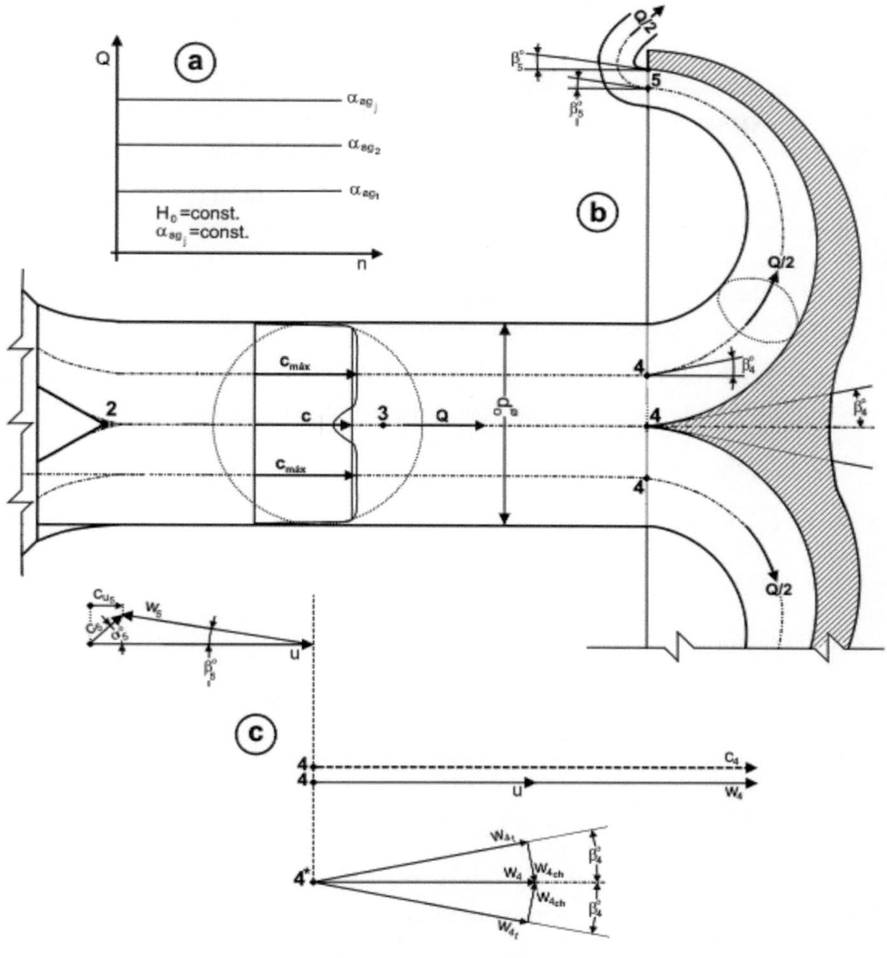

Fig. 7.4 Model for equating and studying the behavior of a *PHIT*

- **Flows with internal and surface energy losses**: to add together all internal losses.
- **Flows with internal, surface, and external energy losses**: to analyze all losses, from entry to the delivery of energy in the axis that will be transformed into mechanical work.

7.2.2 Flows Without Internal and Surface Energy Losses

One should observe in Fig. 7.4c that, due to angle β_4 at the entrance of the flow to the shells, the average shocks, represented by $w_{4_{ch}}$, are canceled out, since they are in opposite directions. Thus, theoretical equations can be developed to determine:

- $Y_{i_{tr}}$ (J/kg)—the theoretical average internal specific work of the flow at the exit of the shells;
- $P_{i_{tr}}$ (kW)—the corresponding power;
- η_{i_t}—the theoretical internal efficiency of the *PHIT*; and
- $r_v = u/c_4$—the optimal speed ratio, between:
  - $u = \frac{\pi \cdot D \cdot n}{60}$ (m/s)—the tangential speed of the rotor; and
  - $c_4 = \frac{4 \cdot Q}{\pi \cdot d_o^2}$—the average of the flow at the entrance of their shells.

Assuming there are no losses:

$c_0 = c_4 = \sqrt{2 \cdot g \cdot H_0} = \sqrt{2 \cdot Y_{i_0}}$ (m/s)
and …
$w_4 = w_5$ (m/s)
via …
$\rho = 1000$ (kg/m^3)
so …
$Y_{i_{tr}} = u \cdot (c_{u_4} - c_{u_5}) \rightarrow c_{u_4} = c_4 - w_4 \rightarrow c_{u_5} = u - w_5 \cdot \cos\beta_5$

Thus:

$$Y_{i_{tr}} = u \cdot w_4 \cdot (1 + \cos\beta_5) \text{ (J/kg)} \rightarrow P_{i_{tr}} = Q \cdot u \cdot (c_4 - u) \cdot (1 + \cos\beta_5) \text{ (kW)} \quad (7.1)$$

The specific work and the average internal power of the flow available at the entrance of the *PHIT*, respectively, are:

$$Y_{i_0} = g \cdot H_0 \text{ (J/kg)}$$

and

$$P_{i_0} = Q \cdot g \cdot H_0 = 10^{-3} \cdot Q \cdot Y_{i_0} = Q \cdot \frac{c_4^2}{2} \text{ (kW)} \quad (7.2)$$

The theoretical internal yield of the *PHIT* in function of the geometry of the *RO* shells, is:

$$\eta_{i_t} = \frac{P_{i_{t_r}}}{P_{i_0}} = \frac{2 \cdot u \cdot (c_4 - u) \cdot (1 + \cos\beta_5)}{c_4^2} = 2 \cdot \frac{u}{c_4} \cdot \left(1 - \frac{u}{c_4}\right) \cdot (1 + \cos\beta_5) \quad (7.3)$$

$\eta_{i_{t_{max}}}$ must have $\frac{\partial \eta_{i_t}}{\partial (u/c_4)} = 0$ or $\frac{\partial \eta_{i_t}}{\partial (u/c_4)} = 0$, to determine the maximum theoretical internal yield, which results in the following r_v speed relationship:

$$\frac{u}{c_4} = r_v = 0.5 \quad (7.4)$$

Substituting the value in (7.4) into (7.3), we can obtain the maximum theoretical internal yield of *PHIT* in function of the geometry of the *RO* shells:

$$\eta_{i_{t_{max}}} = 0.5 \cdot (1 + \cos\beta_5) \quad (7.5)$$

The angle at the exit of the shells should be as small as possible. Currently, this angle has been fixed in the range of 4 $4 \leq \beta_5^o \leq 10$, which results in the following range for the theoretical limit of *RO* internal performance:

$$\beta_5 = 10^o \rightarrow 0.9924 \leq \eta_{i_{r_t}} \leq 0.9988 \rightarrow \beta_5 = 4^o \quad (7.6)$$

7.2.3 *Flows with Internal and Surface Energy Losses*

Considering energy losses in the flow from entry section 0 in the *PHIT* to entry section 4 in the shells of the *RO*, represented by a coefficient, $\varphi_{0;\,4}$, that reduces the absolute average velocity at the entrance of the shell as:

$$c_0 = \sqrt{2 \cdot g \cdot H_0} = \sqrt{2 \cdot Y_{i_0}}$$

to …

$$c_4 = \phi_{0;4} \cdot c_0 = \phi_{0;4} \cdot \sqrt{2 \cdot g \cdot H_0} = \phi_{0;4} \cdot \sqrt{2 \cdot Y_{i_0}}$$

Thus:

$$H_0 = \frac{c_4^2}{2 \cdot g \cdot \phi_{0;4}^2} \ (mca)$$

Or:

$$Y_{i_0} = \frac{c_4^2}{2 \cdot \phi_{0;4}^2} \text{ (J/kg)} \tag{7.7}$$

In the flow through the *RO* shells, the losses will be represented by the coefficient, ψ_5, which reduces the relative average velocity at the entrance of the shell from w_4 to $w_5 = \psi_5 \cdot w_4$.

Via these coefficients, and by applying Euler's Equation, we get:

$$Y_{i_{r_5}} = u \cdot (c_{u_4} - c_{u_5}) = u \cdot [c_4 - (u - w_5 \cdot \cos\beta_5)] = u \cdot [c_4 - (u - \psi_5 \cdot w_4 \cdot \cos\beta_5)]$$

Or:

$$Y_{i_{r_5}} = u \cdot \{c_4 - [u - (c_4 - u) \cdot \psi_5 \cdot \cos\beta_5]\} \ (J/kg)$$

And:

$$P_{i_{r_5}} = Q_r \cdot Y_{i_{r_5}} = Q_r \cdot u \cdot \{c_4 - [u - (c_4 - u) \cdot \psi_5 \cdot \cos\beta_5]\} \ (kW) \tag{7.8}$$

In this expression $Q_r = \eta_f \cdot Q$, η_f is the escape yield that will be dealt with in the next chapter.

The internal power available in the flow at the entrance of the *PHIT*, at section 0, can be written as follows:

$$P_{i_0} = Q \cdot g \cdot H = Q \cdot \frac{c_0^2}{2} = Q \cdot \frac{c_4^2}{2 \cdot \phi_{0;4}^2} \ (kW) \tag{7.9}$$

Establishing the quotient between expressions (7.8) and (7.9) we thus obtain the internal yield of the *PHIT*.

$$\eta_i = \frac{P_{i_{r_5}}}{P_{i_0}} = \frac{2 \cdot \varphi_{0;4}^2 \cdot \eta_f \cdot u \cdot \{c_4 - [u - (c_4 - u) \cdot \psi_5 \cdot \cos\beta_5]\}}{c_4^2}$$

Or:

$$\eta_i = 2 \cdot \phi_{0;4}^2 \cdot \eta_f \cdot \frac{u}{c_4} \cdot \left\{1 - \left[\frac{u}{c_4} - \left(1 - \frac{u}{c_4}\right) \cdot \psi_5 \cdot \cos\beta_5\right]\right\} \tag{7.10}$$

To initially determine the maximum internal efficiency, $\eta_{i_{r_{max}}}$, Eq. (7.10) is partially derived in relation to (u/c_4), $\frac{\partial \eta_i}{\partial (u/c_4)}$, canceling out the resulting expression, which will allow us to determine the speed ratio r_v as follows:

$$\frac{\partial \eta_i}{\partial (u/c_4)} = -4 \cdot \phi_{0;4}^2 \cdot \eta_f \cdot \frac{u}{c_4} \cdot \psi_5 \cdot \cos\beta_5$$
$$+ \left\{1 - \left[\frac{u}{c_4} - \left(1 - \frac{u}{c_4}\right) \cdot \psi_5 \cdot \cos\beta_5\right]\right\} \cdot 2 \cdot \phi_2^2 = 0 \rightarrow$$

$$3 \cdot \frac{u}{c_4} \cdot \psi_5 \cdot \cos\beta_5 + \frac{u}{c_4} = 1 + \psi_5 \cdot \cos\beta_5$$

… which results for the speed ratio, as:

$$r_v = \frac{u}{c_4} = \frac{1 + \psi_5 \cdot \cos\beta_5}{3 + \psi_5 \cdot \cos\beta_5} \tag{7.11}$$

Since the obtained velocity r_v ratio neither enters coefficient $\varphi_{0;\,4}$ nor η_f, it can also be used to determine the maximum values of both of $\eta_{i_{max}}$ and $P_{i_{r_{5max}}}$, respectively, via expressions (7.10) and (7.9), by substituting r_v from expression (7.11) in them, respectively resulting in:

$$\eta_{i_{max}} = \frac{P_{i_{r_5}}}{P_{i_0}} = 2 \cdot \phi_{0;4}^2 \cdot \eta_f \cdot r_v \cdot \{1 - [r_v - (1 - r_v) \cdot \psi_5 \cdot \cos\beta_5]\} \tag{7.12}$$

$$P_{i_{r_{5max}}} = 10^{-3} \cdot Q \cdot \eta_f \cdot u \cdot c_u \cdot \{1 - [r_v - (1 - r_v) \cdot \psi_5 \cdot \cos\beta_5]\}\ (kW) \tag{7.13}$$

Taking the r_v from expression (7.4), $r_v = r_{v(7.4)} = 0.5$, with expression (7.11), $r_v = r_{v(7.11)}$, gives $r_{v(7.11)} \leq r_{v(7.4)}$.

For example, with $\beta_5 = 6^o$ and $\psi_5 = 0.98$, the result is:

$$r_{v(7.11)} = \frac{u}{c_4} = \frac{1 + \psi_5 \cdot \cos\beta_5}{3 + \psi_5 \cdot \cos\beta_5} = \frac{1 + 0.98 \cdot \cos 6}{3 + 0.98 \cdot \cos 6} = 0.4968 < 0.5$$

7.2.4 *Flows with Internal, Surface, and External Energy Ventilation Losses*

This analysis will be done by including the external ventilation losses, coming from air friction with the *RO* of the *PHIT*, which will be studied further on in this book.

To determine these losses, Formulas obtained from tests of *PHIT* rotors are of the type:

$$P_{p_v} = 10^{-6} \cdot k_v \cdot \rho_{ar} \cdot n^3 \cdot D_e^5 \text{ (kW)}$$

… with $k_v(1)$ as a constant that is a function of the type of *RO* and the position of its axis. For examples see *NBR* 11,374 and CEI/*IEC* 41.

Substituting …

$$n = \frac{60 \cdot u}{\pi \cdot D_e} \text{ (rpm)} \rightarrow u \text{ (m/s)} \rightarrow$$

$$D_e \ (m)$$

$$c_4 = \frac{4 \cdot Q_1 \cdot \eta_f}{\pi \cdot d_o^2} \text{ (m/s)}$$

and …

$$u = r_v \cdot c_4 = r_v \cdot \frac{4 \cdot Q_1 \cdot \eta_f}{\pi \cdot d_o^2} \rightarrow$$

$$c_4 (\text{m/s}) \rightarrow$$

$$Q_1 \ (\text{m}^3/\text{s}) \rightarrow$$

$$d_o \ (\text{m}) \rightarrow$$

$$\eta_f \ (1)$$

into this formula, we have:

$$P_{p_v} = 10^{-6} \cdot k_v \cdot \rho_{ar} \cdot \left(\frac{60}{\pi}\right)^3 \cdot \frac{4}{\pi} \cdot \left(\frac{D_e}{d_o}\right)^2 \cdot Q_1 \cdot \eta_f \cdot r_v \cdot u^2$$

… and given:

$$k_v^* = 10^{-6} \cdot k_v \cdot \rho_{ar} \cdot \left(\frac{60}{\pi}\right)^3 \cdot \frac{4}{\pi} \cdot \left(\frac{D_e}{d_o}\right)^2 \cdot Q_1 \cdot \eta_f = 1.171 \cdot 10^{-3} \cdot k_v \cdot Q_1 \cdot \eta_f \cdot \left(\frac{D_e}{d_o}\right)^2$$

We have:

$$P_{p_v} = k_v^* \cdot r_v \cdot u^2 \tag{7.14}$$

With this, the efficiency in the *PHIT* axis, considering ventilation losses, without considering mechanical losses, can be obtained as follows:

$$\eta_e = \frac{P_{i_{r_5}} - P_{p_v}}{P_{i_0}} = \frac{P_{i_{r_5}}}{P_{i_0}} - \frac{P_{p_v}}{P_{i_0}}$$

Making substitutions and calculations, we then have:

$$\eta_e = 2 \cdot \phi^2_{0;4} \cdot \eta_f \cdot (r_v - r_v^2 - r_v \cdot \psi_5 \cdot \cos\beta_5 + r_v^2 \cdot \psi_5 \cdot \cos\beta_5 - k_v^* \cdot r_v^3 \qquad (7.15)$$

To determine $\eta_{e_{\max}}$, the partial derivative $\frac{\partial\eta_e}{\partial r_v}$, after the algebra, must give a null result:

$$r_v = \frac{-(1 - \psi_5 \cdot \cos\beta_5) \pm \sqrt{(1 - \psi_5 \cdot \cos\beta_5)^2 + 3 \cdot k_v^* \cdot (1 - \psi_5 \cdot \cos\beta_5)}}{3 \cdot k_v^*} \qquad (7.16)$$

This expression can be applied using the following data, which were taken from a basic design of a 6-jet vertical axis *PHIT*:

- $Q_1 = 5.56$ (m^3/s)
- $\beta_5 = 6^o$
- $\psi_5 = 0.98$
- $\eta_f = 0.995$
- $d_o = 0.264$ (m)
- $D_e = 3.250$ (m)
- $B_e = 0.860$ (m)

With k_v, determined, and with uncertainty that can exceed 40 (%) for whole-cast *PHIT* rotors, from expression $k_v = 0.01 + 12.4 \cdot (B_e/D_e)^4$, we have:

$$k_v = 0.01 + 12.4 \cdot (B_e/D_e)^4 = 0.01 + 12.4 \cdot (0.86/3.83)^4 = 0.0415 \pm 10\% k_v^*$$

$$= 1.171 \cdot 10^{-3} \cdot k_v \cdot Q_1 \cdot \eta_f \cdot \left(\frac{D_e}{d_o}\right)^2$$

$$= 1.171 \cdot 10^{-3} \cdot 00415 \cdot 5.56 \cdot 0.995 \cdot \left(\frac{3.83}{0.86}\right)^2 = 5.33216 \cdot 10^{-3}\ r_v$$

$$= \frac{-(1 - 0.98 \cdot \cos 6) \pm \sqrt{(1 - 0.98 \cdot \cos 6)^2 + 3 \cdot 5.33216 \cdot 10^{-3} \cdot (1 - 0.98 \cdot \cos 6)}}{3 \cdot 5.33216 \cdot 10^{-3}}$$

$$= 0.439$$

This value varies greatly, mainly considering the uncertainties in determining k_v and the power of the *PHIT*. This will be studied further in the next chapters.

7.2.5 Speed and Rotation

The internal hydraulic powers, with and without energy losses, are zero for two tangential *RO* speeds, i.e., $u = 0$ and $u = c_4$(m/s). These speeds, which arise hydraulically, are called, the internal *RO* hydraulic shot speed, internal *RO* hydraulic shot rotation $n_d = \frac{60 \cdot u_d}{\pi \cdot D}$ (rpm), respectively, which would be the highest tangential speeds that the *RO* could reach without considering mechanical losses. The relationship between these and the corresponding nominal u and n results in triggering, or the packing factor:

$$f_{d_n} = f_{e_n} = c_4/u = 1/r_v \tag{7.17}$$

The theoretical maximum value, according to expression (7.4) is:

$$f_{d_{\max}} = 1/0.5 = 2 \tag{7.18}$$

7.3 Analysis, Considerations, and Comments

- It is important to note that the issues and analyses discussed in this chapter are based on the assumption that water flows are constant.
- The equations obtained this far have considered idealized flow models, considering the average kinematic values.
- The specific kinetic energy of the flow $0.5 \cdot c_5^2 (\mathrm{J/kg})$, which is dissipated at the exit of the shells, was not considered in the Equations.
- Euler's Equation, resulting from the Principles of the Conservation of Mass, the Quantity of Movement, and the Momentum of the Quantity of Movement, considers only occurrences at the system and subsystem limits that comprise the *PHIT*, and therefore, does not consider occurrences in their respective interiors. Thus, we have not accounted for the geometry, or what happens to the flow inside, whether the flow is solid or gaseous, neither have we considered the gas index contained in the flowing water.
- These considerations mean that we do not know how, either qualitatively or quantitatively, the energy losses inside the subsystems behave, from the respective inlet and outlet sections.
- This application showed that r_v, considering both the geometry and the energy losses in the split rotor shells, must be less than 0.5 of the lossless r_v to obtain $\eta_{i_{\max}}$.
- As will be studied further, r_v will be preliminarily fixed as a function of the specific kinetic energy available at the *PHIT* inlet, i.e., $Y_0 = 0.5 \cdot c_0^2 \rightarrow c_o = c_{4t} = \sqrt{2 \cdot g \cdot H_{0.}}$

Chapter 8
Geometric Modeling and Theoretical Analyses of the Energetic Losses in a Pelton Hydraulic Impulse Turbine (*PHIT*)

This chapter seeks to:

(a) geometrically model the types of *PHIT*s that will be studied;
(b) characterize each of its components by name;
(c) delimit the space occupied by them; and,
(d) establish criteria for the flow of water as a real fluid between the inlet and outlet sections.

This is done to serve as a foundation and complemented with hypotheses and considerations that will be developed through a routine that allows us to preliminarily quantify

- $Y_{i_{p_j}}$ (J/kg)—the specific internal work lost;
- $P_{i_{p_j}}$ (kW)—the corresponding internal powers;
- Y_{i_j} (J/kg)—the specific work; and
- P_{i_j} (kW)—the respective internal powers …

…. delivered to each of the components that delimit the Real Hydrodynamic Flow Path (RHFP) of the model, and with this, to determine:

- η_{i_j}—its internal efficiency; and
- ϕ_j, ψ_r—and the velocity loss coefficients.

The studies will be complemented with analyses on the transformation of internal hydraulic energy, available from the flow, into mechanical energy on the axis of the *PHIT*. To do this, the external losses P_{e_p} (kW) that occur when operating the *GG* will be studied and quantified, which include the ventilation losses P_{p_v} (kW), coming from the rotation of the *RO* in air, and the mechanical losses P_{p_m} (kW), from the rotating shaft and parts in air and in the bearings. The latter transmit the static and dynamic force of the *GG* that will be studied and designed together with

© The Author(s), under exclusive license to Springer Nature Switzerland AG 2025

Z. de Souza, G. L. Tiago Filho, *Pelton and Crossflow Impulse Hydraulic Turbines*,
https://doi.org/10.1007/978-3-031-87086-6_8

the machine housing. These losses, as was already mentioned and shown in Chap. 7, influence the velocity coefficient, $r_v = u/c_{4t}$. The result is that we will be able to establish a step-by-step algorithm that allows us to conduct preliminary calculations of these losses and to quantify the powers and yields of the *PHIT*s established in the geometric modeling.

8.1 Geometric Modeling

8.1.1 Considerations

The geometries of the components that conduct the flow in the *PHIT* Inlet System (*IS*) have mostly circular cross sections that vary, basically as a function of the positioning and characteristics of the high pressure system (forced conduit), particularly in the lowest part, at the exit of the valve at the entrance of the engine room.

Given the limits of the intended studies, this large variation will be done via three basic models, characterized in Figs. 8.1, 8.2, and 8.3, which have the following components in the *IS* characterized by numbers in the midline of the flow, i.e., the Average Streamline (*AS*), as follows:

- The Transition Component (*TC*) that conducts the flow from the *PHIT* inlet, section 0 to section $1*$, shown in Fig. 8.1.
- The Slow Control Component (*SCC*) that controls the *PHIT* load. The *SCC* goes from section $1*$ to the section 1 inlet of each Injector (*IN*) as per Fig. 8.1 This load

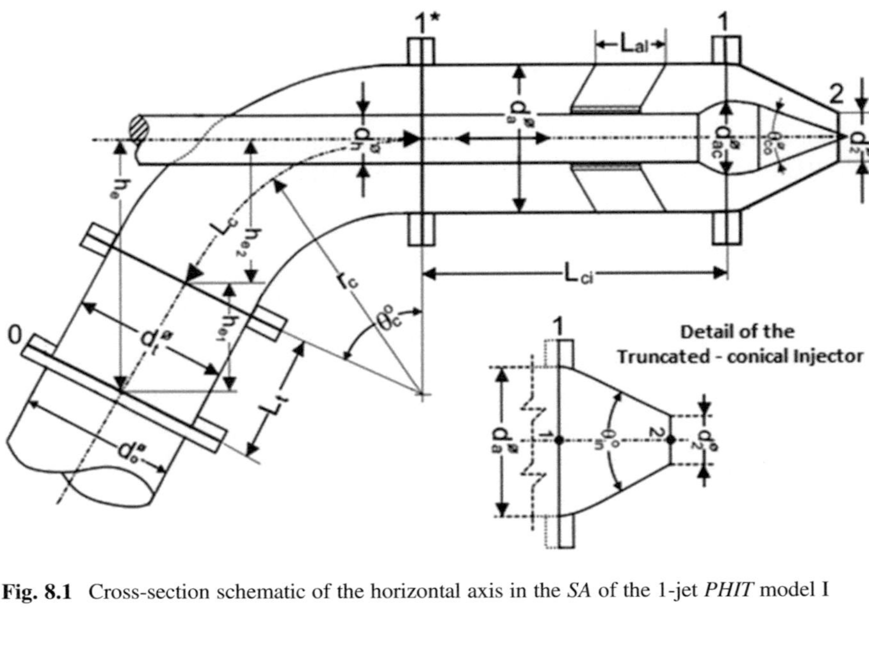

Fig. 8.1 Cross-section schematic of the horizontal axis in the *SA* of the 1-jet *PHIT* model I

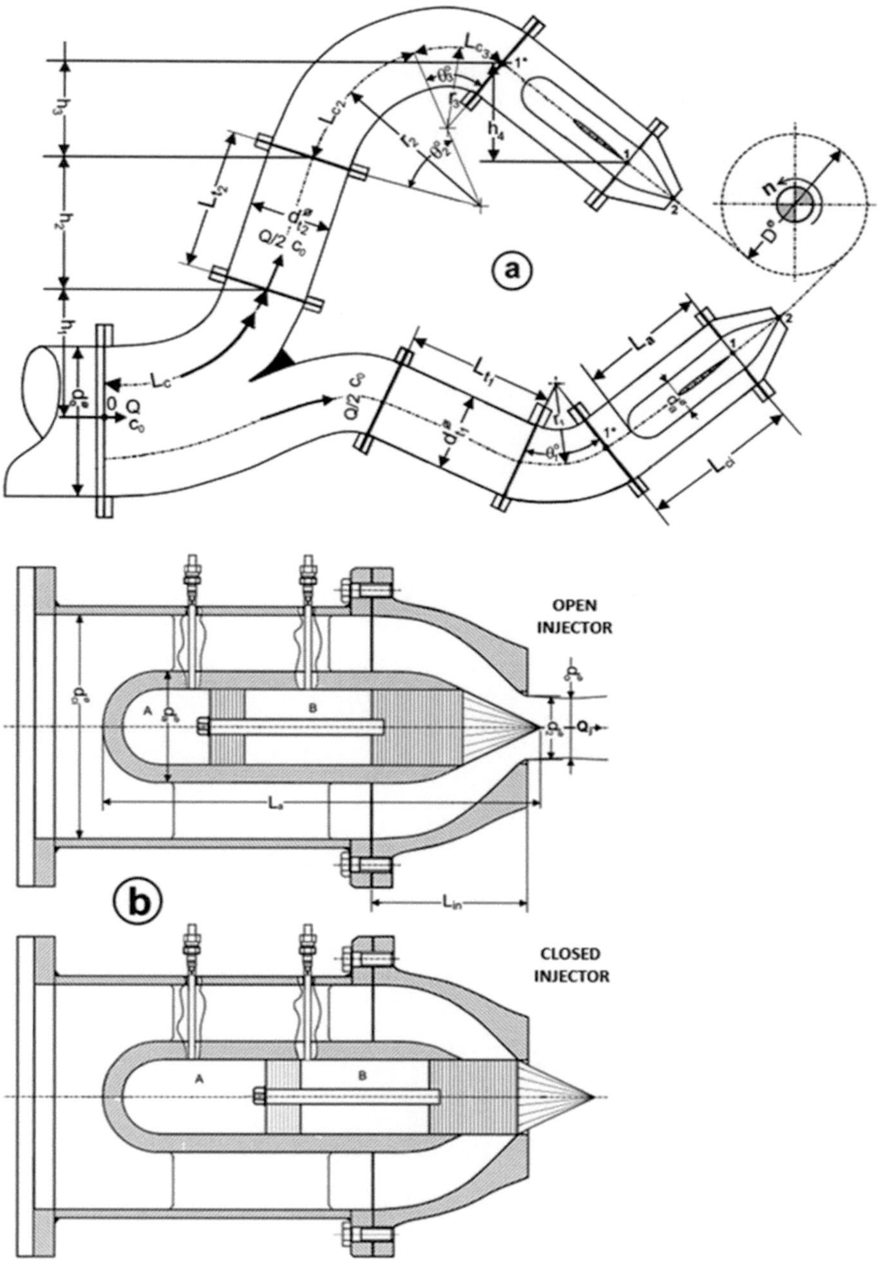

Fig. 8.2 (**a**) Cross-section schematic of the horizontal axis of model II. (**b**) Longitudinal section schematic showing the *SCC* with extreme needle positions

control is made using an internal cylindrical-conical component submitted to alternating translational movement.

- The Injector (*IN*) that conducts the flow between section 1 and section 2 of each flow coming from the *TC*, transforming the energy available at the inlet section 1 (minus losses) into kinetic energy at outlet section 2, as per Fig. 8.1.
- The Admission System (*SA*) that coincides with the flow from inlet section 0 to section 2 at the *IN* outlet, which, for calculation purposes, must consider the greatest energy losses.
- The Quick Control Component (*QCC*) that quickly diverts *RO* run from the *PHIT*. This component, called a Deflector (*DF*), which was shown in Fig. 8.3a, b, and which was schematized in Fig. 1.5 of Chap. 1, and laid out greater detail in Figs. 8.8, and in 8.11, 8.12 and 8.13.

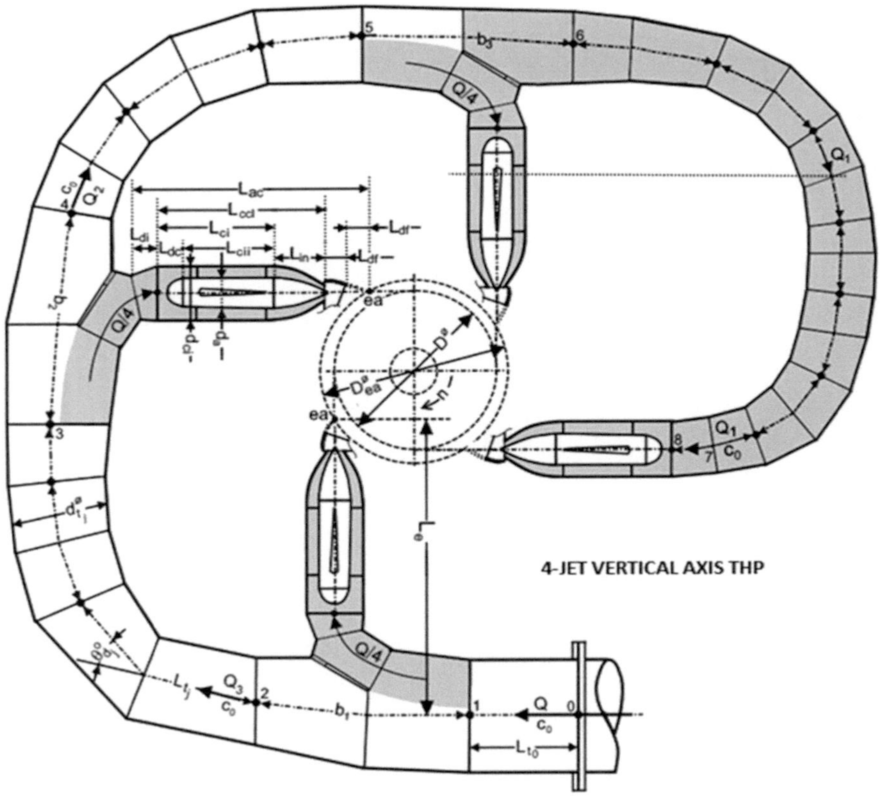

Fig. 8.3 (**a**) Cross-section schematic of the vertical axis of Model III from *SA* to the 4-jet *PHIT*. (**b**) Cross-section schematic to the vertical axis of model III from the *SA* to the 5-jet *PHIT*. (**c**) Cross-section schematic to the vertical axis of model III from the *SA* to the 6-jet *PHIT*

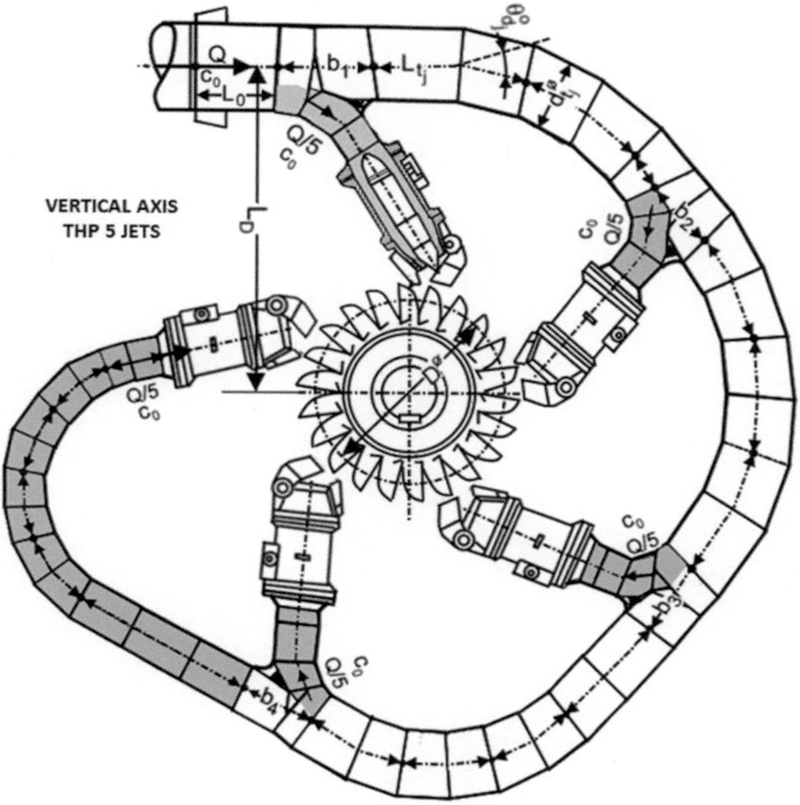

Fig. 8.3 (continued)

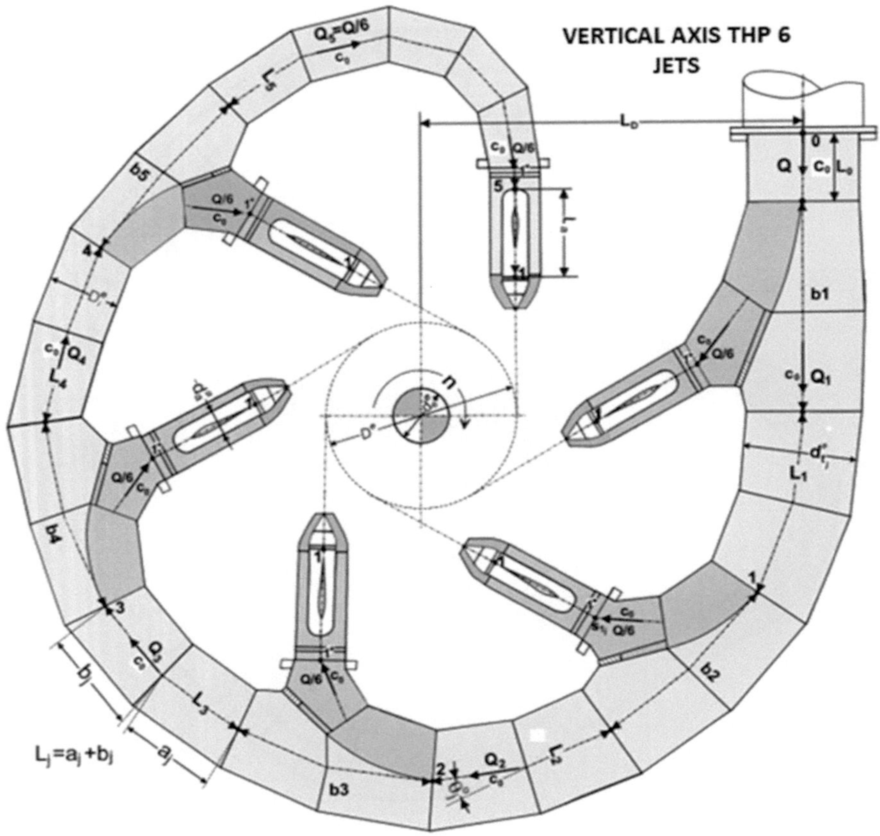

Fig. 8.3 (continued)

- The Free Flow (*FF*) that conducts airflow from the outlet section of the *IN* in 2, to the inlet section of shell 4, with minimum diameter in section 3.
- The *Rotor* Shells (*RS*), which conduct the split flow between sections 4 and 5.
- The Flow Outlet, (*FO*) which carries the section flow past the shell outlet 6 to the Downstream Level (*DL*).

The *PHIT SA* comprises all the components that transform the available pressure and kinetic energy at the inlet at section 0 into available kinetic energy in section 2 (*CT* + *CCL* + *IN*),whether it is a 1-jet or 2-jet *PHIT*.

8.1.2 *Model I—Horizontal Axis 1-Jet PHIT*—Fig. 8.1—Sections **0** to **2**

In this model, the *SA* in Fig. 8.1 comprises fixed tubes and curves, and a movable needle with a cylindrical rod with a conical tip on one side, which can be moved with translation movement, allowing for variations in the cross section of the *IN*, thus controlling the flow. There is, therefore, a need for electrical load compatibility available at the exit of the *EG*.

With reference to the figure, the control is performed via alternate translation inside the flow of the needle comprising a rod and a conical tip. Alternate translation can be done manually for Micro-Pelton Generator Groups (*MPGG*s) or for larger *GG*s using a conical cylinder system, also called a needle, which was schematized in detail in Figure 8.2b.

According to the schematic in Fig. 8.1, and in the detailed drawings of the *IN* in the *SA*, the main parts are the diameters, the lengths, and the angles that are geometrically represented and symbolized. Remember that their geometries, and the quality of the surfaces (materials and average roughness) in contact with the flow, are what are being considered here, for the purpose of this study and the respective calculations for all the models.

8.1.3 *Model II—Horizontal Axis 2-Jet* PHIT—*Fig. 8.2—Sections 0 to 2*

In this model, the flow is divided in two at the beginning via a bifurcation of rounded corners, as shown in Figure 8.2a.

The hypothesis that will be adopted here in this study is that the flow $Q(\mathrm{m/s^3})$ provided by the flow in section 0, at the entrance of the *PHIT* is divided in half, thereby flowing to each injector $Q/2 = Q_1(\mathrm{m/s^3})$.

Via this model, power ratings can reach up to 55 (MW), e.g., the *Netuno Central—Trentin, Haute-Adige*—Italy.

Both in this model, and in the vertical axis models that will be analyzed, the rod and needle schematized in Fig. 8.1 were replaced by conical cylinder systems, the needles, schematized, in some detail, in Figure 8.2b.

In this figure, the longitudinal section along the *SCC* axis shows how the alternating translation movement is processed, which controls the load via flow variations, positioning the needle after the *QCC* activates (the *DF* is not shown in the figure). This positioning variation is obtained by varying the pressures in cylindrical chamber *A* and chamber *B*.

As we will see, the double regulation of the *PHIT* permits this variation when the action must be doubled, starting with the jet deflection via the *DE*, and ending with the needle displacement, all within preestablished limits of over-elevation pressure (water hammer). The part for the *DF* action is shown in the schematic drawings in Fig. 1.5 in Chap. 1.

8.1.4 *Model III—Vertical Axis of a 4-, 5-, and 6-Jet* PHIT—*Figures 8.3a, b, c Respectively—Sections 0 to 2*

When the number of jets is greater than 2, the technical and cost problems related to preventing the water flowing at the exit of the shells from falling on the *RO*, thereby impairing operation, has forced manufacturers to position the axis of the *PHIT* in the vertical position. For this axle position, mostly 4-, 5-, and 6-jet models are manufactured. Their geometries and internal dimensions are given, in cross section to the axis (views), in Fig. 8.3a, b, c.

Figure 8.3a shows a normal section to the axis of a 4-jet *PHIT*, geometrically naming the internal dimensions of the flow, which, since they are dimensionless, can be preliminarily considered in the calculations for energy losses.

Figure 8.3b, c similarly show the 5- and 6-jet models.

The 5-jet *PHIT* will be modeled based on data from the *Jostedal* Power Station, in Oslo, Norway, at:

- $Q = 28.5$ (m^3/s)
- $H = 1130$ (mca)
- $n = 428.6$ (rpm)
- $P = 288$ (MW)

The 6-jet *PHIT* will be modeled based on data from one of eight *PHITS* operating at the *São Carlos* Power Station in Antioquia, Colombia, each rated at 170.4 (MW), at:

- $Q = 33.37$ (m^3/s)
- $H = 578$ (mca)
- $n = 300$ (rpm)

8.2 Velocity at the Entrance of the *PHIT*

Both the Brazilian standards *NBR* 6412 – 9581 and CEI/*IEC* 41 – 60,193 use section 0 as a reference for the *PHIT* entry, from Figs. 8.1, 8.2, and 8.3, located immediately after the valve installed at the end of the penstock.

This section is normally equal to the valve inlet, and the diameter is the same as the end of the penstock, and must be calculated by imposing the condition of minimum energy losses.[1]

If the diameter is unknown, one can start by sizing the *SA* and the *HT* in general, then the *PHIT*, using the *Bondschu* Formula, as:

$$d_{cf} = 1.27 \cdot \frac{Q^{0.1286}}{(H_L + h_s)^{0.1429}} \text{ (m)} \tag{8.1}$$

Where:

- Q (m^3/s)—the outflow;
- H_L(m)—the vertical distance between the maximum upstream level flow to the lowest point of the penstock's internal diameter;
- h_s (m)—the corresponding water hammer.

For *GCH* inventory studies, *Eletrobrás* recommends:

$$d_{cf} = 0.71 \cdot \frac{P_{eM}^{0.43}}{H_M^{0.65}} \text{ (m)} \tag{8.2}$$

Where:

$P_{eM}(kW)$—the shaft power for the head $H_M = (H_L + h_s)\ (m)$

From section 0 onwards, the flow should have minimum internal losses so that kinetic energy is maximum at output section 2 of the *IN*, i.e.:

$$E_{i_2} = Y_{i_2} = \left(Y_{i_0} - \sum\nolimits_0^2 Y_{i_{P_{min_j}}}\right)_{\max}$$

Therefore, the average absolute velocity, c_o, in the *TC* should be the smallest possible. Since the flow is turbulent, losses occur next to the walls of the conduit in the so-called boundary layer, resulting in more turbulence the smaller the respective diameters, which is why, in principle, the diameter at the entrance of the *TC* is taken as d_o equal to d_{cf}, i.e., expressions (8.1) or (8.2), which resulted here, in adopting a constant mean absolute velocity across all sections of the *TC*, calculated by:

$$c_{cf} = \frac{4 \cdot Q}{\pi \cdot d_{cf}^2} \text{ (m/s)} = constant\ in\ all\ stretches\ of\ the\ TC \tag{8.3}$$

Given this hypothesis, the following expression determines the respective internal diameters of the corresponding sections:

[1] *Souza, Z. e outros - Centrais Hidrelétricas - Capítulo 7 - 6.1.2 - pg. 282 - Ed. Interciência - RJ - BR.*

$$d_{t_j} = \sqrt{\frac{4 \cdot Q_j}{\pi \cdot c_{cf}}} \text{ (m)} \tag{8.4}$$

Given these considerations, the specific energy of the pressure at the *PHIT* inlet results in:

$$\frac{p_o}{\rho} = g \cdot H - \frac{c_o^2}{2} \text{ (J/kg)} \tag{8.5}$$

Not considering the losses, all specific energy available at the entrance of the *PHIT* should be transformed into average kinetic energy for the minimum section of the flow at the entrance of the *RO* shells:

$$0.5 \cdot c_e^2 = 0.5 \cdot c_{4t}^2 = Y_{i_0} = g \cdot H \text{ (J/kg)} \rightarrow c_e = c_{4t} = \sqrt{2 \cdot g \cdot H} \text{ (m/s)} \tag{8.6}$$

To provide data on large *PHITS*, we will consider a data compilation made by Pierre Henry,[2] transcribed in Table 8.1.

This table used the following conventions:

- $d_o = D$ (m)—the primitive diameter;
- $D_o = d_2 = d_{in}$ (m)—the jet diameter at the exit of the *IN*;
- $D_2 = d_0$ (m)– the minimum jet diameter;
- $B = B$ (m)– the maximum internal width of the shells;
- $z_r = z_c$—the number of *RO* shells;
- $z_o = z_j$—the number of jets.

Table 8.2 was formatted, and the values were calculated with expressions:

$$\begin{gathered} P_h = 9.81 \cdot Q \cdot H (\text{kW}); \quad \eta_t = 100 \cdot \frac{P_e}{P_h} (\%); \quad Q_1 = \frac{Q}{z_j} (\text{m}^3/\text{s}); \\ n_{qA_1} = 3 \cdot n \cdot \frac{Q_1^{0.5}}{H^{0.75}}; \quad c_{o_t} = c_{4_t} = \sqrt{2 \cdot g \cdot H} (\text{m/s}); \\ c_4 = \frac{4 \cdot Q_1}{\pi \cdot d_o^2} (\text{m/s}); \quad \phi_{o;4} = c_4 / c_o; \quad u = \frac{\pi \cdot D \cdot n}{60}; \quad r_{v_r} = u / c_4; \\ r_v = u / c_{4t}; \quad d_{cf} = 1.27. \frac{Q^{0.4286}}{(H_L + h_s)^{0.1429}} (\text{m}); \quad c_{cf} = \frac{4 \cdot Q}{\pi \cdot d_{cf}^2} (\text{m/s}) \end{gathered} \tag{8.7}$$

[2] Turbomachines Hidraulique, Henry P., Presses Polytechniques et Universitaires Romandes, Lausanne, ISBN 2-88-74-209-9 - 1992.

Table 8.1 PHIT characteristics according to Pierre Henry's Tables

Numbers	H	Q	n	P_e	$D_1 = D$	$D_0 = d_2$	$D_2 = d_0$	B	$z_r = z_c$	$z_0 = z_j$
No.	m	m^3/s	rpm	MW	m	m	m	m	–	–
1	1088.00	5.77	600.00	55.00	2.26	0.21	0.17	0.53	21.00	2.00
2	915.00	19.50	428.60	156.30	2.70	0.26	0.20	0.72	20.00	5.00
3	1130.00	28.50	428.60	288.00	3.18	0.28	0.22	0.73	22.00	5.00
4	1214.40	24.31	500.00	263.40	2.85	0.23	0.19	0.57	23.00	6.00
5	1032.00	24.40	450.00	224.80	2.95	0.26	0.22	0.64	22.00	5.00
6	648.00	18.00	428.60	102.80	2.36	0.25	0.21	0.65	21.00	5.00
7	652.00	18.83	428.60	110.00	2.42	0.25	0.20	0.61	22.00	6.00
8	578.00	33.37	300.00	170.40	3.25	0.34	0.26	0.80	22.00	6.00
9	381.00	22.65	276.90	76.06	2.84	0.31	0.24	0.79	20.00	6.00
10	255.00	33.27	200.00	74.50	3.21	0.42	0.32	1.00	19.00	6.00
11	350.00	49.76	225.00	153.00	3.33	0.46	0.37	1.11	20.00	6.00

1—C. Naturno - Italy; 2—C. L'Eau d'Olle - France; 3—C. Jostedal - Norway; 4—C. Sellrain-Silz - Austria; 5—C. Guavio - Colombia; 6—C. La Bâtiaz - Switzerland; 7—C. Pigai Aoos - Greece; 8—C. San Carlos - Colombia; 9—C. Devil Canyon - United States;10—C. Restituición - Peru; 11—C. San Agaton - Venezuela

Table 8.2 Calculated values with PHIT characteristics

Withdrawals from the Pierre Henry tables							
Number	P_h	η_t	Q_1	ηqA1	c_{4t}	c_4	$\varphi_{o;4}$
No.	MW	%	m³/s	–	m/s	m/s	c_4/c_{4t}
1	61.59	89.3	2.89	16.14	146.11	131.71	0.90
2	175.04	89.3	3.90	15.26	133.99	129.26	0.96
3	315.93	91.2	5.70	15.75	148.90	147.26	0.99
4	115.81	91.0	4.05	14.68	154.36	150.74	0.98
5	247.02	91.0	4.88	16.38	142.29	134.42	0.94
6	114.42	89.8	3.60	19.00	112.76	109.07	0.97
7	120.44	91.3	3.14	17.65	113.10	105.07	0.93
8	189.21	90.1	5.56	18.01	106.49	101.84	0.96
9	84.66	89.9	3.78	18.72	86.46	82.07	0.95
10	83.23	89.5	5.55	22.14	70.73	69.82	0.99
11	170.85	89.6	8.29	24.02	82.87	79.08	0.95

Number	u	r_{vr}	r_v	D/d_o	B/d_o	d_2/d_o	d_{cf}	c_{cf}	φ_o
No.	m/s	u/c_4	u/c_{4t}	–	–	–	m	m/s	c_o/c_e
1	71.00	0.54	0.49	10.76	2.52	1.23	0.96	8.06	0.06
2	60.59	0.47	0.45	13.80	3.67	1.35	1.65	9.13	0.07
3	71.25	0.48	0.48	14.30	3.27	1.28	1.88	10.23	0.07
4	74.61	0.50	0.48	15.41	3.08	1.22	1.74	10.21	0.06
5	69.51	0.52	0.49	13.72	2.98	1.22	1.78	9.76	0.07
6	52.96	0.49	0.47	11.51	3.17	1.22	1.67	8.18	0.07
7	54.20	0.52	0.48	12.39	3.14	1.26	1.71	7.88	0.07
8	51.05	0.50	0.48	12.33	3.03	1.27	2.22	8.64	0.08
9	41.12	0.50	0.48	11.71	3.26	1.29	1.99	7.26	0.08
10	33.62	0.48	0.48	10.09	3.15	1.31	2.49	6.84	0.10
11	39.23	0.50	0.47	9.11	3.04	1.27	2.83	7.93	0.10

The values in Tables 8.1 and 8.2, in italics with a light gray background, were obtained by adopting the hypothesis of linearity between the known characteristics with similar values for n_{qA_1}. Once the values of two of its dependent characteristics are known, e.g., *D/do*, the value of the unknown dependent value can be determined, as follows:

- **Line Number 1—**$n_{qA_{11}} = 16.138$ **results in italic with a gray background.**

  The closest values that contain this specific rotation and its independent characteristics are:

$$\Delta n_{qA_{1_{53}}} = \left(n_{qA_{1_5}} = 16.379\right) - \left(n_{qA_{1_3}} = 15.751\right) = 0.628 \rightarrow$$

$$\Delta (D/d_o)_{53} = [(D/d_o)_5 = 13.721] - [(D/d_o)_3 = 14.301] = -0.58$$

Or…

$$\Delta n_{qA_{1_{15}}} = \left(n_{qA_{1_5}} = 16.379\right) - \left(n_{qA_{1_1}} = 16.138\right) = 0{,}241 \rightarrow$$

$$\Delta(D/d_o)_{15} = \frac{\Delta(D/d_o)_{53} \cdot \Delta n_{qA_{15}}}{\Delta n_{qA_{53}}} = \frac{-0.58 \cdot 0.241}{0.628} = -0.223$$

Thereby giving ...

$$(D/d_o)_1 = (D/d_o)_5 + \Delta(D/d_o)_{15} = 13.721 - 0.223 = 13.498 \rightarrow$$

$$d_{o_1} = 2.26/13.498 = 0.167(\mathrm{m}) \rightarrow$$

$$c_{4_1} = \frac{4 \cdot Q_{1_1}}{\pi \cdot d_{o_1}^2} = \frac{4 \cdot 2.885}{\pi \cdot 0.167^2} = 131.711\ (\mathrm{m/s}) \rightarrow$$

$$\phi_{(04)_1} = \frac{131.711}{146.105} = 0.9015 \rightarrow$$

$$r_{v_1} = \frac{u_1}{c_{4_1}} = \frac{71}{131.711} = 0.539$$

- **Line Number 2—**$n_{qA_{1_2}} = 15.263$**—results in italic with a gray background.**
  The closest values that contain this specific rotation and its independent characteristics are:

$$\Delta n_{qA_{1_{34}}} = \left(n_{qA_{1_3}} = 15.751\right) - \left(n_{qA_{1_4}} = 14.677\right) = 1.074 \rightarrow$$

$$\Delta(D/d_o)_{34} = \left[(D/d_o)_3 = 14.301\right] - \left[(D/d_o)_4 = 15.405\right] = -1.104.$$

Or ...

$$\Delta n_{qA_{1_{23}}} = \left(n_{qA_{1_2}} = 15.263\right) - \left(n_{qA_{1_3}} = 15.751\right) = -0.488 \rightarrow$$

$$\Delta(D/d_o)_{24} = \frac{\Delta n_{qA_{23}} \cdot \Delta(D/d_o)_{34}}{\Delta n_{qA_{34}}} = \frac{-0.488 \cdot (-1.104)}{1.074} = -0.502$$

Thereby giving ...

$$(D/d_o)_2 = (D/d_o)_3 + \Delta(D/d_o)_{24} = 14.301 - 0.502 = 13.793 \rightarrow$$

$$d_{o_2} = \frac{D}{(D/d_o)_2} = \frac{2.7}{13.793} = 0.196\ (\mathrm{m}) \rightarrow$$

$$c_{4_2} = \frac{4 \cdot Q_{1_2}}{\pi \cdot d_{o_2}^2} = \frac{4 \cdot 3.9}{\pi \cdot 0.196^2} = 129.260\ (\mathrm{m/s}) \rightarrow$$

$$\phi_{(04)_2} = \frac{129.260}{133.986} = 0.9647 \rightarrow$$

$$r_{v_{r_2}} = \frac{u_2}{c_{4_2}} = \frac{60.591}{129.260} = 0.469$$

- **Line Number 9—$n_{qA_{19}} = 18.716$—results in italic with a gray background.** The closest values that contain this specific rotation and its independent characteristics are:

$$\Delta n_{qA_{168}} = \left(n_{qA_{1_6}} = 18.996\right) - \left(n_{qA_{1_8}} = 18.005\right) = 0.991 \rightarrow$$

$$\Delta(D/d_o)_{68} = \left[(D/d_o)_6 = 11.512\right] - \left[(D/d_o)_8 = 12.325\right] = -0.813$$

Or …

$$\Delta n_{qA_{196}} = \left(n_{qA_{1_9}} = 18.716\right) - \left(n_{qA_{16}} = 18.9960 = -0.244 \rightarrow\right.$$

$$\Delta(D/d_o)_{96} = \frac{-0.244 \cdot (-0.818}{0{,}991} = 0.201$$

Thereby giving …

$$(D/d_o)_9 = (D/d_o)_6 + \Delta(D/d_o)_{96} = 11.512 + 0.201 = 11.713 \rightarrow$$

$$d_{o9} = D_9/11.713 = 2.836/11.7130 = 0.242\ (m) \rightarrow$$

$$c_{4_9} = \frac{4 \cdot Q_{1_9}}{\pi \cdot d_{o_9}^2} = \frac{4 \cdot 3.775}{\pi \cdot 0.242^2} = 82.072\ (m/s) \rightarrow$$

$$\phi_{(04)_9} = \frac{c_{4_9}}{c_{4_t}} = \frac{82.072}{86.459} = 0.9493 \rightarrow$$

$$r_{v_9} = \frac{u_9}{c_{4_9}} = \frac{41.118}{82.072} = 0.501$$

With the values obtained in Table 8.2, the graphs in Figs. 8.4 and 8.5 were then drawn.

In Fig. 8.4 the $r_v = u/c_{4t}$ values, in function of $n_{qA_1} = 3 \cdot n \cdot \frac{Q_1^2}{H^{0.75}}$, which is for large *PHIT*s, are between the following range:

$$0.470 \leq r_v \leq 0.486$$

or with two decimals …

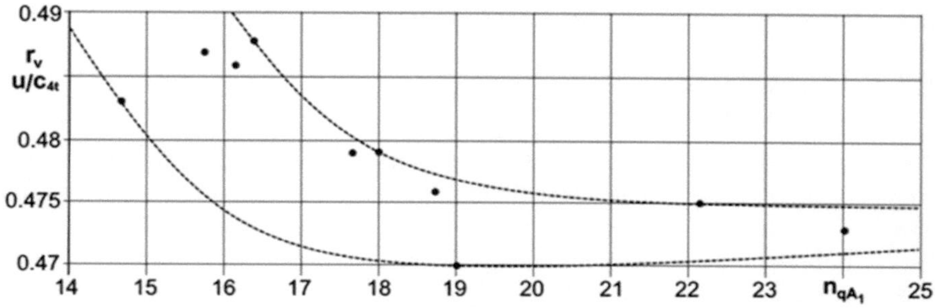

Fig. 8.4 r_v function in relation to n_{qA_1} for a *PHIT* with values from Table 8.2

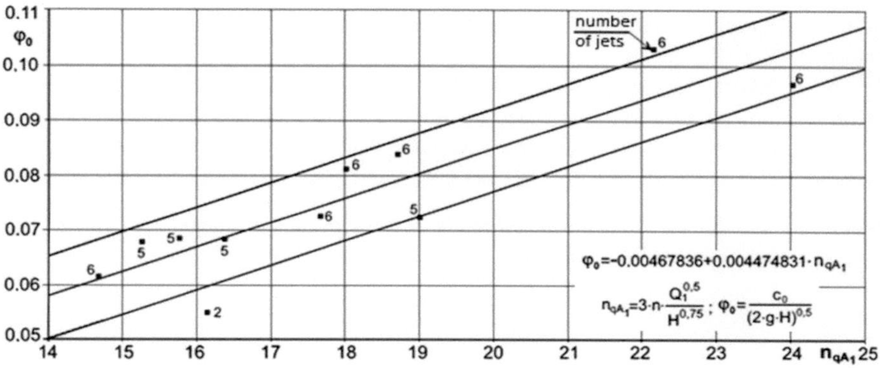

Fig. 8.5 ϕ_0 function in relation to n_{qA_1} for a *PHIT* with values from Table 8.2

$$0.47 \leq r_v < 0.49 \tag{8.8}$$

In this graph, point $r_v = 0.452 \rightarrow n_{qA_1} = 15.263$ was not considered, because it falls outside the currently used range.

The graph in Fig. 8.5 shows the field $\phi_0 = f\left(n_{qA_1}\right)$. These values can be used in preliminary studies, particularly when the penstock characteristics next to the engine room are not available, and by extension the internal diameter at the valve outlet.

Values from these tables, and from other tables that will be obtained from them, will be used in the next chapter.

8.3 Energy Losses in the *PHIT* Transition Component

8.3.1 Considerations

The energy losses in the *TC* (taken here as having the greatest loss) depend on the arrangement of these components in the geometry of each *PHIT*, and can, in principle, be determined preliminarily, based on a system that will be developed for each of the models presented here.

However, we should note that the most probable and reliable values for the flow losses in the parts that comprise the *TC* in the *PHIT*, along with their errors and respective uncertainty ranges, can only be obtained via tests with models or prototypes. These tests and trials must be carried out by accredited laboratories, or on site, within the standards and criteria that have been agreed upon by the respective parties.

Here, we fist seek to calculate the values of these losses, so that the preliminary hydrodynamic studies can be carried out.

To obtain calculations that with lower experimental variance values, one can consult the expressions and coefficients contained in the following publication:

Memento des Pertes de Charge—Idelcik, I.E—Coeffients de Pertes Charge par Fronttement –Eyrolles—Editeur—Paris—France

… or from other recognized scientific literature publications.

8.3.2 Hypotheses

- Water flows in average steady state with a head, constant available H and equally distributed flows Q for each motor jet at the outlet of the respective *IN*, so:

$$Q_1 = Q/z_j$$

- The basic energy loss in the *TC* must be established for the longest flow path between section 0 and section 1∗.
- The average absolute velocity $c = c_0(m/s)$ of the flow will be taken as being constant in all sections of the *TC*.
- Geometric characteristics will be taken as average values, both on their internal and external surfaces, immersed in the flow. This flow geometry will be called Hydraulic Flow Path (*HFP*).
- The kinematic characteristics of the flow, with respect to the *AS*, when used, must observe the principle of Conservation of Mass, i.e.:
  - $Q = \frac{\dot{m}}{\rho} = A \cdot c \rightarrow Q(\mathrm{m}^3/\mathrm{s})$
  - m(kg/s)

- $\rho = 1000\ (\text{kg/m}^3)$
- $A(\text{m}^2)$
- c (m/s)

- To determine the energy losses of the flow in each of the *TC* sections, these sections, or interconnected parts, can be assumed to have connected tubes, curves, deflections, and bifurcations, as shown in the following modeling.
- The flow and geometric characteristics of the *SCC* and *IN*, including their energy losses, will be studied in item 8.4.
- We recommend, unless otherwise indicated, to use a safety factor when calculating the specific work in the range $1.0 \leq f_s \leq 1.1$, i.e.:

$$Y_{i_{p_j}} = f_s \cdot Y_{i_{p_{j\,calculated}}}$$

8.3.3 Energy Losses in the Pipes

$$Y_{i_{p_t}} = 0.5 \cdot \lambda_t \cdot c_0^2 \cdot L_t \cdot d_t^{-1} = 0.5 \cdot \lambda_t \cdot (Q/A_t)^2 \cdot L_t \cdot d_t^{-1} = 0.8106 \cdot \lambda_t \cdot Q^2 \cdot L_t \cdot d_t^{-5} \tag{8.9}$$

- $Y_{i_{p_t}}$ (J/kg)—the internal specific energy lost at the end of the straight run of a long pipe L_t(m), diameter d_t(m), with flows through it Q(m^3/s), with constant absolute average speed, c_0(m/s).
- λ_t—the coefficient of energy losses in a straight section of a pipe which can be determined, preliminarily, via the Colebrook-White expression, as a function of the average roughness of the pipe wall that delimits the flow, and its Reynolds Number, as follows:

$$\lambda_t = \frac{1325}{\left[ln\left(\frac{\varepsilon_t}{3.7 \cdot d_t} + \frac{5.74}{N_{Re_t}^{0,9}}\right)\right]^2}$$

$$N_{Re_t} = \frac{4 \cdot Q}{\pi \cdot \nu \cdot d_t} \cong 1,26 \cdot 10^6 \cdot \frac{Q}{d_t} 10^{-6} \leq \frac{\varepsilon_t}{d_t} \leq 10^{-2}$$

$$5 \cdot 10^3 \leq N_{Re_t} \leq 10^8 \tag{8.10}$$

- ε_t (mm)—the absolute average roughness of the inner wall of a straight section of pipe, Table 8.3.

Observations:

- This form can also be used for sections of conical trunk tubes, with enlargement up to 14°, considering the average value of the diameters.

Table 8.3 Average absolute roughness

Tube surface material	e_t (mm)
Drawn steel	0.0015
Seamless steel	0.02–0.10
Welded steel sheets	0.04–0.10
Cast iron painted with bitumen	0.10–0.15
Cast iron painted with asphalt	0.12–0.30
PVC	0.0015–0.01

- When flow occurs between concentric tubes (d_t, d_h), form can be used with a hydraulic diameter $d_h = \frac{4 \cdot A_t}{U_t} = (d_t - d_h)$, where A_t(m^2) is the cross-sectional area, and U_t(m) is the wet perimeter.

8.3.4 *Losses in the Circular Cross Section Curves*

For curve angle $\theta < 90^o$—Weissbach's Formula:

$$Y_{i_{Pc}} = \lambda_c \cdot \frac{c_0^2}{2} \text{ (J/kg)} \rightarrow$$

$$\lambda_c = \frac{\theta_c}{90} \cdot \left[0.131 + 1.847 \cdot \left(\frac{d_c}{2 \cdot r_c}\right)^{3.5}\right] \rightarrow r_c = \textit{average radius of the curve} \tag{8.11}$$

For curve angle $\theta < 90^o$:

$$Y_{i_{Pc}} = \lambda_c \cdot \frac{c_0^2}{2} \text{ (J/kg)} \rightarrow$$

$$\lambda_c = 0.682 - 0.2437 \cdot \frac{r_c}{d_c} + 4.67 \cdot 10^{-2} \cdot \left(\frac{r_c}{d_c}\right)^2 - 3.868 \cdot 10^{-3} \cdot \left(\frac{r_c}{d_c}\right)^3 + 1.508 \cdot 10^{-4} \cdot \left(\frac{r_c}{d_c}\right)^4 - 0.2265 \cdot 10^{-5} \cdot \left(\frac{r_c}{d_c}\right)^5 \tag{8.12}$$

8.3.5 *Losses in the Deflections*

For deflections with geometries according to Fig. 8.6.

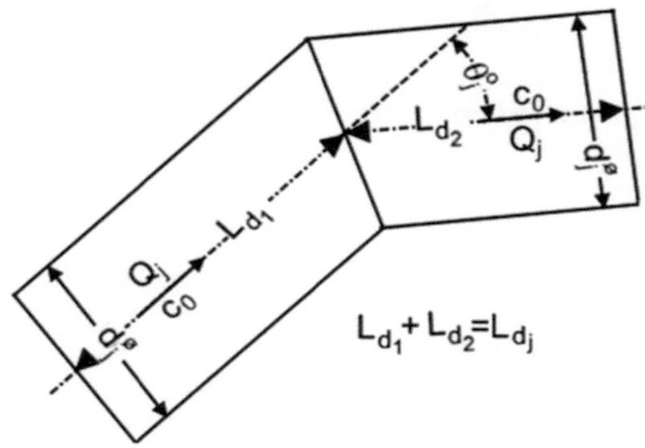

Fig. 8.6 Deflection geometry and the calculation for λ_d

$$Y_{i_{p_d}} = \lambda_d \cdot \frac{c_0^2}{2}\ (\mathrm{J/kg}) \rightarrow$$

$$\lambda_d = 0.9457 \cdot sen^2 0.5\theta_d + 2.05 \cdot sen^4 0.5\theta_d \tag{8.13}$$

The energy losses from flow deflections must be added to the corresponding friction in the pipe of length $L_{d_j} = L_{d_1} + L_{d_j}$, obtained from expressions (8.10) and (8.9).

8.3.6 *Losses in Bifurcations with Sharp Corners and Circular Sections*

The energy losses at bifurcations depend on several factors, especially:

- The internal geometry of the bifurcation parts, whether they be straight or conical.
- The angle of the bifurcation, δ_b^o
- The roughness of the wall materials in contact with the flow.

These and other factors mean that such losses should ideally be obtained on a case-by-case basis, using models that have geometric, kinematic, and dynamic similarity, whenever possible, or with tests using corresponding prototypes with geometric and hydrodynamic similarity.

Figure 8.7 shows a schematic of a longitudinal section with simplified basic geometry, of straight axis sections and bifurcations in welded plates, with geometric and flow characteristics used in the *SA* of a *PHIT* with a vertical axis.

For this, the energy loss coefficients, which are constant in the intervals transcribed in Fig. 8.7, can be used in preliminary studies.

Straight path:

$$Y_{i_{p_{br}}} = \lambda_{br} \cdot \frac{c_0^2}{2}\ (\mathrm{J/kg})$$

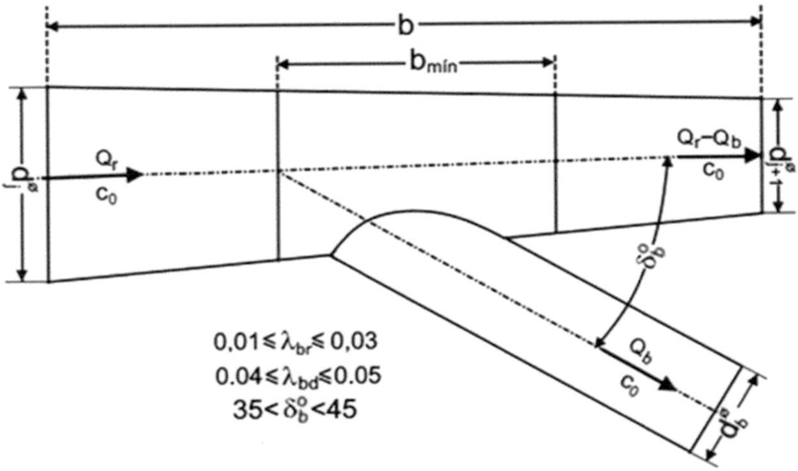

Fig. 8.7 Geometries and characteristics of bifurcations

→

Deviated path:

$$Y_{i_{p_{br}}} \lambda_{bd} \cdot \frac{c_0^2}{2} \text{ (J/kg)} \tag{8.14}$$

8.4 The Transition Component, the Regulating Component, and the Injector

8.4.1 The Transition Component, and the Regulating Component

The schematic in Fig. 8.8 represents, characterizes, and relates the *TC* between section 0 and section 1∗, here, a Deflector (*DF*) and the main parts that make up the *SCC* and the *QCC*. The geometries of these components may differ from manufacturer to manufacturer, particularly with respect to their moving parts, which are called here the moving needle, with its linear and alternating movement system, and the Deflector (*DF*).

The *SCS* normally comprises a differential piston with springs that occupy different volumes depending on the fixed ratios between the diameters of d_a, d_{ci} and the lengths, L_{ci}, L_{dc}.

The crown area, $A_a = \frac{\pi}{4} \cdot \left(d_{ci}^2 - d_a^2\right)$ must be set so that the absolute average speed, c_a, is the smallest possible, with respect to energy losses, while providing for

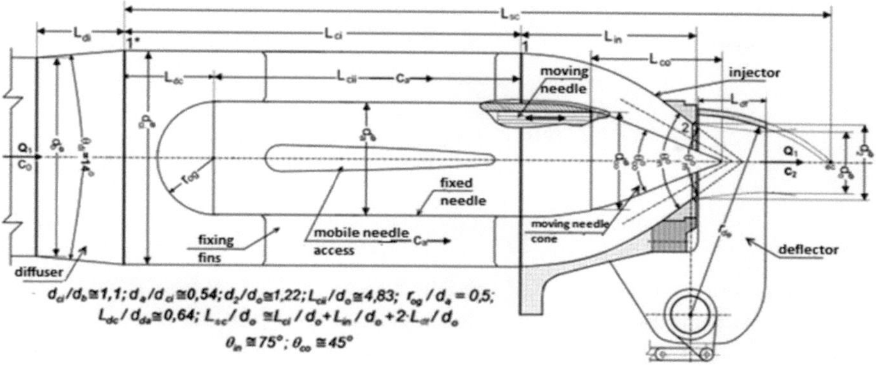

Fig. 8.8 Geometry and relationships between *SCC* and *QCC* characteristics

Table 8.4 Values of c_a/c_t

	$d_a/d_{ci} = 0.50$		$d_a/d_{ci} = 0.58$	
d_{ci}/d_d	c_0/c_a	c_a/c_0	c_0/c_a	c_a/c_0
0.80	0.48	2.08	0.43	2.35
0.90	0.61	1.65	0.54	1.86
1.00	0.75	1.33	0.66	1.51
1.10	0.91	1.10	0.80	1.25
1.20	1.47	0.68	1.30	0.77

the housing of the moving needle. In principle, if possible, it should be equal to the constant used for sizing the *SA*, c_o (m/s), but normally it is larger.

This can be briefly analyzed as follows:

$$Q_1 = \frac{\pi \cdot d_d^2}{4} \cdot c_0 = \frac{\pi \cdot (d_{ci}^2 - d_a^2)}{4} \cdot c_a \rightarrow \frac{c_0}{c_a} = \left(\frac{d_{ci}}{d_d}\right)^2 \cdot \left[1 - \left(\frac{d_a}{d_{ci}}\right)^2\right] \rightarrow d_a = \sqrt{d_{ci}^2 - \frac{4 \cdot Q_1}{\pi \cdot c_a}} \tag{8.15}$$

Table 8.4 was formatted and the values contained therein were calculated with this Equation, varying $\frac{d_{ci}}{d_d}$ from expression (8.15), which allows us to have a first idea of the variations in the ratios of the absolute average speeds being studied, and to preliminarily establish, their limits (in the table in italics):

$$1.0 \leq d_{ci}/d_d < 1,1 \rightarrow 1.1 \leq c_a/c_0 \leq 1.5 \tag{8.16}$$

If a *DI* is needed to connect the output of the bifurcation with the input of the Control System (*SC*) (Cross Section 1∗), the angle of the *DI*, θ_{di}^o, must not exceed 14 (degrees), for reasons related to boundary layer detachment, which then allows us to establish an expression inferiorly limiting the length of the deflector, L_{di}, as follows:

$$0.5 \cdot (d_{ci} - d_d) = L_{di} \cdot tg(0.5 \cdot \theta_{di}) \rightarrow$$

$$L_{di_{min}} = \frac{d_{ci} - d_d}{2 \cdot tg(0.5 \cdot \theta_{di_{max}})} = \frac{d_{ci} - d_b}{2 \cdot tg(0.5 \cdot 14)} \rightarrow$$

$$L_{di_{min}} = 4.072 \cdot (d_{ci} - d_d) \tag{8.17}$$

The angles of the needle cone, θ_a^o, and the injector, θ_{in}^o, are set from θ_a^o which must be small, given the low pressures in the region of the tip of the cone. These low pressures cause wear (cavitation) in this region over the short term, which is why a two-part cone with a removable tip has been developed by designers. The most widely used angles for *PHIT*s that must operate with medium and large heads are around $\theta_a \cong 44^o$ and $\theta_{in} \cong 70^o$.

Generally, in preliminary studies, these angles can be fixed in the following intervals:

$$40^o \leq \theta_a \leq 60^o \rightarrow 60^o \leq \theta_{in} \leq 90 \tag{8.18}$$

8.4.2 The Injector

The *IN* geometry aims to increase the average flow velocity, thus transforming pressure energy into kinetic energy. Since fluid in a liquid state is practically incompressible, e.g., water, the simplest geometric model is a schematized trunk-conical shape, as shown in Fig. 8.9.

Given the principle of Conservation of Energy and average value permanent flows with dynamic characteristics, the following Equation can be developed for modeling this:

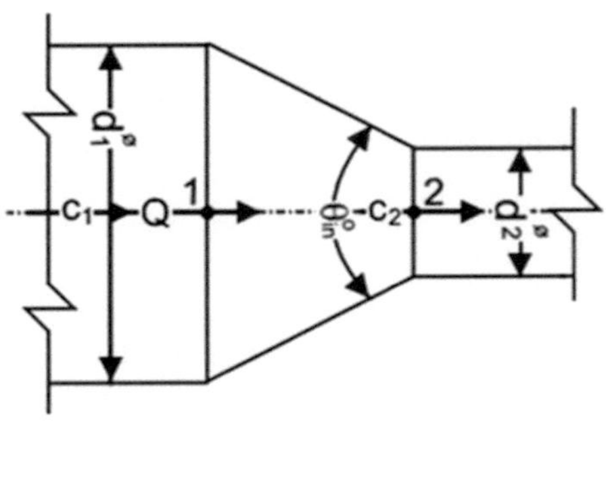

Fig. 8.9 Schematic section of an *IN* in conical section

$$Y_{i_1} = \frac{p_1}{\rho} + \frac{c_1^2}{2} = \frac{p_2}{\rho} + \frac{c_2^2}{2} + Y_{i_{Pin}}$$

to …

$$p_2/\rho = p_{atm}/\rho = 0 \rightarrow Y_{i_1} = \frac{p_1}{\rho} + \frac{c_1^2}{2} = \frac{c_2^2}{2} + Y_{i_{Pin}}$$

… or the specific inner work lost in the *IN* can be determined by:

$$Y_{i_{Pin}} = Y_{i_1} - \frac{c_2^2}{2} = Y_{i_1} - Y_{i_2} \text{ (J/kg)} \tag{8.19}$$

In this expression, Y_{i_2} (J/kg) is the specific work at *IN* output 2.

Considering that c_1^* (m/s) is a theoretical average absolute velocity that in section 1 of the flow, transforms specific work Y_{i_1} (J/kg) into specific kinetic energy, $Y_{i_1} = e_{c_1} = \frac{c_{1*}^2}{2}$ (J/kg), we can obtain equivalent velocity $c_{1*} = \sqrt{2 \cdot Y_{i_1}}$ (m/s).

This equivalent speed allows us to introduce a speed loss coefficient, $\phi_{in} = \phi_1 = \frac{c_2}{c_{1*}}$.

On the other hand, c_2 can be determined once the internal geometry of the *IN* is known for each established flow case $Y_{i_1} = g \cdot H_1$ *e* Q) via:

$$c_2 = \frac{4 \cdot Q}{\pi \cdot d_2^2} \text{ (m/s)}$$

Thus, the speed loss coefficient $\phi_{in} = \phi_1$ can be determined experimentally.

The cited bibliography recommends that ϕ_1 can be preliminarily fixed, for preliminary studies of *IN*s for *PHITs* of the type outlined in Fig. 8.10, in the following range:

$$0.986 \leq \phi_1 \leq 0.998 \tag{8.20}$$

Tests carried out on models with surface roughness, $\varepsilon_t < 0.01$ (mm) with a geometry similar to Fig. 8.10, allow us to establish the following mathematical/statistical expression to determine its head loss coefficient, with reference to c_1(m/s):

$$\lambda_{in} = 0.0025 \cdot \left[1 - \left(\frac{A_2}{A_1}\right)^2\right] \cdot \cot g\left(\frac{\theta_m}{2}\right) \rightarrow Y_{i_{Pin}} = \lambda_{in} \cdot \frac{c_1^2}{2} \text{ (J/kg)} \tag{8.21}$$

The specific work lost in the *IN*, can also be determined as follows:

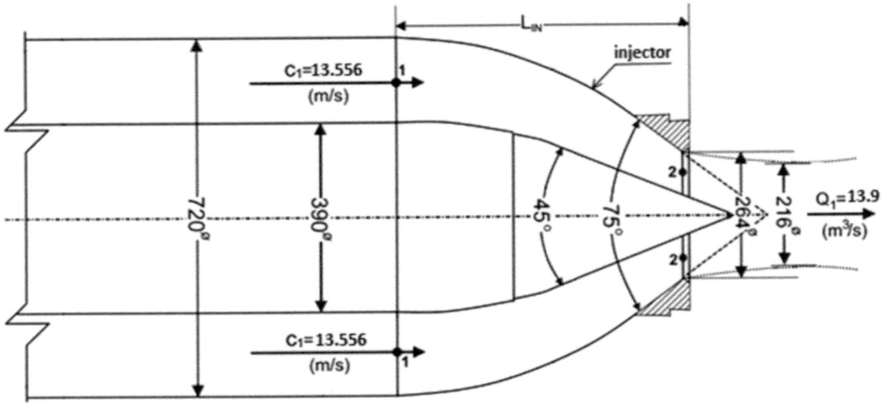

Fig. 8.10 Scaled diagram of a typical *IN* with a needle

$$Y_{i_{p_{in}}}=Y_{i_1}-Y_{i_2}=\frac{c_{1*}^2}{2}-\frac{(\phi_1\cdot c_{1*})^2}{2}=\left(1-\phi_1^2\right)\cdot\frac{c_{1*}^2}{2} \tag{8.22}$$

Then, the internal yield of the *IN* is:

$$\eta_{i_{in}}=1-\frac{Y_{i_{p_{in}}}}{Y_{i_1}}=1-\left(1-\phi_1^2\right)=\phi_1^2 \tag{8.23}$$

Considering the longitudinal section if the *IN* with constant geometric and hydrodynamic characteristics, as per the diagram in Fig. 8.10, the steps below can be used to determine the values of the characteristics studied in this item.

Calculate:

1. The area normal to the flow between the cylinders:

$$A_1=\frac{\pi}{4}\cdot\left(0.72^2-0.392\right)=0.287691\ \left(\mathrm{m}^2\right)$$

2. The flow:

$$Q_1=c_1\cdot A_1=13.556\cdot 0.287691=3.9\ (\mathrm{m/s})$$

3. The maximum flow velocity:

$$c_{\max}=\frac{4\cdot Q}{\pi\cdot d_o^2}=\frac{4\cdot 3.9}{\pi\cdot 0.216^2}=106.431\ (\mathrm{m/s})$$

4. The normal area to the flow at the exit of the *IN* considering, $c_1=c_{\max}$:

$$A_2 = \frac{Q}{c_{\max}} = \frac{3.9}{106.431} = 0.036643\ \left(\mathrm{m}^2\right)$$

5. The average angle of the flow at the *IN*:

$$\theta_m = 0.5 \cdot (75 + 45) = 60^o$$

6. The pressure loss coefficient at the *IN*:

$$\lambda_{in} = 0.0025 \cdot \left[1 - \left(\frac{A_2}{A_1}\right)^2\right] \cdot \cot g\left(\frac{\theta_m}{2}\right)$$
$$= 0.0025 \cdot \left[1 - \left(\frac{0.036643}{0.287691}\right)^2\right] \cdot \cot g\left(\frac{60}{2}\right) = 0.0042599$$

7. The specific internal pressure drop of the flow at the *IN*:

$$Y_{i_{p_{in}}} = \lambda_{in} \cdot \frac{c_1^2}{2} = 0.0042599 \cdot \frac{13.556^2}{2} = 0.3914\ (\mathrm{J/kg})$$

8. The internal power lost from the flow at the *IN*:

$$P_{i_{p_{in}}} = Q \cdot Y_{i_{p_{in}}} = 3.9 \cdot 0.3914 = 1.527\ (\mathrm{kW})$$

9. The coefficient of kinetic energy losses at the *IN*:

$$\phi_1 = \sqrt{1 - \frac{2 \cdot Y_{i_{p_{in}}}}{c_1^2}} = \sqrt{1 - \frac{2 \cdot 0.3914}{13.556^2}} = 0.9979$$

10. The specific internal work at the output of the *IN*:

$$Y_{i_2} = \frac{c_{\max}^2}{2} = \frac{106.432^2}{2} = 5663.885\ (\mathrm{J/kg})$$

11. The internal power at the *IN* output:

$$P_{i_2} = Q \cdot Y_{i_2} = 3.9 \cdot 5663.885 = 22089.153\ (\mathrm{kW})$$

12. The internal power at the *IN* input:

$$P_{i_1} = P_{i_2} + P_{i_{p_{IN}}} = 22089.153 + 1.527 = 22090.680\ (\mathrm{kW})$$

13. The *IN* internal yield:

$$\eta_{i_{IN}} = 1 - \frac{P_{i_{P_{in}}}}{P_{i_1}} = 1 - \frac{1{,}527}{22090{,}680} = 0.9999 = 99.99\ (\%)$$

Since $\eta_{i_n} = \phi_1^2 = 0.9978$, the difference to the one thousandth comes from approximations made in the calculations.

8.5 The Quick Control Component

8.5.1 Considerations

The *QCC*, herein referred to as the Deflector (*DF*), is a component that partially or totally diverts the jet from the *RO* shells of the *PHIT* within any certain time, as established by the Project Team (*PT*), with partial or total load rejection in the *GG*, and also, when necessary, to maintain a partial load available after the *QCC* is activated.

The geometry for the *DF* operation and its positioning, considering a vertical view and the flow axis line, is shown in the diagrams in Fig. 1.5 from Chap. 1.

From these geometries and positioning, the section diagram in Fig. 8.11, shows the two main parts i.e., the *DF* and the fixed axis in the *IN*, if the movement system is located above the flow axis line, which will be called, *upper drive* and *top-down*

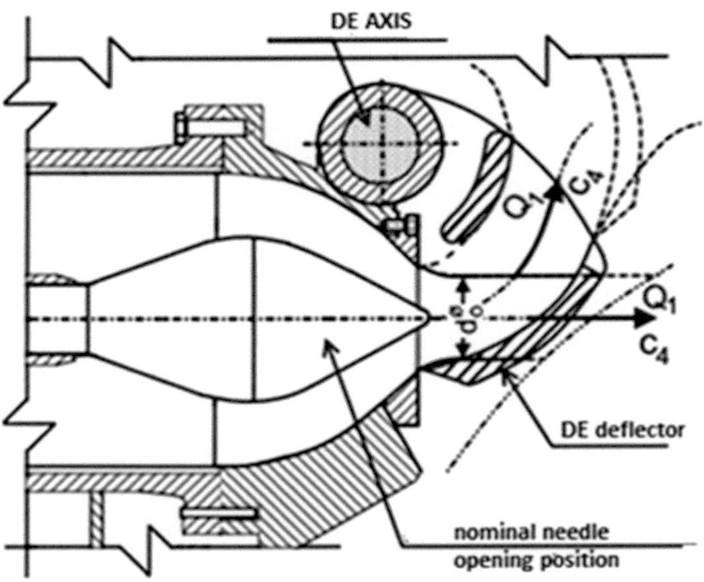

Fig. 8.11 *DE* cutaway diagram with *upper drive* and *top-down displacement*

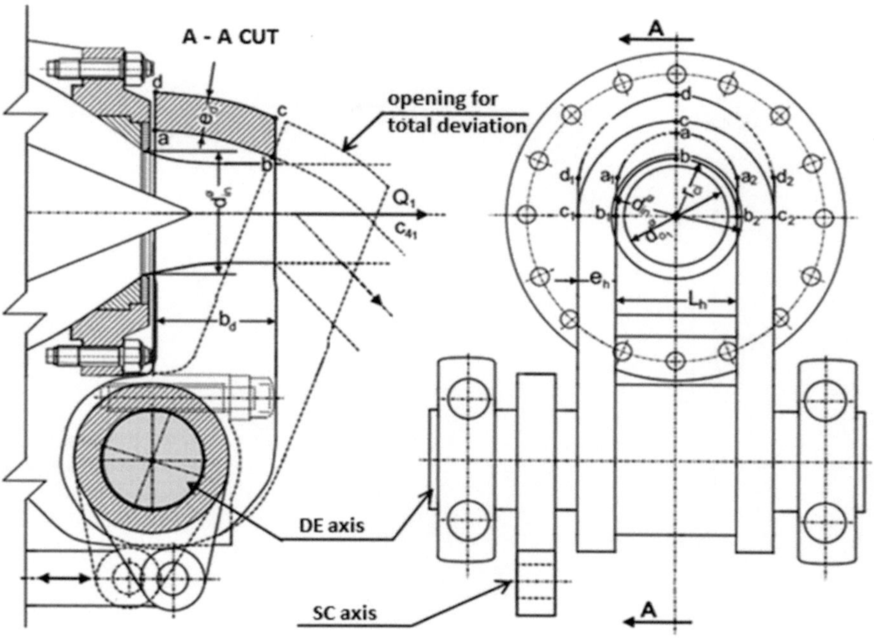

Fig. 8.12 Section view of a *DF* with a *bottom drive* and *top-down displacement*

displacement, i.e., the upper position, opening without jet deflection, and the lower position, for total deflection or eventual partial deflection.

Figure 8.12 shows a section view of another type of *DF* geometry with a *bottom drive* and *top-down displacement*. In this geometry, the *DF* displacement axis is housed at the lower part of the *IN*.

To visualize and better understand the *DF* projections drawn in Fig. 8.12, we schematized a projection of it in Fig. 8.13a, along with the forces that must be maintained for the total deflection of the maximum jet to occur.

Whatever the geometry established for the *DF*, the flow through the inner surface, over time $t_d(s)$, in which it deflects the entire jet, is chaotic with shocks and large fluctuations, substantially altering the average cross section, and practically covering the entire internal surface, which is called the wet surface.

8.5.2 Geometry and Equation

Considering the constant geometry in Fig. 8.13, we will show how an Equation can be made to obtaining a first approximation of the driving force, F_m, and the motor or driving moment, M_m, to keep the *DF* deflecting the jet completely, which is absolutely necessary for specifying the Dual Command System (*DCS*) that will command the double adjustment (deflector/needle) of the *PHIT*.

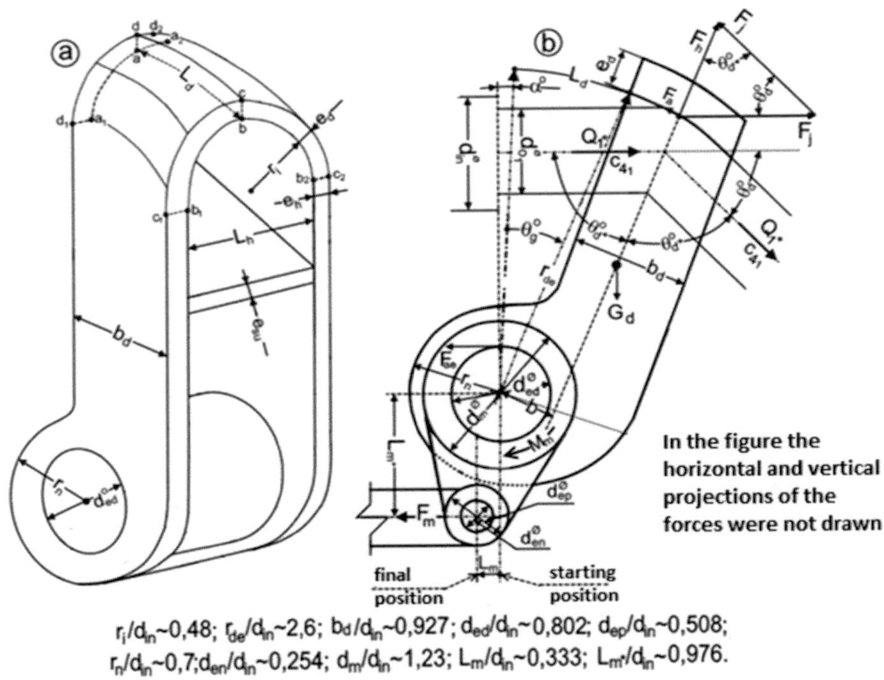

Fig. 8.13 (**a**) Schematic overview of the *DF* geometry. (**b**) Geometry, characteristics, and forces for the *DF* when positioned for full jet deflection

In the lower part of Fig. 8.13, there are values for some relations of these characteristics for this geometry, which can be used to start drafting compatibility between drawings and structural calculations for the *DF* and for fixing it in the *PHIT* structure.

The Equation of forces from the flow must have the following conditions as a starting point:

$$Q_{1*} = \frac{Q_{\max}}{z_j}\ \left(m^3/s\right) \rightarrow H_1 = (1 + h_s) \cdot H_{\max}\ (m) \rightarrow h_s - water\ hammer \quad (8.24)$$

Once these conditions have been established, the following Equation can be established:

- Average absolute velocity of the flow in the diameter of the jet d_{0_1}

$$c_{4_1} = \phi_{0;4} \cdot \sqrt{2 \cdot g \cdot H_1}\ (\mathrm{m/s}) \rightarrow 0.97 \leq \phi_{0;4} \leq 0.99 \rightarrow g = 9.81\ \left(\mathrm{m/s^2}\right) \quad (8.25)$$

- Average speed for turning the *DF* to fully deflect the jet:

$$c_g = \frac{180}{\pi \cdot t_d} \cdot r_{de} \cdot \theta_g^o \text{ (m/s)} \tag{8.26}$$

- Jet thrust force:

$$F_j = \rho \cdot Q_{1*} \cdot c_{4_1} \text{ (N)} \tag{8.27}$$

- Hydraulic force resulting from the flow over the *DF* and its projections:

$$F_h = 2 \cdot F_j \cdot \cos\theta_d \rightarrow F_{h_x} = F_h \cdot \cos\theta_{d*} \rightarrow F_{h_y} = F_h \cdot \sin\ \theta_{d*} \tag{8.28}$$

- Frictional force of the flow on the inner surface of the *DF*:

$$F_{a_h} = \mu_{aa} \cdot F_h \text{ (N)} \rightarrow \mu_{aa} \cong 0.01\text{–water/steel friction coefficient} \tag{8.29}$$

- Force resulting from flow and friction on the *DE*:

$$F_{r_d} = \sqrt{F_h^2 + F_{a_h}^2} \text{ (N)} \tag{8.30}$$

- Angle θ_{d*}^o that F_h forms with F_j:

$$\theta_{d*} = 0.5 \cdot \left(180 - \theta_d^o\right) \ (degrees) \tag{8.31}$$

- Friction force between the *DF* and its axis:

$$F_{ae} = \mu_e \cdot \left(F_{h_y} + F_{ah_y} - G\right) \ \text{(N)} \rightarrow \mu_{e_{est}} \cong 0.15 \rightarrow \mu_{e_{din}} \cong 0.06 \tag{8.32}$$

- Motor strength:

$$F_m = F_{h_x} - F_{ah_x} - F_{ae} \text{ (N)} \tag{8.33}$$

- Motor or driving moment:

$$M_m = F_m \cdot L_{m*} \text{ (N} \cdot \text{m)} \tag{8.34}$$

- Motor work:

$$\tau_m = F_m \cdot L_m \text{ (J)} \tag{8.35}$$

8.6 Free Flow Losses

In the stretch of the flow between the outlet of *IN* 2 and the outlet of the rotor shell 5, the flow absorbs surrounding air through the external surface via penetrating shocks and turbulence, thus transforming into a two-phase flow, increasing the dissolved gas content.

This absorption results in a small reduction in the average specific mass of the flow, which of course, will correspond to an energy loss.

A prediction of the value for this mass reduction or leakage water, both theoretically and practically, is highly uncertain, which is why, in the calculations that will be developed here, leakage performance, η_f, will be introduced as follows:

$$\eta_f = 1 - \frac{Q_f}{Q} \rightarrow Q_f = (1 - \eta_f) \cdot Q \tag{8.36}$$

For the purpose of calculating this loss, which also depends on the step between the shells, we recommend considering a value for this yield in the following range:

$$0.99 \leq \eta_f < 1.0 \tag{8.37}$$

Thus, the corresponding internal energy loss results in:

$$P_{i_{Pf}} = Q_f \cdot Y_{i_0} = (1 - \eta_f) \cdot Q \cdot Y_{i_0} = Q \cdot Y_{i_{Pf}} \text{ (kW)} \rightarrow Y_{i_{Pf}} = (1 - \eta_f) \cdot Y_{i_0} \text{ (J/kg)} \tag{8.38}$$

8.7 Losses in the Shells

In addition to the losses mentioned in the previous subitem, there are losses due to friction on the flow surfaces along the walls of the shells (boundary layer), with the surrounding air due to turbulence. In this flow, the boundary layers elongate in the direction of flow, according to the modeling shown in Figs. 7.2 and 7.3a.

These losses, concentrated in the shells by the flow, are difficult to determine, reduce the average relative velocity of the flow, and are predicted by introducing a dimensionless coefficient, $\psi_r = \psi_5 = \frac{w_5}{w_4}$.

Once this coefficient is established, we can determine the average flow velocity, $w5$, at the exit edge of the shell, the specific work, and internal power lost, along with the corresponding internal efficiency, as follows:

$$w_5 = \psi_5 \cdot w_4 (\text{m/s}) \rightarrow$$

$$Y_{i_{Pco}} = \frac{w_4^2}{2} - \frac{w_5^2}{2} = (1 - \psi_5^2) \cdot \frac{w_4^2}{2} \text{ (J/kg)} \rightarrow$$

$$P_{i_{Pco}} = Q \cdot Y_{i_{Pco}} \text{ (kW)} \rightarrow$$

$$\eta_{i_{co}} = 1 - \frac{P_{i_{P_{co}}}}{P_{i_4}} = 1 - \left(1 - \psi_5^2\right) = \psi_5^2 \tag{8.39}$$

The ψ_5 values should be within the following range:

$$0.96 \leq \psi_5 \leq 0.99 \tag{8.40}$$

The lowest values for shell surfaces with $\varepsilon > 0.01$ (mm), and for a low power *PHIT*, are 0.99, or even more for hydraulically smooth shell surfaces, $\varepsilon \cong 0.001$ (mm), and medium and high powered *PHIT*s.

It is important to note that these standards, in principle, do not consider the flow energy at the exit of the shells as being losses, unless agreed upon between the respective parties:

$$Y_s = 0.5 \cdot c_5^2 \ (\text{J/kg}) \rightarrow P_s = 10^3 \cdot Q \cdot Y_s \ (\text{kW}) \tag{8.41}$$

One should also consider that increases in d_o (large flows, low heads, larger wet surfaces) increase friction, while reductions in ψ_5, on the contrary, reduce friction, but increase turbulence, with possible cavitation and loss increases from the escape.

8.8 External Ventilation Losses

Exhaustive tests in accredited laboratories have been carried out to determine these losses, as well as their statistical Equations, and the results have been transcribed into Standards, which allow them to be used to prove contractual clauses, both in models and in tests.

Here, we will first transcribe the constant in Brazilian standard NB-228 from *ABNT* (registered in INMETRO as *NBR* 11,374 – *NBR* 3), which recommends the following formula for calculating the external ventilation losses in *PHIT RO* without housings:

$$P_{P_v} = 10^{-6} \cdot \rho_{ar} \cdot k_v \cdot n^3 \cdot d_{ex}^5 \ (kW) \tag{8.42}$$

Where:

- ρ_{air} (kg/m^3)—is the specific mass of air (1.32 kg/m^3 to 20 ° C);
- n (rpm)—is the *PHIT* rotor rotation;
- d_{ex} (m)—is the outside diameter of *PHIT RO* shell crown;
- k_v—is the ventilation loss coefficient that can be obtained from the following formulas, for cases without housings.

- **Integrally cast rotors:**

$$k_v = 0.01 + 12.40 \cdot \left(\frac{B_e}{d_{ex}}\right)^4 \tag{8.43}$$

- **Rotors with detachable shells:**

$$k_v = 0.35 + 0.25 \cdot \frac{B_e}{d_{ex}} \tag{8.44}$$

Where:

$B_{em}(m)$—is the maximum outer shell thickness.

Losses for similar rotors inside housings can be determined from losses obtained from formula (8.43) or (8.44), multiplied by the coefficient $k_{vp}/k_{vpo} < 1$ obtained from Figs. 52 and 53 on pages 70 and 71 of the aforementioned standard.

Standard CEI/*IEC* 41—3a edition—1991—11—pages 280/282 recommends using the following expressions to determine the *RO* ventilation losses of a *PHIT* with housings, with a degree of uncertainty of 50 (%), as per Figs. 8.14 and 8.15.

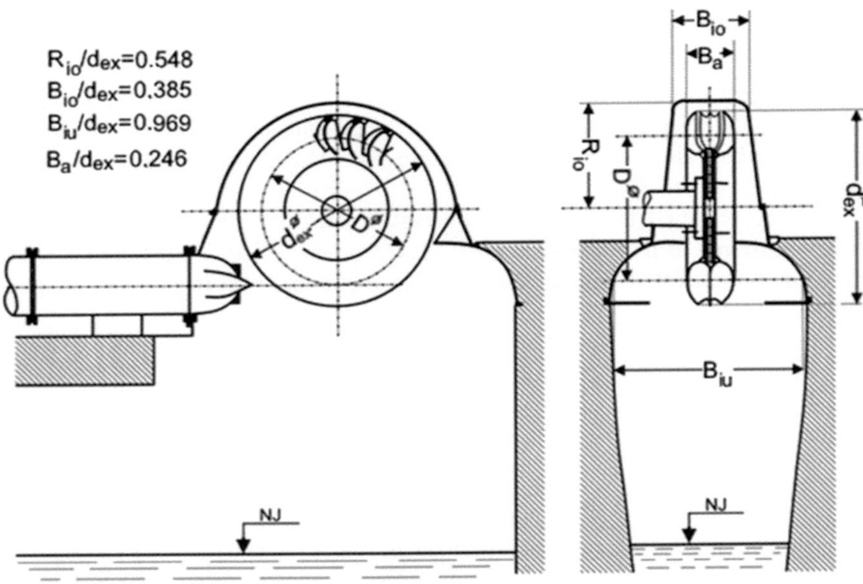

Fig. 8.14 Copy of Fig. 58 from the CEI/*IEC* 41 Standard

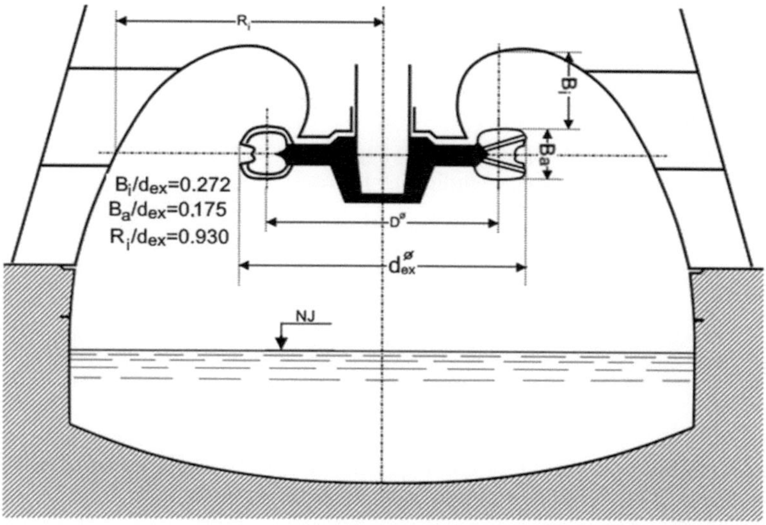

Fig. 8.15 Copy of Fig. 59 of the CEI/*IEC* 41 Standard

- **Losses for Horizontal Axis Rotors**—Fig. 8.14

$$P_{p_v}=15\cdot 10^{-3}\cdot n^3\cdot d_{ex}^5\cdot\left(\frac{B_a}{d_{ex}}\right)^{\frac{1}{4}}\cdot\left(\frac{B_{io}}{d_{ex}}\right)^{\frac{3}{4}}\cdot\left(\frac{B_{pu}}{d_{ex}}\right)^{\frac{5}{4}}\cdot\left(\frac{R_{io}}{d_{ex}}\right)^{\frac{7}{4}}\cong 1.12945\cdot 10^{-3}\cdot n^3\cdot d_{ex}^5(\mathrm{kW})$$

$$\rightarrow$$

$$n\left(s^{-1}\right), d_{ex}; R_{io}; B_a; B_{io}; B_{iu}\ (m) \tag{8.45}$$

According to this standard, the ventilation losses can be determined from the relations of the expressions obtained allowing for the proportionalities from Figs. 8.14 and 8.15, resulting in the respective numerical coefficients.

- **Losses for Vertical Axis Rotors**—Fig. 8.15

$$P_{p_v}=22\cdot 10^{-3}\cdot n^3\cdot d_{ex}^5\cdot\left(\frac{B_a}{d_{ex}}\right)^{2/3}\cdot\left(\frac{B_i}{d_{ex}}\right)^{4/3}\cdot\left(\frac{R_i}{d_{ex}}\right)\cong 4.62607\cdot 10^{-3}\cdot n^3\cdot D_e^5\ (kW)\rightarrow$$

$$n\left(s^{-1}\right), d_{ex}; R_i; B_a; B_i; (m) \tag{8.46}$$

8.9 Mechanical Losses

8.9.1 Horizontal Axis

The mechanical power lost in the rotating support system of the horizontal axis *PHIT* can be obtained, in preliminary studies, as follows:

- **Power lost due to friction in the bearings:**

$$P_{p_{am}} = 10^{-3} \cdot k_m \cdot G \cdot u_m \text{ (kW)} \tag{8.47}$$

Where:

- $G(N)$—is the sum of the weights of the *RO*, the axle, and the devices installed on it;
- u_m (m/s)—is tangential speed of the shaft in the plain bearing and in the center bearing of the rotating component (balls, rollers, bushings);
- k_m – is bearing type coefficient: plain bearing $= 0.002 \leq k_m \leq 0.003$;
- $k_m = 0.001$—is for a rolling bearing.

- **Power lost in gaskets (seals):**

$$P_{p_g} = 10^{-3} \cdot k_g \cdot A_g \cdot p_g \cdot u \text{ (kW)} \tag{8.48}$$

Where:

- $A_g = \pi \cdot d_g \cdot L_g$ (m^2)—is the friction surface of the gasket;
- d_g—is the diameter of the gasket;
- L_g—is the length of the gasket;
- p_g (N/m^2)—is the packing pressure, 10^6 to $2 \cdot 10^6$ (N/m^2);
- u (m/s)—is the tangential velocity of the shaft at the location of the packings;
- k_g—is the gasket coefficient, fixed in the range $0.001 \leq k_g \leq 0.005$.

- **Power lost by ventilation in disks installed on the shaft:**

$$P_{p_d} = 10^{-9} \cdot n^3 \cdot d_d^5 \cdot (1 + 1.8 \cdot B_d/d_d) \text{ (kW)} \tag{8.49}$$

Where:

- n (rpm)—is the disk rotation;
- d_d (m)—is the disk diameter;
- B_d (m)—is the disk width.

8.9.2 Vertical Axis

The mechanical power lost to friction in the rotating support system of the vertical axis *PHIT* occurs in the thrust bearing, and the CEI/*IEC* Standard 41—3a edition—1991—11—page. 288 recommends that one obtain this using the following formulas:

- **Plain thrust bearing:**

$$P_{p_{am}} = 58 \cdot 10^{-3} \cdot \mu \cdot n^2 \cdot d_m^3 \cdot L_m \cdot \delta^{-1} \ \text{(kW)} \tag{8.50}$$

- **Thrust bearing with pads**

$$P_{p_{am}} = 3.6 \cdot 10^{-3} \cdot k_s \cdot z_s \cdot \mu \cdot n^2 \cdot d_m^2 \cdot L_c \cdot b \cdot \delta^{-1} \ \text{(kW)} \tag{8.51}$$

Where:

- μ (Pa · s)—is the dynamic viscosity of the oil;
- n (rps)—is the rotation;
- d_m (m)—is the bearing diameter;
- L_m (m)– is the axial length of the bearing;
- δ (m)—is the full diametric span;
- k_s—is a constant that depends on the pad geometry, which can be adopted, in first approximation between 4 and 8;
- z_s—is the number of pads;
- L_c (m)—is the pad circumference length;
- $L_c = 0.8 \cdot \pi \cdot d_m \cdot z_s^{-1}$;
- b (m)—is the axial length of the pad.

- **Total mechanical power lost**

$$P_{p_m} = P_{p_{am}} + P_{p_g} + P_{p_d} \tag{8.52}$$

8.10 Powers and Yields

In this subitem, we will offer a synthesis on the conceptualization and form of the powers and yields that will be used in this study, considering the specific mass of water, $\rho = 1000\ (kg/m^3)$.

- **Internal power available in the flow or hydraulic fluid at the *PHIT* inlet:**

$$P_{i_0} = P_h = g \cdot Q \cdot H_0 = Q \cdot Y_{i_0}\ (kW) \tag{8.53}$$

- **Internal power at the *RO* output of the *PHIT*:**

$$P_{i_r} = Q_r \cdot Y_{i_r}\ (\mathrm{kW}) \tag{8.54}$$

  – **Internal yield of the *PHIT*:**

$$\eta_i = \frac{P_{i_r}}{P_{i_0}} = \frac{Q_r \cdot Y_{i_r}}{Q \cdot Y_{i_0}} = \frac{\eta_f \cdot Y_{i_r}}{Y_{i_0}} = 1 - \frac{\sum_{i=0}^{5} P_{i_{pi}}}{P_{i_0}} \tag{8.55}$$

- **The *PHIT* internal power considering kinetic flows at the *RO* outlet as energy losses:**

$$P_{i_{rs}} = Q_r \cdot \left(Y_{i_r} - 0.5 \cdot c_5^2\right) = Q_r \cdot \left(Y_{i_r} - Y_{i_{rs}}\right)\ (\mathrm{kW}) \tag{8.56}$$

  – **The internal yield of *PHIT* considering flow kinetics at the *RO* outlet as energy losses:**

$$\eta_{i_s} = \frac{P_{i_{rs}}}{P_{i_0}} = \frac{Q_r \cdot Y_{i_{rs}}}{Q \cdot Y_{i_0}} = \frac{\eta_f \cdot Y_{i_{rs}}}{Y_{i_0}} = 1 - \frac{\sum_{i=0}^{6} P_{i_{p_i}}}{P_{i_0}} \tag{8.57}$$

- **The external power of the *PHIT***

  This power does not include the flow power lost at the *RO* output.

$$P_{ex} = P_{i_r} - P_{p_v}\ (kW) \tag{8.58}$$

  – The external yield of *PHIT*:

$$\eta_{ex} = \frac{P_{ex}}{P_{i_0}} \tag{8.59}$$

- **External power of the *PHIT* considering the kinetic energy of the flow at the *RO* outlet as a loss:**

$$P_{ex_s} = P_{ex} - P_{p_v}\ (\mathrm{kW}) \tag{8.60}$$

– **External yield of the *PHIT* considering the kinetic energy of the flow at the exit of the *RO* as a loss:**

$$\eta_{ex_s} = \frac{P_{ex_s}}{P_{i_0}} \tag{8.61}$$

- **Power on the *PHIT* shaft:**
  This power does not include the power lost in the *PHIT* rotating system.

$$P_e = P_{ex} - P_{p_m}(\mathrm{kW}) \tag{8.62}$$

– **Efficiency on the shaft or outside the *PHIT*:**

$$\eta_e = \frac{P_e}{P_{i_0}} = \eta_i \cdot \eta_{ex} \tag{8.63}$$

- **Power on the *PHIT* axis considering the kinetic energy of the flow at the *RO* outlet as a loss:**

$$P_{e_s} = P_{ex_s} - P_{p_s}\ (\mathrm{kW}) \tag{8.64}$$

– **Yield on the *PHIT* axis considering the kinetic energy of the flow at the *RO* outlet as a loss:**

$$\eta_{e_s} = \frac{P_{e_s}}{P_{i_0}} \tag{8.65}$$

– **The Mechanical Performance of the *PHIT*:**

$$\eta_e = \frac{P_m}{P_e} \tag{8.66}$$

– **The total yield of the *PHIT*:**

$$\eta_t = \frac{P_m}{P_h = P_{i_0}} = \frac{P_e}{P_{i_0}} \cdot \frac{P_m}{P_e} = \eta_e \cdot \eta_m \tag{8.67}$$

– **The mechanical performance of the *PHIT* considering the kinetic energy of the flow at the rotor outlet as a loss:**

$$\eta_{m_s} = \frac{P_m}{P_{e_s}} \tag{8.68}$$

- ***GG* power with the *PHIT*:**

$$P_{gg} = \eta_{el} \cdot P_e \ (\text{kW}) \tag{8.69}$$

  - ***GG* yield with the *PHIT*:**

$$\eta_{gg} = \frac{P_{gg}}{P_{i_0}} = \eta_t \cdot \eta_{el} \ (\text{kW}) \tag{8.70}$$

- ***GG* power with *PHIT* considering the kinetic energy of the flow at the *RO* outlet as a loss:**

$$P_{gg_s} = \eta_{el} \cdot P_{e_s} \ (\text{kW}) \tag{8.71}$$

  - ***GG* yield with the *PHIT* considering the kinetic energy of the flow at the *RO* outlet as a loss:**

$$\eta_{gg_s} = \frac{P_{gg_s}}{P_{i_0}} \tag{8.72}$$

8.11 Algorithms for Determining Energy Losses in the *SA* of the *PHIT*

8.11.1 Objective

The purpose of establishing these algorithms and their validations is not only to numerically show their use, but also to make a comparison between the value of the internal losses of the specific work, Y_{i_p}, and the velocity coefficients $\varphi_{0;\ 1}$ between the models under study, with regards to the *SA* of *THP* only.

Therefore, in the validation, the imposition of constant values for the following input characteristics will be adopted as a criterion for all models, as well as for the necessary geometric quantities obtained respectively from Figs. 8.1, 8.2a, and 8.3c, drawn to scale, with reference to the diameters in meters, fixed for each of the figures.

For all three models, the following input data will be considered:

- $H = H_0 = 578$ (mca)
- $Q_1 = 5.562$ (m3/s)
- $n = 300$ (rpm)
- $c_0 = 8.645$ (m/s)
- $g = 9.81$ (m/s2)

- $\rho = 1000\ (\text{kg/m}^3)$
- $\varepsilon_t = 0.02\ (\text{mm})$

8.11.2 Model I—Fig. 8.1

8.11.2.1 Complementary Input Data

- $z_j = 1$
- $L_{tu} = 0.747$
- $h_e = 1.538$
- $h_{e_1} = 0.659$
- $h_{e_2} = 0.879$
- $r_c = 1.538$
- $d_1 = d_t = 0.905$
- $d_h = 0.209$
- $d_c = 0.439$
- $L_a = 0.404$
- $L_h = 1.745$
- $L_c = 1.745\ (\text{m})$
- $\theta_c = 65^\circ$
- $f_s = 1.0$
- $\lambda_r = 0.01$

8.11.2.2 Script and Calculations

1. The flow at the inlet:

$$Q = Q_1 \cdot z_j = 5.562 \cdot 1.0 = 5.562\ \left(\text{m}^3/\text{s}\right)$$

2. The specific internal work of the flow available at the inlet:

$$Y_{i_0} = g \cdot H_0 = 9.81 \cdot 578 = 5670.180\ (\text{J/kg})$$

3. The internal flow power available at the inlet:

$$P_{i_0} = Q \cdot Y_{i_0} = 5.562 \cdot 5670.18 = 31537.541\ (\text{kW})$$

4. The average static pressure energy at the inlet:

$$\frac{p_o}{\rho} = Y_{i_0} - \frac{c_o^2}{2} = 5670.18 - \frac{8.645^2}{2} = 5632.812\ (\text{J/kg})$$

5. The inlet pipe diameter:

$$d_t = d_0 = \sqrt{\frac{4 \cdot Q}{\pi \cdot c_0}} = \sqrt{\frac{4 \cdot 5.56}{\pi \cdot 8.645}} = 0.905 \text{ (m)}$$

6. The Reynolds number in the tube:

$$N_{Re_t} \cong 1.26 \cdot 10^6 \cdot \frac{Q_1}{d_t} = 1.26 \cdot 10^6 \cdot \frac{5.562}{0.905} = 7.372 \cdot 10^6$$

7. The energy loss coefficient in the tube:

$$\lambda_t = \frac{1.325}{\left[ln\left(\frac{\varepsilon_t}{3.7 \cdot d_t} + \frac{5.74}{N_{Re_t}^{0.9}}\right)\right]^2} = \lambda_t = \frac{1.325}{\left[ln\left(\frac{\varepsilon_t}{3.7 \cdot d_t} + \frac{5.74}{N_{Re_t}^{0.9}}\right)\right]^2} = 0.00954$$

8. The specific internal work of the flow lost in the tube:

$$Y_{i_{p_t}} = f_s \cdot 0.8106 \cdot \lambda_t \cdot Q^2 \cdot L_t \cdot d_t^{-5} = 1 \cdot 0.8106 \cdot 0.00954 \cdot 5.562^2 \cdot 0.747 \cdot 0.905^{-5}$$
$$= = 0.294 \text{ (J/kg)}$$

9. The internal flow power lost in the tube:

$$P_{i_{p_t}} = Q_1 \cdot Y_{i_{p_t}} = 5.562 \cdot 0.294 = 1.635 \text{ (kW)}$$

10. The internal yield of the tube:

$$\eta_{i_t} = 1 - \frac{P_{i_{p_t}}}{P_{i_0}} = 1 - \frac{1.635}{31537.541} = 0.9999 = 99.99\ (\%)$$

11. The energy loss coefficient on the curve:

$$\lambda_c = \frac{\theta_c}{90} \cdot \left[0.131 + 1.847 \cdot \left(\frac{d_c}{2 \cdot r_c}\right)^{3.5}\right] = \frac{65}{90} \cdot \left[0.131 + 1.847 \cdot \left(\frac{0.905}{2 \cdot 1.745}\right)^{3.5}\right]$$
$$= 0.1046$$

12. The specific internal work of the flow lost in the curve:

$$Y_{i_{p_c}} = f_s \cdot \lambda_c \cdot \frac{c_0^2}{2} = 1.0 \cdot 0.1046 \cdot \frac{8.645^2}{2} = 3.909 \text{ (J/kg)}$$

13. The internal flow power lost in the curve:

$$P_{i_{P_c}} = Q_1 \cdot Y_{i_{P_c}} = 5.562 \cdot 3.909 = 21.742\ (\mathrm{kW})$$

14. The hydraulic diameter of the flow in the tube with an inner rod:

$$D_h = (d_1 - d_h) = (0.905 - 0.29) = 0.615\ (\mathrm{m})$$

15. The Reynolds number of the flow in the tube with an inner rod:

$$N_{Re_h} \cong 1.26 \cdot 10^6 \cdot \frac{Q_1}{d_h} = 1.26 \cdot 10^6 \cdot \frac{5.562}{0.615} = 11.391 \cdot 10^6$$

16. The energy loss coefficient in the tube with an inner rod:

$$\lambda_h = \frac{1.325}{\left[ln\left(\frac{\varepsilon_t}{3.7 \cdot d_h} + \frac{5.74}{N_{Re_h}^{0.9}}\right)\right]^2} = \frac{1.325}{\left[ln\left(\frac{0.02}{3.7 \cdot 615} + \frac{5.74}{\left(11.391 \cdot 10^6\right)^{0.9}}\right)\right]^2} = 0.01528$$

17. The specific internal work of the flow lost in the tube with an internal rod:

$$Y_{i_{P_h}} = 0.8106 \cdot f_s \cdot \lambda_h \cdot Q_1^2 \cdot L_h \cdot d_h^{-5} = 0.8106 \cdot 1 \cdot 0,01528 \cdot 5562^2 \cdot 1.784 \cdot 0.615^5$$
$$= 7.770\ (\mathrm{J/kg})$$

18. The internal flow power lost in the tube with an internal rod:

$$P_{i_{P_h}} = Q_1 \cdot Y_{i_{P_h}} = 5.562 \cdot 7.770 = 43.214\ (\mathrm{kW})$$

19. The specific internal work of the flow lost from the difference in water column heights, h_e:

$$Y_{i_{P_{he}}} = g \cdot h_e = 9.81 \cdot 1.538 = 15.088\ (\mathrm{J/kg})$$

20. The internal power of the flow lost by the difference in water column heights, h_e:

$$P_{i_{P_{he}}} = Q_1 \cdot Y_{i_{P_{ne}}} = 5.562 \cdot 15.088 = 83.919\ (\mathrm{kW})$$

21. The sum, between section 0 and section 1 of the internal losses of specific work coming from the differences in heights:

$$Y_{i_{P0;1}} = Y_{i_{P_t}} + Y_{i_{P_c}} + Y_{i_{P_h}} + Y_{i_{P_{he}}} = 0.294 + 3.909 + 7.770 + 15.088 = 27.061\ (\mathrm{J/kg})$$

22. The total internal loss of available power between section 0 and section 1 arising from the difference in heights:

$$P_{i_{p0;\,1}} = Q_1 \cdot Y_{i_{pc0;\,1}} = 5.562 \cdot 27.061 = 150.513\ (\mathrm{kW})$$

23. The specific internal work available from the flow at the IN entrance:

$$Y_{i_1} = Y_{i_0} - Y_{i_{p0;\,1}} = 5670.180 - 27.061 = 5643.119\ (\mathrm{J/kg})$$

24. The internal performance of the system between section 0 and section 1:

$$\eta_{i0;\,1} = 1 - \frac{P_{i_{p0;\,1}}}{P_{i_0}} = 1 - \frac{150.513}{5643.119} = 0.9733 = 97.33\ (\%)$$

25. Energy loss coefficient between section 0 and section 1:

$$\phi_{0;\,1} = \sqrt{\frac{Y_{i_1}}{Y_{i_0}}} = \sqrt{\frac{5643.119}{5670.180}} = 0.9976$$

26. The specific internal work of the run lost by the difference in the water column heights h_{e_1}:

$$Y_{i_{ph_{e1}}} = g \cdot h_{e_1} = 9.81 \cdot 0.659 = 6.465\ (\mathrm{J/kg})$$

27. The internal power of the flow lost by the difference in the water column heights h_{e_1}:

$$P_{i_{ph_{e1}}} = Q_1 \cdot Y_{i_{pn_{e1}}} = 5.562 \cdot 6.465 = 35.948\ (\mathrm{kW})$$

28. The internal flow specific work available at the entrance of the curve:

$$Y_{i_{ec}} = Y_{i_0} - \left(Y_{i_{ptu}} + Y_{i_{ph_{e1}}}\right) = 5670.180 - (0.290 + 6.465) = 5663.425\ (\mathrm{J/kg})$$

29. The internal flow power available at the entrance of the curve:

$$P_{i_{ec}} = Q_1 \cdot Y_{i_{ec}} = 5.562 \cdot 5663.425 = 31499.970\ (\mathrm{kW})$$

30. The average static pressure energy of the flow at the entrance to the curve:

$$\frac{p_{e_c}}{\rho} = Y_{i_{ec}} - \frac{c_o^2}{2} = 5663.425 - \frac{8.645^2}{2} = 5626.057\ (\mathrm{J/kg})$$

31. The internal yield of the curve:

$$\eta_{i_c} = 1 - \frac{P_{i_{pc}}}{P_{i_{ec}}} = 1 - \frac{21.742}{31499.97} = 0.9993 = 99.93\ (\%)$$

32. The specific internal work of the flow lost by the difference in water column heights h_{e_2}:

$$Y_{i_{p_{he2}}} = g \cdot h_{e_2} = 9.81 \cdot 0.879 = 8.623\ (\mathrm{J/kg})$$

33. The internal power of the flow lost by the difference in water column heights:

$$P_{i_{p_{he2}}} = Q_1 \cdot Y_{i_{pn2}} = 5.562 \cdot 8.623 = 47.961\ (\mathrm{kW})$$

34. The internal flow power available at the exit of the curve:

$$P_{i_{sc}} = P_{i_{ec}} - \left(P_{i_{pc}} + P_{i_{p_{h_2}}}\right) = 31499.97 - (21.742 + 47.961) = 31430.267\ (\mathrm{kW})$$

35. The specific internal work of the flow at the exit of the curve:

$$Y_{i_{sc}} = \frac{P_{i_{sc}}}{Q_1} = \frac{31430.267}{5.562} = 5650.893\ (\mathrm{J/kg})$$

36. The average flow velocity at the exit of the curve:

$$c_{s_c} = \frac{4 \cdot Q_1}{\pi \cdot \left(d_1^2 - d_h^2\right)} = c_{s_c} = \frac{4 \cdot 5.562}{\pi \cdot \left(0.905^2 - 0.209^2\right)} = 9.134\ (\mathrm{m/s})$$

37. The average static pressure energy of the flow at the exit of the curve:

$$\frac{p_{e_{sc}}}{\rho} = Y_{i_{ec}} - \frac{c_{s_c}}{2} = 5663.425 - \frac{9.134^2}{2} = 5621.710\ (\mathrm{J/kg})$$

38. The internal flow yield between section 0 and section 1:

$$\eta_{i_{0;1}} = 1 - \frac{P_{i_{sh}}}{P_{i_{sc}}} = 1 - \frac{43.214}{31430.267} = 0.9986 = 99.86\ (\%)$$

39. The internal flow power available at the *IN* entrance:

$$P_{i_1} = P_{i_{sc}} - P_{i_{p_h}} = 31430.267 - 43.214 = 31387.053\ (\mathrm{kW})$$

40. The specific work of the flow available at the entrance of the *IN*:

$$Y_{i_1} = \frac{P_{i_1}}{Q_1} = \frac{31375.817}{5.562} = 5641.103\ (\text{kW})$$

41. The average flow velocity at the *IN* inlet:

$$c_1 = \frac{4 \cdot Q_1}{\pi \cdot \left(d_1^2 - d_c^2\right)} = c_1 = \frac{4 \cdot 5.562}{\pi \cdot \left(0.905^2 - 0.439\right)} = 11.307\ (\text{m/s})$$

42. The average static pressure energy of the flow at the *IN* inlet:

$$\frac{p_{e_1}}{\rho} = Y_{i_1} - \frac{c_1}{2} = 5641.103 - \frac{11.307^2}{2} = 5577.179\ (\text{J/kg})$$

8.11.3 *Model II—Fig. 8.2^a^*

8.11.3.1 Input Data

- $z_j = 2$
- $h_1 = h2 = 1.010$
- $h_3 = 0.763$
- $h_4 = 0.788$
- $L_{t_1} = L_{t_2} = 1.132$
- $L_c = 1.602$
- $L_{c_2} = 1.299$
- $L_{c_3} = 0.644$
- $L_{t_3} = 0.523$
- $L_a = 1.132$
- $d_a = 0.427$
- $r_c = 1.329$
- $r_1 = 0.615$
- $r_2 = 1.329$
- $r_3 = 0.615(m)$
- $\lambda_r \cong 0.01$
- $\lambda_b \cong 0.05$
- $\theta_1 = 60^\circ$
- $\theta_2 = 56^\circ$
- $\theta_3 = 60^\circ$
- $f_s = 1$

The upper stretch was taken as a reference.

We calculated:

$$A_a = \frac{\pi}{4} \cdot \left(d_5^2 - d_a^2\right) = \frac{\pi}{4} \cdot \left(0.905^2 - 0.472^2\right) = 0.468287\ \left(\mathrm{m}^2\right)$$

$$c_a = \frac{Q_5}{A_a} = \frac{5.562}{0.468287} = 11.876\ (\mathrm{m/s})$$

8.11.3.2 Script—Calculations

1. The specific internal work of the flow available at the inlet:

$$Y_{i_0} = g \cdot H_0 = 9.81 \cdot 578 = 5670.180\ (\mathrm{J/kg})$$

2. The internal flow power available at the inlet for each jet:

$$P_{i_{0_1}} = Q_1 \cdot Y_{i_0} = 5.562 \cdot 5670.18 = 31537.541\ (\mathrm{kW})$$

3. The average static pressure energy of the flow at the inlet:

$$\frac{p_0}{\rho} = Y_{i_0} - \frac{c_o^2}{2} = 5670.18 - \frac{8.645^2}{2} = 5632.812\ (\mathrm{J/kg})$$

4. The internal specific work lost from the flow at the top *r* of the bifurcation:

$$Y_{i_{p_r}} = \lambda_r \cdot \frac{c_0^2}{2} \cong 0.01 \cdot \frac{8.645^2}{2} = 0.374\ (\mathrm{J/kg})$$

5. The internal flow power lost at the top *r* of the bifurcation:

$$P_{i_{p_r}} = Q_1 \cdot Y_{i_{p_r}} = 5.562 \cdot 0.374 = 2.080\ (\mathrm{kW})$$

6. The diameter of the tubes:

$$d_{t_1} = d_{t_2} = d_{t_3} = \sqrt{\frac{4 \cdot Q_1}{\pi \cdot c_0}} = \sqrt{\frac{4 \cdot 5{,}562}{\pi \cdot 8{,}645}} = 0.905\ (\mathrm{m})$$

7. The internal specific work lost from the flow at the deflection *d* part of the bifurcation:

$$Y_{i_{p_d}} = \lambda_d \cdot \frac{c_0^2}{2} \cong 0.05 \cdot \frac{8.645^2}{2} = 1.868\ (\mathrm{J/kg})$$

8. The internal power of the flow lost in the deflection d part of the bifurcation:

$$P_{i_{P_d}} = Q_1 \cdot Y_{i_{P_d}} = 5.562 \cdot 1.868 = 10.390 \text{ (kW)}$$

9. The internal yield of the bifurcation:

$$\eta_{i_b} = 1 - \frac{P_{i_{P_r}} + P_{i_{P_d}}}{P_{i_0}} = 1 - \frac{2.08 + 10.39}{31537.541} = 0.9996 = 99.96\ (\%)$$

10. The Reynolds number of the flow in the tube L_{t_2}:

$$N_{Re_{t_2}} \cong 1.26 \cdot 10^6 \cdot \frac{Q_1}{d_{t_2}} = 1.26 \cdot 10^6 \cdot \frac{5.56}{0.905} = 7.741 \cdot 10^6$$

11. The energy loss coefficient of the flow in the tube L_{t_2}:

$$\lambda_{t_2} = \frac{1.325}{\left[ln\left(\frac{\varepsilon_t}{3.7 \cdot d_{t_2}} + \frac{5.74}{N_{Re_{t_2}}^{0.9}}\right)\right]} = \frac{1.325}{\left[ln\left(\frac{0.02}{3.7 \cdot 905} + \frac{5.74}{(7741000)^{0.9}}\right)\right]^2} = 0.009925$$

12. The specific internal work of the flow lost in the tube L_{t_2}:

$$\begin{aligned} Y_{i_{P_{t_2}}} &= 0.8106 \cdot f_s \cdot \lambda_{t_2} \cdot Q_1^2 \cdot L_{t_2} \cdot D_{t_2}^{-5} \\ &= = 0.8106 \cdot 1 \cdot 0.009925 \cdot 5.562^2 \cdot 1.132 \cdot 0.905^{-5} = = 0.464 \text{ (J/kg)} \end{aligned}$$

13. The internal power of the flow lost in the tube L_{t_2}:

$$P_{i_{P_{t_2}}} = Q_1 \cdot Y_{i_{P_{t2}}} = 5.562 \cdot 0.464 = 2.581 \text{ (kW)}$$

14. The internal yield of the tube L_{t_2}:

$$\eta_{i_{t_2}} = 1 - \frac{P_{i_{P_{t_2}}}}{P_{i_0}} = 1 - \frac{2.581}{31527.541} = 0.9999 = 99.99\ (\%)$$

15. The coefficient of energy loss of the flow in the first part of the curve:

$$\lambda_2 = \frac{\theta_2}{90} \cdot \left[0.131 + 1.847 \cdot \left(\frac{d_{t_2}}{2 \cdot r_2}\right)^{3.5}\right] = \frac{56}{90} \cdot \left[0.131 + 1.847 \cdot \left(\frac{0.905}{2 \cdot 1.329}\right)^{3.5}\right] = 0.0265$$

16. The specific internal work of the flow lost in the first part of the curve:

$$Y_{i_{p_2}} = \lambda_2 \cdot \frac{c_0^2}{2} = 0.0265 \cdot \frac{8.645^2}{2} = 0.990\ (\mathrm{J/kg})$$

17. The internal power of the flow lost in the first part of the curve:

$$P_{i_{p_2}} = Q_1 \cdot Y_{i_{p_2}} = 5.562 \cdot 0.990 = 5.506\ (\mathrm{kW})$$

18. The energy loss coefficient in the second part of the curve:

$$\lambda_3 = \frac{\theta_3}{90} \cdot \left[0.131 + 1.847 \cdot \left(\frac{d_{t_2}}{2 \cdot r_3}\right)^{3.5}\right] = \frac{60}{90} \cdot \left[0.131 + 1.847 \cdot \left(\frac{0.905}{2 \cdot 0{,}615}\right)^{3.5}\right] = 0.50804$$

19. The specific internal work of the flow lost in the second part of the curve:

$$Y_{i_{p_3}} = f_s \cdot \lambda_3 \cdot \frac{c_0^2}{2} = 1 \cdot 0.50804 \cdot \frac{8.645^2}{2} = 18.984\ (\mathrm{J/kg})$$

20. The internal power of the flow lost in the second part of the curve:

$$P_{i_{p_3}} = Q_1 \cdot Y_{i_{p_3}} = 5.562 \cdot 18.984 = 105.589\ (\mathrm{kW})$$

21. The specific internal work of the flow lost across the curve:

$$Y_{i_{p_c}} = Y_{i_{p_2}} + Y_{i_{p_3}} = 0.990 + 18.984 = 19.974\ (\mathrm{J/kg})$$

22. The internal power lost throughout the curve:

$$P_{i_{p_c}} = Q_1 \cdot Y_{i_{p_c}} = 5.562 \cdot 19.974 = 111.095\ (\mathrm{kW})$$

23. The specific internal work of the flow lost in the tube L_{t_3}:

$$\begin{aligned} Y_{i_{p_{t_3}}} &= 0.8106 \cdot f_s \cdot \lambda_{t_3} \cdot Q_1^2 \cdot L_{t_3} \cdot D_{t_3}^{-5} \\ &= 0.8106 \cdot 1 \cdot 0.009925 \cdot 5.562^2 \cdot 0.523 \cdot 0.905^{-5} = = 0.241\ (\mathrm{J/kg}) \end{aligned}$$

24. The internal power of the flow lost in the tube L_{t_3}:

$$P_{i_{p_{t_3}}} = Q_1 \cdot Y_{i_{p_{t_3}}} = 5.562 \cdot 0.214 = 1.341\ (\mathrm{kW})$$

25. The Reynolds number of the flow in the tube with the inner needle:

$$N_{Re_a} \cong 1.26 \cdot 10^6 \cdot \frac{Q_1}{d_h} = 1.26 \cdot 10^6 \cdot \frac{5.562}{0.478} = 14.656 \cdot 10^6$$

26. The energy loss coefficient in the tube with internal needle:

$$\lambda_a = \frac{1.325}{\left[ln\left(\frac{\varepsilon_t}{3.7 \cdot d_h} + \frac{5.74}{N_{Re_a}^{0.9}}\right)\right]^2} = \frac{1.325}{\left[ln\left(\frac{0.02}{3.7 \cdot 478} + \frac{5.74}{\left(14.656 \cdot 10^6\right)^{0.9}}\right)\right]^2} = 0.010517$$

27. The specific internal work of the flow lost in the tube with an internal needle:

$$\begin{aligned} Y_{i_{Pa}} &= 0.8106 \cdot f_s \cdot \lambda_a \cdot Q_1^2 \cdot L_{h_2} \cdot d_h^{-5} \\ &= 0.8106 \cdot 1 \cdot 0.0105517 \cdot 5.562^2 \cdot 1.132 \cdot 0.478^5 = 11.964\ (\text{J/kg}) \end{aligned}$$

28. The internal flow power lost in the tube with a needle:

$$P_{i_{Pa}} = Q_1 \cdot Y_{i_{Pa}} = 5.562 \cdot 11.964 = 66.542\ (\text{kW})$$

29. The specific internal work of the flow lost by the difference in water column heights, $h_e = h_1 + h_2 + h_3 - h_4 = 1.01 + 1.01 + 0.763 - 0.788 = 1.995\ (m)$:

$$Y_{i_{Phe}} = g \cdot h_e = 9.81 \cdot 1.995 = 19.571\ (\text{J/kg})$$

30. The internal power of the flow lost by the difference in water column heights, $h_e = 0.975\ (m)$:

$$P_{i_{Phe}} = Q_1 \cdot Y_{i_{Pne}} = 5.562 \cdot 19.571 = 108.854\ (\text{kW})$$

31. The total internal loss of specific work available between section 0 and section 1:

$$\begin{aligned} Y_{i_{Pct}} &= Y_{i_{Pr}} + Y_{i_{Pt_2}} + Y_{i_c} + Y_{i_{Pa}} + Y_{i_{Phe}} \\ &= 0.374 + 0.464 + 19.974 + 11.964 + 19.571 = 52.347\ \ (\text{J/kg}) \end{aligned}$$

32. The total internal loss of available power in the flow between section 0 and section 1:

$$P_{i_{Pct}} = Q_1 \cdot Y_{i_{Pct}} = 5.562 \cdot 52.347 = 291.154\ (\text{kW})$$

33. The specific internal work available from the flow at the entrance of each *IN*:

$$Y_{i_1} = Y_{i_0} - Y_{i_{Pct}} = 5670.180 - 52.237 = 5617.943\ (\mathrm{J/kg})$$

34. The internal power available at the input of each *IN*:

$$P_{i_1} = Q_1 \cdot Y_{i_1} = 5.562 \cdot 5617.943 = 31246.999\ (\mathrm{kW})$$

35. The internal yield between section 0 and section 1:

$$\eta_{i0;1} = 1 - \frac{P_{i_{Pct}}}{P_{i_0}} = 1 - \frac{291.154}{31246.999} = 0.9907 = 99.07\ (\%)$$

36. Coefficient of energy loss of the flow between section 0 and section 1:

$$\phi_{0;1} = \sqrt{\frac{Y_{i_1}}{Y_{i_0}}} = \sqrt{\frac{5617.945}{5670.180}} = 0.9954$$

37. The specific internal work of the flow lost by the difference in water column heights, $h_1 = 1.010\ (m)$:

$$Y_{i_{Ph_1}} = g \cdot h_1 = 9.81 \cdot 1.01 = 9.908\ (\mathrm{J/kg})$$

38. The internal power of the flow lost by the difference in water column heights, $h_1 = 1.010\ (m)$:

$$P_{i_{Ph_1}} = Q_1 \cdot Y_{i_{Ph_1}} = 5.562 \cdot 9.908 = 55.108\ (\mathrm{kW})$$

39. Internal flow specific work available at the pipe inlet h_2:

$$Y_{i_{eh_2}} = Y_{i_0} - \left(Y_{i_{Pr}} + Y_{i_{Ph_1}}\right) = 5670.180 - (0.374 + 9.908) = 5659.898\ (\mathrm{J/kg})$$

40. The internal flow power available at the pipe inlet h_2:

$$P_{i_{eh_2}} = Q_1 \cdot Y_{i_{eh_2}} = 5.562 \cdot 5659.898 = 31480.353\ (\mathrm{kW})$$

41. The static pressure energy at the pipe inlet h_2:

$$\frac{p_{eh2}}{\rho} = Y_{i_{eh_2}} - \frac{c_o^2}{2} = 5659.898 - \frac{8.645^2}{2} = 5622.530\ (\mathrm{J/kg})$$

42. The specific internal work of the flow lost from the difference in water column heights $h_2 = 1.010\ (m)$:

$$Y_{i_{Ph_2}} = g \cdot h_2 = 9.81 \cdot 1.01 = 9.908\ (\mathrm{J/kg})$$

43. The internal power of the flow lost from the difference in water column heights, $h_2 = 1.010\ (m)$:

$$P_{i_{Ph_2}} = Q_1 \cdot Y_{i_{Ph_2}} = 5.562 \cdot 9.908 = 55.108\ (\mathrm{kW})$$

44. Internal flow specific work available at the entrance of curve 2:

$$Y_{i_{ec_2}} = Y_{i_{eh_2}} - \left(Y_{i_{Pt_2}} + Y_{i_{Ph_2}}\right) = 5659.898 - (0.464 + 9.908) = 5649.526\ (\mathrm{J/kg})$$

45. The internal flow power available at the entrance of curve 2:

$$P_{i_{ec}} = Q_1 \cdot Y_{i_{ec}} = 5.562 \cdot 5649.526 = 31422.664\ (\mathrm{kW})$$

46. The average static pressure energy of the flow at the entrance of the curve:

$$\frac{p_{e_c}}{\rho} = Y_{i_{ec}} - \frac{c_o^2}{2} = 5649.526 - \frac{8.645^2}{2} = 5612.158\ (\mathrm{J/kg})$$

47. The specific internal work of the flow lost from the difference in elevation, $h_3 = 0.763$ (m):

$$Y_{i_{Ph_3}} = g \cdot h_3 = 9.81 \cdot 0.763 = 7.485\ (\mathrm{J/kg})$$

48. The internal power of the flow lost from the difference in water column heights, $h_3 = 0.763$ (m):

$$P_{i_{Ph_3}} = Q_1 \cdot Y_{i_{Ph_3}} = 5.562 \cdot 7.485 = 41.632\ (\mathrm{kW})$$

49. The specific internal work available in the flow at the exit of curve 3:

$$Y_{i_{sc_3}} = Y_{i_{ec_2}} - \left(Y_{i_{Pc}} + Y_{i_{Ph_3}}\right) = 5649.526 - (19.974 + 7.485) = 5622.067\ (\mathrm{J/kg})$$

50. The static pressure energy of the flow at the exit of curve 3:

$$\frac{p_{e_{sc}}}{\rho} = Y_{i_{sc}} - \frac{c_0^2}{2} = 5622.067 - \frac{8.645^2}{2} = 5584.699\ (\mathrm{J/kg})$$

51. The static pressure energy of the flow at the *IN* inlet:

$$\frac{p_{e_1}}{\rho} = Y_{i_1} - \frac{c_1}{2} = 5627.381 - \frac{11.187^2}{2} = 5564.807\ (\mathrm{J/kg})$$

8.11.4 Model III—Fig. 8.3c

8.11.4.1 Input Data

$$z_j = 6$$

Constant values in Table 8.1. θ_j^o was measured in Fig. 8.3c, calculated by proportion, $d_a = 0.472$ (m), and the rest observe the step items. The following was also calculated:

$$d_0 = \sqrt{\frac{4 \cdot Q_0}{\pi \cdot c_0}} = \sqrt{\frac{4 \cdot 33.37}{\pi \cdot 8.645}} = 2.217$$

$$d_h = d_5 - d_a = 0.905 - 0.427 = 0.478\ (\mathrm{m})$$

$$A_a = \frac{\pi}{4} \cdot \left(d_5^2 - d_a^2\right) = \frac{\pi}{4} \cdot \left(0.905^2 - 0.427^2\right) = 0.50006\ \left(\mathrm{m}^2\right)$$

$$c_a = \frac{Q_5}{A_a} = \frac{5.56}{0.50006} = 1.187\ (\mathrm{m/s})$$

8.11.4.2 Script and Calculations

1. The specific internal work of the flow available at the inlet:

$$Y_{i_0} = g \cdot H_0 = 9.81 \cdot 578 = 5670.18\ (\mathrm{J/kg})$$

2. The internal flow power available at the inlet:

$$P_{i_0} = Q_0 \cdot Y_{i_0} = 33.37 \cdot 5670.18 = 189213.907\ (\mathrm{kW})$$

3. Identify each section between section 0 and section 1 with the highest energy loss named in column *a* in Table 8.5.
4. The flow in each section:

Table 8.5 Characteristics taken from Fig. 8.3

a	b	c	d	e	f	g	h	i	j
Caract.	Q_j	d_j	L_j	***Lr*1**	θ_j	Yi_{pr1}	***Pi* pr1**	***Nre*. 10^6**	**Δtr**
Unid.	m³/s	m	m	m	Degrees	J/kg	kW	–	–
0.0000	33.37	2.22	–	–	–	–	–	–	–
			1.24	–	0	–	–	189.60	0.01
1.0000	27.81	2.02	3.41	3.79	14.5	0.37	10.40	173.06	0.01
2.0000	22.25	1.81	3.32	3.66	18	0.37	8.32	154.82	0.01
3.0000	16.68	1.57	3.30	3.25	15	0.37	6.24	134.12	0.01
4.0000	11.12	1.28	3.32	3.52	7	0.37	4.16	109.46	0.01
5.0000	5.56	0.91	7.16	3.45	21	0.37	2.09	77.41	0.01
					34				
					34				
Agulha	5.56	0.48	1.13	–	0	–	–	14.66	0.01
Σ	–	–	22.88	17.67	34	1.87	31.21	–	–

k	l	m	n	o	p	q
Yi_{rg1}	Pi_{ptr}	$\lambda d1$	Yi_{pr1}	Pi_{pr1}	Yi_{br}	***P*1/*P***
J/kg	kW	–	J/kg	kW	J/kg	J/kg
–	–	–	–	–	5670.18	5632.81
0.18	6.01	0.00	0.00	0.00	5670.00	5632.63
0.55	15.21	0.02	0.58	16.18	5668.50	5631.13
0.60	13.30	0.02	0.91	20.24	5666.62	5629.25
0.71	11.78	0.02	0.62	10.41	5664.91	5627.54
0.91	10.09	0.00	0.13	1.48	5663.50	5626.13
2.93	16.31	0.03	1.26	6.99	5651.78	5614.41
		0.10	3.58	19.89		
		0.10	3.58	19.89		
11.96	66.47	0.00	0.00	0.00	5639.82	5577.25
17.83	139.16	–	10.66	95.08	–	–

$$Q_j = Q_{j-1} - Q_0/z_j = Q_{j-1} - 33.37/6 = Q_{j-1} - 5.562\ \left(\mathrm{m}^3/\mathrm{s}\right)$$

… transcribing the values in column *b*.

5. The average diameter of each section between section 0 and section 1:

$$d_j = \sqrt{\frac{4 \cdot Q_j}{\pi \cdot c_0}} = \sqrt{\frac{4 \cdot Q_j}{\pi \cdot 8.645}} = 0.38377 \cdot \sqrt{Q_j}\ (\mathrm{m})$$

… transcribing the values in column *c*.

6. The average scale coefficient:

$$k = \sum_0^5 \left(\frac{d_j}{d_{j_{fig.8.3}}} \right) = 0.4225 \text{ (m/cm)}$$

The values for $d_{j_{fig.3}}$ (cm) were obtained from Fig. 8.3c, and plotted to scale.

7. The length of the tubes:

$$L_j = k \cdot L_{j_{fig.8.3}}$$

were calculated from measuring the sections in tubes $L_{j_{fig.8.3}}$ (cm) in Fig. 8.3c, plotted to scale, and then transcribing the respective values in (m) in column *d*, as well as the sum in the last line of the column.

8. The length values L_{r_j} from the *ACL* of the bifurcations in the *r* direction read in (cm) in Fig. 8.3, and transcribing the values in column *e* in (m), as well as the sum in the last line of this column.
9. The angles θ_j^o of the directions of the pipes measured in Fig. 8.3c, and transcribing the values in column *f*.
10. The specific internal work lost from the flow in the upper part *r* of the bifurcations:

$$Y_{i_{Pr_j}} = \lambda_r \cdot \frac{c_0^2}{2} \cong 0.01 \cdot \frac{8.645^2}{2} = 0.374 \; (J/kg) = const$$

... transcribing the values in column *g* and the sum in the last line of column *g*.

11. The internal powers of the flow lost in the upper part *r* of the bifurcations:

$$P_{i_{Pr_j}} = Q_j \cdot Y_{i_{Pr_j}} = 0.374 \cdot Q_j \; (kW)$$

... transcribing the values in column *h* and the sum in the last line of this column.

12. The Reynolds numbers of the flow in the tube:

$$N_{Re_{t_j}} \cong 1,26 \cdot 10^6 \cdot \frac{Q_j}{d_j}$$

... transcribing the values in column *i*.

13. The energy loss coefficients of the flow in the sections:

$$\lambda_{t_j} = \frac{1.325}{\left[ln \left(\frac{\varepsilon_{tu}}{3.7 \cdot d_{t_j}} + \frac{5.74}{N_{Re\,tj}^{0.9}} \right) \right]} = \frac{1.325}{\left[ln \left(\frac{0.02}{3.7 \cdot d_t} + \frac{5.74}{N_{Re_j}{}^{0.9}} \right) \right]^2}$$

...transcribing the values in column *j*.

14. The specific internal work of the flow lost in the tube sections:

$$Y_{i_{P_{tj}}} = 0.8106 \cdot f_s \cdot \lambda_{t_j} \cdot Q_j^2 \cdot L_{t_j} \cdot d_j^{-5}\ (J/kg)$$

… transcribing the values in column *k* and their and their sum in the last line of this column.

15. The internal powers of the flow lost in the tube sections:

$$P_{i_{P_{tj}}} = Q_j \cdot Y_{i_{P_{tj}}}\ (kW)$$

…transcribing the values in column 1 and their sum in the last line of this column.

16. The energy loss coefficients of the flow arising from the deflections of the tube sections:

$$\lambda_{d_j} = 0.9457 \cdot \sin^2 0.5\theta_{d_j} + 2.05 \cdot \sin^4 0.5\theta_{d_j}$$

…transcribing the values in column *m*.

17. The specific internal work of the flow lost in the deflections of the tube sections:

$$Y_{i_{P_{d_j}}} = \lambda_{d_j} \cdot f_s \cdot \frac{c_0^2}{2} = \lambda_{d_j} \cdot 1 \cdot \frac{8.645^2}{2} = 37.368 \cdot \lambda_{d_j}\ (\text{J/kg})$$

… transcribing the values in column n and their sum in the last line of this column.

18. The internal powers of the flow lost in the deflections of the tube sections:

$$P_{i_{P_{d_j}}} = Q_j \cdot Y_{i_{P_{d_j}}}\ (\text{kW})$$

…transcribing the values in the column and their sum in the last line of this column.

19. The total internal specific work of the flow lost between section 0 and section 1:

$$Y_{i_{P_{ct}}} = \sum\nolimits_{j=0}^{j} Y_{i_{P_{rj}}} + \sum\nolimits_{j=0}^{j} Y_{i_{P_{tj}}} + \sum\nolimits_{j=0}^{j} Y_{i_{P_{d_j}}}$$
$$= 1.870 + 17.827 + 10.662 = 30.359\ (\text{J/kg})$$

20. The specific internal work of the flow available at the *IN* entrance:

$$Y_{i_1} = Y_0 - Y_{i_{P_{ct}}} = 5670.180 - 30.359 = 5639.821\ (\text{J/kg})$$

21. The total internal power of the flow lost in the *TC* + *SCC*:

$$P_{i_{Pct}} = \sum_{j=0}^{j} P_{i_{Pr_j}} + \sum_{j=0}^{j} P_{i_{Pt_j}} + \sum_{j=0}^{j} P_{i_{Pd_j}} = P_{i_{Pct}}$$
$$= 31.206 + 139.158 + 95.083 = 265.447 \text{ (kW)}$$

22. The internal flow power available at the *IN* entrance:

$$P_{i_1} = P_0 - P_{i_{Pct}} = 189157.205 - 265.447 = 188891.758 \text{ (J/kg)}$$

23. The internal performance of *TC* + *SCC*:

$$\eta_{i_{0;1}} = 1 - \frac{P_{i_{Pct}}}{P_{i_0}} = 1 - \frac{265.447}{189157.205} = 0.9986 = 99.86\ (\%)$$

24. The *TC* velocity coefficient:

$$\phi_{0;1} = \frac{c_1}{c_0} = \sqrt{\frac{2 \cdot Y_{i_1}}{2 \cdot Y_{i_0}}} = \sqrt{\frac{2 \cdot 5639.821}{2 \cdot 5670.180}} = 0.9973$$

25. The specific internal works of the flow entering each bifurcation in the r direction between section 0 and section 0:

$$Y_{i_{bi_j}} = Y_{i_{bi_{j-1}}} - \sum \left(Y_{i_{Pr}} + Y_{i_{Pt}} + Y_{i_{Pd}}\right)_j \text{ (J/kg)}$$

… transcribing the values in column *p*.

26. The average static pressure energy of the flow at the entrance of each bifurcation in the r-direction of the *r*, *TC* + *SCC*:

$$\frac{p_j}{\rho} = Y_{i_{j-1}} - \frac{c_0^2}{2} = Y_{i_{j-1}} - \frac{8,645^2}{2} = Y_{i_{j-1}} - 37.368 \text{ (J/kg)}$$

… transcribing the values in column *q*.
For the needle substitute C_o for $C_a = 11.187$ (m/s).

8.11.5 Comments

- The graph in Fig. 8.16 shows the distribution of the specific pressure energy, p/ρ (J/kg) as a function of the length of the *ACL* between section 0 and section 1. The greatest drop occurs in the section containing the needle, where the flow cross section is strongly reduced, which produces an increase in specific kinetic energy, $c^2/2$ (J/kg).
- The lowest internal efficiency between section 0 and section 1 occurs in MI:

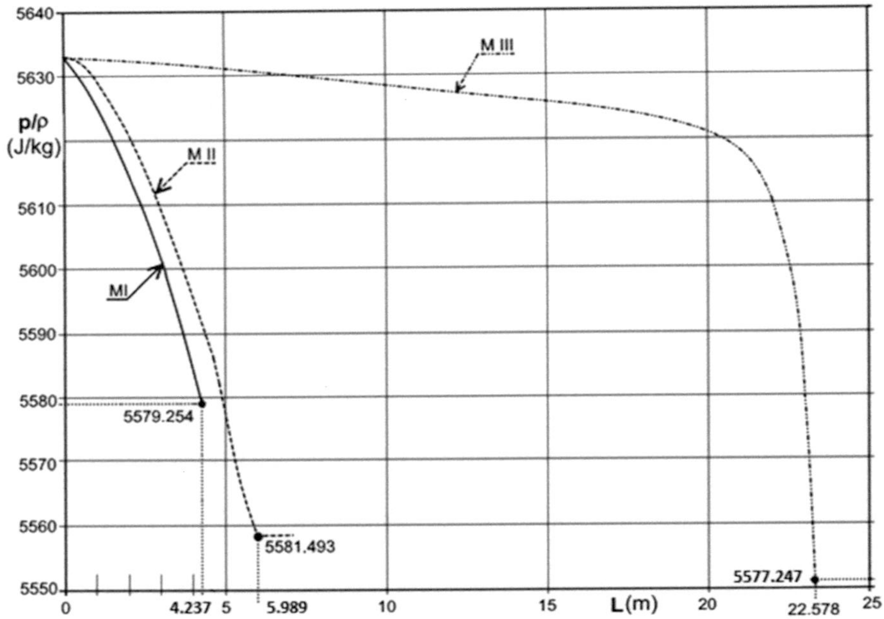

Fig. 8.16 Graphs with functions $p/\rho = f(L)$ for the models studied

$\eta_{i_{0;\,1_{MI}}} = 97.33$ (%), in other words, energy losses of 2.67 (%)

The highest internal efficiency occurs in MIII:

$\eta_{i_{c0;\,1_{MIII}}} = 99.86$ (%), or a loss of only 0.14 (%).

This is mainly due to the specific energy loss arising from the difference in elevations and the necessary curve at the entrance that is crossed by part of the rod of the load control system used here.

- The velocity coefficients between section 0 and section 1 that relate the specific energies available in the flow at the *IN* inlet with those at the *PHIT* $\phi_{0;\,1}$inlet, observe the assumptions, and have little variation, i.e., $0.9954 \leq \phi_{ct} \leq 0.9976$.
- The issue of the average flow velocity between section 0 and section 1, and consequently its internal diameter, must be studied on a case-by-case basis considering the minimum cost of energy losses, in the same way that was done for the penstocks in the *HPP*.
- The other internal losses (*IN*, free flow, shells) are independent of the modeling, i.e., whether the axis is horizontal or vertical.

8.12 Algorithm and Validation for Determining the Energy Losses at the *PHIT IN* Input

8.12.1 Considerations and Input Data

- The algorithm in this item will be developed for internal losses as a continuation of MIII, item 8.9., however it is also valid for continuation of MI and MII.
- Thus, the algorithm and its validation have already been developed and validated until step 27 of subitem 8.9.4, and its continuity should start at step 28.
- The complementary data for the validation will be:
  - the loss coefficients

    $\phi_{in} = \phi_1 = 0.99$
    $\psi_5 = 0.98$
    $\eta_{f_3} = 1.0$
    $\eta_{f_c} = 0.995$

  - the speed ratio, $r_v = 0.465$,
  - the *RO* rotation,
  - $n = 300$ (rpm) $= 5$ (Hz),
  - the average angle at the exit of the shells, $\beta_5 = 6^o$,
  - the specific mass of air at $20^oC - \rho = 0.132$ (kg/m^3),
  - $\eta_m = 0.97$,
  - $\eta_{el} = 0.98$,
  - and the following geometric quantities will have their values justified in Chap. 9:

    $d_{ex} = 3.8$,
    $b_a = 0.8$,
    $B_i = 1.100$,
    $R_i = 3.7$ (m)

See Fig. 8.15.

8.12.2 Script and Calculations

27. The theoretical average absolute velocity of the flow at the entrance of the *IN*:

$$c_{1*} = \sqrt{2 \cdot Y_{i_1}} = \sqrt{2 \cdot 5639.821} = 106.206 \text{ (m/s)}$$

28. The average specific internal work of the flow lost in each *IN*:

$$Y_{i_{p_{in}}} = 0.5 \cdot \left(1 - \phi_1^2\right) \cdot c_{1*}^2 = 0.5 \cdot \left(1 - 0.99^2\right) \cdot 106.206^2 = 112.233\ (\mathrm{J/kg})$$

29. The average internal flow power lost at each injector:

$$P_{i_{p_{in}}} = Q_1 \cdot Y_{i_{p_{in}}} = 5.56 \cdot 112.233 = 624.016\ (\mathrm{kW})$$

30. The internal yield of each *IN*:

$$\eta_{i_{in}} = 1 - \frac{P_{i_{p_{in}}}}{P_{i_1}} = 1 - \frac{624.016}{5.56 \cdot 5639.821} = 0.9801 = 98.01\ (\%)$$

31. The average specific internal work of the flow available at the exit of each *IN*:

$$Y_{i_2} = Y_{i_1} - Y_{i_{p_{in}}} = 5639.821 - 112.233 = 5527.588\ (\mathrm{J/kg})$$

32. The average internal power of the flow available at the output of each *IN* and inlet of the *RO* shells:

$$P_{i_2} = P_{i_4} = Q_1 \cdot Y_{i_2} = 5.56 \cdot 5527.588 = 30733.389\ (\mathrm{kW})$$

33. The average absolute velocity of the flow at the exit of each *IN* and entry into the *RO* shells:

$$c_2 = c_4 = \sqrt{2 \cdot Y_{i_2}} = \sqrt{2 \cdot 5527.588} = 105.144\ (\mathrm{m/s})$$

34. The tangential velocity at the pitch diameter of the *RO*:

$$u = r_v \cdot c_4 = 0.465 \cdot 105.144 = 48.892\ (\mathrm{m/s})$$

35. The average relative velocity at the entrance of the *RO* shells:

$$w_4 = c_4 - u = 105.144 - 48.892 = 56.252\ (\mathrm{m/s})$$

36. The average relative velocity at the exit of the *RO* shells:

$$w_5 = \psi_5 \cdot w_4 = 0.98 \cdot 56.252 = 55.127\ (\mathrm{m/s})$$

37. The average internal specific work lost from the flow in the *RO* shell:
38. $Y_{i_{p_r}} = 0.5 \cdot \left(w_4^2 - w_5^2\right) = 0.5 \cdot \left(56.252^2 - 55.127^2\right) = 62.651\ (\mathrm{J/kg})$

39. The average internal flow power lost in the *RO* shell:

$$P_{i_{Pir}} = Q_1 \cdot \eta_{f_c} \cdot Y_{i_{p_r}} = 5.56 \cdot 0.995 \cdot 62.651 = 346.598 \text{ (kW)}$$

40. The *RO* internal yield:

$$\eta_{i_r} = 1 - \frac{P_{i_{p_r}}}{P_{i_4}} = 1 - \frac{346.598}{5.56 \cdot 5527.588} = 0.9887 = 98.87 \text{ (\%)}$$

41. The internal efficiency of the rotor in relation to the P_{i_0}:

$$\eta_{i_{r0}} = 1 - \frac{P_{i_{p_r}}}{\frac{P_{i_0}}{z_j}} = 1 - \frac{346{,}598}{5{,}56 \cdot 5670{,}18} = 0.9890 = 98.90 \text{ (\%)}$$

42. The average absolute velocity of the flow at the exit of the *RO* shell:

$$\begin{aligned} c_5 &= \sqrt{u^2 - w_5^2 - 2 \cdot u \cdot w_5 \cdot \cos \beta_5} \\ &= \sqrt{48.892^2 - 55.127^2 - 2 \cdot 48.892 \cdot 55.127 \cdot \cos 6} = 8.271 \text{ (m/s)} \end{aligned}$$

43. The average internal specific work of the flow available at the shell outlet of the *RO* jet:

$$Y_{i_{5_1}} = 0.5 \cdot c_5^2 = 0.5 \cdot 8.271^2 = 34.205 \text{ (J/kg)}$$

44. The average internal flow power available at the shell outlet of the *RO* jet:

$$P_{i_{5_1}} = P_{i_1} = Q_1 \cdot Y_{i_5} = 5.56 \cdot 34.205 = 190.180 \text{ (kW)}$$

45. The average internal power available at the output *j* of the *RO* on the *PHIT*:

$$P_{i_5} = z_j \cdot Q_1 \cdot Y_{i_{5_1}} = 6 \cdot 5.56 \cdot 34.205 = 1141.079 \text{ (kW)}$$

46. The internal yield of available energy not used by *RO* of the *PHIT*:

$$\eta_i = 1 - \frac{z_j \cdot P_{i_5}}{P_{i_0}} = 1 - \frac{1141.079}{189157.205} = 0.9940 = 99.40 \text{ (\%)}$$

47. The average internal specific work of the flow lost between the entry of the *PHIT* and the exit of a jet from the rotor:

$$Y_{i_p} = Y_{i_{Pct}} + Y_{i_{Pin}} + Y_{i_{p_r}} = 30.545 + 112.233 + 62.651 = 205.429 \text{ (J/kg)}$$

48. The average internal specific power of the flow lost between *PHIT* inlet and *RO* shells outlet:

$$P_{i_p} = z_j \cdot \left(P_{i_{p_{ct}}} + P_{i_{p_{in}}} + P_{i_{p_r}}\right) = 6 \cdot (265.447 + 624.016 + 346.598)$$
$$= 7416.366 \text{ (kW)}$$

49. The average internal power of the *PHIT*:

$$P_i = P_{i_0} - P_{i_p} = 189157.205 - 7416.366 = 181740.839 \text{ (kW)}$$

50. The internal yield of the *PHIT*:

$$\eta_i = \frac{P_i}{P_{i_0}} = \frac{18174.839}{189157.205} = 0.9608 = 96.08\ (\%)$$

51. The diameter of the jet:

$$d_o = \sqrt{\frac{4 \cdot Q_1}{\pi \cdot c_2}} = \sqrt{\frac{4 \cdot 5.56}{\pi \cdot 105.144}} = 3.113 \text{ (m)}$$

52. The pitch diameter of the *RO*:

$$D = \frac{u}{\pi \cdot n} = \frac{48.892}{\pi \cdot 5} = 3.113 \text{ (m)}$$

53. The *RO* internal yield:

$$\eta_{i_r} = 1 - \frac{P_{i_{p_{in}}}}{P_{i_o}/z_j} = 1 - \frac{624.016}{5.56 \cdot 5670.18} = 0.9998$$

54. Ventilation losses using IEC/*IEC* 41 formula for a vertical axis *RO PHIT* with housings:

$$P_{p_v} = 22 \cdot 10^{-3} \cdot n^3 \cdot d_{ex}^5 \cdot \left(\frac{B_a}{d_{ex}}\right)^{2/3} \cdot \left(\frac{B_i}{d_{ex}}\right)^{4/3} \cdot \left(\frac{R_i}{d_{ex}}\right) = 22 \cdot 10^{-3} \cdot 5^3 \cdot 3.8^5$$
$$\cdot \left(\frac{0.8}{3.8}\right)^{2/3} \cdot \left(\frac{1.1}{3.8}\right)^{4/3} \cdot \left(\frac{3.7}{3.8}\right) = 143.646 \text{ (kW)}$$

55. The shaft power:

$$P_e = P_i - P_{p_v} = 181740.839 - 143.646 = 181453.547 \text{ (kW)}$$

56. The yield on the *PHIT* axis:

$$\eta_e = \frac{P_e}{P_{i_0}} = \frac{181453.547}{189157.205} = 0.9593 = 95.93\ (\%)$$

57. The mechanical power of the *PHIT*:

$$P_m = P_e \cdot \eta_m = 181453.547 \cdot 0.97 = 176009.941\ (\text{kW})$$

58. The total yield of the *PHIT*:

$$\eta_t = \frac{P_e}{P_{i_0}} \cdot \frac{P_m}{P_e} = \eta_e \cdot \eta_m = 0.9593 \cdot 0.97 = 0.9305 = 93.05\ (\%)$$

59. The electrical power of the *PHIT*:

$$P_{el} = P_m \cdot \eta_{el} = 176009.941 \cdot 0.98 = 172489.742\ (\text{kW})$$

60. The yield of the *GG* with *PHIT*s:

$$\eta_{gg} = \frac{P_{el}}{P_{io}} = \eta_t \cdot \eta_{el} = \frac{172489.749}{189157.205} = 0.9305 \cdot 0.98 = 0.9199 = 91.19\ (\%)$$

8.12.3 Analysis and Comments

- The results obtained that the greatest losses are concentrated in the energy transformation that occurs in the *IN*, due to several factors, where large variations of the absolute average velocity of the flow occur.
- The kinetic energy that the *RO* cannot use comes from the impossibility of $\beta_5 = 0$, so this angle, as was mentioned, must be as small as possible, normally >4o. This limitation will be analyzed in detail in the next chapters.
- Most Norms do not consider the losses arising from the difference in heights between the outlet of the shell flow and the downstream level, given the high heads used by *PHIT*s.
- CEI/*IEC* 41 warns that the calculated ventilation losses are subject errors that can reach 50 (%). Here, $P_{p_V} = 143.646$ (kW) it can reach 215,469 (kW), probably for externally rough *RO* with shells fixed to the disk by screws. If the formulas contained in NB - 228 were used, it would result in:

  – $k_{vf} = 0.01 + 12.40 \cdot \left(\frac{B_e}{d_{ex}}\right)^4 = 0,01 + 12,40 \cdot \left(\frac{0.8}{3.8}\right)^4 = 0.03436$
  – $P_{p_{vf}} = 10^{-6} \cdot \rho_{ar} \cdot k_{vf} \cdot n^3 \cdot d_{ex}^5 = 10^{-6} \cdot 0.132 \cdot 0.03436 \cdot 300^3 \cdot 3.8^5 = 97.031\ (kW)$

  for cast *RO*s in *PHIT*s without housings.

- $k_{vd} = 0.035 + 0.25 \cdot \frac{B_e}{d_{ex}} = 0.08763$
- $P_{P_{vd}} = 10^{-6} \cdot \rho_{ar} \cdot k_{vd} \cdot n^3 \cdot d_{ex}^5 = 10^{-6} \cdot 0.132 \cdot 0.08763 \cdot 300^3 \cdot 3.8^5 = 247.462\,(\text{kW})$

 for *RO*s with detachable shells, without housings.

These results show that the formulas provide values that serve only for preliminary studies. Better values can be obtained in the laboratory by testing geometrically similar models and trying to transpose them to prototypes.

- In the application, the formulas were not used to calculate the mechanical losses due to the fact that they depend on quantities that will only be obtained after the preliminary structural design.
- Considering the studies carried out, and the data available for public use, the following intervals can be established for the speed coefficients, the yields, and the *PHIT* speed ratio, with the lowest values always being for small power low quality manufacturing. We also recommended consulting the values in Table 8.1, for setting the r_v particularly for medium and large powers particularly, referencing the closest value to the corresponding specific rotation, n_{qA}.

 - $0.960 \leq \phi_{0;\,2} \leq 0.997$
 - $0.96 \leq \eta_i \leq 0.99$
 - $0.95 \leq \eta_e \leq 0.92$
 - $0.96 \leq \eta_m \leq 0.98$
 - $0.88 \leq \eta_t \leq 0.96$
 - $0.96 \leq \eta_{el} \leq 0.98$
 - $0.85 \leq \eta_{GG} \leq 0.9$
 - $0.46 \leq r_v < 0.5$.

(8.61)

The following question has been left unanswered should one want to try to solve it.

Why do the values of $Y_{i_{Pr_j}}$ *remain the same, while the values of* $P1_{Pr_j}$ *differ in* Table 8.3*?*

Chapter 9
The Preliminary Hydrodynamic Calculation of the Flow in a Pelton Hydraulic Impulse Turbine (*PHIT*) Rotor

The main objective of this chapter is to present a procedure that obtains a first approximation of the geometric physical limits of the Real Hydraulic Flow Path (*RHFP*) in the Rotor Shells (*RS*) of the *PHIT*. The complexity of studying these physical limits, between each half of the flow and the respective inner surface of the shells, must occur with minimal friction and turbulence (shocks) in a very short time, i.e., in the order of a tenth of a second. Even when using computational resources, the results can be far from representing real flows. To improve this, exhaustive laboratory studies on homologous models (which are geometrically similar, with the same specific rotation) should be conducted. Given these facts, the chapter will be limited to developing studies that allow us to graphically or analytically obtain the relationships between flow characteristics and the *PHIT* geometry, using the Principles of Mechanics applied to Hydrodynamics (Euler's Equation) and coefficients, preferably dimensionless ones. Having done this, we will be able to develop and validate algorithms that will reproduce scale drawings of the *PHIT RO*, particularly the shells. These algorithms, and the respective software that will result from them, may be used to prepare the structural calculations, the respective Preliminary Project, and manufacture of models, which, via tests, will establish the Executive Project and the Manufacturing Project according to the contracts agreed upon by the parties involved.

9.1 Specific Rotation

Specific rotations are expressions that establish the quantitative relationship between the fluid dynamic energy of the flows and the mechanics in the Turbomachine *TM*, based on the Principles of Mechanics (Euler's Equation for *TM*) and the Theory of

© The Author(s), under exclusive license to Springer Nature Switzerland AG 2025

Z. de Souza, G. L. Tiago Filho, *Pelton and Crossflow Impulse Hydraulic Turbines*,
https://doi.org/10.1007/978-3-031-87086-6_9

Similarity,[1] which conceptualizes homologous Hydraulic Turbomachine *HTM* as follows:

Homologous *HTM*s are geometrically similar, and have the same specific rotation, but since they do not have the same N_{R_e}, they will not have the same yield.

In the specific case of a *PHIT*, not considering pressure variations of the flow in the rotors, the efficiency is practically the same when homologous, making the specific rotation fundamental for studying the flow in the shells.

An expression for the specific rotation starts from Euler's Equation, which in its dimensionless[2] form can be written as follows:

$$\psi = 2 \cdot (1 - \phi \cdot \cot g\beta_4) \rightarrow \psi = \frac{2 \cdot g \cdot H}{u^2} \rightarrow \phi = \frac{c_m}{u} \tag{9.1}$$

Considering the Similarity Theory, two *TM*s must have the same dimensionless coefficients:

$$\psi = \frac{2 \cdot g \cdot H}{u^2} = \frac{2 \cdot g \cdot H}{(\pi \cdot D \cdot n)^2} \rightarrow \phi = \frac{c_m}{u} = \frac{4 \cdot Q}{\pi^2 \cdot D^3 \cdot n}$$

Thus:

$$\frac{Q_1}{D_1^3 \cdot n_1} = \frac{Q_2}{D_2^3 \cdot n_2}$$

Manipulating the equalities, one initially obtains:

$$\frac{H_1}{H_2} \cdot \left(\frac{n_2}{n_1}\right)^2 = \left(\frac{D_1}{D_2}\right)^2$$

and

$$\frac{Q_1}{Q_2} \cdot \frac{n_2}{n_1} = \left(\frac{D_1}{D_2}\right)^3$$

And after transformations:

[1] Souza, Zulcy—Projeto de MF—Tomo I—pg. 75—Editora Interciência—ISNM 978–85–7193-258-6—Rio de Janeiro—BR—2011.

[2] Souza, Zulcy—Projeto de MF—Tomo I—pgs. 51 e 73—Editora Interciência—ISNM 978–85–7193-258-6—Rio de Janeiro—BR—2011.

$$\frac{n_1^{2/3} \cdot Q_1^{1/3}}{H_1^{1/2}} = \frac{n_2^{2/3} \cdot Q_2^{1/3}}{H_2^{1/2}} \rightarrow \frac{n_1 \cdot Q_1^2}{H_1^{3/4}} = \frac{n_2 \cdot Q^2}{H_2^{3/4}}$$

This equality is called specific rotation in the Technical System Units (*TSU*), with:

- n(rpm)
- Q(m^3/s)
- H (mca)

$$n_q = n \cdot \frac{Q^{1/2}}{H^{3/4}} \left(\frac{\text{rot}}{\text{min}}\right) \tag{9.2}$$

Physically, n_q corresponds, theoretically (without considering any losses):

The rotation n (rpm), of a *PHIT* operating under Q (m^3/s) and H (mca) starts to operate subject to a head $H = 1$ (mca), injecting the jets into the respective rotor shells, with a flow of water $Q = 1.0$ (m^3/s) that would maintain the same rotation n (rpm).

In International System of Units (*SI*) terms, the specific rotation, with:

- n (rpm)
- Q (m^3/s)
- Y (J/kg)

… results in the following expression:

$$n_{qA} = 10^3 \cdot n \cdot \frac{Q^{1/2}}{Y^{3/4}} \left(\frac{\text{rot}}{seg}\right) \equiv (\text{rps}) \tag{9.3}$$

The expression n_{qA} can be expressed as a function of n_q, resulting in:

$$n_{q_A} = 3 \cdot n \cdot \frac{Q^{1/2}}{H^{3/4}} = 3 \cdot n_q \ (\text{rps}) \tag{9.4}$$

The International Electrotechnical Commission (*IEC*) Standards recommends using the following formula for specific rotation with n (rps), Q (m^3/s), E (J/kg) $\rightarrow$ $E\left(\frac{\text{J}}{\text{kg}}\right) \equiv \left[\frac{\text{N}\cdot\text{m}}{\text{kg}}\right] \equiv \left[\frac{\text{kg}\cdot\text{m}}{\text{s}^2} \cdot \frac{\text{m}}{\text{kg}}\right] \equiv \left(\frac{\text{m}^2}{\text{s}^2}\right)$.

$$N_{QE} = n \cdot \frac{Q^{1/2}}{E^{3/4}} = 10^{-3} \cdot n_{qA} = 3 \cdot n_q \ (\text{rps}) \tag{9.5}$$

For a *PHIT*, the specific rotation is usually determined by the jet, by entering a flow rate in the formula:

$$Q_1 = Q/z_j \rightarrow z_j = number\ of\ jets \tag{9.6}$$

9.2 Characteristic Relations

9.2.1 Reference Diameter and Jet Diameter

The reference diameter, DDD (m), of the THP runner is defined as the circumference tangent and perpendicular to the jet axis coming from the nozzle at its throat diameter, dod_odo (m) (Fig. 7.3a).

The diameter, D, can be expressed as a function of the characteristics linked to the *RO* as follows:

$$u = \frac{\pi \cdot D \cdot n}{60} \rightarrow D = \frac{60 \cdot u}{\pi \cdot n}\ (\mathrm{m}) \tag{9.7}$$

With … u (m/s) and n (rpm).

The diameter of the jet d_o can be expressed as a function of characteristics linked to the flow in a jet, considering the volumetric losses from the two-phase flow, η_{v_j}, between the exit section of the *IN* 2 and the entrance section of the edges of the shells at 4, according to:

$$Q_r = Q_1 \cdot \eta_{v_j} = \frac{\pi \cdot d_o^2}{4} \cdot c_3\ \left(\mathrm{m}^3/\mathrm{s}\right) \rightarrow c_3 \cong c_4 \cong \phi_{0;4} \cdot \sqrt{2 \cdot g \cdot H}\ (\mathrm{m/s})$$

After substitutions we get:

$$d_o = 2 \cdot \left(\frac{\eta_{v_j} \cdot Q_1}{\pi \cdot c_4}\right)^{0.5} = \left(\frac{4 \cdot \eta_{v_j} \cdot Q_1}{\pi \cdot \phi_{0;4} \cdot \sqrt{2 \cdot g \cdot H}}\right)^{0.5} (m) \rightarrow g\ \left(\mathrm{m/s^2}\right) \text{ and } H_o(\mathrm{mca}) \tag{9.8}$$

The relationship between these two characteristics, which, as will be seen, is fundamental for initially designing a *PHIT*, particularly the *RO*, can now be established:

$$\begin{aligned}\frac{D}{d_o} &= \frac{30}{\pi^{0.5}} \cdot \left(\frac{c_4}{\eta_{v_j} \cdot Q_1}\right)^{0.5} \cdot \frac{u}{n} = 16.926 \cdot \left(\frac{c_4}{\eta_{v_j} \cdot Q_1}\right)^{0.5} \cdot \frac{u}{n} \\ &= \frac{30 \cdot (2 \cdot 9.81)^{0.25}}{\pi^{0.5}} \cdot \left(\frac{\phi_{0;4}}{\eta_{v_j} \cdot Q_1}\right)^{0.5} \cdot H^{0.25} \cdot \frac{u}{n} = 35.622 \cdot \left(\frac{\phi_{0;4}}{\eta_{v_j} \cdot Q_1}\right)^{0.5} \cdot H^{0.25} \cdot \frac{u}{n}\end{aligned} \tag{9.9}$$

9.2.2 The Specific Rotation and the Diameter Ratio

Expression (9.4) applied to $z_j = 1$ can be related to other characteristics, as follows:

$$n_{qA_1} = 3 \cdot n \cdot \frac{Q_1^{1/2}}{H^{3/4}} \rightarrow n = \frac{60 \cdot u}{\pi \cdot D} \rightarrow Q_1 = \frac{\pi \cdot d_o^2}{4} \cdot c_4 \rightarrow c_4 = \phi_{0;4} \cdot c_{4t} \rightarrow H = \frac{c_{4t}^2}{2 \cdot g}$$

Substituting and simplifying, we get:

$$n_{qA_1} = \frac{90 \cdot (2 \cdot g)^{0.75}}{\pi^{0.5}} \cdot \phi_{0;4}^{1.5} \cdot \frac{u}{c_{4t}} \cdot \left(\frac{D}{d_0}\right)^{-1}$$

With …

$$r_v = \frac{u}{c_{4t}}$$

and …

$$k_{n_{qA_1}} = \frac{90 \cdot (2 \cdot 9.81)^{0.75}}{\pi^{0.5}} \cong 473.36$$

We have:

$$n_{qA_1} = 473.36 \cdot \phi_{0;4} \cdot r_{v_t} \cdot \left(\frac{D}{d_o}\right)^{-1} \tag{9.10}$$

This expression serves as a foundation for preliminarily establishing the link between the transformation of hydraulic energy into mechanical energy.

In this expression, a limiter appears, which is the diameter of jet, d_o, which is linked to the free flow losses, albeit superficially, in item 8.5 in Chap. 8, since the greater its value, the greater the volumetric losses or leakage. The value increases with increasing Q_1, and decreases with increasing c_{4t}, i.e., with increasing H.

Such considerations have limited $d_o < 250$ (mm), and higher values are used exceptionally and mainly by limiting the number of jets to 6 and limiting combinations of Q and H values, as can be seen from the values shown in Table 8.1.

Depending on the specific rotation, a link can also be established with the number of jets, z_j, as follows:

$$n_{q_{A_1}} = 3 \cdot n \cdot \frac{(Q/z_j)^{1/2}}{H^{3/4}} = \rightarrow z_j = \frac{9}{n_{qA_1}^2} \cdot \frac{Q}{H^{1.5}} \cdot n^2 \rightarrow n = \frac{H^{0.75}}{3 \cdot Q^{0.5}} \cdot z_j^{0.5} \cdot n_{qA_1} \tag{9.11}$$

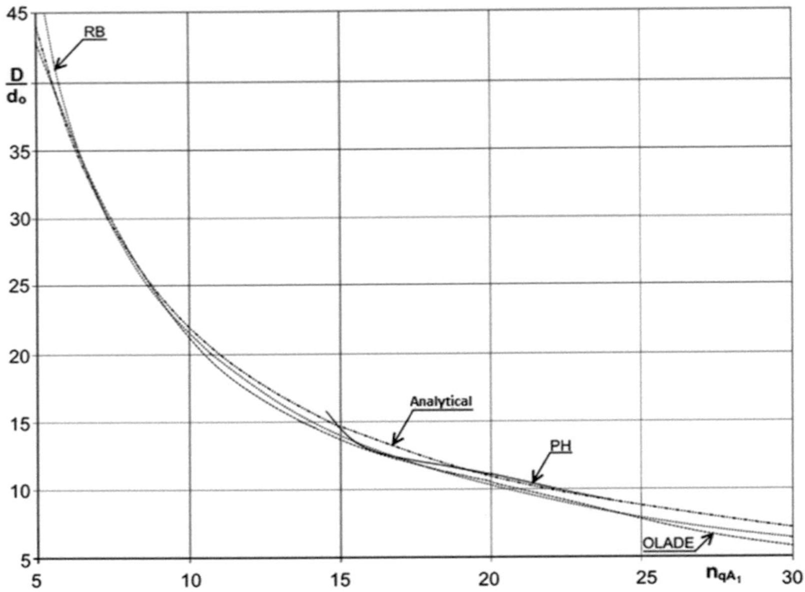

Fig. 9.1 Graph of function $D/d_o = f\left(n_{q_{A_1}}\right)$

In the graph from Fig. 9.1, the functions $D/d_o = f\left(n_{q_{A_1}}\right)$, were represented using calculated data from indications from four authors listed in the bibliography here, resulting in the following equations:

- Latin American Energy Organization—(*OLADE*)

$$\frac{D}{d_o} = \frac{228}{n_{qA_1}} - 1.12 \tag{9.12}$$

- (*RB*)—Richard Bran—Turbomachines

$$\frac{D}{d_o} = 90.738 - 12.715 \cdot n_{qA_1} - 0.021 \cdot n_{qA_1}^2 + 0.00021 \cdot n_{qA_1}^3 \tag{9.13}$$

- (*PH*)—Pierre Henry—Hydraulic Turbomachines

$$\frac{D}{d_o} = 213.868 - 32.017 \cdot n_{qA_1} + 1.703 \cdot n_{qA_1}^2 - 0.03046 \cdot n_{qA_1}^3$$

$$\rightarrow$$

$$14 \leq n_{qA_1} \leq 19 \; \frac{D}{d_o} = 20.283 - 0.462 \cdot n_{qA_1}$$

$$\rightarrow$$

$$19 \leq n_{qA_1} \leq 24 \tag{9.14}$$

- Analysis from expression (9.10) with: $\varphi_{0;\,4} = 0.99$ and $r_v = 0.48$

$$\frac{D}{d_0} = \frac{224.941}{n_{qA_1}} \tag{9.15}$$

Analyzing the four drawings in Fig. 9.1, shows that the greatest variation occurs with $n_{qA_1} = 30$, for one n_{qA_1}, decreasing this variation to nearly 0 for $n_{qA_1} \cong 7, 5$.

To obtain the (*PH*) function, we used the n_{qA_1} and D/d_o values from Table 8.1, which allowed us to plot the 11 pairs of values in Fig. 9.2. Then, the dashed line was drawn, and with the help of the "Tablecurve 2D" software computer program, the two valid equations in the respective ranges, n_{qA_1}, were obtained, indicated in the figure.

These average functions were transcribed in Fig. 9.1, and should preferably be used for high power *PHIT*s with more than one jet, while always observing the

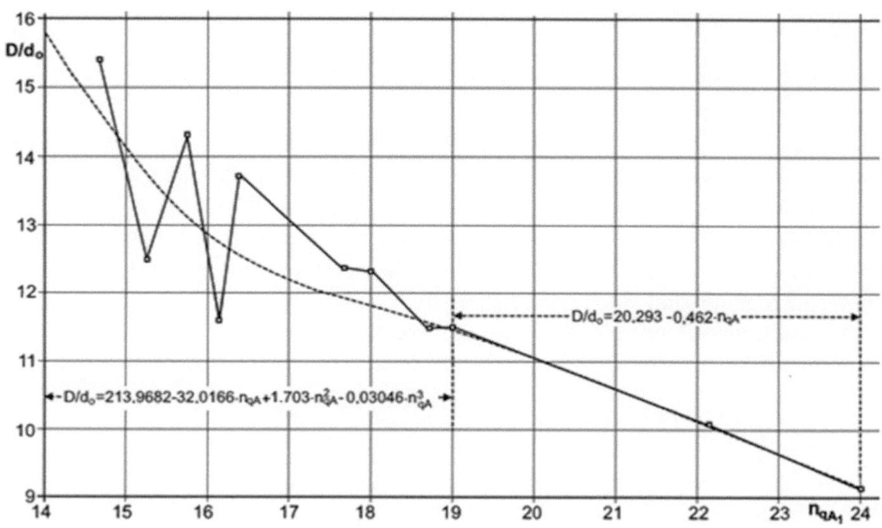

Fig. 9.2 Graph of function $D/d_o = f\left(n_{q_{A_1}}\right)$—Pierre Henry—Table 8.1

deflections that occur, particularly in the range $14 \leq n_{qA_1} \leq 19$, in which case, the value from the dashed line is preferrable.

9.3 Fundamentals for Calculating the Hydrodynamic Design of a *PHIT*

The Basic Project of a *PHIT* starts given values established according to contracts between parties, fixed withing specific clauses, that include:

- the maximum flow—Q_{max}
- the nominal flow—Q
- the minimum flow—Q_{min}
- the corresponding drops—H_{max}, H, H_{min}.

The *PT* uses the so-called nominal values, Q (m^3/s), $H = H_0$ and topographic characteristics of the site as references for the preliminary hydrodynamic calculations, particularly for the Engine Room, the penstock, and the levels downstream the river.

Using these elements, the *PT* preliminarily fixes:

- η_m
- $\phi_{0;\,2} \cong \phi_{0;\,4}$
- r_v

… and using expressions (9.9) and (9.10) they can establish an interval for:

- n(rpm)
- D
- d_0 (m)

… and via analysis, their preliminary values.

To fix η_t, one can use the graph in Fig. 9.3, which was drawn with points taken from page 245 of the book *Máquinas de Fluxo*—Bran/Zulcy de Souza, and 11 points from Table 8.2, which were calculated from the values in Table 8.1.

In Fig. 9.3, the values between the average curve, $\eta_{t_m} = f(n_{A_1})$ and the upper curve are recommended for high power *PHIT*s, and between the average and lower curve for medium and small power *PHIT*s.

The concentration of points allows us to establish the best interval:

$$14 < n_{qA_1} < 20 \tag{9.16}$$

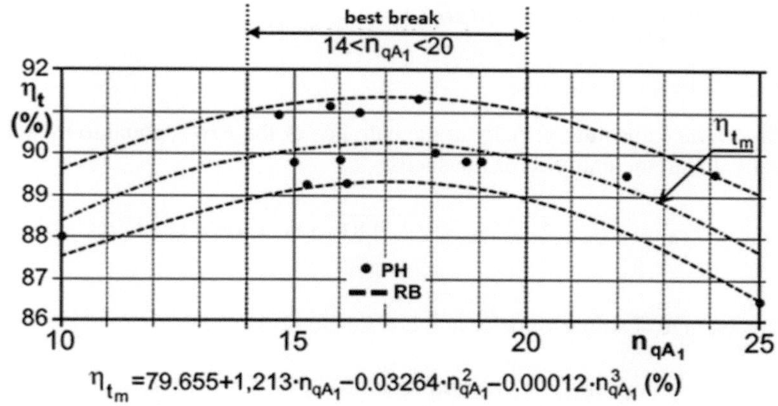

Fig. 9.3 The *PHIT* Total yield– Richard Bran/Zulcy de Souza; Pierre Henry

9.4 Initial Features

9.4.1 Objective and Data

- Develop and validate an algorithm that allows for the following characteristics for developing an initial study of a Basic *PHIT* Project.
  - n (rpm)
  - $c4$ (m/s)
  - n_{qA_1}
  - D
  - d_o (m)
- Therefore, the data must be definitively fixed as:
  - $H = H_0$(mca),
  - Q (m^3/s); and
  - the frequency f_s (Hz) of the *EG* system.
    To validate and compare the results, we fixed the following:
  - $Q = 33.37$(m^3/s)
  - $H = H_0 = 578$(mca)
  - $f = 60$ (Hz).
- $r_v = 048$, $\phi_{0;\ 2} \cong \phi_{0;\ 4} = 0.98$ will be fixed within the limits established in Chaps. 7 and 8.

9.4.2 Algorithm and Validation

Calculate and store the results:

1. The average absolute velocity at the entrance of the *PHIT*, equal to the theoretical value at the entrance of the shells:

$$c_0 = c_{4t} = \sqrt{2 \cdot g \cdot H_0} = \sqrt{2 \cdot 9.81 \cdot 578} = 106.491\ (\mathrm{m/s})$$

2. The average absolute velocity at the entrance of the shells:

$$c_4 = \phi_{0;4} \cdot c_{4t} = 0.98 \cdot 106.491 = 104.361\ (\mathrm{m/s})$$

3. The tangential velocity of the *RO* at the pitch diameter:

$$u = r_v \cdot c_{4t} = 0.48 \cdot 106.491 = 51.116\ (\mathrm{m/s})$$

4. Format Table 9.1, placing the types of *PHIT*s used, with their number of jets, z_j and the *H*—horizontal axis and *V*—vertical axis positions.
   Calculate and position the values in the respective columns:
5. In column 2 the flows:

$$Q_{1_i} = \frac{Q}{z_{j_i}} = \frac{33.37}{z_{j_i}}\ (\mathrm{m^3/s})$$

6. In column 3 the diameters:

$$d_{o_i} = \sqrt{\frac{4 \cdot Q_{1_i}}{\pi \cdot c_4}} = \sqrt{\frac{4 \cdot Q_{1_i}}{\pi \cdot 104{,}361}} = 0.1105 \cdot \sqrt{Q_{1_i}}$$

Table 9.1 Number of z_j

1	2	3
Z_{ji}	Q_{ji}	d_{oi}
–	m³/s	m
1H	33.37	0.63
2H	16.69	0.42
4V	8.34	0.31
5V	6.67	0.28
6V	5.56	0.26

Table 9.2 EG poles

PHIT rotations		
1	2	3
6.00	600.00	36.01
8.00	450.00	27.01
10.00	360.00	21.61
12.00	300.00	18.01
14.00	257.14	15.43
16.00	225.00	13.51

Depending on the d_{o_i} value, choose one or more to continue the steps. In this case, to validate the algorithm, the values in *italics* were chosen and stored:

- $z_j = 6$, $Q_1 = 5.562$ (m^3/s)
- $d_o = 0.261$.

  This value, $d_o = 0.261$ is greater than $d_o < 250$ (m), which will probably result in a greater leakage loss, which can reduce the η_1. However, using $d_o < 0.250$ (m), there will be an increase in D, and consequently an increase in the dimensions of the rotor and its cost. In these cases, technical-financial analysis is recommended, so that the decision can be taken by the *PT*.

7. Format Table 9.2, placing the number of pole pairs of the *EG*, z_{pp}, in column 1, and since it is a large *PHIT*, start with $z_{\mathrm{pp}} = 6$, and follow, preferably, with even numbers.

   Calculate and position the values in the respective columns.
8. In column 2 the rotations:

$$n_i = 3600/z_{pp_i} \text{ (rpm)}$$

9. In column 3 the specific rotations:

$$n_{qA_i} = 3 \cdot n_i \cdot \frac{Q_1^{0.5}}{H_0^{0.75}} = 3 \cdot n_i \cdot \frac{5.562^{0.5}}{578^{0.75}} = 0.06002 \cdot n_i$$

10. Considering the highest yield range, $14 < n_{qA_1} < 20$ the values written in *italics* were chosen and stored:

- $z_{\mathrm{pp}} = 12$, $n = 300$ (rpm)
- $n_{qA_1} = 18, \eta_{t_{\max}} = 0.91$

11. Calculate the pitch diameter:

$$D = \frac{60 \cdot u}{\pi \cdot n} = \frac{60 \cdot 51.116}{\pi \cdot 300} = 3.254 \text{ (m)}$$

12. Determine the probable average total yield:

$$\begin{aligned}\eta_{t_m} &= 79.655 + 1.213 \cdot n_{qA_1} - 0.03264 \cdot n_{qA_1}^2 - 0.00012 \cdot n_{qA_1}^3 \\ &= 79.655 + 1.213 \cdot 18 - 0.03264 \cdot 18^2 - 0.00012 \cdot 18^3 = 90.22\ (\%).\end{aligned}$$

9.4.3 Exit

Table 9.3 with data, and Table 9.4 with the result of the calculations.

9.5 Usual Ranges for the Relative Shell Dimensions

9.5.1 The Relationship Between Dimensions

In Fig. 9.4, the main dimensions of the interior of the inner surface of the shells are named in the upper part, shown in section, and in the lower part, their intervals are referenced to the diameter of the jet d_o, along with the diameters of the circumferences of the *RO*:

- *D*—The diameter jet center primitive
- D_{e_a}—The outer diameter of the shell edge

Table 9.3 Data

H	Q	f	$\varphi_{0;4}$	r_v
mca	m^3/s	Hz	–	–
578.00	33.37	60.00	0.98	0.48

Table 9.4 Results

z_j	Q_1	z_{pp}	n	n_{qA1}	$D_{/do}$	c_{4t}
jatos	m^3/s	pp	rpm	1/s	–	m/s
6.00	5.56	12.00	300.00	18.00	12.47	106.49

c_4	d_o	D	u	$\eta_{t\,m}$
m/s	m	m	m/s	%
104.36	0.26	3.25	51.99	90.2

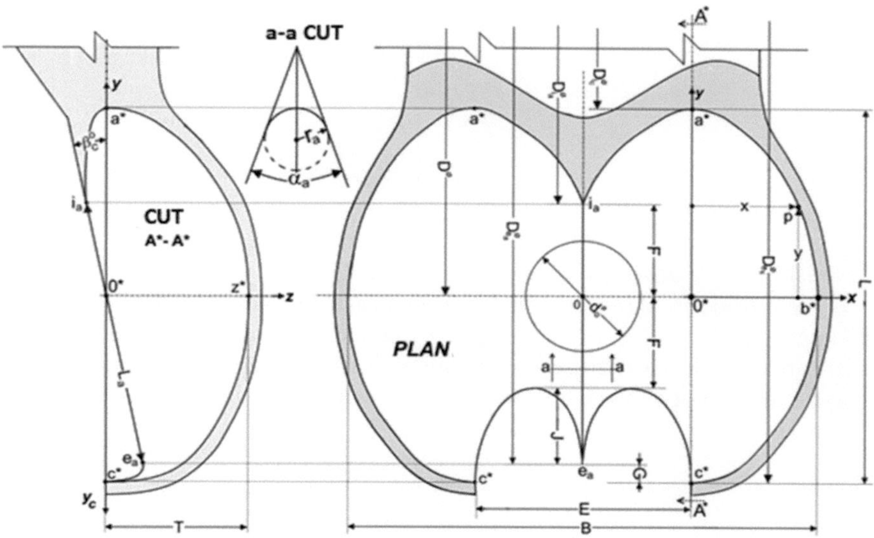

Fig. 9.4 Names and interval relationships of the shell dimensions and diameters

- D_{i_a}—The inner diameter of the shell edge
- D_{e_c}—The outer diameter of the inner surface of the shell
- D_{i_c}—The inner diameter of the inner surface of the shell
- The coordinate system, in view y_p, x_p, and in cutaway $A* - A*$, y_c, z_c.

- $2.5 < B/d_o \leq 3.5$
- $2.4 \leq L/d_o \leq 3.4$
- $0.85 \leq T/d_o \leq 1.4$
- $1.05 \leq E/d_o \leq 1.5$
- $0.32 \leq J/d_o \leq 0.45$
- $0.85 \leq F/d_o \leq 1.0$
- $0 \leq G/d_o \leq 0.15$
- $3 \leq \beta_a^o \leq 16$
- $\alpha_a^o/2 = 20^o \rightarrow r_a = 4.5$ (mm)
- $\alpha_a^o/2 = 15^o \rightarrow r_a = 3.5$ (mm)
- $\alpha_a^o/2 = 10^o \rightarrow r_a = 2.5$ (mm)
- $D_{e_c} = D + 2 \cdot (F + J + G)$
- $D_{i_c} = D_{e_c} - 2 \cdot L$
- $D_{e_a} = D_{e_c} - 2 \cdot G$
- $D_{i_a} = D_{e_a} - 2 \cdot (2 \cdot F + J)$

To establish the ratios of the main dimensions of the inner surface of the shells, we recommend, initially using the value of n_{qA_1} in Table 8.1 or the value to be fixed for the ratio B/d_o or T/d_o in the bibliography, then starting from the relation:

$$r_B = \frac{(B/d_o)_{\text{fixed}}}{(B/d_o)_j}$$

or

$$r_T = \frac{(T/d_o)_{\text{fixed}}}{(T/d_o)_j}$$

… which is important in fixing the number of shells, z_c.

In these ratios, j is the value of the maximum, average, or minimum ratios, shown in Fig. 9.4. Lines $c * b * a*$ and $c * z * a*$ can, for the purpose of first drawing the inner surface of the shell that will internally limit the structural calculation, one can consider the arcs of ellipses, obtained from the equation of the ellipse:

$$\frac{x^2}{a^2} + \frac{y^2}{b^2} = 1 \rightarrow y = b \cdot \sqrt{1 - \frac{x^2}{a^2}}$$

as follows:

- **For the blueprint, row $c * b * a*$** – $b = 0 * \bar{b}* = 0.5 \cdot L \rightarrow a = 0 * \bar{a}* = 0.5 \cdot (B - E)$, i.e.:

$$y_p = 0.5 \cdot L \cdot \sqrt{1 - \left(\frac{2 \cdot x_p}{B - E}\right)^2} \tag{9.17}$$

- **For the A*-A* cut, line $c * z * a*$** – $b = 0 * \bar{b}* = 0.5 \cdot L \rightarrow b = 0 * \bar{z}* = T$, i.e.:

$$y_c = 0.5 \cdot L \cdot \sqrt{1 - \left(\frac{x_c}{T}\right)^2} \tag{9.18}$$

Once the dimensions are obtained, these will serve as a first drawing of the limits of the inner surface of the shell projected in the views shown in Fig. 9.4, which, once corrected and complemented, will serve as a basis drawing the inner surface of the bipartite shells, which will be detailed in Chap. 10. This three-dimensional surface differs from manufacturer to manufacturer and currently depends on the computational behavior of the flow dynamics from the inlet to the outlet of the *RO*, complemented by tests in specialized laboratories. However, in general, it can be summarized in two different models.

In the first model, Fig. 9.5a, the projection of the midline, *ab* of the wet surface is parallel to the dividing edge of the shell *AB* of the flow, while in the second model it is inclined, as shown in Fig. 9.5b. This inclination, adopted by some manufacturers,

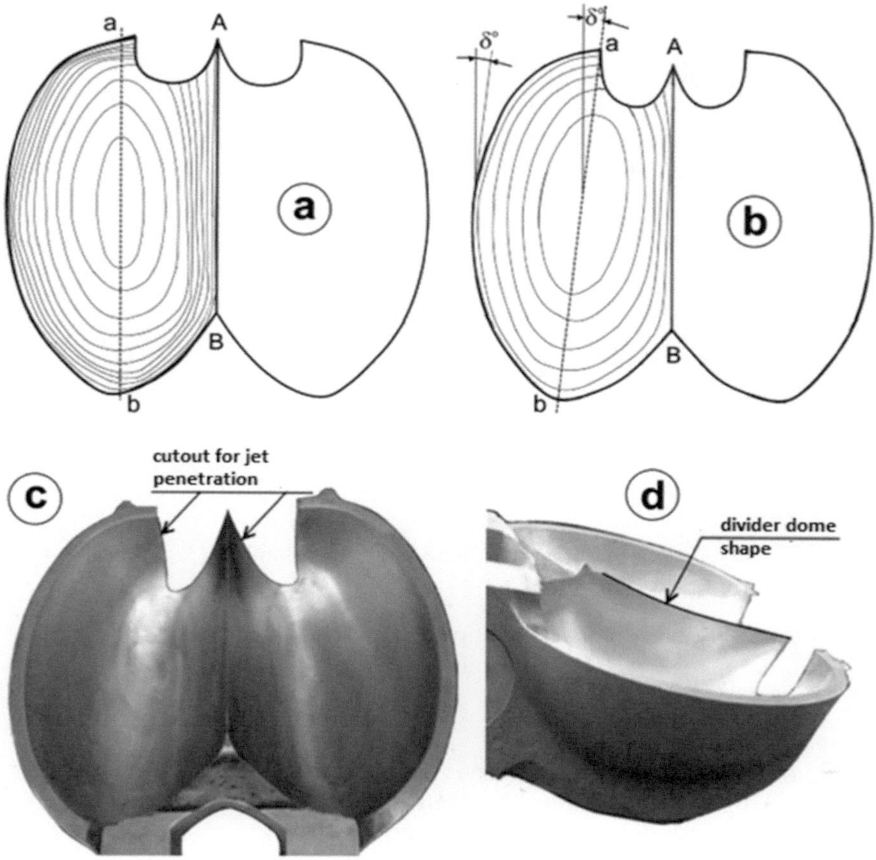

Fig. 9.5 Position, in plan view, of the midline ab of the casing and the shape of its water jet splitter. (**a**) right casing; (**b**) left casing; (**c**) cutout for jet passage; (**d**) water jet divider

facilitates the flow exit, when the level lines are closer, which promotes greater proximity between the shells.

However, we should note that shells that are very close together are difficult to fix on the disk, and the *RO* casting process should be used to make a single piece.

Photos *c* and *d* in Fig. 9.5 are from the *PHIT* shell used at the *Cubatão* Hydroelectric Power Plant, in São Paulo, Brazil, which was donated to *UNIFEI*, and manufactured by VOITH, Germany.

In Fig. 9.5d, the edge of the dividing edge of the jet is shown, with a bulge, when the most usual is to be a straight line. If such bulging has not occurred due to wear and tear, it may also be purposeful, aiming to meet the front distribution of absolute flow velocities.

9.5.2 Algorithm and Validation

9.5.2.1 Input and Validation Data

Establish and validate steps to obtain the shell dimensions for the *PHIT* based on the following characteristics:

- $n_{qA_1} = 18$
- $d_o = 0.266$ (m)
- $D = 3.257$ (m)
- $r_B = \frac{(B/d_o)_{\text{fixed}} = 3.034}{(B/d_o)_{\text{max.}}}$

9.5.2.2 Script and Calculations

1. Calculate the ratio:

$$r_B = \frac{(B/d_o)_{\text{fixed}}}{(B/d_o)_{\text{max.}}} = \frac{3.034}{3.5} = 0.8669$$

2. Calculate the other ratios by multiplying their maximum value by r_B or, if less than the minimum, take the minimum value of the range and then store the results:

   - $B = 3.034 \cdot 266 = 807$ (mm)
   - $L = r_B \cdot (L/d_o)_{\text{max}} \cdot d_0 = 0.8669 \cdot 1.0 \cdot 266 = 761$ (mm)
   - $T = 0.8669 \cdot 1.4 \cdot 266 = 323$ (mm)
   - $G = 0.8669 \cdot 0.15 \cdot 266 = 35$ (mm)
   - $E = 0.8669 \cdot 1.5 \cdot 266 = 346$ (mm)
   - $F = 0.8669 \cdot 1.0 \cdot 266 = 231$ (mm)
   - $J = 0.8669 \cdot 0.45 \cdot 266 = 104$ (mm)

3. Choose a computer program to draw graphics, which has a range of uncertainties (accuracy/precision) for the drawings, and obtain drawing values of at least ±0.1 (mm), e.g., AutoCAD.
4. Open the program, choose the scale, mark point 0 on the worksheet, and from that point, mark two orthogonal axes, starting Fig. 9.6a
5. Draw the circumference diameter $d_o = 266$ (mm), the center at 0,
6. From the horizontal axis that passes through 0 downwards, mark $F = 231$ (mm) + $J = 104$ (mm) vertically and in sequence, and $F = 231$ (mm) upwards.
7. At the lower end of J, draw a horizontal line that meets the vertical axis at e_a, i.e., the lower end of the dividing edge of the shell.
8. Draw a horizontal line through e_a, marking $G = 35$ (mm) downwards, and draw a horizontal line through the lower end.

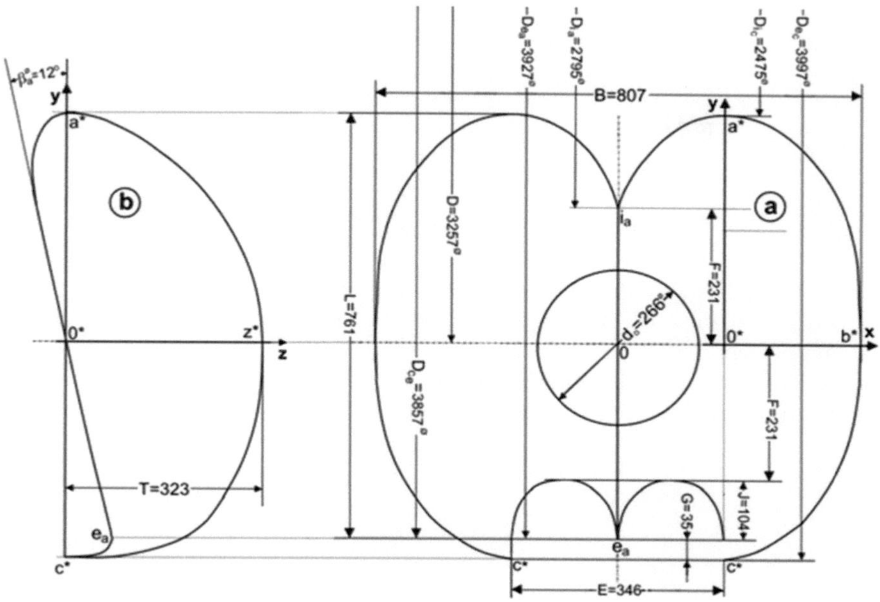

Fig. 9.6 Scale drawing of the section view of the shell. (**a**) elevation view; (**b**) plan view

9. From the vertical axis that passes through 0, on the straight line that passes through the lower end of G, mark $0.5 \cdot B = 0.5 \cdot 807 = 403.5$ (mm) to the right and to the left, obtaining point $c*$, and mark the horizontal segments at the bottom $E = 846$ (mm), and $B = 2 \cdot 0.406.5 = 807$ (mm) in the upper one.
10. From $c*$ draw the vertical y axis, and passing through 0, the horizontal x axis which are centered at $0*$
11. From $c*$, in the vertical direction, mark $L = 761$ (mm).
12. Calculate, store, and mark the diameters:

$$D_{e_c} = D + 2 \cdot (F + J + G) = 3257 + 2 \cdot (231 + 104 + 35) D_{i_c}$$

$$D_{i_c} = D_{e_c} - 2 \cdot L = 3997 - 2 \cdot 761 = 2475 \text{ (mm)}$$

$$D_{e_a} = D_{e_c} - 2 \cdot G = 3997 - 2 \cdot 35 = 3927 \text{ (mm)}$$

$$D_{i_a} = D_{e_a} - 2 \cdot (2 \cdot F + J) = 3927 - 2 \cdot (2 \cdot 231 + 104) = 2795 \text{ (mm)}$$

13. Draw, with a straight line and circle arcs, the cutout, with the right part between the points e_a and $c*$:
14. Draw the ellipse arc $(c * b * \ * a)*$:

$$y = 0.5 \cdot (B - E) \cdot \sqrt{1 - \frac{4}{L^2} \cdot x^2} = 0.5 \cdot (807 - 346) \cdot \sqrt{1 - \frac{4}{761^2} \cdot x^2} = 230.5$$

$$\cdot \sqrt{1 - \left(\frac{x}{380.5}\right)^2} \text{ (mm)} \rightarrow$$

$$0 \leq x \text{ (mm)} \leq 230.5 \rightarrow$$

$$0 \leq y \text{ (mm)} \leq 392$$

15. Fold Fig. 9.6a to the right, obtaining the horizontal projection of the inner surface of the shell, placing, the values of the corresponding dimensions in millimeters:
16. Referencing Fig. 9.6a, draw Fig. 9.6b $T = 318$ (mm) on the same scale:
17. Draw the ellipse arc ($c * z * a*$) on the cut:

$$y = T \cdot \sqrt{1 - \frac{4}{L^2} \cdot z^2} = 323 \cdot \sqrt{1 - \left(\frac{z}{380.5}\right)^2} \rightarrow$$

$$0 \leq z \text{ (mm)} \leq 323 \rightarrow$$

$$0 \leq y \text{ (mm)} \leq 392.$$

9.6 Relative Average Trajectory

9.6.1 Conceptual and Graphical Method

To make the Basic Project it absolutely important to study the Relative Average Trajectory (*RAT*) of the fluid particles, particularly in the flow through the *RO* shells, which occurs partly when in contact with the internal surfaces, partly inside the water, and partly in contact with the surrounding air.

This *RAT* is nothing more than the average displacement of a water particle in contact with the *PHIT RO* shells, from input edge 4 to output edge 5.

For this to occur, this particle must always be in contact with the inner surface of the shell so that it can transmit all the jet energy.

For this to happen, the direction of its theoretical average relative speed, w (m/s) (not considering the effects of turbulence and friction) remains constant.

Thus, to obtain the *RAT* of a flow with a jet of diameter of d_o (m) in Pelton shells with primitive diameters D (m), we will draw the horizontal projection of the jet in the vertical view, along with diameter D, and the external diameter of the dividing edge of the shell D_{e_a} (m), in this view, as shown in Fig. 9.7

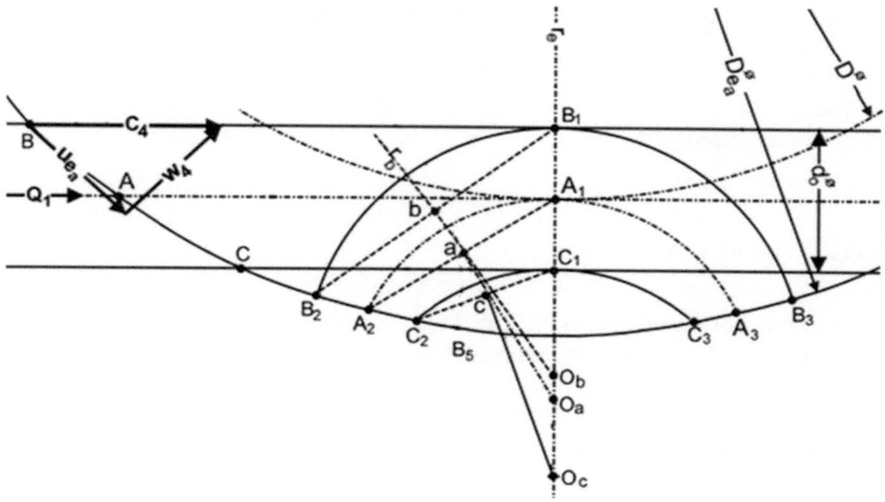

Fig. 9.7 Graphic method for obtaining the *RAT* as arcs of circles

In this graph, under a dynamic aspect, in the same time t (s) that the water particle moves horizontally from B to B_1, with the speed c_4 (m/s), in contact with the inner surface of the shell of the *RO*, which rotates with n (rpm), it moves in the arc from B to B_2 with the tangential speed:

$$u_{e_a} = \frac{\pi \cdot D_{e_a} \cdot n}{60} \ (\mathrm{m/s})$$

therefore …

$$t = \frac{c_4}{B\bar{B}_1} = \frac{u_{e_a}}{\mathrm{arc}(BB_2)} \ (\mathrm{s})$$

… which determines the arc length between B and B_2 via the expression:

$$\mathrm{arc}(BB_2) = \frac{u_{e_a}}{c_4} \cdot B\bar{B}_1 \ (\mathrm{m})$$

Being that the *RAT* that starts at B is limited by the diameter of the flow of d_o, it therefore has to pass through B_1 and leave the inner surface of the shell at B_3. The *RAT*, in first approximation, can be drawn as the arc of the circumference that passes by points B_2; B_1; B_3.

To draw this arc, just connect points B_2 and A_1 with a straight line, then through their midpoint b, draw r_b that determines point O_b on r_e that passes through B_1, the center of the circle arc B_2; B_1; B_3, which is taken as the corresponding *RTM*.

RTM, A_2, A_1, A_3, and C_2, C_1, C_3 were all drawn using the same method.

In the graph of the figure, the average velocity triangle for the particle at point B was drawn, showing the direction of the relative velocity w_4, parallel to the direction of tangent line r_t to the *RTM*.

The length of the circle arc C_2, C_1, C_3, in this drawing, is called the theoretical step between the shells, $pc*$.

9.6.2 *Analytical Method*

The objective of the analytical method, in general, is to generate functions $y = f(x)$ for the *RTM*, and consequently, to obtain the 3D geometry of the inner surface of the shells, starting by using information on the characteristics obtained from the basic hydrodynamic calculation of the *PHIT* via Computational Fluid Dynamics (*CFD*), and complementing this with tests and analysis in laboratory modeling.

Normally, this method uses a pre-established function. Here, the arcs of circles will be used considering the results from the graphic method described in Fig. 9.7, and thus, its basic equation is that of a circle.

$$y = \sqrt{r^2 - x^2} \tag{9.19}$$

9.6.3 *Algorithm and Validation*

9.6.3.1 Input and Validation Data

- D_{e_a}
- D
- d_o(m)
- c_4 (m/s)
- n (rpm)

Validation will be done for:

- $D_{e_a} = 3.770$ (m)
- $D = 3.196$ (m)
- $d_o = 0.261$ (m)
- $c_4 = 104.361$ (m/s)
- $n = 300$ (rpm)

… which will be used to develop of the steps in the layout schematized in Fig. 9.8.

9.6.3.2 Script and Validation

Calculate:

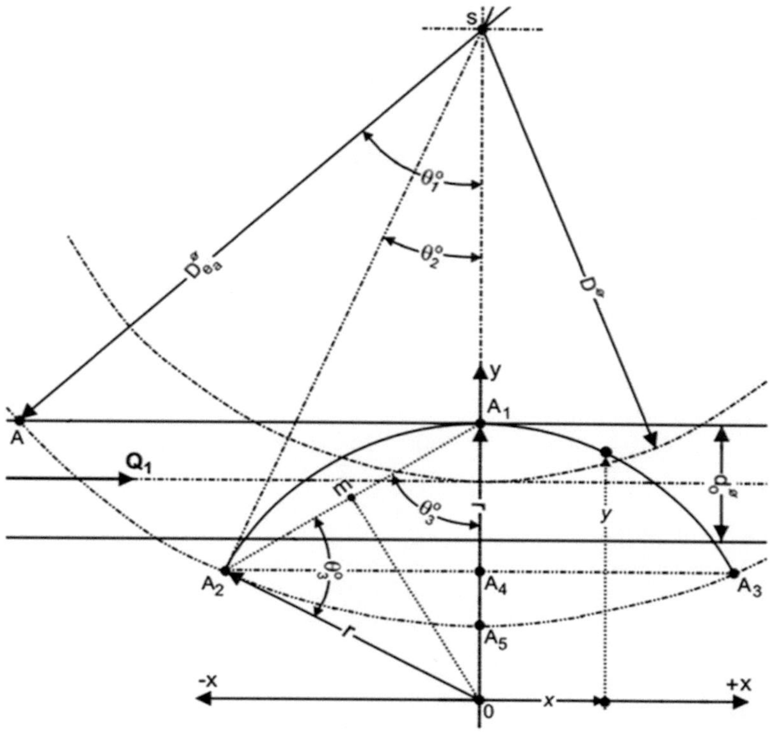

Fig. 9.8 Graph for the *RAT* Equation

1. The length of the segment:

$$\overline{AA_1} = 0.5 \cdot \sqrt{D_{e_a}^2 - (D - d_o)^2} = 0.5 \cdot \sqrt{3.77^2 - (3.196 - 0.261)^2} = 1.183\ (\mathrm{m})$$

2. The tangential velocity at the outer diameter of the dividing edge of the shell:

$$u_{e_a} = \frac{\pi \cdot n}{60} \cdot D_{e_a} = \frac{\pi \cdot 300}{60} \cdot 3.77 = 59.219\ (\mathrm{m/s})$$

3. The angle covered by the arc (*AA*5):

$$\theta_1 = \arcsin \frac{2 \cdot A\bar{A}_1}{D_{e_a}} = \arcsin \frac{2 \cdot 1.183}{3.77} = 38.872^o$$

4. The length of the arc (*AA*5):

$$\text{arc}(AA_5) = \frac{\theta_1 \cdot D_{e_a} \cdot \pi}{360} = \frac{38.872 \cdot 3.77 \cdot \pi}{360} = 1.279 \text{ (m)}$$

5. The length of the arc (*AA*2):

$$\text{arc}(AA_2) = \frac{u_{e_a}}{c_4} \cdot \overline{AA_1} = \frac{59.219}{104.361} \cdot 1.183 = 0.671 \text{ (m)}$$

6. The length of the arc (*A*2*A*5):

$$\text{arc}(A_2A_5) = \text{arc}(AA_5) - \text{arc}(AA_2) = 1.279 - 0.671 = 0.608 \text{ (m)}$$

7. The angle covered by the arc (*A*2*A*5):

$$\theta_2 = \frac{\text{arc}(A_2A_5)}{D_{e_a}} \cdot \frac{360}{\pi} = \frac{0.608}{3.77} \cdot \frac{360}{\pi} = 18.481^\circ$$

8. The length of the segment:

$$\overline{SA_4} = 0.5 \cdot D_{e_a} \cdot \cos\theta_2 = 0.5 \cdot 3.77 \cdot \cos 18.481 = 1.788 \text{ (m)}$$

9. The length of the segment:

$$\overline{A_1A_4} = \overline{SA_4} - 0.5 \cdot (D - d_o) = 1.788 - 0.5 \cdot (3.196 - 0.261) = 0.320 \text{ (m)}$$

10. The length of the segment:

$$\overline{A_2A_4} = 0.5 \cdot D_{e_a} \cdot \sin\,\theta_2 = 0.5 \cdot 3.77 \cdot \sin\,18.481 = 0.598 \text{ (m)}$$

11. The angle:

$$\theta_3 = \arctan\frac{\overline{A_2A_4}}{\overline{A_1A_4}} = \arctan\frac{0.598}{0.320} = 61.848^\circ$$

12. The length of the segment:

$$\overline{A_1A_2} = \frac{\overline{A_2A_4}}{\sin\,\theta_3} = \frac{0.597}{\sin\,61.848} = 0.677 \text{ (m)}$$

13. The length of the radius of the circle:

$$r = \frac{0.5 \cdot \overline{A_1A_2}}{\cos\theta_3} = \frac{0.5 \cdot 0.677}{\cos 61.848} = 0.717 \text{ (m)}$$

14. The *RTM* Equation, $y = \sqrt{r^2 - x^2} = \sqrt{0.717^2 - x^2} = \sqrt{0.5147 - x^2}\ (m)$, with the limit values of the ordinate abscissa in the following intervals,

$$-0.598 \leq x \text{ (m)} = \pm\overline{A_2A_4} \text{ (m)} \leq 0.598$$

and

$$\overline{0A_1} = r = 0.717 \geq y \text{ (m)} \geq \overline{0A_4} = r \cdot \sin\left(0.5 \cdot \theta_3^{\circ}\right) = 0.368$$

It is recommended to use numbers with at least five significant digits in the calculations.

9.7 Number and Step Between the Shells

9.7.1 Considerations

The actual number of shells, z_c and the actual step between shells, p_c, must be as large as possible, so that:

- A change in speed n (rpm) in the *PHIT* is within the limits of the load frequency change, normally less than $\pm 2a\ 3$ (%).
- The entire flow rate Q (m^3/s) supplied by the flow to the shells, together with its average kinetic energy, $0.5 \cdot \rho \cdot Q \cdot c_4^2 = 500 \cdot Q \cdot c_4^2\ (W) = 0.5 \cdot Q \cdot c_4^2\ (kW)$, subtracting the transformed energy losses is $[\sum(escape + friction + outlet + ventilation)]$ and delivered as power on to the *PHIT* axis, p_c (*kW*).
- That the flow outlet of the preceding shell does not hit the back of the preceding shell. In this aspect, the relation T/d_o is fundamentally important for fixing the z_c as the real step p_c between them.

To ensure this, one needs to position most shells, z_c between their first point, point e (input), and the last point s (output), in contact with each jet at the outer diameter D_{e_a} of the dividing edge of the *RO* shells of the *PHIT* (Fig. 9.9).

The graph in Fig. 9.9 shows a schematic cross-section axis drawing of the *RO* of a *PHIT* projecting for only one C_1 shell.

Taking the design characteristics of this *PHIT* as:

- $Q_1 = 5.56$ (m^3/s)
- $H = 578$ (m)
- $n = 300$ (rpm)
- $z_j = 6$
- $D_{e_a} = 3.770$ (m)

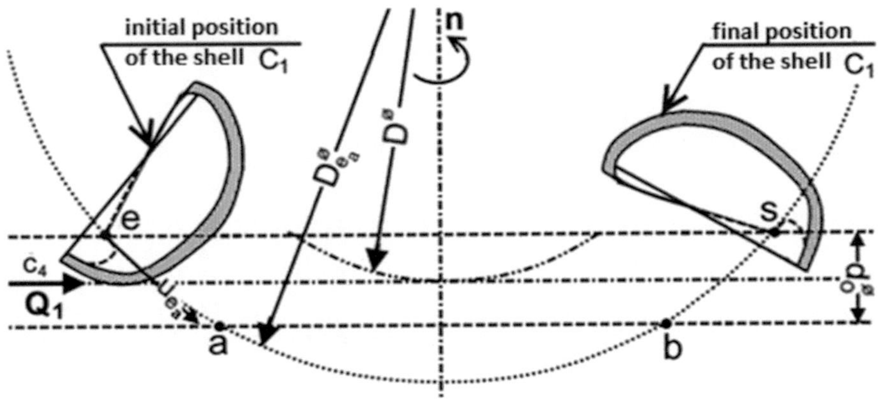

Fig. 9.9 *PHIT RO* Graph showing the extreme positions of the C_1 shell relative to the jet

- $D = 3.196$ (m)
- $d_o = 0.261$ (m)
- $c_4 = 104.361$ (m/s)
- $u_{e_a} = 59.219\ (\mathrm{m/s})$

The question is:

What would happen, theoretically, with this PHIT if it were built and operationalized with a single C_1 shell?

Initially, it would only receive energy during the flow of the jet on its inner surface between its initial and final position during the average time:

$$t = \frac{\overline{es}}{c_4} = \frac{\sqrt{D_{e_a}^2 - (D - d_o)^2}}{c_4} = \frac{\sqrt{3.77^2 - (3.196 - 0.261)^2}}{104.361} = 0.0226728\ (\mathrm{s})$$

In this time, t, the *RO* travels:

$$\mathrm{arc}(es) = u_{e_a} \cdot t = 59.219 \cdot 0.0226728 = 1.343\ (\mathrm{m})$$

This value corresponds to an angle:

$$\theta = \arcsin \frac{2 \cdot \mathrm{arc}(es)}{D_{e_a}} \cdot \frac{180}{\pi} = \arcsin \frac{2 \cdot 1.343}{3.77} \cdot \frac{180}{\pi} = 40.821^{\circ}$$

This result shows that in arc(se) $= \theta * = 360 - \theta = 360 - 40.821 = 319.179^o$ of the *RO*, the *PHIT* would not receive energy, with very large variations in rotation (*frequency* $> 3\%$).

By contrast, considering that each jet submitted to Head H receives a corresponding energy per jet, then the flow, $Q_1^* = \frac{vol(eabs)}{t}$ is much smaller than Q_1, therefore, a lot of power is lost:

$$P_p = g \cdot \left(Q_1 - Q_1^*\right) \cdot H \text{ (kW)}$$

One may acquire the calculations for the values and percentages, as well as comment, by sending a request to the author at zulcysouza@gmail.com.

Now, to guarantee our third need, we need to conduct a structural calculation and the respective executive and manufacturing projects of the *RO*, which will allow us to study the flow behavior in computer operational simulations. This will be very valuable in finally defining the geometry of the *PHIT RO*, particularly the shells, including the cutout, and the back of the shells.

Regarding the preliminary numerical determination of the number of shells, z_c and the respective step p_c between shells, three methods will be studied and analyzed:

- The Graphic Method
- The Theoretical Analytical Method
- The Statistical Method

9.7.2 Graphic Method

9.7.2.1 Methodology

To justify this method, the analysis will be developed by sequencing figures with the not-to-scale schematics, representing C_1 shells; C_2 …, separated by the pitch p_c, from an *RO* of a *PHIT*, which rotates at *n* (rpm). These shells receive energy from the flow Q_1 (m^3/s) in one of the jets with a diameter of $d_o =$ (mm), subject to a Head *H* (mca), which provides the *RO* shells with an average specific kinetic energy at $0.5 \cdot c_4^2$ (J/kg).

- In the schematic in Fig. 9.10A, shell C_2, moves in time *t* (*s*), with the lower end of its dividing edge of the jet from the initial position, point *a* to point *b*, traveling arc *ab*, with the tangential velocity $u_a = r_v \cdot c_4$ (m/s), where $C_4 = const.$ (m/s) is the average flow velocity. With this speed, C_4, and at the same time *t*, the stream of the flow moves horizontally, from point *a* to point *d*, always maintaining contact with the inner surface of shell C_2, therefore, covering the distance $\bar{ad} = \frac{ar(ab)}{r_v}$ (m).

  Thus, as we already saw, any point of the water/shell on *ar*(*ab*) can be determined, or part *bd* of the *PHIT*, represented here by line segment $\overline{bd}$. In Fig. 9.10B, a cross section of the jet is shown, divided into two parts by a plane containing a horizontal line originating at point *b*, which divides Q_1 (m^3/s) into

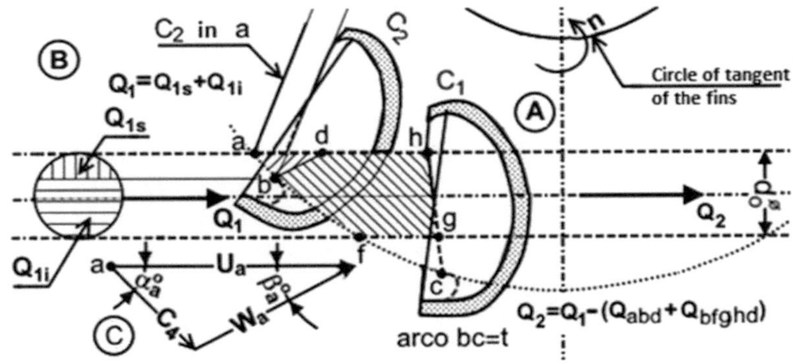

Fig. 9.10 Jet displacement from *a* to *d* and shell rotation C_2 between *a* to *b* at the same time *t*

two flows, $Q_{1s} < Q_{1i}$ (m^3/s). The upper part of the jet (vertical dashes) limited by the plane passing through the straight line superiorly delimits the part of the jet flow where flowing volume (*abd*) (dashes to the right) delivers energy to shell C_2 through the internal surfaces in time *t* (*s*). At the same time, the lower part of the jet delivers energy to shell C_1 through the internal surfaces, with energy coming from the volume flow (*bfghe*) (dashes to the left). In this case, with only two shells, the flow used to transform energy delivered to the *RO* by a jet can be calculated by:

$$Q_r = \frac{vol(abd) + vol(bfghd)}{t} = Q_{abd} + Q_{bfghd}\ \left(\mathrm{m}^3/\mathrm{s}\right)$$

… giving, therefore, the flow of the jet that does not participate in this energy transformation (available flow of the jet in the *RO* output, $Q_2 = Q_1 - Q_r\,(m^3/s)$. In Fig. 9.8C, the triangle with the average speeds at point *a* was drawn, which allows us to observe that the relative speed, w_a (m/s) is centrifugal.

- In the schematic in Fig. 9.11A, the *RO* rotating with *n* (rpm) with the same two shells, and now with shell C_2, rotating in a time, $t* > t\ (s)$, with point *a* at the upper end of the jet, rotates until position *b* at the lower end. During this same rotation and time, shell C_1 moves from p_{c*} = arc (*BC*) always observing that $\bar{ad} = \frac{ar(ab)}{r_v}$ (m). In this case, everything happens as was described in the schematic in Fig. 9.10A, and one should observe that the energy of the flow increases, delivered by the volumes to the inner surfaces of the shells in Fig. 9.10B, which are shell C_2, volume (*abd*), and shell C_1, volume (*bgd*) respectively, give the flow generating energy as $Q_{v*} = \frac{vol(abd)+vol(bgd)}{t*}$ less than Q_1, leaving the following flow to generate energy, $Q_2 = Q_1 - Q_{v*}\ (m^3/s)$.
- In the schematic of Fig. 9.12, shell C_3 enters more, which moves from *a* to *b* with tangential velocity u_a = (m/s),while the upper jet particle, with the velocity c_4 = (m/s), moves through segment c_4 = (m/s).

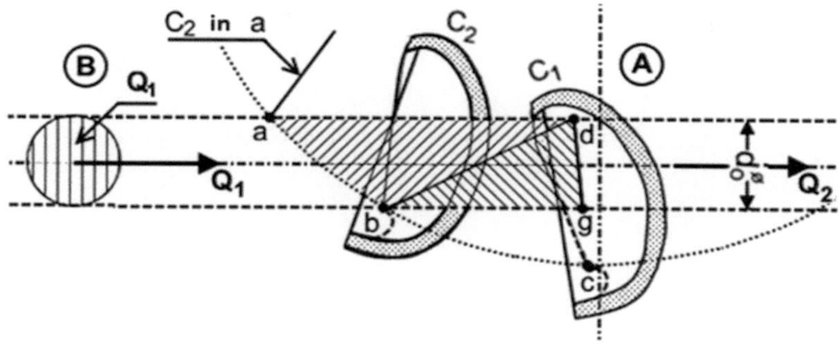

Fig. 9.11 Jet displacement from a to d and shell C_2 rotation between a and b at the same time $t* > t\,(s)$

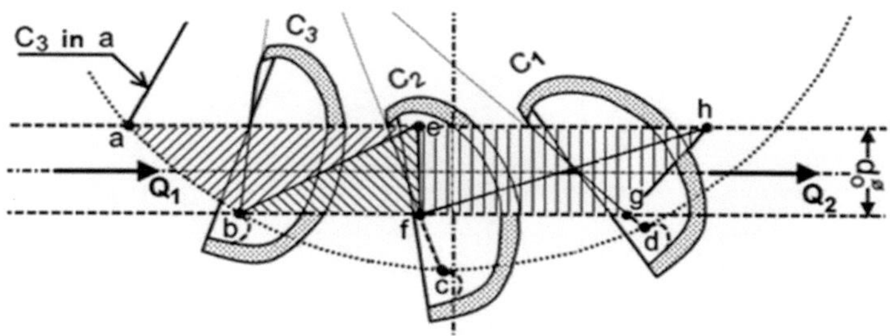

Fig. 9.12 Jet displacement with *RO* rotation with three shells

In this case, the energy delivered by the volumes to the inner surfaces of the shells for C_3 volume (*abe*), C_2 volume (*bfe*), and C_1 volume (*fghe*), respectively, i.e. flow, $Q_{v**} = \frac{vol(abe)+vol(bef)+vol(fghe)}{t*} > Q_{v*}$ is still smaller than Q_1, leaving the following flow to generate energy $Q_2 = Q_1 - Q_{v**}$ (m^3/s).

- Finally, the diagram in Fig. 9.13 shows that the *RO* needs to entirely cover the jet of 4 shells entering shell C_5 in the upper fillet of the jet, while this fillet exits at *f* in shell C_1 at the same time.

Figure 9.14, shows the full distribution of the entire flow Q_1 of the jet, always for shell 4 of the *RO* when it rotates at n (rpm) and receives the available hydraulic energy H_0, minus the losses between section 0; 4, which will be transformed into mechanical energy on the axis of the *PHIT*.

The numerical application below shows that this graphical method allows us to obtain z_c* and p_c*, since, as we will see, dimensional structural problems and flow problems at the exit of the shell jet must be considered when obtaining a real integer number of shells, z_c, and the corresponding real step, p_c.

Fig. 9.13 RO covering the entire jet

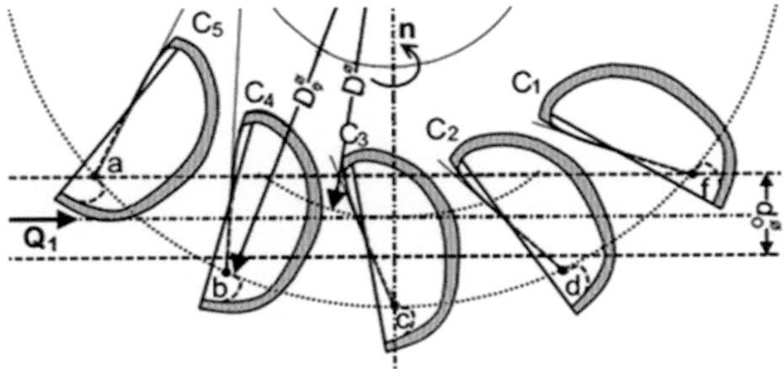

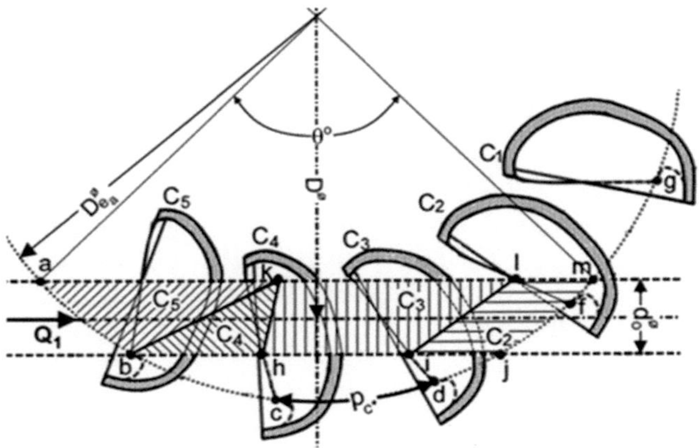

Fig. 9.14 Distribution of all the Q_1 flow on the 4 *RO* shells

9.7.2.2 Application

- **Input data**
  - $D_{e_a} = 3{,}770$ (m)
  - $D = 3.196$ (m)
  - $d_0 = 3.196$ (m)
  - $c_4 = 104.361$ (m/s)

Symbology in Fig. 9.14.

- Steps and Calculations

  Calculate:

1. The line segment:

$$\overline{am} = \sqrt{D_{e_a}^2 - (D - d_o)^2} = \sqrt{3.77^2 - (3.196 - 0.261)^2} = 2.366\ (m)$$

2. The angle of reach of segment $\bar{am}$:

$$\theta = 2 \cdot \arcsin \frac{\overline{am}}{D_{e_a}} = 2 \cdot \arcsin \frac{2.366}{3.77} = 77.74^{\circ}$$

3. The theoretical number of shells:

$$z_{c*} = \frac{z_{c_{af}}}{\theta} \cdot 360 = \frac{4}{77.74} \cdot 360 = 18.5$$

4. The theoretical step between the shells:

$$p_{c*} = \frac{\pi \cdot D_e}{z_{c*}} = \frac{\pi \cdot 3.77}{18.5} = 0.640\ (m) \rightarrow \alpha_{c*} = \frac{360}{18.5} = 19.459^{\circ}$$

9.7.3 Theoretical Analytical Method

9.7.3.1 Considerations

To explain this method for obtaining the theoretical step, $pc*$; $\alpha c*$ we will use Fig. 9.15, the basic drawing of which has already been shown in previous subitems (see the Item 9.6 drawing of the *PHIT*.

The basis for the analytical method is that the time t (s) that particle c of the flow in the jet with average absolute velocity c_4 (m/s) takes to travel straight segment $a\bar{a}_1$ (m) (Fig. 9.5) is the same as the outer end from the dividing line of the shell traversing the arc of circumference, arc (aa_2) (m), with the tangential speed of the rotor u_{e_c} (m/s), i.e.:

$$\text{arc}(aa_2) = \frac{u_{e_a}}{c_4} \cdot \overline{aa_1}\ (\text{m})$$

9.7.3.2 Algorithm and Validation

- **Input and validation data**
  - $D_{e_a} = 3.770$ (m)
  - $D = 3.196$ (m)
  - $d_0 = 0.261$ (m)
  - $c_4 = 104.361$ (m/s)
  - $u_{e_a} = 59.29$ (m/s)

Symbology Fig. 9.15.

- Steps and Calculations

  Calculate:

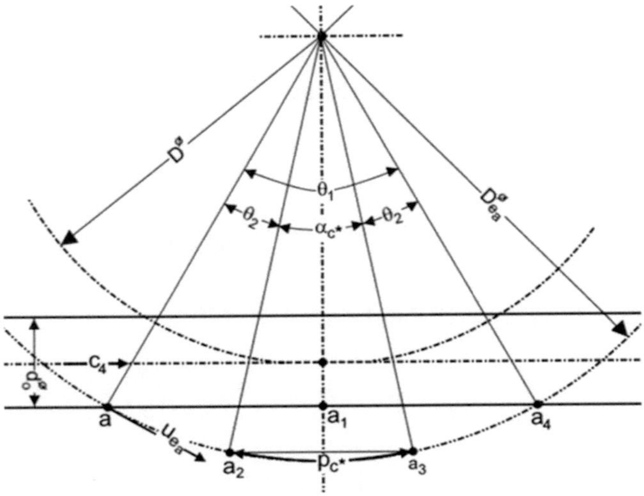

Fig. 9.15 Graph for scripting the analytical method

1. The line segment:

$$\overline{aa_1} = 0.5 \cdot \sqrt{D_{e_a}^2 - (D + d_o)^2} = 0.5 \cdot \sqrt{3.77^2 - (3.196 + 0.261)^2} = 0.752 \text{ (m)}$$

2. The line segment:

$$\overline{aa_4} = 2 \cdot \overline{aa_1} = 2 \cdot 0.752 = 1.504 \text{ (m)}$$

3. The angle:

$$\theta_1 = 2 \cdot \arcsin \frac{\overline{aa_4}}{D_{e_a}} = 2 \cdot \arcsin \frac{1.504}{3.77} = 47.024^\circ$$

4. The arc:

$$\text{arc}(aa_2) = \frac{u_{e_a}}{c_4} \cdot \overline{aa_1} = \frac{59.219}{104.361} \cdot 0.752 = 0.427 \text{ (m)}$$

5. The angle:

$$\theta_2 = \frac{2 \cdot \text{arc}(aa_2)}{D_{e_a}} = \frac{2 \cdot 0.427}{3.77} = 0.226375 \ (rd) = 0.226375 \cdot \frac{180}{\pi} = 12.970^\circ$$

6. The theoretical step between shells in degrees:

$$\alpha_{c*} = \theta_1 - 2 \cdot \theta_2 = 47.024 - 2 \cdot 12.97 = 21.083^\circ$$

7. The theoretical number of shells:

$$z_{c*} = \frac{360}{\alpha_{c*}} = \frac{360}{21.083} = 17.075 \; shells$$

8. The length of the circumference D_{e_a}:

$$C_{D_{ea}} = \pi \cdot D_{e_a} = \pi \cdot 3.77 = 11.844 \text{ (m)}$$

9. The theoretical pitch between shells in meters:

$$p_{c*} = \frac{C_{D_{ea}}}{z_{c*}} = \frac{11.844}{17.075} = 0.694 \text{ (m)} \rightarrow \alpha_{c*} = \frac{360}{z_{c*}} = \frac{360}{17.075} = 21.083^\circ$$

9.7.4 Statistical Method

This method is based on expressions obtained from the analyses on the number of *PHIT RO* shells that have shown efficient operational performance, whether in laboratory tests, or via the *PHIT* installed on site, whether inside or outside of the warranty period.

The graph in Fig. 9.16 shows the result obtained with a plot of 36 points taken from micro, small and large *RO*.

- **Input and validation data**
  - $Q_1 = 5.562$ (m3s)
  - $H = 578$ (m)

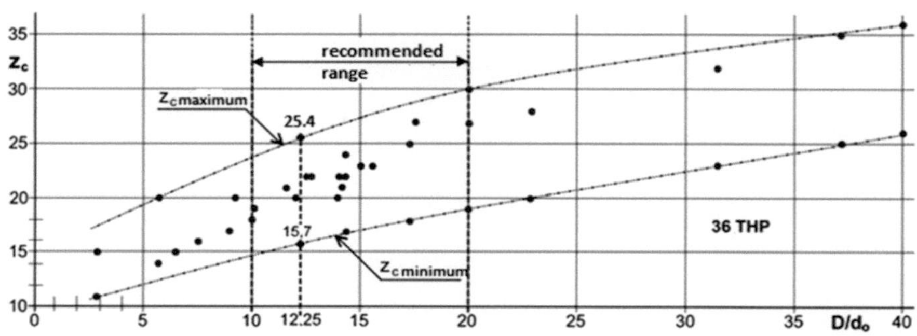

Fig. 9.16 Graph of expressions of the limit functions of the number of shells for 36 *RO* in a *PHIT*

- $n = 300$ (rpm)
- $D_{e_a} = 3.770$ (m)
- $D = 3.196$ (m)
- $d_o = 0.261$

Calculate:

$$D/d_o = 3.196/0.261 = 12.25$$

… which corresponds, in the graph, to the following limits for the actual numbers of shells:

$$z_{c_{\min}} = 15.7$$

and

$$z_{c_{\text{mam}}} = 25.4$$

… or considering the integer values:

$$z_{c_{\min}} = 16 \rightarrow z_{c_{\text{mam}}} = 25$$

… thus the corresponding steps are obtained:

$$p_{c_{\text{mim}}} = \frac{\pi \cdot D_{e_a}}{z_{c_{\max}}} = \frac{\pi \cdot 3.77}{25} = 0.476\ (\text{m}) \rightarrow \alpha_{a_{\min}} = \frac{360}{25} = 14.4^{\circ}$$

and

$$p_{c_{\max}} = \frac{\pi \cdot D_{e_a}}{z_{c_{\min}}} = \frac{\pi \cdot 3.77}{16} = 0.740\ (\text{m}) \rightarrow \alpha_{c_{\max}} = \frac{360}{16} = 35^{\circ}$$

9.7.5 Real Step

If the values obtained in the graphical method, with the limits established in Fig. 9.16 as a reference for first fixing the real pitch between the shells p_c(m), α_c(*degrees*),and consequently from the number of shells of the RO, z_c, which resulted, in:

$$p_{c_{\min}} = 0.476\ (\text{m}) \rightarrow \alpha_{c_{\min}} = 14.4^{\circ} \rightarrow z_{c_{\min}} = 10\ \textit{shells}$$

and

$$p_{c_{\max}} = 0.740\ (\text{m}) \rightarrow \alpha_{c_{\max}} = 35^{\circ} \rightarrow z_{c_{\max}} = 25\ \textit{shells}$$

… **are** related to those obtained by the two previous methods, then:

- Graphic Method

$$r_{p_{cg_{\min}}} = \frac{p_{c_{\min}}}{p_{c*_g}} = \frac{0.476}{0.640} = 0.744 \rightarrow r_{p_{cmam}} = \frac{p_{c_{\max}}}{p_{c*_g}} = \frac{0.740}{0.640} = 1.156$$

- Analytical Method

$$r_{p_{ca_{\min}}} = \frac{p_{c_{\min}}}{p_{c*_a}} = \frac{0.476}{0.694} = 0.686 \rightarrow r_{p_{cg_{\max}}} = \frac{p_{c_{\max}}}{p_{c*_a}} = \frac{0.740}{0.694} = 1.066$$

These results are expected, since the absolute average velocity of flow c_4 is greater than the tangential one, u_{e_a} of the *RO* in D_{e_a}, or here $\frac{c_4}{u_{ea}} = \frac{104.361}{59.219} = 1.762$.

Currently, we can position $z_c = \frac{\pi \cdot D_{ea}}{p_c}$ by taking a value for the real p_c in the interval, e.g.:

If the theoretical values obtained by these two methods are taken as initial references for fixing the relationship r_{p_c} between the real step, p_c, and the theoretical step, $pc*$, among the shells, this relationship would be between the one currently recommended in the bibliography, as:

$$0.65 \leq r_{p_c} = p_c/p_{c*} \leq 0.80 \tag{9.20}$$

9.8 The Flows Directed to each Shell in the Active Positions

9.8.1 Considerations and Equation

The graphic method indirectly showed how to obtain the number of shells of the *PHIT RO* in operation that receive the jet energy, called here *active shells*, through a sequence of not-to-scale figures, representing shells C_1, C_2 ..., supposedly separated by pitch p_c, from a *PHIT RO* rotating at *n* (rpm). These shells receive energy from a flow Q_1 (m^3/s) in one of their jets, subjected to a head *H* (mca), which gives rise to a cylindrical jet with a diameter d_o(mm), and an average specific kinetic energy at $0.5 \cdot c_4^2$ (J/kg).

To establish the equation, we will initially consider the time $t_o(s)$ that, with uniform movement, (Fig. 9.17) it takes the end of an *RO* shell to move from position *a* to position *b* with tangential velocity $u_{e_a} = \frac{\pi \cdot D_{e_a} \cdot n}{60}$ (m/s). This time can be determined as long as the arc length (*ab*) is known, via expression:

$$t_{ab} = \frac{\text{arc}(ab)}{u_{e_a}} \ (s) \tag{9.21}$$

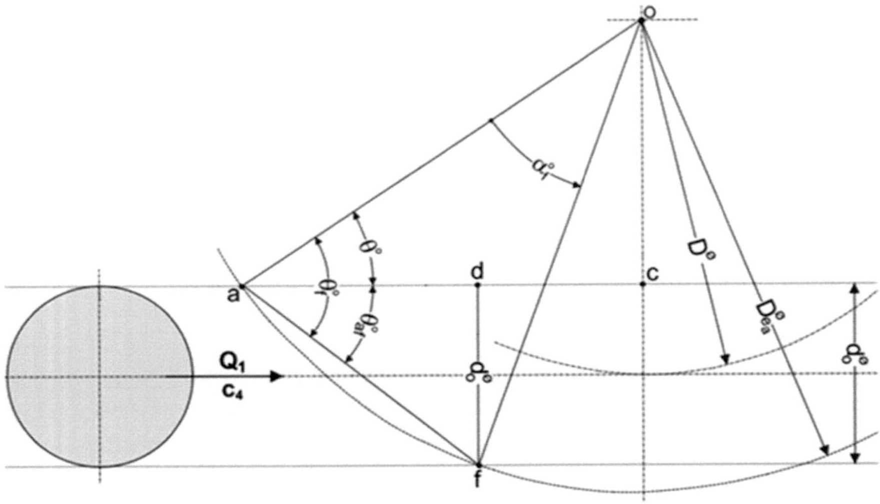

Fig. 9.17 Diagram for determining the angle α_f

During this time, as was imposed in Fig. 9.10, where this shell receives the flow $Q_{1_{ab}} = \frac{vol(abd)}{t_{ab}}$ (m/s), which can also be determined if the cross section of the jet is determined, A_{j_1}, in Fig. 9.10, represented with vertical crosshatches, i.e., $Q_{1_s} = A_{j_{1_s}} \cdot c_4$ (m/s).

To determine the angle of coverage of the jet α_f° when the shell rotates along arc (*af*), in Fig. 9.17, receiving flow Q_1 (m^3/s), we have:

The right triangle (*Oac*) as:

$$\overline{ac} = 0.5 \cdot \sqrt{D_{e_a}^2 - (D - d_o)^2} \rightarrow \theta = \arccos\left(\frac{2 \cdot \overline{ac}}{D_{e_a}}\right) \quad (9.22)$$

From the equilateral triangle (*Oaf*) we get:

$$\overline{af} = D_{e_a} \cdot \sin\left(0.5 \cdot \alpha_f\right) \rightarrow \theta_f = 90 - 0.5 \cdot \alpha_f \quad (9.23)$$

From the right triangle (*afd*) we have:

$$d_o = \overline{af} \cdot \sin\, \theta_{af} = \overline{af} \cdot \sin\, \left(\theta - \theta_f\right) = \overline{af} \cdot \sin\, \left(90 - 0.5 \cdot \alpha_f - \theta\right) \quad (9.24)$$

Combining these last three expressions, we get:

$$D_{e_a} \cdot \sin(0.5 \cdot \alpha_f) = \frac{do}{\sin(90 - 0.5 \cdot \alpha_f - \theta)} \rightarrow \sin(0.5 \cdot \alpha_f) \cdot \sin\left(90 - 0.5 \cdot \alpha_f - \theta = \frac{d_o}{D_{e_a}}\right) \tag{9.25}$$

Considering the trigonometric expression, sin $a \cdot$ sin $b = 0,5 \cdot [\cos(a - b) - \cos(a + b)]$ and setting $a = 0.5 \cdot \alpha_f$ and $b = 90 - 0.5 \cdot \alpha_f - \theta$ in the expression (9.25), we have:

$$\sin(0.5 \cdot \alpha_f) = 0.5 \cdot [\cos(0.5 \cdot \alpha_f - 90 + 0.5 \cdot \alpha_f + \theta) - \cos(0.5 \cdot \alpha_f + 90 - 0.5 \cdot \alpha_f - \theta)]$$

or

$$\cos(\alpha_f - 90 + \theta) = \cos(90 - \theta) + \frac{2 \cdot d_o}{D_{e_a}} \rightarrow \alpha_f = (90 - \theta) - \text{arc}\left[\cos(90 - \theta) + \frac{2 \cdot d_o}{D_{e_a}}\right] \tag{9.26}$$

Once the jet/shell coverage angle is determined, α_f, it is possible, as will be shown, to obtain the flow distribution Q_1 in the active rotor shells.

Therefore, in Fig. 9.18, one must know the positions of the activity arcs (aj). With these, once the angles $\alpha_j^\circ < \alpha_f^\circ$ have been fixed, and the areas of the jet $A_j = area\ (nmpqn)$ cross-sections have been determined, the corresponding flow rates can be obtained, $Q_{aj} = A_j \cdot c_4$, as will be shown in sequence, observing the symbology from the graphs in Fig. 9.18.

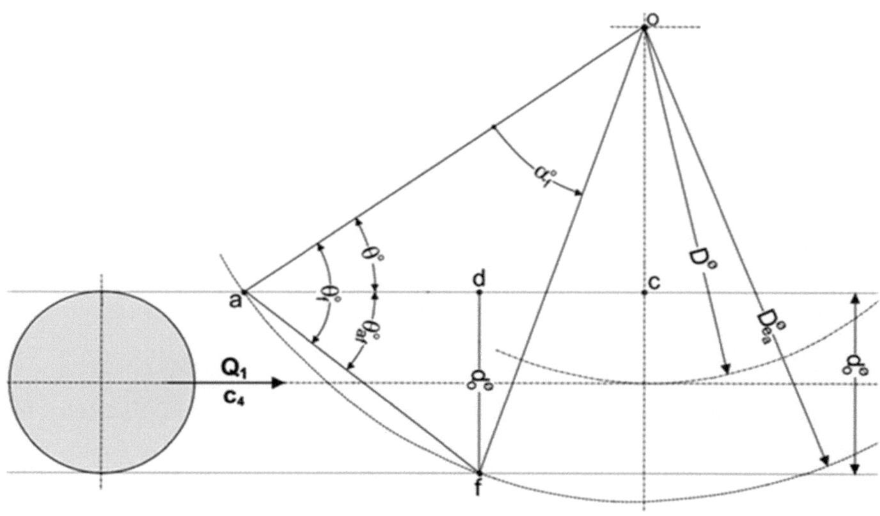

Fig. 9.18 Diagram for equating the directed flow for each shell in its active position. (**a**) jet trajectory; (**b**) and (**c**) cross sections

- Figure 9.18b—$h_j < d_o/2$

$$y_j = 0.5 \cdot d_o - h_j \rightarrow x_j = 0.5 \cdot d_o \cdot \cos\delta_j = 0.5 \cdot \sqrt{d_o^2 - (d_o - 2 \cdot h_j)^2}$$

or

$$\delta_j = \arccos\sqrt{1 - \left(1 - \frac{2 \cdot h_j}{d_o}\right)^2} \quad (^\circ) \tag{9.27}$$

The isosceles triangle (*opm*) allows us to write:

$$\beta_j = 100 - 2 \cdot \delta_j$$

... and the area (*opqn*) of this sector, as:

$$A_{\beta_j} = \frac{\beta_j}{360} \cdot A_o \rightarrow A_o = \frac{\pi \cdot d_o^2}{4} \tag{9.28}$$

The area of the isosceles triangle (*opm*):

$$A_{\Delta_j} = y_j \cdot x_j = 0.5 \cdot (0.5 \cdot d_o - h_j) \cdot \cos\delta_j \tag{9.29}$$

Resulting in the cross section of the jet:

$$A_j = A_{\beta_j} - A_{\Delta_j} \tag{9.30}$$

- Figure 9.18c $h_j < d_o/2$

$$y_j = h_j - 0.5 \cdot d_o \rightarrow x_j = 0.5 \cdot d_o \cdot \cos\delta_j = 0.5 \cdot \sqrt{d_o^2 - (2 \cdot h_j - d_o)^2}$$

or

$$\delta_j = \arccos\sqrt{1 - \left(\frac{2 \cdot h_j}{d_o} - 1\right)^2} \quad (^\circ) \tag{9.31}$$

This expression is formally equal to expression (9.27), since algebraically, the square of a number is always positive.

The isosceles triangle (*opm*) allows us to write:

$$\beta_j = 180 + 2 \cdot \delta_j$$

… and the area (*opqn*) in this sector, as:

$$A_{\beta_j} = \frac{\beta_j}{360} \cdot A_o \rightarrow A_o = \frac{\pi \cdot d_o^2}{4} \tag{9.32}$$

The area of the isosceles triangle (*opm*):

$$A_{\Delta_j} = y_j \cdot x_j = 0.5 \cdot \left(h_j - 0.5 \cdot d_o\right) \cdot \cos \delta_j \tag{9.33}$$

The cross section of the jet, is then:

$$A_j = A_{\beta_j} + A_{\Delta_j} \tag{9.34}$$

Once the angle α_f has been subdivided into at least seven parts, this equation will allow us to determine the points for the drawing functions $Q_{1_f} = f\left(t_{o_j}\right)$, and $A_j = f\left(t_{oj}\right)$, with a reasonable degree of uncertainty, as will be shown in the next subitem, using an algorithm and respective validation, thus achieving the goal of this item, which was:

To determine the flow directed to each shell in its active position.

9.8.2 Algorithm and Validation

The objective of this subitem is to establish and validate an algorithm based on knowledge of the following *PHIT* characteristics $(H; Q_1; c_4; n; z_c; p_c; D_{e_a}; D; d_o)$ which provides:

The number of active shells, the areas, and the jet flow rates delivered to each shell, while they are active.

9.8.2.1 Input and Validation Data

- $H = 500$ (m)
- $Q_1 = 2.470$ (m3/s)
- $N = 450$ (rpm)
- $c_4 = 97.065$ (m/s)
- $z_c = 18$ *shells*
- $p_c = 0.4304$ (m) – 20°

- $D = 1.980$ (m)
- $d_o - 0.780$ (m)
- $D_{e_a} = 2.466$ (m)

9.8.2.2 Script and Calculations

Using the symbology from Figs. 9.17 and 9.18, calculate:

1. The line segment:

$$\overline{ac} = 0.5 \cdot \sqrt{D_{e_a}^2 - (D - d_o)^2} = 0.5 \cdot \sqrt{2.466^2 - (1.98 - 0.18)^2} = 0.8428 \text{ (m)}$$

2. The angle:

$$\theta^\circ = \arccos\left(\frac{2 \cdot \overline{ac}}{D_{e_a}}\right) = \arccos\left(\frac{2 \cdot 0.8428}{2.466}\right) = 46.880^\circ$$

3. The angle:

$$\alpha_f = (90 - \theta) - \text{arc}\left[\cos(90 - \theta) + \frac{2 \cdot d_o}{D_{e_a}}\right]$$
$$= (90 - 46.880) - \text{arc}\left[\cos(90 - 46.880) + \frac{2 \cdot 0,18}{2.466}\right] = 14.273^\circ$$

4. The angles of the interval of functions:

$$\Delta = \frac{\alpha_f}{j} = \frac{14.273}{8} = 1.78412$$

   In the validation we will use seven intervals of 2° and 1 interval of 0.273°.
5. With the data and calculated values, draw Fig. 9.19.
6. The coverage angle with the flow $Q_1$and the respective arc length:

$$\alpha_{Q_1} = p_c - \alpha_f = 20 - 14.273 = 5.727\ (^\circ) = 123.25 \text{ (mm)}$$

7. The arc span where the shells receive flow:

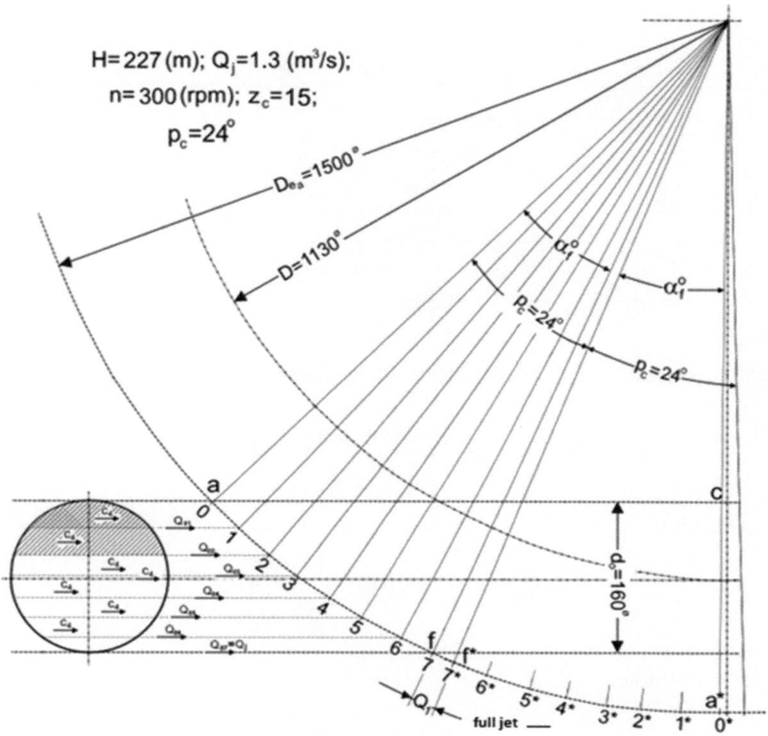

Fig. 9.19 Initial to-scale drawing to obtain the functions of the Q_1 distribution in *PHIT* active shells

$$\mathrm{arc}(aa*) = \mathrm{arc}(00*) = \frac{(p_c + \alpha_f) \cdot D_{e_a} \cdot \pi}{360} = \frac{(20 + 14.273) \cdot 2466 \cdot \pi}{360} = 737.55\ (\mathrm{mm})$$

8. To continue the script, the graphic reference will be Fig. 9.20, with a detailed scale of Fig. 9.19, filling in lines from Table 9.5, as follows:
9. In the characteristics and unit line, the point numbers in arc (*af*) or 08 from Fig. 9.19:
10. In line 1, the angles α_{0j}° :
11. In line 2, the arcs:

$$\mathrm{arc}(0J) = \frac{\pi \cdot D_{e_a}}{360} \cdot \alpha_{0j} = \frac{\pi \cdot 2466}{360} \cdot \alpha_{0j} = 21.51991 \cdot \alpha_{0j}\ (\mathrm{mm})$$

12. In line 3, the displacement times of the lower end of the dividing edge of the shell with the tangential velocity:

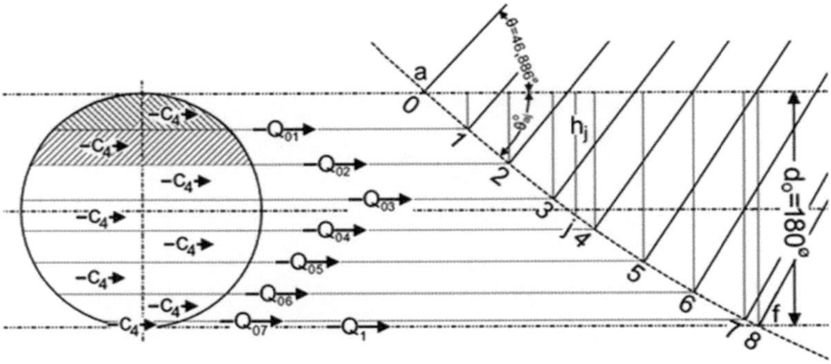

Fig. 9.20 Detailed drawing of Fig. 9.19 for clarifying the steps in sequence

$$u_{e_a} = \frac{\pi \cdot D_{e_a} \cdot n}{60} = \frac{\pi \cdot 2.446 \cdot 450}{60} = 57.6325 \text{ (m/s)}$$

Therefore:

$$10^3 \cdot t_{0j} = \frac{\text{arc}(0j)}{u_{e_a}} = \frac{\text{arc}(0j)}{57.6325}$$

For example:

$$10^3 \cdot t_{01} = \frac{\text{arc}(01)}{u_{e_a}} = \frac{43.04}{57.6325} = 0.746\ (s)$$

Or:

$$t_{01} = 0.746 \cdot 10^{-3}\ (s)$$

13. On line 4, the segments:

$$\overline{0j} = D_{e_a} \cdot \sin\left(0.5 \cdot \alpha_{0j}\right) = 2466 \cdot \sin\left(0.5 \cdot \alpha_{0j}\right) \text{ (mm)}$$

14. On line 5, the angles:

$$\theta_{0j} = 90 - \theta - 0.5 \cdot \alpha_{0j} = 90 - 46.88 - 0.5 \cdot \alpha_{0j} = 43.12 - 0.5 \cdot \alpha_{0j}\ (^\circ)$$

15. In line 6, the heights of the sectors:

$$h_j = \overline{0j} \cdot \sin\ \theta_{0j}$$

16. In line 7, the angles:

Table 9.5 Calculation of characteristics

Line	Caracteristics	Unit	1	2	3	4	5	6	7	8
1		degrees	2.0	4.0	6.0	8.0	10.0	12.0	14.0	14.3
2	arc(0j)	mm	43.04	86.08	129.12	172.16	215.20	258.24	301.28	307.15
3	$10^{3\cdot} t_{0J}$	s	0.75	1494.00	2240.00	2987.00	3.74	4.48	5.23	5.33
4	0_j	mm	43.04	86.06	129.06	172.02	214.93	257.77	300.53	306.36
5	θ_{0j}	degrees	42.1	41.1	40.1	39.1	38.1	37.1	36.1	36.0
6	h_j	mm	28.87	56.60	83.18	108.54	132.68	155.56	177.16	180.00
7	j	degrees	42.8	21.8	4.4	11.9	28.3	46.8	75.6	90.0
8	β_j	degrees	94.4	136.4	171.3	203.8	236.6	273.5	331.1	360.0
9	$A_{\beta j}$	cm^2	66.76	96.44	121.09	144.04	167.26	193.34	234.07	254.47
10	$A_{\Delta j}$	cm^2	22.43	15.51	3.40	9.07	18.79	22.46	10.86	0.00
11	A_j	cm^2	44.33	80.93	124.49	153.11	186.05	215.80	244.93	254.47
12	Q_j	m^3/s	0.43	0.79	1.21	1.49	1.81	2.10	2.38	2.47

$$\delta_j = \arccos\sqrt{1-\left(1-\frac{2\cdot h_j}{d_o}\right)^2}$$
$$= \arccos\sqrt{1-\left(1-\frac{2\cdot h_j}{180}\right)^2} = \arccos\sqrt{1-\left(1-\frac{h_j}{90}\right)^2}\ (^\circ)$$

17. In line 8, the angles covered by the sectors of the circles:

$$h_j < 0.5\cdot d_o < 9\,0\ (\mathrm{mm}) \rightarrow \beta_j = 180 - 2\cdot\delta_j \cos\sqrt{1-\left(1-\frac{h_j}{90}\right)^2}\ (^\circ)$$

and

$$h_j > 0.5\cdot d_o > 90\ (\mathrm{mm}) \rightarrow \beta_j = 180 + 2\cdot\delta_j\ (^\circ)$$

18. The transverse surface of the jet:

$$A_o = \frac{\pi\cdot d_o^2}{4} = \frac{\pi\cdot 18^2}{4} = 254.469\ (\mathrm{cm}^2)$$

19. In line 9, the transverse surfaces covered by the sectors of the circles:

$$A_{\beta_j} = \frac{\beta_j}{360}\cdot A_0 = \frac{\beta_j}{360}\cdot 254.469 = 0.70686\cdot\beta_j\ (\mathrm{cm}^2)$$

20. In line 10, the surfaces of isosceles triangles (*oab*), for:

$$h_j < 0.5\cdot d_o = 90\ (\mathrm{mm}) \rightarrow A_{\Delta_j} = 0.5\cdot(0.5\cdot d_o - h_j)\cdot\cos\delta_j = 0.5\cdot(90 - h_j)\cdot\cos\delta_j$$

and for:

$$h_j > 0.5\cdot d_o = 90\ (\mathrm{mm}) \rightarrow A_{\Delta_j} = 0.5\cdot(h_j - 0.5\cdot d_o)\cdot\cos\delta_j = 0.5\cdot(h_j - 90)\cdot\cos\delta_j$$

21. In line 11, the transverse surfaces of the jet, for:

$$h_j < 0.5\cdot d_o = 90\ (\mathrm{mm}) \rightarrow A_j = A_{\beta_j} - A_{\Delta_j}$$

and for:

$$h_j > 0.5\cdot d_o = 90\ (\mathrm{mm}) \rightarrow A_j = A_{\beta_j} + A_{\Delta_j}$$

22. In line 12, the flows accumulated in the segments of the cross sections of the jet:

$$Q_j = c_4 \cdot A_j \cdot 10^{-4} = 0.0097065 \cdot A_j = 0.0097065 \cdot A_j$$

23. With data in the table supplemented by:

$$\alpha_{08*} = 20\ (^{\circ})$$

$$10^3 \cdot t_{08*} = 7{,}475\ (s)$$

$$\text{arc}(08*) = \text{arc}(08) + \text{arc}(88*) = 307.15 + 123.25 = 430.40\ (\text{mm})$$

The functions were drawn:

$$Q_j = f[\text{arc}(0j)]$$

... made available by the jet for 1 shell (Fig. 9.21a), and for 4 active shells (Fig. 9.21b).

24. The calculations performed in the previous steps, show that the jet $Q_{j_i} = Q_{1_i}$ provides flows covering arc AC (Fig. 9.22) at 737.55 (mm), with average absolute speed $c_4 = 97.065$ (m/s),with a maximum value at $Q_{jet_{max}} = 2.47$ (m^3s) in the *ab* arc at 123.5 (mm). While this occurs, the shell, rotating with tangential velocity $u_{e_a} = 57.633$ (m/s), receives flows in a larger arc $AE = \frac{c_4}{u_{ea}} \cdot 737.55 = \frac{97.065}{57.633} \cdot 737.55 = 1242.18$ (mm), but with an inverse relationship, $\frac{u_{ea}}{c_4} = \frac{57.633}{97.065} = 0.5938$, with a maximum value at $Q_{shell_{\max}} = \frac{u_{ea}}{c_4} \cdot Q_j = 0.5938 \cdot 2.47 = 1.467\ (m^3/s)$, covering $ara * b * = \left(\frac{123.5}{0.5938}\right) = 207.98$ (mm).

Given these characteristics, and the values in lines 2 and 12 of Table 9.5, the values transcribed in Table 9.6 were calculated, which allowed us to draw Fig. 9.22.

The graphs in Fig. 9.22, represented by the equal areas *ABCA* and *ADEA*, show how the transfer of energy from the jet to the shell of the *PHIT RO* occurs. This is important since, even though these energies are the same, their impact on the shell is smaller, since $Q_{i_{\max}} < Q_{c_{\max}}$.

25. Considering the average values of the shells shown in Fig. 9.4, the following characteristics were calculated in (mm):

- $T = 1.3 \cdot do = 1.30 \cdot 180 \cong 203$
- $B = 3 \cdot do = 3 \cdot 180 = 540$
- $L = 2.9 \cdot do = 2.9 \cdot 180 = 522$
- $E = 1.3 \cdot do = 1.30 \cdot 180 = 234$
- $J = 0.39 \cdot do = 0.39 \cdot 180 \cong 70$
- $F = 0.983 \cdot do = 0.983 \cdot 180 = 177$

Figure 9.23 was drawn with these values and $\alpha_a = 12^{\circ}$, in scale, and using a respective cutaway of *AA*, along with the ellipses with the equations:

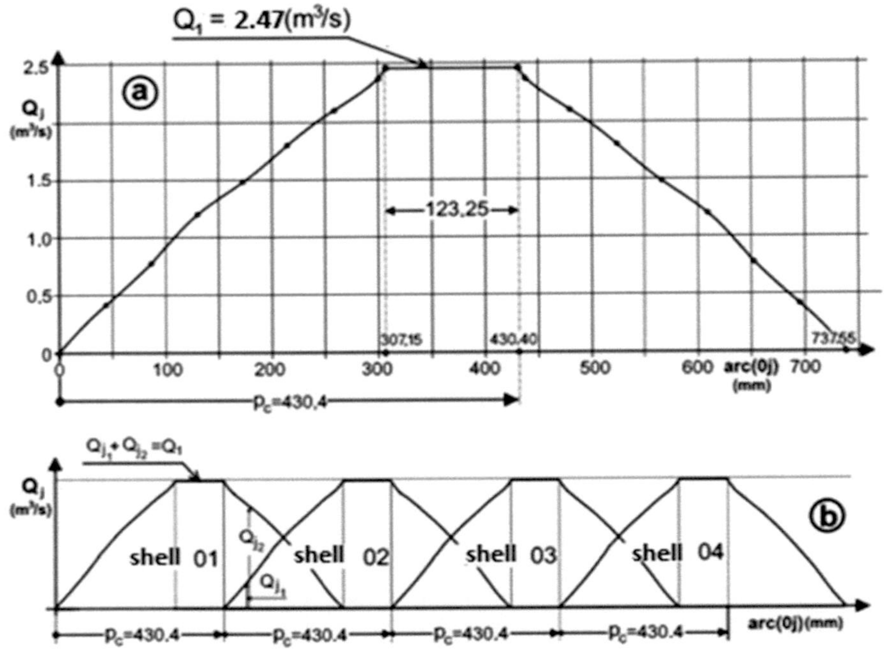

Fig. 9.21 Graphs with drawing of the functions $Q_j = f[\text{arc}(0j)]$. (**a**) for one casing; (**b**) for four casings

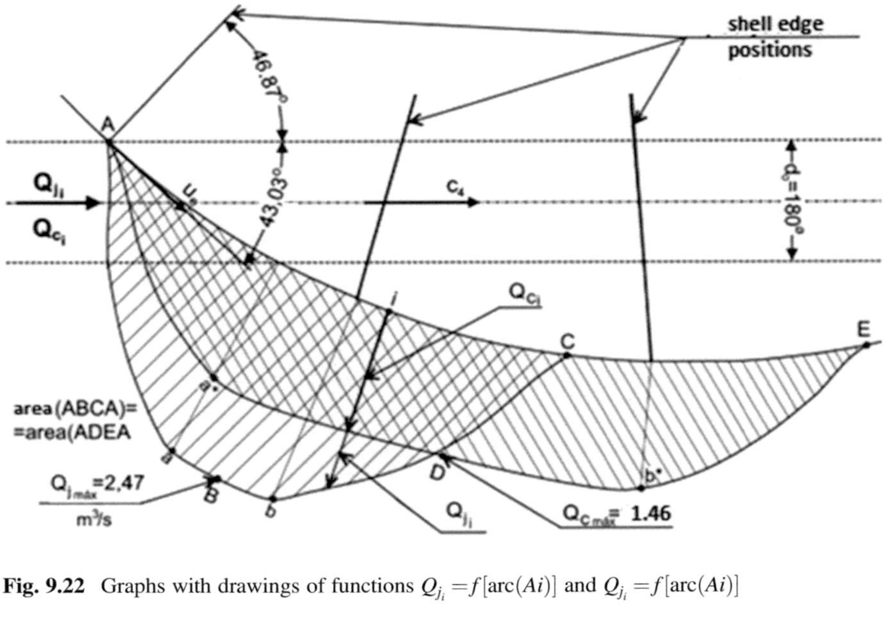

Fig. 9.22 Graphs with drawings of functions $Q_{j_i} = f[\text{arc}(Ai)]$ and $Q_{j_i} = f[\text{arc}(Ai)]$

Table 9.6 Calculation of tracing characteristics in Fig. 9.21

Caracteristics	Unit	1	2	3	4	5	6	7	8
arc(A_i) to C	mm	43.04	86.08	129.12	172.16	215.20	258.24	301.28	307.15
arc(A_i) to E	mm	72.48	144.96	217.45	289.93	362.41	434.89	507.38	517.26
	m^3/s	0.43	0.79	1208.00	1.49	1.81	2.10	2.38	2.47
	m^3/s	0.26	0.47	0.72	0.88	1.07	1.24	1.41	1.47

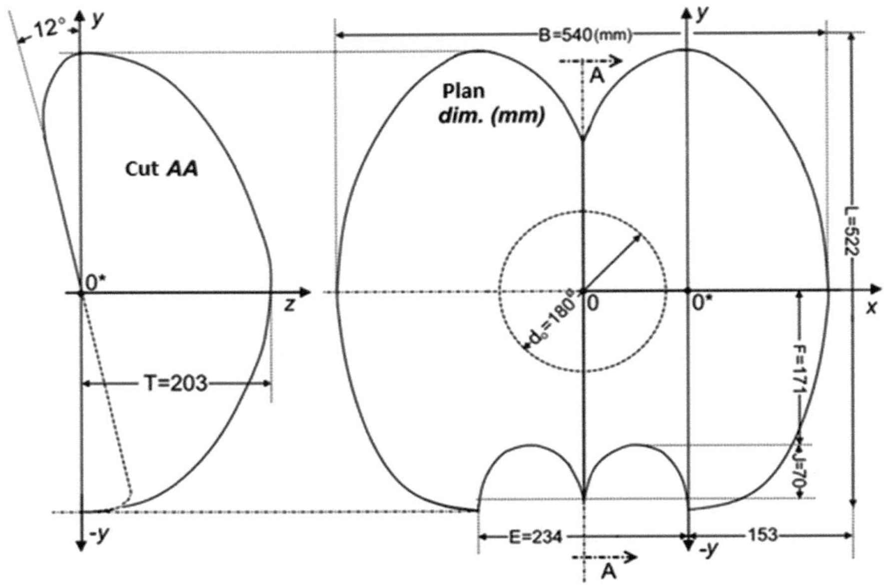

Fig. 9.23 To-scale drawing of the section view of the shell

$$y = 0.5 \cdot L \cdot \sqrt{1 - \left(\frac{2 \cdot x}{B - E}\right)^2} = 261 \cdot \sqrt{1 - \left(\frac{x}{153}\right)^2} \text{ (mm) with } 0 \leq x(\text{mm}) \leq 153$$

and

$$y = 0.5 \cdot L \cdot \sqrt{1 - \left(\frac{2 \cdot z}{T}\right)^2} = 261 \cdot \sqrt{1 - \left(\frac{z}{203}\right)^2} \text{ (mm) with } 0 \leq z(\text{mm}) \leq 203$$

26. Figure 9.24 shows a to-scale drawing of the internal limits of the ellipse on the right side of the shell view, with the complements in arcs of circumferences, and the dividing edge of the shell. Figure 9.25 shows the limits corresponding to the *AA* cut.

9.9 Slope of the Dividing Edge of the Shell

9.9.1 Considerations

By imposing the condition of maximum transfer of kinetic energy available in the jets delivered to the active shells of the *PHIT RO*, the inclination of the dividing edge of each of its shells must satisfy the following proposition:

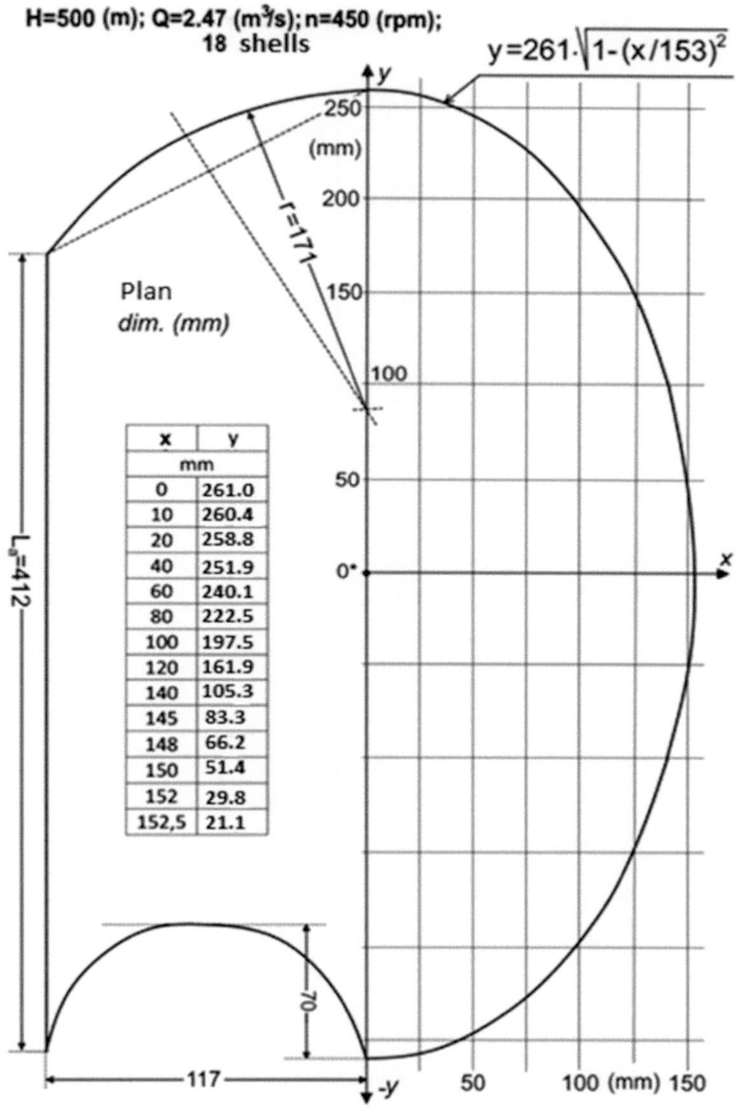

Fig. 9.24 To-scale drawing of the right of the shell

The dividing edge must be perpendicular to the middle position of each shell between the first and last contact with the full jet.

This proposition can be obtained via a graphic process, which will allow us to elaborate and validate an algorithm, provided that the following geometric characteristics of the *PHIT RO*, and the jet that transmits energy, are available:

z	y
mm	
0	261.0
10	260.7
20	259.7
30	258.1
40	255.9
50	253.0
60	249.3
80	239.9
100	227.1
120	210.5
140	189.0
160	160.6
180	120.7
190	92.0
200	44.7
201	36.6
203	0

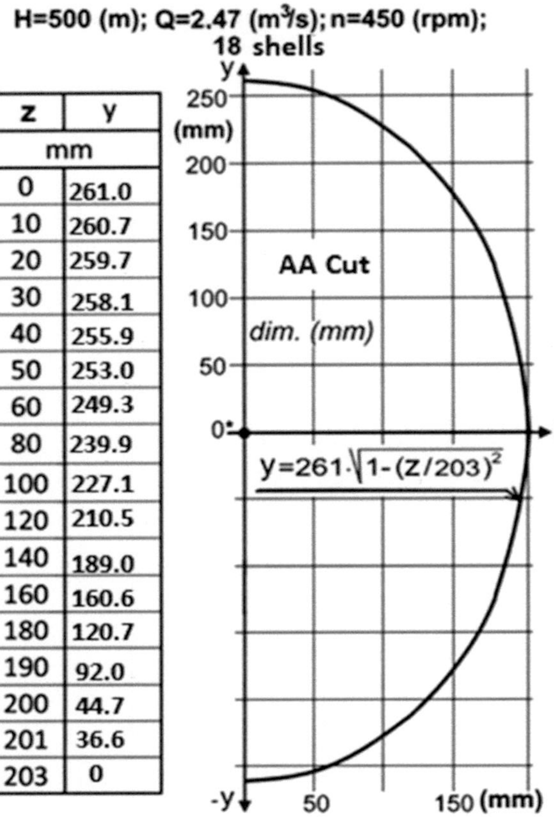

Fig. 9.25 Medium cutaway to-scale drawing of the shell

- D_o
- D
- D_{i_a}
- D_{e_a}
- p_c
- α_c.

With these characteristics, from an orthogonal system of axes centered at O, Fig. 9.26 begins by drawing the corresponding circles, then marking point B_1 where the vertical axis meets the horizontal tangent to circle D, which represents the jet axis, and two other horizontals at points B and B_2, and their limits in the drawing.

At the lower limit of the jet, mark point A at the meeting point with circumference D_{e_a}.

Move point A to a position A_1 by drawing its *RAT* in the form of a circular arc passing through points A_1, B_2, and A_3.

Connect $A_1$with O, and from this, mark the step $\alpha_c^o = \text{arch}A_1, A_2 = p_c$ (mm) obtaining, D_{e_a} on point A_2, which must belong to the dividing edge of the shell in its last full contact with the jet.

Once this is done, the sought-after position of the dividing edge of the shell, which observes the established proposition, can be obtained by observing the following.

First, connect point O with A_1 via a line segment, and starting from O, in a first attempt, mark an angle to the left of the segment, with side $\beta_1 = \alpha_c$ with its vertex at point A_1, meeting the horizontal axis at R.

Then, draw a circle with center at O and radius $r = \bar{OR}$ with the tangents to this circle passing through $A_1$and A_2, marking the meeting point of tangent A_2R with the *RAT* with an X, which, in general, is not on the upper limit of the jet, showing that the dividing edge of the shell is not fully receiving the jet.

For this to happen, the *RO* must be rotated in the opposite direction, with the center at O, up to position X_1, which belongs to the upper limit of the jet, thus establishing the direction of the shell edge between its initial position, tangent to the circumference of radius r that passes through A_1 and a final one that passes through A_2 and X_1, also tangent to r, forming angle θ^o.

If the line corresponding to angle $0, 5 \cdot \theta^{\circ}$ is normal to the horizontal axis passing through O and R, i.e., $\beta = 90^{\circ}$, the proposition is met. If not, as shown in the Figure, $\beta < 90^{\circ}$, angle β_1 must be changed, and the drawing must be redone until this is met,

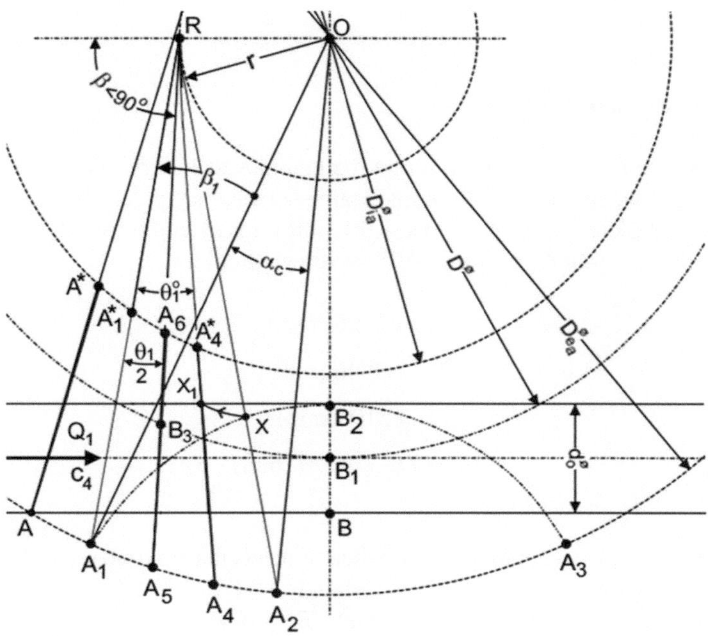

Fig. 9.26 Graphical acquisition of the inclination of the dividing edge of the *PHIT* shell

when the position of the dividing edge of the shells will be defined by line segment $\overline{A_5A_6}$, and point B_3 defined when it meets circle D.

9.9.2 *Algorithm and Validation*

9.9.2.1 Input Data and Values for Validation

- $D_{i_c} = 2226$ (mm)
- $D_{i_a} = 2946$ (mm)
- $D = 3196$ (mm)
- $D_{e_a} = D_{e_c} = 3770$ (mm)
- $d_o = 206$ (mm)
- $B = 3.02 \cdot d_o$
- $E = 1.65 \cdot d_o$
- $F = 0.85 \cdot d_o$
- $G = 0$
- $J = 0.385 \cdot d_o$
- $T = 1.22 \cdot d_o$
- $z_c = 22$
- $\alpha_c^o \cong 16.36^o$
- $p_c \cong 538$ (mm)
- $\beta_c = 10^o$

9.9.2.2 Script and Calculations

1. Choose a computer program to plot the graphics and carry out the work (scale plotting), which has a range of uncertainties (accuracy/precision), and that can obtain the plots values to at least, ±0.1 (mm), for example, AutoCAD.
2. With the center at O, start Fig. 9.27 by drawing circles:

$$D_{i_c} = 2226 \text{ (mm)}$$

$$D_{i_a} = 2946 \text{ (mm)}$$

$$D = 3196 \text{(mm)}$$

$$D_{e_a} = D_{e_c} = 3770 \text{ (mm)}$$

3. On the vertical axis, mark point B_1 where it meets the circumference D, drawing the jet axis through it.
4. From the axis of the jet, point B_1, draw horizontals distant by $0.5 \cdot d_o = 0.5 \cdot 206 = 103$ (mm), obtaining the limits of the jet in the drawing,

and points B_2, B, and where the lower limit of the jet meets the circumference, D_{e_a}, point A.

5. Over D_{e_a} mark point A_1, then connect this point with O.
6. From A_1, draw a horizontal line that meets D_{e_a} at point A_3.
7. From A_1, draw the *RAT* in contact with the inner surface of the shell, as a circle that passes through points $A_1B_2A_3$. The center, O_1, that passes through these three points is obtained by joining A_1 with B_2, determining the midpoint M, and from there, draw a normal, which finds the vertical axis of the coordinates at the sought after point O_1.
8. Connect A_1 with O, marking $\alpha_c^o = 16.63^\circ$ with the vertex at O, to the right the step α_c^o, obtaining D_{e_a} over point A_2.

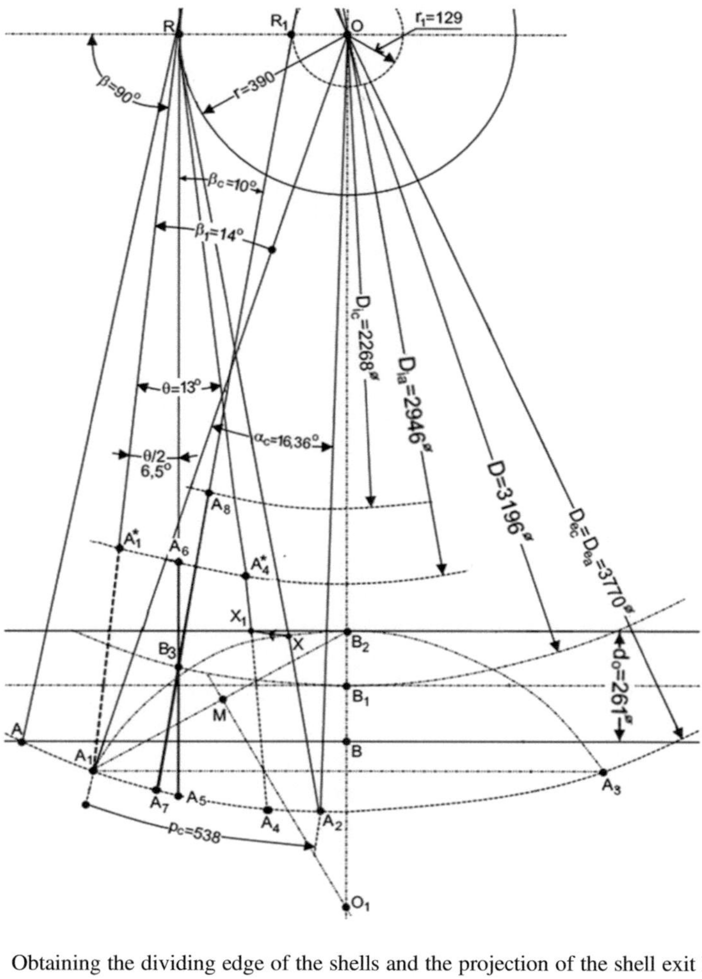

Fig. 9.27 Obtaining the dividing edge of the shells and the projection of the shell exit line

9. With the vertex at A_1, mark angle β_1 to the left of $\overline{AO}$ to start the trials, adopt $\beta_1 \cong \alpha_c = 16^\circ$ following the script, resulting in $\beta > 90^\circ$. The trial with $\beta_1 = 14^\circ$, resulted $\beta_1 \cong 90^\circ$ after observing the sequence.
10. Mark angle $\beta_1 = 14^\circ$ from the direction of line A_1O to the left obtaining point R on the horizontal axis.
11. On the to-scale drawing, measure the radius r_1, as $r_1 = 390$ (mm).
12. With the center at O and the radius $r = 390$ (mm) draw a circle.
13. From A_2, draw a tangent to the circle radius r which meets the RAT at point X.
14. With center at O and radius OX, draw a circular arc that meets the upper projection of the jet at point X_1.
15. From X_1, draw a tangent to the circle r, which in its extension meets circle D_{e_a} at point A_4.
16. Measure the θ angle formed by the tangents to the circle r with vertices at A_1 and A_4, in this case, $\theta = 13^\circ$.
17. Draw the median line of angle $\theta = 13^\circ$ which meets the circles D_{i_a}, D, and D_{e_a} respectively at points A_6, B_3, and A_5.
18. Measure the angle β that forms the average line drawn with the horizontal axis that passes through the OR. If this angle is practically equal to $\beta \cong 90^\circ$, as was the case with $\beta_1 = 14^\circ$, attempt to continue. If not, return to step 9l, changing the value of β_1 until this occurs, and then continue.
19. From point B_3 of the average normal, mark an angle fixed in the data, $\beta_c = 10^\circ$ to the right, with a side that meets the horizontal axis OR at point R_1, then drawing with the center at O and the radius measured on the scale r_1 when $r_1 = 129$ (mm) a circle with a center at O. In this circle radius r_1, all projections of the lines of the maximum longitudinal sections of the shells will be tangent to it.
20. Mark the projection of the length of the dividing edge of the shell $\overline{A_5A_6}$.
21. Mark and calculate the projection of the maximum section length of the shell longitudinal section:

$$L = \overline{A_7A_8} = 0.5 \cdot (D_{e_c} - D_{i_c}) = 0.5 \cdot (3770 - 2268) = 751 \text{ (mm)}$$

22. Calculate the length of the dividing edge of the shells:

$$L_a = \overline{A_5A_6} = 0.5 \cdot (D_{e_a} - D_{i_a}) = 0.5 \cdot (3770 - 2946) = 412 \text{ (mm)}$$

23. Calculate limiting dimensions of the inner surface of the shells in section view:

- $B = 3.02 \cdot d_o = 3.02 \cdot 261 = 790$ (mm)
- $E = 1.3 \cdot d_o = 1.3 \cdot 261 = 340$ (mm)
- $F = 0.85 \cdot d_o = 0.85 \cdot 261 = 222$ (mm)
- $G = 0$
- $J = 0.385 \cdot d_o = 0.385 \cdot 261 = 100$ (mm)
- $T = 1.22 \cdot d_o = 1.22 \cdot 261 = 318$ (mm)

- $y = \pm 0.5 \cdot L \cdot \sqrt{1 - \left(\frac{4 \cdot x}{B}\right)^2} = \pm 0.5 \cdot 751 \cdot \sqrt{1 - \left(\frac{4 \cdot x}{790}\right)^2} = \pm 375.5 \cdot \sqrt{1 - \left(\frac{x}{197.5}\right)^2}$ (mm)
- $y = \pm 0.5 \cdot L \cdot \sqrt{1 - \left(\frac{z}{T}\right)^2} = \pm 0.5 \cdot 751 \cdot \sqrt{1 - \left(\frac{2 \cdot z}{318}\right)^2} = \pm 375.5 \cdot \sqrt{1 - \left(\frac{z}{159}\right)^2}$ (mm)
- $z = \pm T \cdot \sqrt{1 - \left(\frac{4 \cdot x}{B}\right)^2} = \pm 0.5 \cdot 318 \cdot \sqrt{1 - \left(\frac{4 \cdot x}{790}\right)^2} = \pm 375.5 \cdot \sqrt{1 - \left(\frac{x}{197.5}\right)^2}$ (mm) − *ladodireito*.

24. In the cutaway, draw the lines that delimit the internal surface view, of the cutaways in the maximum sections, and of the jet, to scale (Fig. 9.28).
25. The diameter of the shaft calculated by torsion, from the expression is:

$$d_e = 0.118 \cdot \sqrt[3]{\frac{P_e}{n}} = 0.118 \cdot \sqrt[3]{\frac{170400}{300}} = 0.977\ (m)$$

$d_e = 1000$ (mm) has been adopted here in this preliminary calculation. The circumference corresponding to this diameter was drawn in Fig. 9.29, to show that the maximum space available for drawing the disk that connects the shells to the axis is calculated by the difference as:

$$d_d = D_{i_c} - d_e = 2.268 - 1.000 = 1.268\ (\mathrm{m}) = 1388\ (\mathrm{mm})$$

26. In Fig. 9.29, a preliminary drawing was made to scale of the horizontal projection of the *RO* of the vertical axis being studies. This layout was based on drawings form Figs. 9.26 and 9.27, and values used in the algorithm validation.

9.10 Geometry of the Bottom Edge of the Dividing Edge of the Shell

Figure 9.30 shows a section in the central part (middle view) of a *PHIT RO* shell at the instant that the edge is normal to the direction of the jet. There, its main characteristics were named and limited with respect to the lower end of the dividing edge of the flow, which must be observed in the design and manufacturing so that the flow exit occurs without clashing with the shell that precedes it.

Thus, the edge of all shells must have at least the angle formed by tangent direction, *t*, towards the outside with the *RAT* from the lower end of the edge that divides the jet, to prevent jet shocks on the back of the shell that precedes it.

On the other hand, to avoid a noticeable deflection so that jet particles (those that are not in contact with the shell, but very close to the *IN*) continue their straight path to avoid the preceding shell, this means that the soffit and extrados angle, α° will have to be as small as possible, as shown in Fig. 9.31a, cutaway *xy*. If a large α° is used, as shown in the *xy* section of Fig. 9.31c, there will likely be a noticeable reduction in volumetric yield, due to water losses from strong jet deflection. In the

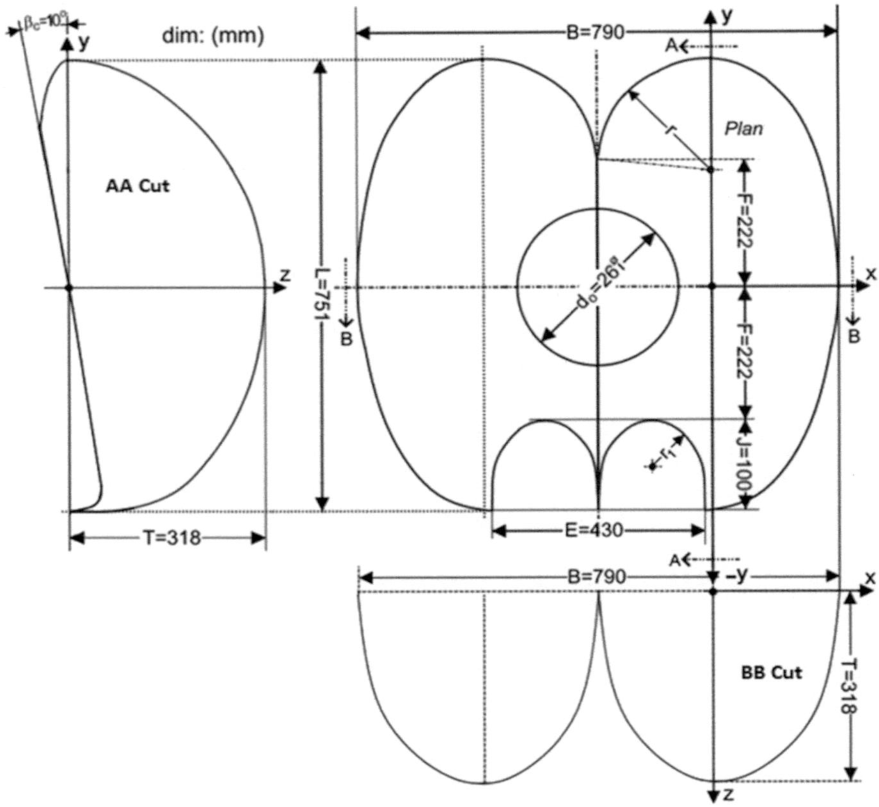

Fig. 9.28 Shell designs

$x * y*$ cutaways, parallel to xy that pass through the inner limits of the cutout, or point f in Fig. 9.31c, angle α° increases, and should be in the range shown in the Figure.

In this cut, machining must be done on the soffit of the shell to avoid shocks, which is normally done manually and controlled through templates that are part of the shell manufacturing project.

As an exercise, we recommend that scholars use the data and results of the algorithm validation to draw, in scale, the cutaways corresponding to Figs. 9.30 and 9.31.

The photographs in Fig. 9.32 show the parts that normally receive manual machining and control with templates, particularly the inner surfaces of the shells and the outer surfaces (back) limited by the tangents of the *RAT*, observing the tests and results of the corresponding models.

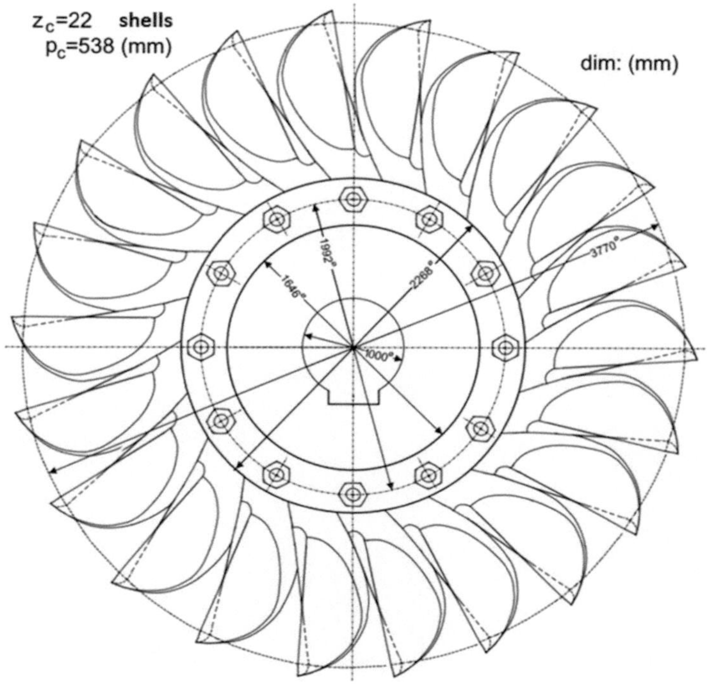

Fig. 9.29 To-scale horizontal projection of a *PHIT RO* used to validate the algorithm

Fig. 9.30 Cutaway through the middle view of the shell at the instant when the edge is in the position normal to the jet

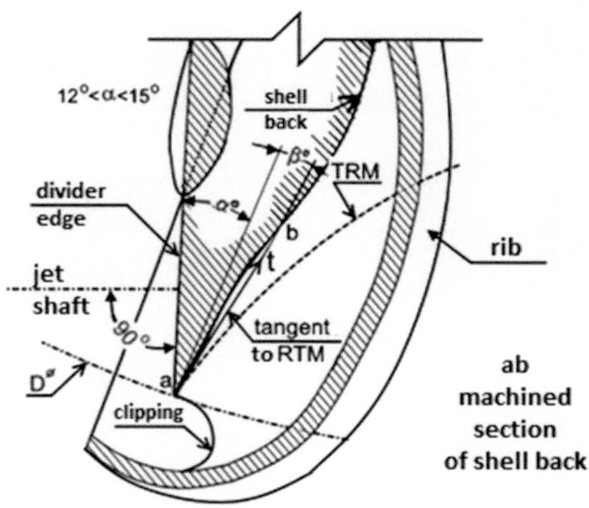

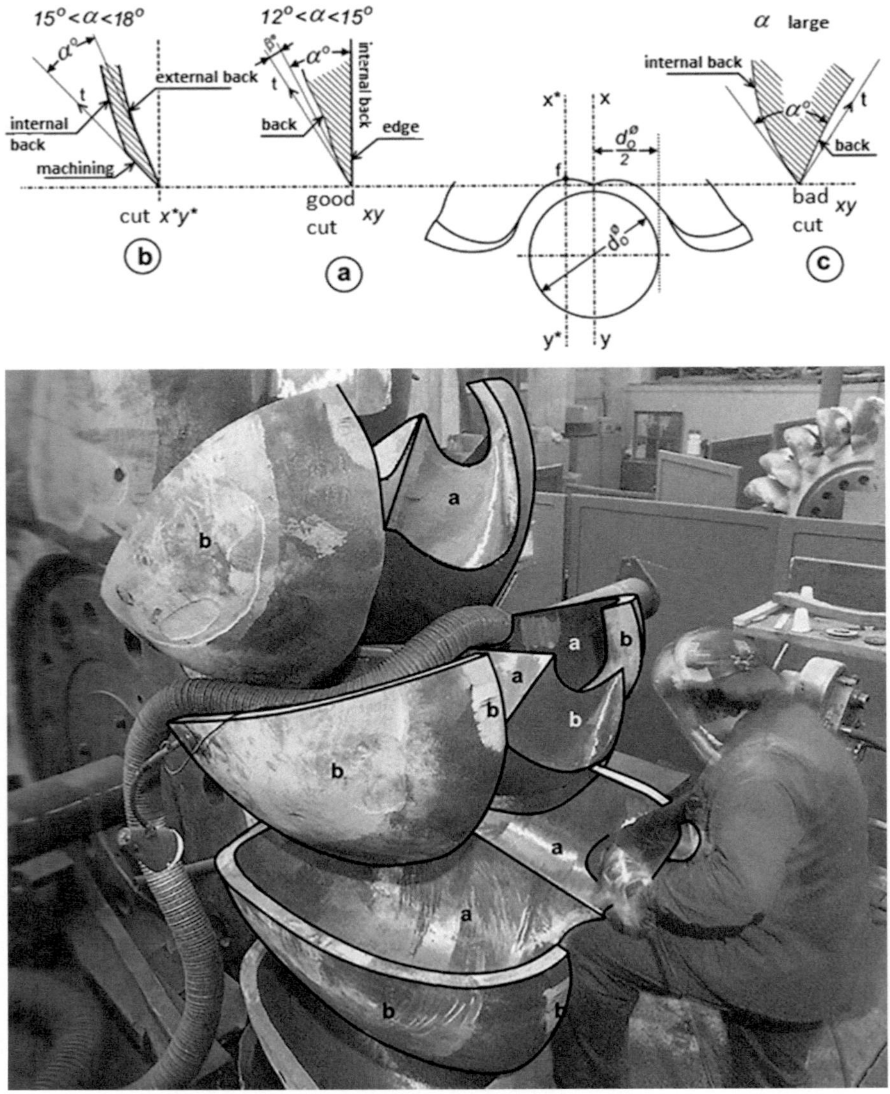

Fig. 9.31 Sections with usual angles. (**a**) good; (**b**) desirable; (**c**) bad

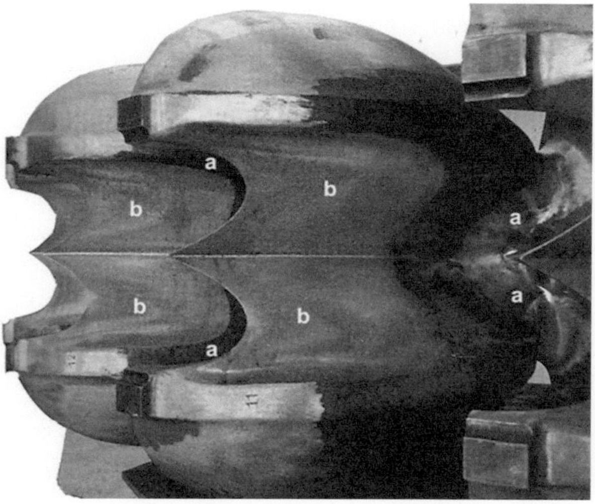

Fig. 9.32 Machining on shell surfaces: a—internal. b—external

Chapter 10
The Hydrodynamic Geometric Drawings for the Shell Surface

The main objective of this chapter is to present the details of the hydrodynamic geometric layout of the inner surface of the shells. For this purpose, everything established in the previous chapters will be used as a hydrodynamic foundation, and the previously established geometry, in its smallest details, e.g., the dimensionless dimensions, will be used as a starting point. We should, therefore, expect that the result, once it has been complemented with the structural calculations of the shell, will allow us to develop an algorithm to build a model or prototype, along with templates, that can be sent to a 3D printer (machine), in this case, initially the inner surface of the shell. The collaboration of engineer Mr. Thiago Soares Correa was fundamentally important in doing this, particularly with respect to mapping the reference shell, in checking and interpolating the values, in drawing the 3D, in making the drawings and coordinates availability of for manufacture, and for prototyping the model or prototype in a 3D printer.

10.1 General Considerations and Fundamentals

- The analysis of the water flow inside the *PHIT RO* shells is very complex, since the flow depends not only on the impact line of the jet but also on the variation of the shape of the cross section of the jet, from its entry in line at the shell division, until the exit, which must occur without collisions with the backs of the shells that precede it. Furthermore, the analysis also depends on the trajectories of the water particles in contact with the inner surfaces of the shells, which vary in length, thus resulting in energy losses. Every water particle that penetrates each shell takes a certain amount of time to travel to the surface from the entrance on edge 4 to exit at edge 5 (Fig. 10.1). For each shell position, and for each water particle, there is a time lag and displacement of the space occupied between the outgoing and incoming flow. These problems occur with variations to the cross section of the flow in the shells, which is three-dimensional, and with variations to the relative

© The Author(s), under exclusive license to Springer Nature Switzerland AG 2025
Z. de Souza, G. L. Tiago Filho, *Pelton and Crossflow Impulse Hydraulic Turbines*,
https://doi.org/10.1007/978-3-031-87086-6_10

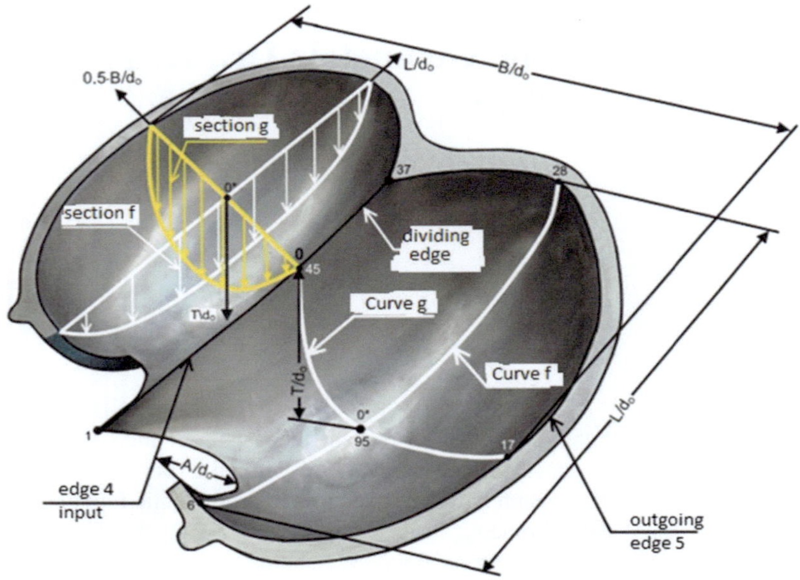

Fig. 10.1 Photograph of a shell showing the fundamentals and conventions that will be used to develop the studies

velocity *w* from zero along the inner surface of the shells to the outer surface in contact with air. Thus, even considering average conditions, this study is very difficult to equate a stationary shell, let alone with all of them in motion, to a small degree of uncertainty, for modeling and visualization in the laboratory.

- Given the issues, and in addition to what has already been studied, an inductive method will be presented here in this book that will allow us to obtain a preliminary internal surface for a shell, which once identified, can be modeled via a 3D machine printer, and serve as a basis for defining the geometry, under the hydrodynamic and structural aspects of the *RO* and the axis of the *PHIT*, either initially at the modeling stage or later for testing or modifying the corresponding prototypes, when and where applicable.
- These foundations were obtained from mapping the inner surface of a shell located at the Hydrodynamic Laboratory for Small Hydroelectric Power Plants (*LHPCH*) at *UNIFEI*, which was photographed and transposed to Fig. 10.1, that characterizes the curves, medium, symbols, conventions, and coordinate axes, with the variables that will be used in digital and analog model.

10.2 Hypotheses for Drawing the Inner Surface of Shells Using the Inductive Method

- The inductive method referenced the coordinates, conventions, and dimensionless relationships from the constants in Fig. 10.2a, b.
  - The following were fixed as maximum limits in the three normal directions (see Fig. 9.4):

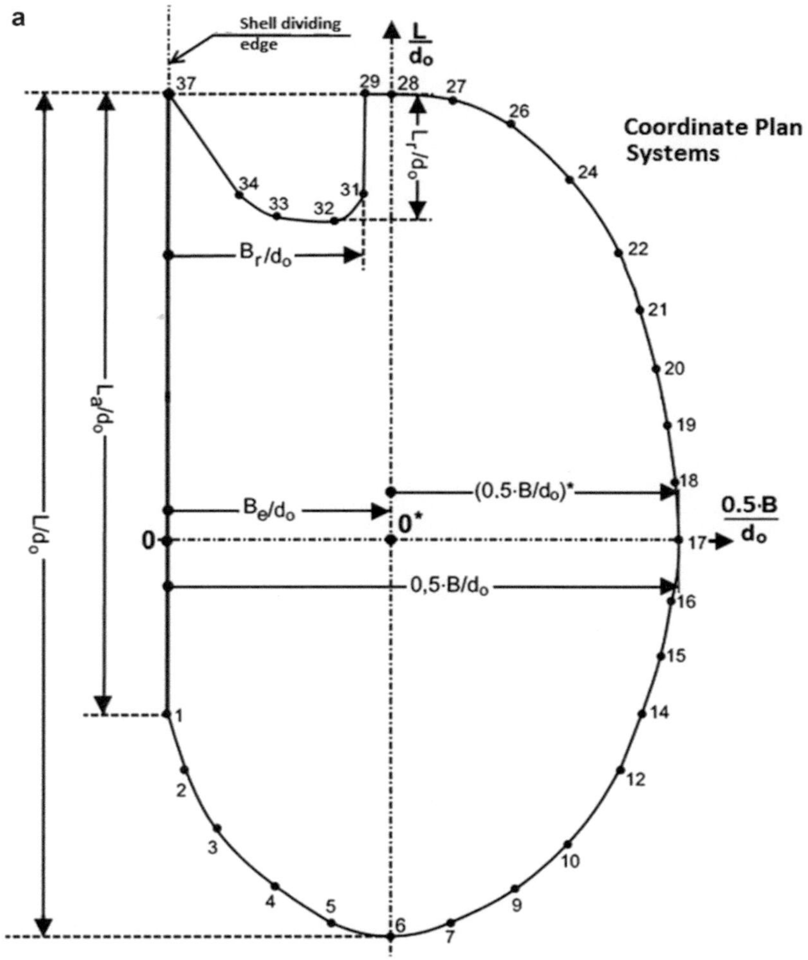

Fig. 10.2 (**a**) View of the shell boundaries, the coordinate systems, and the dimensionless relationships between the features. (**b**) Detail of the shell cutout in a vertical projection, and its characteristics

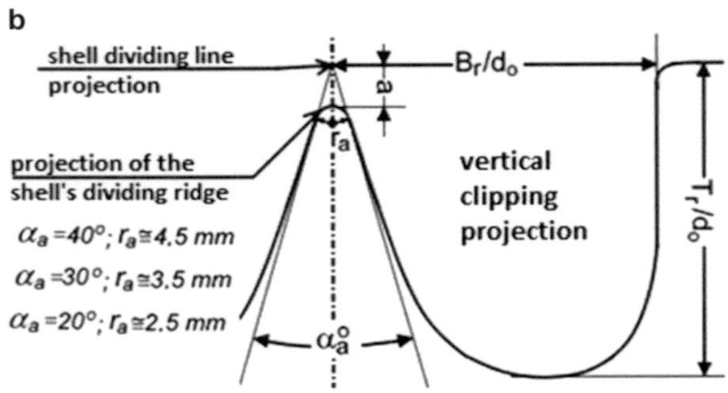

Fig. 10.2 (continued)

- $$L/d_o = 2.61$$
- $$0.5 \cdot B/d_o = 1.564$$
- $$T/d_o = 1.0 \tag{10.1}$$

– The following were calculated with the coefficients fixed in Fig. 10.2a, b:

- $$L_r/d_o = 0.532 \cdot L/d_o$$
- $$L_a/d_o = 0.732 \cdot L/d_o$$
- $$B_r = 0.386 \cdot (0.5 \cdot B/d_o)$$
- $$B_e/d_o = 0.437 \cdot (0.5 \cdot B/d_o) \tag{10.2}$$

If different ratios are fixed (see recommended limits in Fig. 9.4), use geometrically similar coefficients, defined below, for the characteristics of expressions (10.2), and for all values of the coordinates shown in Table 10.1:

$$f_L = \frac{(L/d_o)_{fix}}{L/d_o} = \frac{(L/d_o)_{fix}}{2.61}$$

$$f_B = \frac{(0.5 \cdot B/d_o)_{fix}}{0.5 \cdot B/d_o} = \frac{(0.5 \cdot B/d_o)_{fix}}{1.564}$$

$$f_T = \frac{(T/d_o)_{fix}}{T/d_o} \frac{(T/d_o)_{fix}}{1.0} \tag{10.3}$$

- The following is an example that uses these coefficients:
  Establish …

$$(0.5 \cdot B/d_o)_{fix} = 1.5$$

and …

$$d_o = 0.239\ (m)$$

then…

$$0.5 \cdot B_{max} = 1.5 \cdot 0.239 = 0.358(m)$$

This value or any corresponding value from Table 10.1 can be obtained considering:

$$f_B = \frac{0.5 \cdot B/d_o)_{fix}}{0.5 \cdot B/d_o} = \frac{1.5}{1.552} = 0.966$$

Therefore:

$$0.5 \cdot B_{max} = (0.5 \cdot B/d_o)_{ref} \cdot f_B \cdot d_o = 1.552 \cdot 0.966 \cdot 0.239 = 0.358\ (m)$$

… then generalizing to the corresponding coordinates from Table 10.1 as follows:

Table 10.1 Surface Coordinates

Inner Shell				Inner Shell			
Num.	0.5·B/do	L/d$_o$	T/d$_o$	Num.	0.5·B/do	L/d$_o$	T/d$_o$
1	0.0000	−0.5450	−0.0285	77	0.6790	−0.3790	0.9750
2–2*	0.0610	−0.7160	0.0000	78	0.8590	−0.3790	0.9240
3–3*	0.1530	−0.8910	0.0000	79	1032.0000	−0.3790	0.7800
4–4*	0.3260	−1.0850	0.0000	80	1219.0000	−0.3790	0.5600
5–5*	0.5060	−1.1930	0.0000	81	1392.0000	−0.3790	0.2800
6–6*	0.6790	−1.2250	0.0000	82	0.0610	−0.1850	0.2530
7–7*	0.8590	−1.1750	0.0000	83	0.1530	−0.1850	0.5390
8–8*	1.0320	−1.0900	0.0000	84	0.3260	−0.1850	0.8300
9–9*	1.0670	−1.0690	0.0000	85	0.5060	−0.1850	0.9650
10–10*	1.2190	−0.9310	0.0000	86	0.6790	−0.1850	0.9950
11–11*	1.2520	−0.8910	0.0000	87–87*	0.8590	−0.1850	0.9450
12–12*	1.3720	−0.7150	0.0000	88–88*	1032.0000	−0.1850	0.8200
13–13*	1.3920	−0.6790	0.0000	89–89*	1219.0000	−0.1850	0.5980
14–14*	1.4500	−0.5300	0.0000	90–90*	1392.0000	−0.1850	0.3460
15–15*	1.4990	−0.3590	0.0000	91–91*	0.0610	0.0000	0.2650
16–16*	1.5360	−0.1820	0.0000	92–92*	0.1530	0.0000	0.5500
17–17*	1.5640	0.0000	0.0000	93–93*	0.3260	0.0000	0.8500
18–18*	1.5550	0.1730	0.0000	94–94*	0.5060	0.0000	0.9750
19–19*	1.5270	0.3590	0.0000	95–95*	0.6790	0.0000	1000.0000
20–20*	1.4830	0.5380	0.0000	96–96*	0.8590	0.0000	0.9500
21–21*	1.4280	0.7250	0.0000	97–97*	1032.0000	0.0000	0.8300
21*–21**	1.3920	0.8110	0.0000	98–98*	1219.0000	0.0000	0.6100
22–22*	1.3550	0.8850	0.0000	99–99*	1392.0000	0.0000	0.3600
23–23*	1.3110	0.9770	0.0000	100–100*	1545.0000	0.0000	0.0400
24–24*	1.2180	1.1350	0.0000	101–101*	0.0610	0.1730	0.2650
25–25*	1.0970	1.2690	0.0000	102–102*	0.1530	0.1730	0.5450
26–26*	1.0150	1.3180	0.0000	103–103*	0.3260	0.1730	0.8450
27–27*	0.8590	1.3550	0.0000	104–104*	0.5060	0.1730	0.9550
28–28*	0.6790	1.3720	0.0000	105–105*	0.6790	0.1730	0.9750
29–29*	0.5990	1.3750	0.0000	106–106*	0.8590	0.1730	0.9220
30–30*	0.5990	1.2490	0.2900	107–107*	1032.0000	0.1730	0.8100
31–31*	0.5990	1.0700	0.5000	108–108*	1219.0000	0.1730	0.5980
32–32*	0.5060	0.9900	0.5800	109–109*	1392.0000	0.1730	0.3440
33–33*	0.3260	1.0030	0.5600	110–110*	0.0610	0.3590	0.2650
34–34*	0.2160	1.0700	0.4840	111–111*	0.1530	0.3590	0.5330
35–35*	0.1530	1.1450	0.3870	112–112*	0.3260	0.3590	0.8300
36–36*	0.0880	1.2490	0.2800	113–113*	0.5060	0.3590	0.9150
36*-36**	0.0610	1.2900	0.2010	114–114*	0.6790	0.3590	0.9280
37	0.0000	1.3750	0.0260	115–115*	0.8590	0.3590	0.8700
38	0.0000	1.2490	0.0224	116–116*	1032.0000	0.3590	0.7600
39	0.0000	1.0700	0.0173	117–117*	1219.0000	0.3590	0.5600

(continued)

Table 10.1 (continued)

Inner Shell				Inner Shell			
Num.	0.5·B/do	L/d$_o$	T/d$_o$	Num.	0.5·B/do	L/d$_o$	T/d$_o$
40	0.0000	0.8970	0.0124	118–118*	1392.0000	0.3590	0.2950
41	0.0000	0.7250	0.0076	119–119*	0.0610	0.5380	0.2650
42	0.0000	0.5380	0.0022	120–120*	0.1530	0.5380	0.5200
43	0.0000	0.3590	−0.0028	121–121*	0.3260	0.5380	0.7950
44	0.0000	0.1730	−0.0081	122–122*	0.5060	0.5380	0.8620
45	0.0000	0.0000	−0.0130	123–123*	0.6790	0.5380	0.8710
46	0.0000	−0.1850	−0.0183	124–124*	0.8590	0.5380	0.8100
47	0.0000	−0.3790	−0.0238	125–125*	1032.0000	0.5380	0.6890
48–48*	0.5060	−1.0740	0.3190	126–126*	1219.0000	0.5380	0.5000
49–49*	0.6790	−1.0740	0.4680	127–127*	1392.0000	0.5380	0.2300
50–50*	0.8590	−1.0740	0.3300	128–128*	0.0610	0.7250	0.2650
51–51*	1.0320	−1.0740	0.0530	129–129*	0.1530	0.7250	0.4940
52–52*	0.3260	−0.8910	0.3310	130–130*	0.3260	0.7250	0.7300
53–53*	0.5060	−0.8910	0.6120	131–131*	0.5060	0.7250	0.7900
54–54*	0.6790	−0.8910	0.7250	132–132*	0.6790	0.7250	0.8000
55–55*	0.8590	−0.8910	0.6300	133–133*	0.8590	0.7250	0.7220
56–56*	1.0320	−0.8910	0.4100	134–134*	1032.0000	0.7250	0.5900
56*-56**	1.2190	−0.8910	0.0700	135–135*	1219.0000	0.7250	0.4000
57–57*	0.1530	−0.7160	0.2400	136–136*	1392.0000	0.7250	0.1240
58–58*	0.3260	−0.7160	0.5500	137–137*	0.0610	0.8970	0.2600
59–59*	0.5060	−0.7160	0.7870	138–138*	0.1530	0.8970	0.4580
60–60*	0.6790	−0.7160	0.8610	139–139*	0.3260	0.8970	0.6400
61–61*	0.8590	−0.7160	0.7700	140–140*	0.5060	0.8970	0.6800
62–62*	1.0320	−0.7160	0.6000	141–141*	0.6790	0.8970	0.6750
63–63*	1.2190	−0.7160	0.3330	142–142*	0.8590	0.8970	0.5950
64–64*	0.0610	−0.5450	0.1550	143–143*	1032.0000	0.8970	0.4580
65–65*	0.1530	−0.5450	0.4000	144–144*	1219.0000	0.8970	0.2650
66–66*	0.3260	−0.5450	0.6900	145–145*	0.0610	1.0700	0.2500
67–67*	0.5060	−0.5450	0.8720	146–146*	0.1530	1.0700	0.4000
68–68*	0.6790	−0.5450	0.9320	147–147*	0.6790	1.0700	0.4970
69–69*	0.8590	−0.5450	0.8670	148–148*	0.8590	1.0700	0.4280
70–70*	1.0320	−0.5450	0.7200	149–149*	1032.0000	1.0700	0.2800
71–71*	1.2190	−0.5450	0.4750	150–150*	1219.0000	1.0700	0.0950
72–72*	1.3920	−0.5450	0.1500	151–151*	0.0610	1.2490	0.2150
73–73*	0.0610	−0.3790	0.2120	152–152*	0.6790	1.2490	0.2800
74–74*	0.1530	−0.3790	0.5020	153–153*	0.8590	1.2490	0.2000
75–75*	0.3260	−0.3790	0.7800	154–154*	1.0320	1.2490	0.0800
76–76*	0.5060	−0.3790	0.9430	–	–	–	–

Note: The coordinates of column Num. On the right with 1 or 2 asterisks have reversed signs ie: Those with more (+); minus (−) and as minus (−); plus (+). Except the coordinates of the dividing edge of the shell

- $$0.5 \cdot B_j = (0.5 \cdot B/d_o)_{ref_j} \cdot f_B \cdot d_o$$

- $$L_j = L_{ref_j} \cdot f_L \cdot d_o$$

- $$T_j = T_{ref_j} \cdot f_T \cdot d_o$$

- $$A_{fix_j} = A_{ref_j} \cdot f_A \cdot d_o$$

- $$F_{fix_j} = F_{ref_j} \cdot f_F \cdot d_o$$

- $$G_{fix_j} = G_{ref_j} \cdot f_G \cdot d_o$$

- $$J_{fix_j} = J_{ref_j} \cdot f_J \cdot d_o.$$

- Figure 10.3 shows a view of the geometry of the right half of the shell with numbered points, and the respective longitudinal and transverse sections that will be used, as well as the direction of the normal axes.
- The complete view of the shell (Fig. 10.4a) was first obtained by rotating the right half by 180 (degrees), with its dividing edge as the axis, then with each of its cross sections, as shown in Figure 10.4b.

The longitudinal parts were obtained via symmetric transfer, for example: section d to section $d**$. The view was completed by numbering the corresponding points ($95 \rightarrow 95*; 118 \rightarrow 118*\ldots$).

- Table 10.1 contains the three most likely orthogonal dimensionless coordinates for each point on the inner surface of the shell, up to three decimal places.
- In Table 10.1 the values of T/d_o ($1d*, 47e, \ldots, 37$) in *italics*, corresponding to the dividing edge of the shell. These values must be calculated and replaced once $\boldsymbol{\alpha}_c^o$, L_a have been established, and coordinate point $T/d_o = 0$ is determined via the expressions given in Fig. 10.4c.

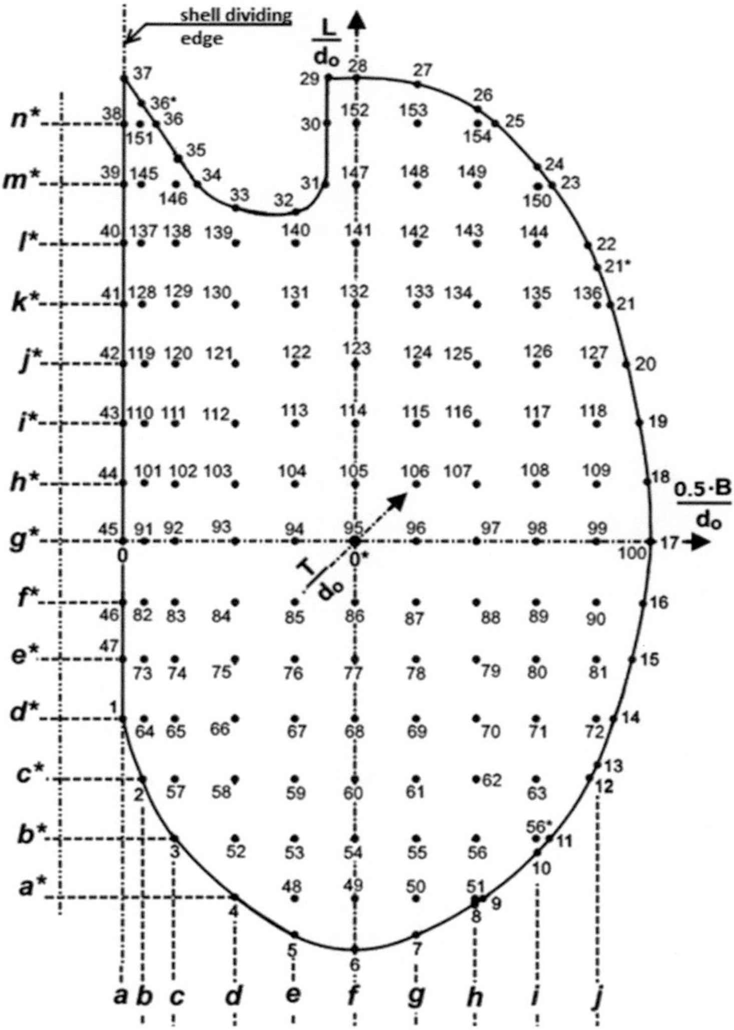

Fig. 10.3 View of half of the inner surface of the shell, with numbered points and coordinates, and the names of the longitudinal and transverse normal sections

- The correspondence of these coordinates, for most computer programs is:
- $$x=0.5\cdot B/d_o$$
- $$y=L/d_o$$
- $$z=-T/d_o$$

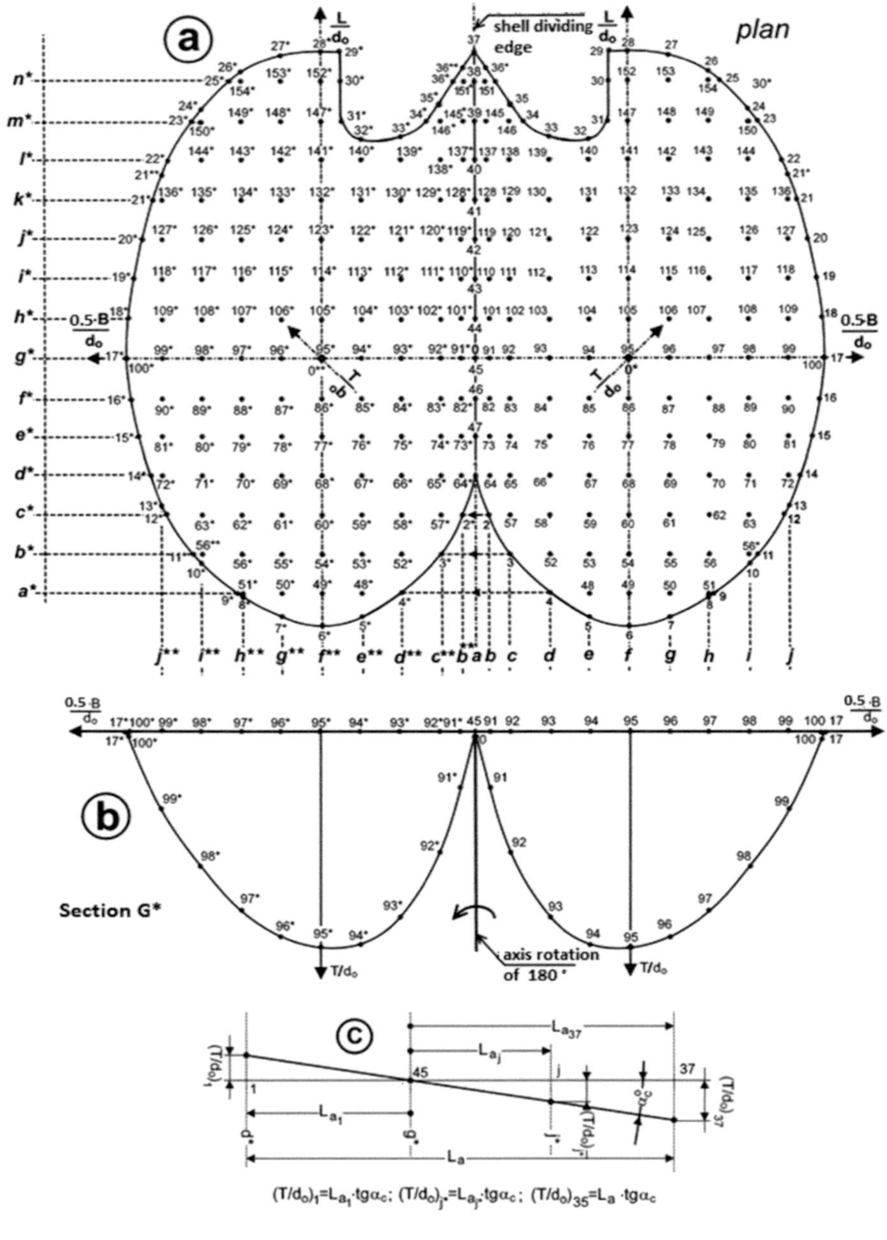

Fig. 10.4 (**a**) View of the inner surface of the shell. (**b**) Folding of section $g*$. (**c**) Partition edge

- The equations were run using constant equations from the computer program "Table Curve 2D v. 5.01" with:
  - R^2 as the Coefficient of Determination: $\mathrm{COD} = R^2 = \frac{e_m^2}{n-1}$;
  - The root mean square error, as: $e_m^2 = \frac{\sum_{i=1}^{n}(x_i - \bar{x})^2}{n}$;
  - x_i as the point coordinate; and
  - $\bar{x}$ as the most probable value of the n coordinates considered in each normal section.

If one should like to obtain the residuals, we recommend transporting the values of the respective coordinates to the program.

10.3 Table of Coordinates, Equation, and Graphs of the Shell Limits for $L/d_o x 0.5 \cdot B/d_o$

In this item, all graphic and analytical elements will be provided to obtain the limits of the inner surface of the shell. With these limits, templates can be built to control the manufacturing of the model or prototypes of the shells, since these are integral parts of the rotors, and to control their internal and external adjustments so that the angles at their respective outlets are practically shockless in the predicted regions.

- *Dimensional Coordinates of the Shell Boundaries in blueprint*—Table 10.2.

  Graphs and equations of limit curves between numbers 1 and 29; 29*—Fig. 10.5.

Table 10.2 Coordinates of the limits of the Shell in Plan

Num.	L/d_o	0.5·B/do	Num.	L/d_o	0.5·B/do
1	−0.5450	0.0000	16–16*	−0.1820	1.5360
2–2*	−0.7160	0.0610	17–17*	0.0000	1.5640
3–3*	−0.8910	0.1530	18–18*	0.1730	1.5550
4–4*	−1085.0000	0.3260	19–19*	0.3590	1.5270
5–5*	−1193.0000	0.5060	20–20*	0.5380	1.4830
6–6*	−1225.0000	0.6790	21–21*	0.7250	1.4280
7–7*	−1175.0000	0.8590	21*-21**	0.8110	1.3920
8–8*	−1090.0000	1.0320	22–22*	0.8850	1.3550
9–9*	−1067.0000	1.0690	23–23*	0.9770	1.3110
10–10*	−0.9310	1.2190	24–24*	1.1350	1.2180
11–11*	−0.8910	1.2520	25–25*	1.2680	1.0970
12–12*	−0.7150	1.3720	26–26*	1.3180	1.0150
13–13*	−0.6790	1.3920	27–27*	1.3550	0.8590
14–14*	−0.5300	1.4500	28–28*	1.3720	0.6790
15–15*	−0.3590	1.4990	29–29*	1.3750	0.5990

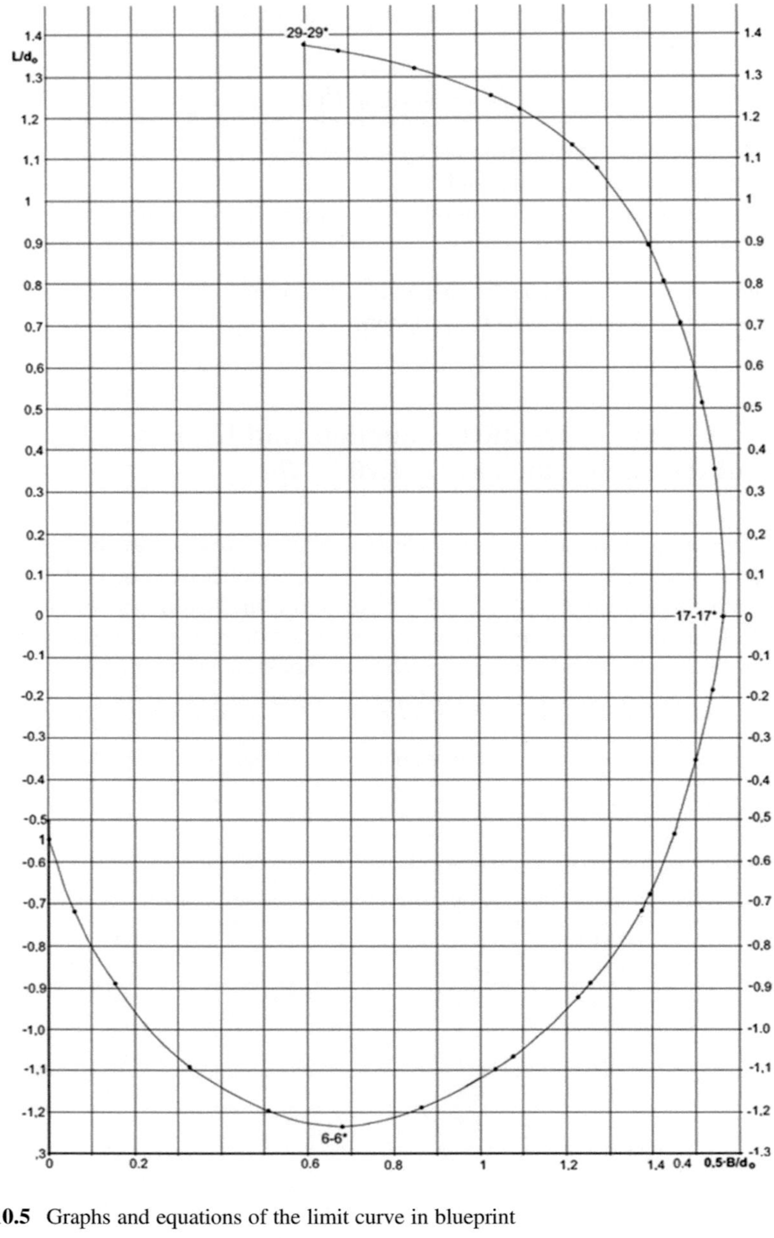

Fig. 10.5 Graphs and equations of the limit curve in blueprint

10.4 Projection of Half of the Shell Cutaway on the Vertical Plane 0.5 · $B/d_o xT/d_o$

The coordinates in the 0.5 · B/d_o and T/d_o directions are projected in true magnitude on the plane that is normal to them (vertical view), and are therefore fundamentally important in manufacturing the templates, which will be used in the manufacturing or quality controls of the resulting product, allowing us to set manufacturing protocols. Thus, the graph with constant coordinates in Fig. 10.6 is quite useful.

10.5 Coordinate Tables, Equations, and Graphs of the Normal Longitudinal Sections for $T/d_o xL/d_o$

- The study that will be developed in this item seeks to provide elements that allow one to control the manufacturing process, either directly or by creating a machine program, and also to control product quality, and to set the respective

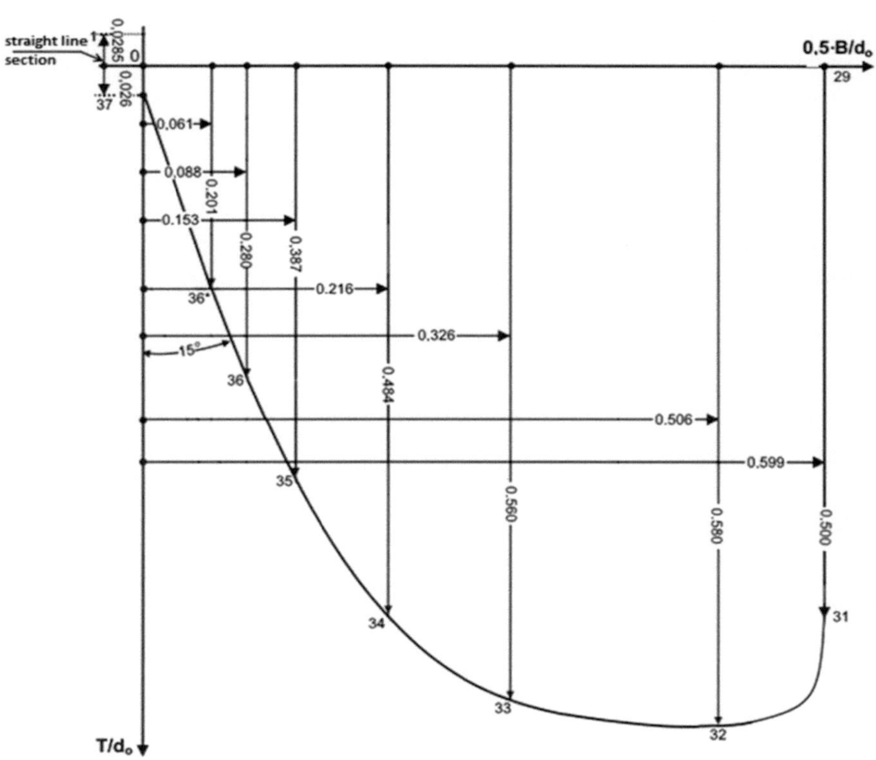

Fig. 10.6 Projection and coordinates of half of the cutaway on plane 0.5 · Bd_oxL/d_oo

manufacturing protocols, with respect to the normal longitudinal sections on the $T/d_o xL/d_o$ plane.

10.5.1 Section A

- *Dimensionless Coordinates*—Table 10.3.
- *Equation, Experimental Variance, and Angle*

$$\frac{T}{d_o} = 0.028385 \cdot \frac{L}{d_o} - 0.013029 \rightarrow R^2 = 0.9999 \rightarrow tg\alpha_d = \frac{d(T/d_o)}{d(L/d_o)} = 0.028385$$
$$\rightarrow \alpha_d \cong 1.63^o$$

- Graph—Fig. 10.7.

10.5.2 Section B

- *Dimensionless Coordinates*—Table 10.4.
- *Graph*—Fig. 10.8.
- *Experimental Equations and Variances.*

Table 10.3 Section a in the longitudinal normal plane T/doxL/do

Num.	T/d_o	L/d_o	Num.	T/d_o	L/d_o
1	−0.0285	−0.5450	42	0.0022	0.5380
47	−0.0238	−0.3790	41	0.0076	0.7250
46	−0.0183	−0.1850	40	0.0124	0.8970
45	−0.0130	0.0000	39	0.0173	1.0700
44	−0.0081	0.1730	38	0.0224	1.2490
43	−0.0028	0.3590	37	0.0260	1.3750

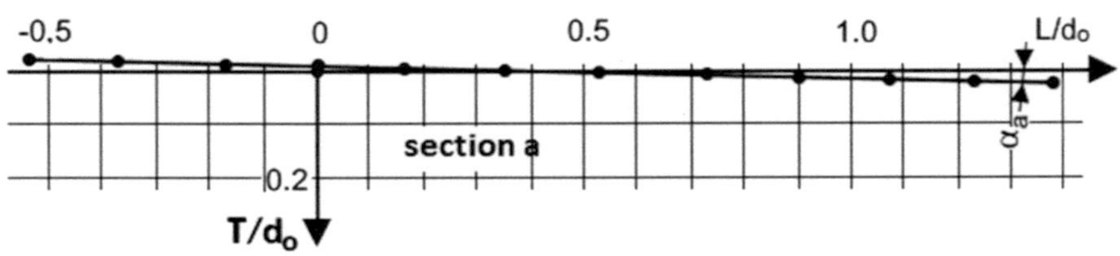

Fig. 10.7 Graph of longitudinal normal section *a* (dividing edge of the shell)

Table 10.4 Section b in the longitudinal normal plane $T/d_o x L/d_o$

Num.	T/d_o	L/d_o	Num.	T/d_o	L/d_o
2–2*	0.0000	−0.7160	119–119*	0.2650	0.5380
64–64*	0.1400	−0.5450	128–128*	0.2650	0.7250
73–73*	0.2120	−0.3790	137–137*	0.2600	0.8970
82–82*	0.2530	−0.1850	145–145*	0.2450	1.0700
91–91*	0.2650	0.0000	151–151*	0.2150	1.2490
101–101*	0.2650	0.1730	36*-36**	0.2010	1.3040
110–110*	0.2650	0.3590	–	–	–

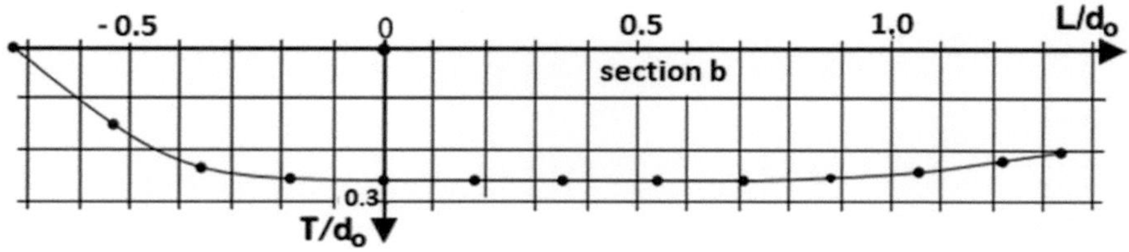

Fig. 10.8 Graph of longitudinal normal section *b*

Numbers 2;2*—91; 91*→Interval:

$$-0.716 \leq \frac{L}{d_o} \leq 0 \frac{T}{d_o} = \frac{0.264925 - 0.516821 \cdot \left(\frac{L}{d_o}\right)^2}{1 - 0.691027 \cdot \left(\frac{L}{d_o}\right)^2} \rightarrow R^2 = 0.9999$$

Numbers 91; 91*—128; 128*→ Interval:

$$0 \leq \frac{L}{d_o} \leq 0.725 \ \frac{T}{d_o} = 0.265 \rightarrow R^2 = 0.9999$$

Numbers 128; 128*—36; 36*→ Interval:

$$0.725 \leq \frac{L}{d_o} \leq 1.304 \frac{T}{d_o} = 0.437994 - 0.594797 \cdot \frac{L}{d_o} + 0.709827 \cdot \left(\frac{L}{d_o}\right)^2 - 0.301451 \cdot \left(\frac{L}{d_o}\right)^3 R^2 = 0.9999$$

10.5.3 Section C

- *Dimensionless Coordinates*—Table 10.5.
- *Graph*—Fig. 10.9.
- *Experimental Equation and Variance.*

$$\frac{T}{d_o} = 0.551192 - 0008970 \cdot \frac{L}{d_o} - 0.216310 \cdot \left(\frac{L}{d_o}\right)^2 + 0.337952 \cdot \left(\frac{L}{d_o}\right)^3 - 0.234940 \cdot \left(\frac{L}{d_o}\right)^4 \rightarrow R^2 = 0.9998.$$

10.5.4 Section D

- *Experimental Equation and Variance.*
- *Graph*—Fig. 10.10.
- *Experimental Equation and Variance* (Table 10.6).

Table 10.5 Section c in the longitudinal normal plane T/doxL/do

Num.	T/d_o	L/d_o	Num.	T/d_o	L/d_o
3–3*	0.0000	−0.8910	111–111*	0.5330	0.3590
57–57*	0.2800	−0.7160	120–120*	0.5200	0.5380
65–65*	0.4190	−0.5450	129–129*	0.4940	0.7250
74–74*	0.5020	−0.3790	138–138*	0.4580	0.8970
83–83*	0.5390	−0.1850	146–146*	0.4000	1.0700
92–92*	0.5500	0.0000	35–35*	0.3870	1.1450
102–102*	0.5450	0.1730	–	–	–

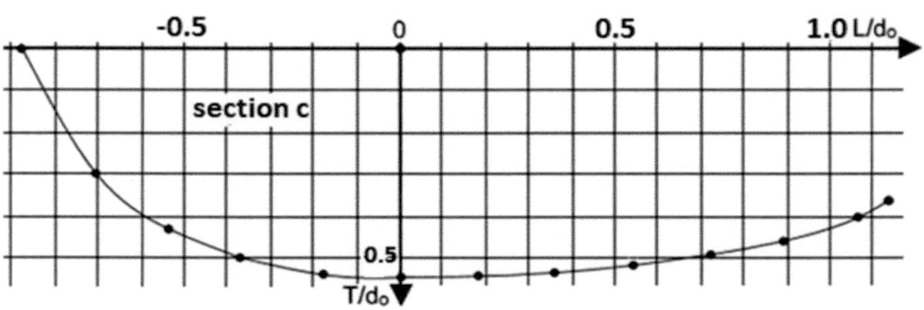

Fig. 10.9 Graph of longitudinal normal section *c*

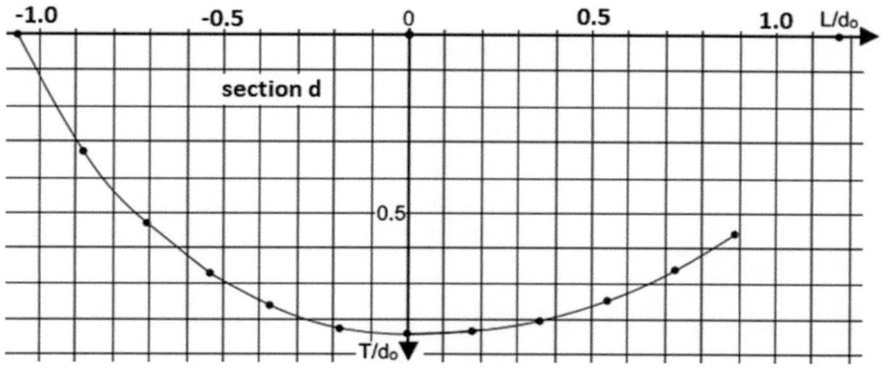

Fig. 10.10 Graph of longitudinal normal section *d*

Table 10.6 Section d in the longitudinal normal plane T/doxL/do

Num.	T/d_o	L/d_o	Num.	T/d_o	L/d_o
4–4*	0.0000	−1.0850	103–103*	0.8450	0.1730
52–52*	0.3310	−0.8910	112–112*	0.8300	0.3590
58–58*	0.5500	−0.7160	121–121*	0.7950	0.5380
66–66*	0.6900	−0.5450	130–130*	0.7300	0.7250
75–75*	0.7800	−0.3790	139–139*	0.6400	0.8970
84–84*	0.8300	−0.1850	33–33*	0.5600	1.0030
93–93*	0.8500	0.0000	–	–	–

$$\frac{T}{d_o} = 0.847444 + 0.024784 \cdot \frac{L}{d_o} - 0.272459 \cdot \left(\frac{L}{d_o}\right)^2 + 0.284257 \cdot \left(\frac{L}{d_o}\right)^3 - 0.230959 \cdot \left(\frac{L}{d_o}\right)^4 - 0.122097 \cdot \left(\frac{L}{d_o}\right)^5 \rightarrow R^2 = 0.9995$$

10.5.5 Section E

- *Dimensionless Coordinates*—Table 10.7.
- *Graph*—Fig. 10.11.
- *Experimental Equation and Variance*

$$\frac{T}{d_o} = \frac{0.972669 + 0.156628 \cdot \frac{L}{d_o} - 0.552386 \cdot \left(\frac{L}{d_o}\right)^2}{1 + 0.208161 \cdot \frac{L}{d_o} - 0.216656 \cdot \left(\frac{L}{d_o}\right)^2} \rightarrow R^2 = 0.9998$$

Table 10.7 Section and in the longitudinal normal plane T/doxL/do

Num.	T/d_o	L/d_o	Num.	T/d_o	L/d_o
5–5*	0.0000	−1.1930	94–94*	0.9750	0.0000
48–48*	0.3190	−1.0740	104–104*	0.9550	0.1730
53–53*	0.6120	−0.8910	113–113*	0.9150	0.3590
59–59*	0.7870	−0.7160	122–122*	0.8720	0.5380
67–67*	0.8720	−0.5450	131–131*	0.7900	0.7250
76–76*	0.9430	−0.3790	140–140*	0.6800	0.8970
85–85*	0.9650	−0.1850	32–32*	0.5800	0.9900

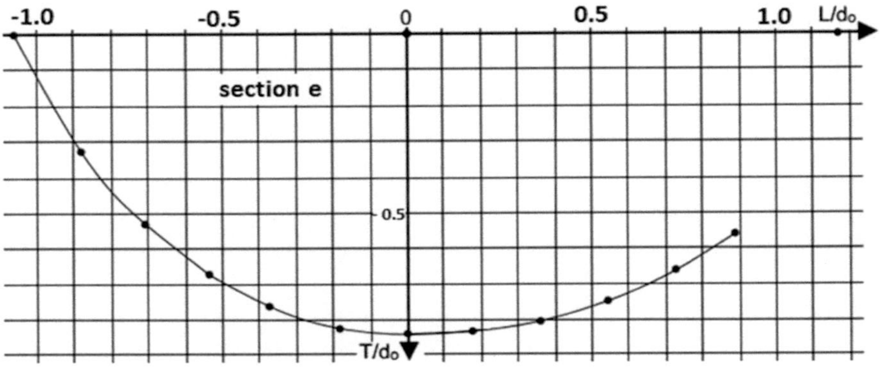

Fig. 10.11 Graph of longitudinal normal section *e*

10.5.6 Section F

- *Coordinates*—Table 10.8.
- *Graph*—Fig. 10.12.
- *Experimental Equation and Variance*

$$\frac{T}{d_o} = \frac{0.993637 + 0.083488 \cdot \frac{L}{d_o} - 0.583757 \cdot \left(\frac{L}{d_o}\right)^2}{1 + 0.150681 \cdot \frac{L}{d_o} - 0.299166 \cdot \left(\frac{L}{d_o}\right)^2} \rightarrow R^2 = 0.9994$$

Table 10.8 Section f in the longitudinal normal plane T/doxL/do

Num.	T/d_o	L/d_o	Num.	T/d_o	L/d_o
6–6*	0.0000	−1.2250	105–105*	0.9750	0.1730
49–49*	0.4680	−1.0740	114–114*	0.9280	0.3590
54–54*	0.7250	−0.8910	123–123*	0.8710	0.5380
60–60*	0.8610	−0.7160	132–132*	0.8000	0.7250
68–68*	0.9320	−0.5450	141–141*	0.6750	0.8970
77–77*	0.9750	−0.3790	147–147*	0.4970	1.0700
86–86*	0.9950	−0.1850	152–152*	0.2800	1.2490
95–95*	1.0000	0.0000	28–28*	0.0000	1.3750

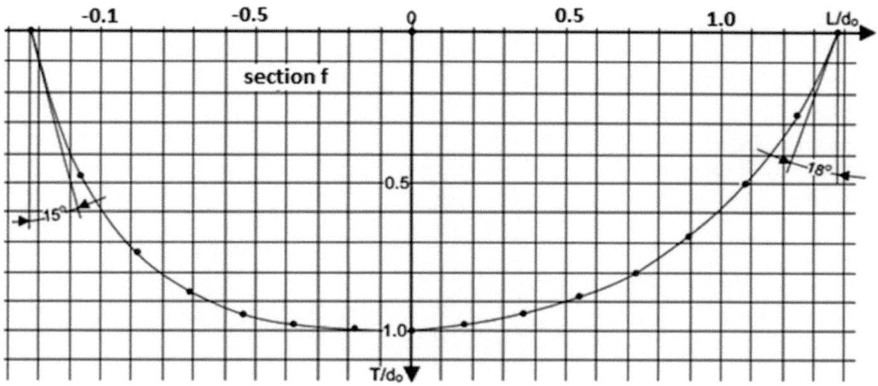

Fig. 10.12 Graph of longitudinal normal section *f*

10.5.7 Graph of Sections A to F—Fig. 10.13

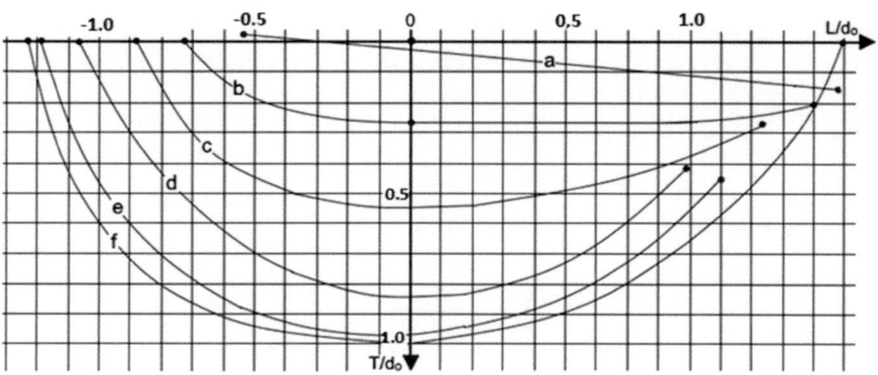

Fig. 10.13 Graph of the longitudinal normal section from *a* to *f*

10.5.8 Section G

- *Dimensionless Coordinates*—Table 10.9.
- *Graph*—Fig. 10.14.
- *Experimental Equation and Variance*

$$\frac{T}{d_o} = \frac{0.999512 + 0.105460 \cdot \frac{L}{d_o} - 0.589885 \cdot \left(\frac{L}{d_o}\right)^2}{1 + 0.188244 \cdot \frac{L}{d_o} - 0.281315 \cdot \left(\frac{L}{d_o}\right)^2} \rightarrow R^2 = 0.9995$$

Table 10.9 Section g in the longitudinal normal plane T/doxL/do

Num.	T/d_o	L/d_o	Num.	T/d_o	L/d_o
7–7*	0.0000	−1.1930	106–106*	0.9230	0.1730
50–50*	0.3300	−1.0740	115–115*	0.8700	0.3590
55–55*	0.6300	−0.8910	124–124*	0.8100	0.5380
61–61*	0.7700	−0.7160	133–133*	0.7220	0.7250
69–69*	0.8670	−0.5450	142–142*	0.5950	0.8970
78–78*	0.9240	−0.3790	148–148*	0.4280	1.0700
87–87*	0.9450	−0.1850	153–153*	0.2000	1.2490
96–96*	0.9500	0.0000	27–27*	0.0000	1.3550

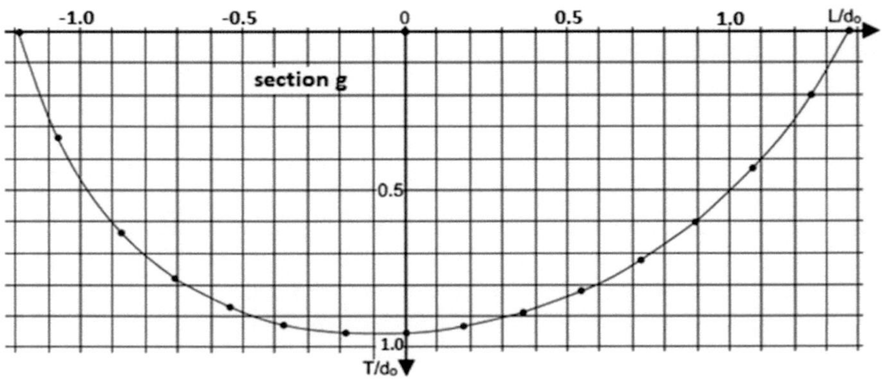

Fig. 10.14 Graph of longitudinal normal section *g*

10.5.9 Section H

- *Dimensionless Coordinates*—Table 10.10.
- *Chart*—Fig. 10.15.
- *Experimental Equation and Variance*

$$\frac{T}{d_o} = \frac{0.827351 + 0.127069 \cdot \frac{L}{d_o} - 0.577158 \cdot \left(\frac{L}{d_o}\right)^2}{1 + 0.227863 \cdot \frac{L}{d_o} - 0.204790 \cdot \left(\frac{L}{d_o}\right)^2} \rightarrow R^2 = 0.9997$$

10.5.10 Section I

- *Dimensionless Coordinates*—Table 10.11.
- *Graph*—Fig. 10.16.
- *Experimental Equation and Variance*

Table 10.10 Section h in the longitudinal normal plane T/doxL/do

Num.	T/d_o	L/d_o	Num.	T/d_o	L/d_o
8–8*	0.0000	−1.0900	107–107*	0.8100	0.1730
51–51*	0.0530	−1.0740	116–116*	0.7600	0.3590
56–56*	0.4100	−0.8910	125–125*	0.6890	0.5380
62–62*	0.6000	−0.7160	134–134*	0.5900	0.7250
70–70*	0.7200	−0.5450	143–143*	0.4590	0.8970
79–79*	0.7800	−0.3790	149–149*	0.2800	1.0700
88–88*	0.8200	−0.1850	154–154*	0.0800	1.2490
97–97*	0.8300	0.0000	26–26*	0.0000	1.3180

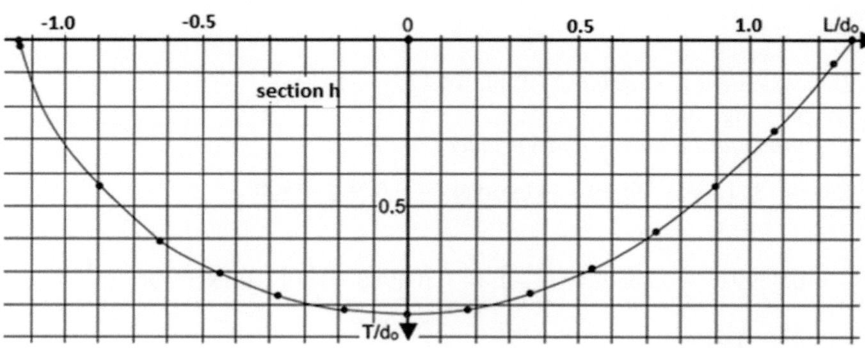

Fig. 10.15 Graph of longitudinal normal section *h*

Table 10.11 Section i in the longitudinal normal plane T/doxL/do

Num.	T/d_o	L/d_o	Num.	T/d_o	L/d_o
10–10*	0.0000	−0.9310	108–108*	0.5980	0.1730
56*-56**	0.0700	−0.8910	117–117*	0.5600	0.3590
63–63*	0.3330	−0.7160	126–126*	0.5000	0.5380
71–71*	0.4750	−0.5460	135–135*	0.4000	0.7250
80–80*	0.5600	−0.3790	144–144*	0.2650	0.8970
89–89*	0.5980	−0.1850	150–150*	0.0950	1.0700
98–98*	0.6100	0.0000	24–24*	0.0000	1.1350

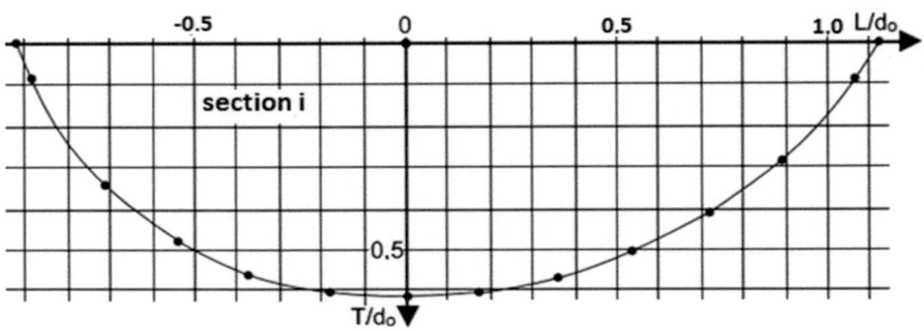

Fig. 10.16 Graph of longitudinal normal section *i*

$$\frac{T}{d_o} = 0.609135 + 0.0261 \cdot \frac{L}{d_o} 0.327879 \cdot \left(\frac{L}{d_o}\right)^2 - 0.135034 \cdot \left(\frac{L}{d_o}\right)^3 + 0.272988 \cdot \left(\frac{L}{d_o}\right)^4 - 0.054965 \cdot \left(\frac{L}{d_o}\right)^5 \rightarrow R^2 = 0.9995$$

10.5.11 Section J

- *Dimensionless Coordinates*—Table 10.12.
- *Graph*—Fig. 10.17.
- *Experimental Equations and Variances.*

Numbers 13; 13 $*$ $-$ 99; 99 $*$ →Interval: $-0{,}679 \leq \frac{L}{d_o} \leq 0$

$$\frac{T}{d_o} = 0.359914 - 0.346084 \cdot \left(\frac{L}{d_o}\right)^2 - 1.704988 \cdot \left(\frac{L}{d_o}\right)^4 + 1.653596 \cdot \left(\frac{L}{d_o}\right)^6 \rightarrow R^2 = 0.9999$$

Numbers 99; 99 $*$ $-$ 21; 11 $*$ → Interval: $-0{,}679 \leq \frac{L}{d_o} \leq 0$

Table 10.12 Section j in the longitudinal normal plane T/doxL/do

Num.	T/d_o	L/d_o	Num.	T/d_o	L/d_o
13–13*	0.0000	−0.6790	109–109*	0.3350	0.1730
72–72*	0.1500	−0.5450	118–118*	0.3100	0.3590
81–81*	0.2800	−0.3790	127–127*	0.2300	0.5380
90–90*	0.3460	−0.1850	136–136*	0.1300	0.7250
99–99*	0.3600	0.0000	21*-21**	0.0000	0.8110

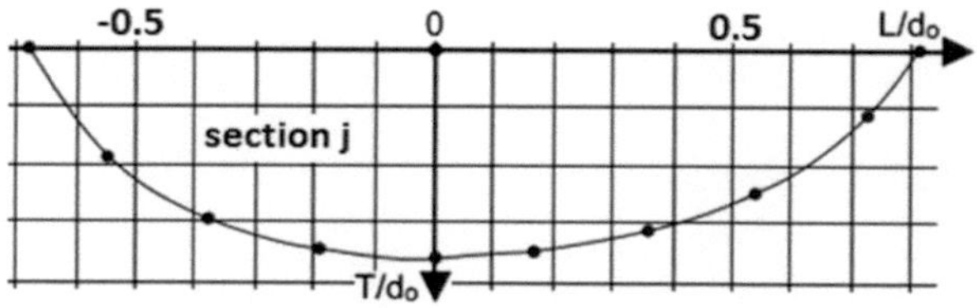

Fig. 10.17 Graph of longitudinal normal section *j*

$$\frac{T}{d_o} = 0.360835 - 0.638900 \cdot \left(\frac{L}{d_o}\right)^2 + 1.236957 \cdot \left(\frac{L}{d_o}\right)^4 - 1.670851 \cdot \left(\frac{L}{d_o}\right)^6$$

$$\rightarrow R^2 = 0.9999$$

10.5.12 Graph of Sections F to J—Fig. 10.18

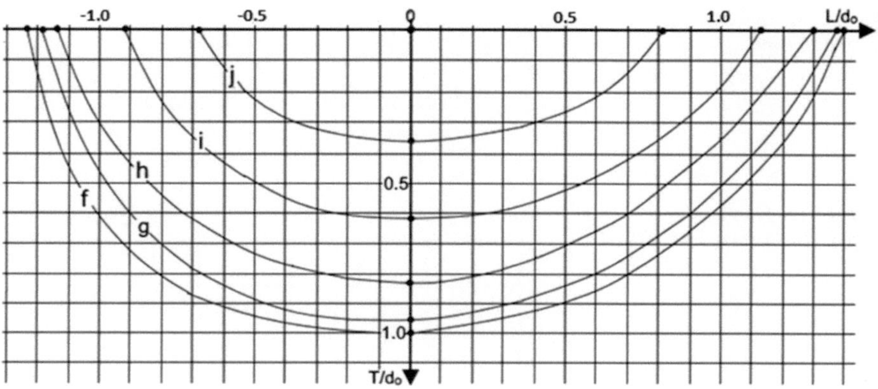

Fig. 10.18 Graph of the longitudinal normal section from *f* to *j*

10.6 Coordinate Tables, Equation, and Graphs of the Normal Transversal Sections for $T/d_o x0.5 \cdot B/d_o$

- As was done in the previous item, the study here in this item seeks to provide elements that allow for controlling manufacturing, either directly or via a machine program, and also to control the product quality and to create the respective manufacturing protocols, now referring to the normal transversal sections in the $T/d_o x0.5 \cdot B/d_o$ plane.

*10.6.1 Section A**

- *Dimensionless Coordinates*—Table 10.13.
- *Graph*—Fig. 10.19.
- *Experimental Equation and Variance*

$$\frac{T}{d_o} = \frac{-0.554640 + 1.508529 \cdot \left(\frac{B}{2 \cdot d_o}\right)^{0.5} - 0.940707 \cdot \frac{B}{2 \cdot d_o}}{1 - 2.063202 \cdot \left(\frac{B}{2 \cdot d_o}\right)^{0.5} + 1.187227 \cdot \frac{B}{2 \cdot d_o}} \rightarrow R^2 = 0.9999$$

Table 10.13 Section a* in the longitudinal normal plane T/dox0.5·B/do

Num.	T/d_o	0.5·B/do	Num.	T/d_o	0.5·B/do
4–4*	0.0000	0.3260	50–50*	0.3300	0.8590
48–48*	0.3190	0.5060	51–51*	0.0530	1.0320
49–49*	0.4680	0.6790	9–9*	0.0000	1.0670

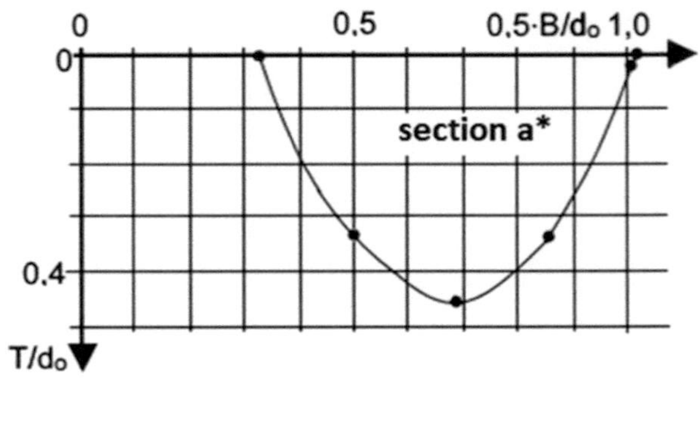

Fig. 10.19 Graph of transversal normal section a*

10.6.2 Section B*

- *Dimensionless Coordinates*—Table 10.14.
- *Graph*—Fig. 10.20.
- *Experimental Equations and Variances.*

Numbers 3; 3 * – 55; 55 * → Interval $-0.153 \le \frac{B}{2 \cdot d_o} \le 0.859$

$$\frac{T}{d_o} = -0.299085 + 1.846553 \cdot \frac{B}{2 \cdot d_o} + 1.015289 \cdot \left(\frac{B}{2 \cdot d_o}\right)^2 - 2.219763 \cdot \left(\frac{B}{2 \cdot d_o}\right)^3 \rightarrow R^2 = 0.9999$$

Numbers 55; 55 * – 11; 11 * → Interval $0.859 \le \frac{B}{2 \cdot d_o} \le 1.252$

$$\frac{T}{d_o} = 5.282881 - 3.153409 \cdot \frac{B}{2 \cdot d_o} - 1.670047 \cdot \left(\frac{B}{2 \cdot d_o}\right)^{-1} \rightarrow R^2 = 0.9999$$

Table 10.14 Section b* in the longitudinal normal plane T/dox0.5·B/do

Num.	T/d$_o$	0.5·B/do	Num.	T/d$_o$	0.5·B/do
3–3*	0.0000	0.1530	55–55*	0.6300	0.8590
52–52*	0.3310	0.3260	56–56*	0.4200	1.0320
53–53*	0.6120	0.5060	56*-56**	0.0700	1.2190
54–54*	0.7250	0.6790	11–11*	0.0000	1.2520

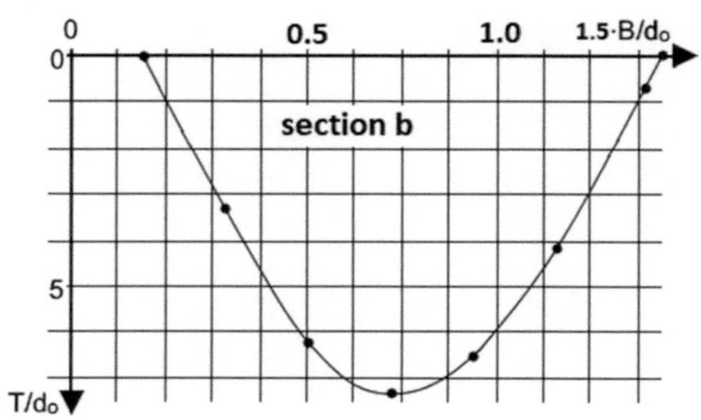

Fig. 10.20 Graph of normal cross section b*

10.6.3 Section C*

- *Dimensionless Coordinates*—Table 10.15.
- *Graph*—Fig. 10.21.
- *Experimental Equations and Variances.*

Numbers 2; 2 * – 60; 60 * → Interval $0.061 \leq \frac{B}{2 \cdot d_o} \leq 0.679$

$$\frac{T}{d_o} = 0.340875 + 3.947214 \cdot \left(\frac{B}{2 \cdot d_o}\right)^2 - 4.047963 \cdot \left(\frac{B}{2 \cdot d_o}\right)^3 - 0.021603 \cdot \left(\frac{B}{2 \cdot d_o}\right)^{-1} \rightarrow R^2 = 0.9999$$

Numbers 60; 60 * – 12; 12 * → Interval $0.679 \leq \frac{B}{2 \cdot d_o} \leq 1.372$

$$\frac{T}{d_o} = 0.951053 - 0.089646 \cdot \left(\frac{B}{2 \cdot d_o}\right)^2 - 0.220647 \cdot \left(\frac{B}{2 \cdot d_o}\right)^4 \rightarrow R^2 = 0.9999$$

Table 10.15 Section c* in the longitudinal normal plane T/dox0.5·B/do

Num.	T/d$_o$	0.5·B/do	Num.	T/d$_o$	0.5·B/do
2–2*	0.0000	0.0610	61–61*	0.7700	0.8590
57–57*	0.2400	0.1530	62–62*	0.6000	1.0320
58–58*	0.5500	0.3260	63–63*	0.3330	0.0000
59–59*	0.7870	0.5060	12–12*	1.2190	1.3720
60–60*	0.8610	0.6790	–	–	–

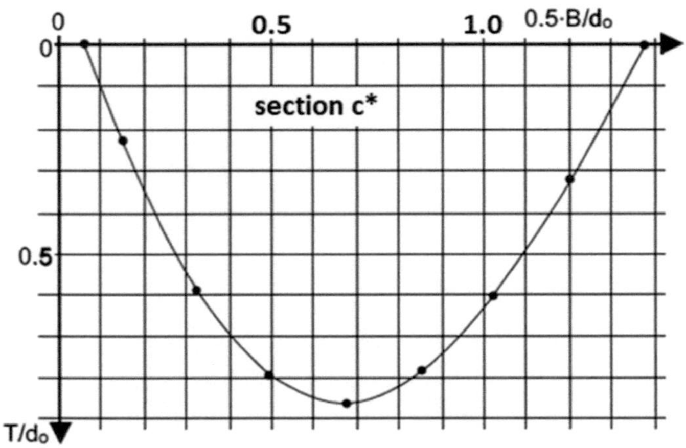

Fig. 10.21 Graph of normal cross section c*

10.6.4 Section D*

- *Dimensionless Coordinates*—Table 10.16.
- *Graph*—Fig. 10.22.
- *Experiment Equation and Variance.*

$$\frac{T}{d_o} = 0.034279 + 3.770324 \cdot \frac{B}{2 \cdot d_o} - 7.529457 \cdot \left(\frac{B}{2 \cdot d_o}\right)^2 + 12.733999 \cdot \left(\frac{B}{2 \cdot d_o}\right)^3 - 14.977994 \cdot \left(\frac{B}{2 \cdot d_o}\right)^4 + 8.785757 \cdot \left(\frac{B}{2 \cdot d_o}\right)^5 - 1.993445 \cdot \left(\frac{B}{2 \cdot d_o}\right)^6 \rightarrow R^2 = 0.9996$$

Table 10.16 Section d* in the longitudinal normal plane T/dox0.5·B/do

Num.	T/d_o	0.5·B/do	Num.	T/d_o	0.5·B/do
1–1*	–0.0285	0.0000	69–69*	0.8670	0.8590
64–64*	0.1550	0.0610	70–70*	0.7200	1.0320
65–65*	0.4000	0.4190	71–71*	0.4750	1.2190
66–66*	0.6900	0.3260	72–72*	0.1500	1.3920
67–67*	0.8720	0.5060	14–14*	0.0000	1.4500
68–68*	0.9320	0.6790	–	–	–

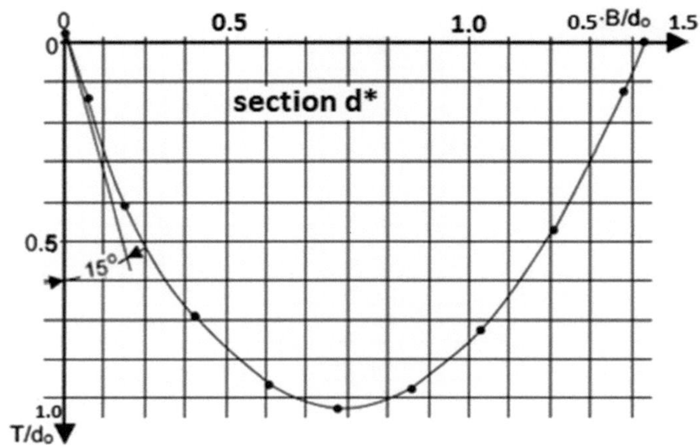

Fig. 10.22 Graph of normal cross section d*

10.6.5 Section E*

- *Dimensionless Coordinates*—Table 10.17.
- *Graph*—Fig. 10.23.
- *Experimental Equations and Variances.*

Numbers 47 – 78; 78 * → Interval $0 \leq \frac{B}{2 \cdot d_o} \leq 0.859$

$$\frac{T}{d_o} = 0.0227706 + 7.343804 \cdot \frac{B}{2 \cdot d_o} - 8.553870 \cdot \left(\frac{B}{2 \cdot d_o}\right)^{1,5} + 2.577104 \cdot \left(\frac{B}{2 \cdot d_o}\right)^{2} - 0.533377 \cdot \left(\frac{B}{2 \cdot d_o}\right)^{0,5} \rightarrow R^2 = 0.9994$$

Numbers 78; 78 * – 15; 15 * → Interval $0.859 \leq \frac{B}{2 \cdot d_o} \leq 1.499$

Table 10.17 Section e* in the longitudinal normal plane T/dox0.5·B/do

Num.	T/d_o	0.5·B/do	Num.	T/d_o	0.5·B/do
47	0.0238	0.0000	78–78*	0.9240	0.8590
73–73*	0.2120	0.0610	79–79*	0.7800	1.0320
74–74*	0.5020	0.1530	80–80*	0.5600	1.2190
75–75*	0.7800	0.3260	81–81*	0.2800	1.3920
76–76*	0.9430	0.5060	15–15*	0.0000	1.4990
77–77*	0.9750	0.6790	–	–	–

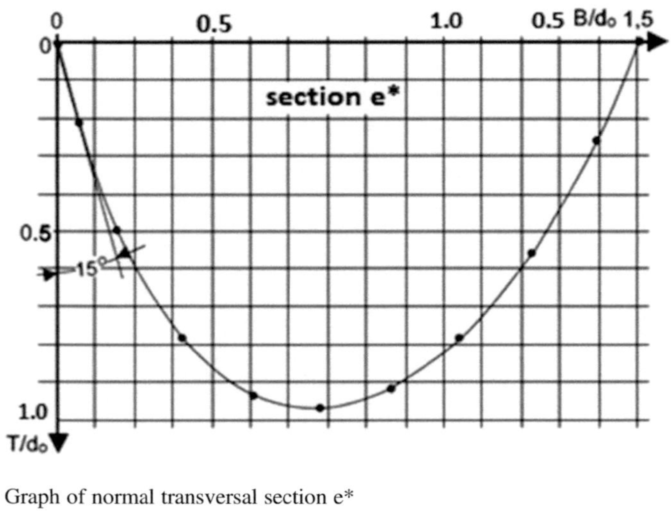

Fig. 10.23 Graph of normal transversal section e*

$$\frac{T}{d_o} = \frac{1.417404 + 0.945119 \cdot \frac{B}{2 \cdot d_o}}{1 - 0.320645 \cdot \frac{B}{2 \cdot d_o} - 0.094270 \cdot \left(\frac{B}{2 \cdot d_o}\right)^2} \rightarrow R^2 = 0.9991$$

*10.6.6 Section F**

- *Dimensionless Coordinates*—Table 10.18.
- *Graph*—Fig. 10.24.
- *Experimental Equation and Variance.*

Table 10.18 Section f* in the longitudinal normal plane T/dox0.5·B/do

Num.	T/d_o	0.5·B/do	Num.	T/d_o	0.5·B/do
47	0.0183	0.0000	87–87*	0.9450	0.8590
82–82*	0.2530	0.0610	88–88*	0.8200	1.0320
83–83*	0.5390	0.1530	89–89*	0.5980	1.2190
84–84*	0.8300	0.3260	90–90*	0.3460	1.3920
85–85*	0.9650	0.5060	16–16*	0.0000	1.5360
86–86*	0.9950	0.6090	–	–	–

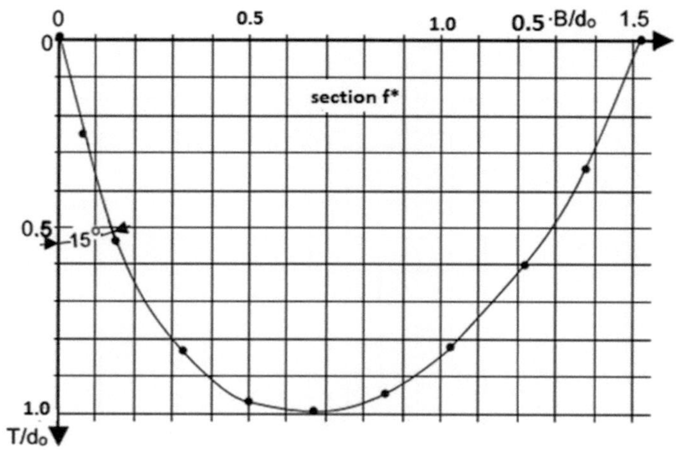

Fig. 10.24 Graph of longitudinal normal section f*

$$\frac{T}{d_o} = 0.016547 + 4.434314 \cdot \frac{B}{2 \cdot d_o} - 8.070254 \cdot \left(\frac{B}{2 \cdot d_o}\right)^2 + 7.951069 \cdot \left(\frac{B}{2 \cdot d_o}\right)^3 - 4.409370 \cdot \left(\frac{B}{2 \cdot d_o}\right)^4 + 0.928988 \cdot \left(\frac{B}{2 \cdot d_o}\right)^5 \rightarrow R^2 = 0.9999$$

10.6.7 Section G*

- *Dimensionless Coordinates*—Table 10.19.
- *Graph*—Fig. 10.25.
- *Experimental Equation and Variance*

Table 10.19 Seção g* no plano normal longitudinal T/dox0.5·B/do

Num.	T/d_o	0.5·B/do	Num.	T/d_o	0.5·B/do
45	0.0130	0.0000	96–96*	0.9500	0.8590
91–91*	0.2650	0.0610	97–97*	0.8300	1.0320
92–92*	0.5500	0.1530	98–98*	0.6100	1.2190
93–93*	0.8500	0.3260	99–99*	0.3600	1.3920
94–94*	0.9750	0.5060	100–100*	0.0400	1.5450
95–95*	1.0000	0.6790	17–17*	0.0000	1.5640

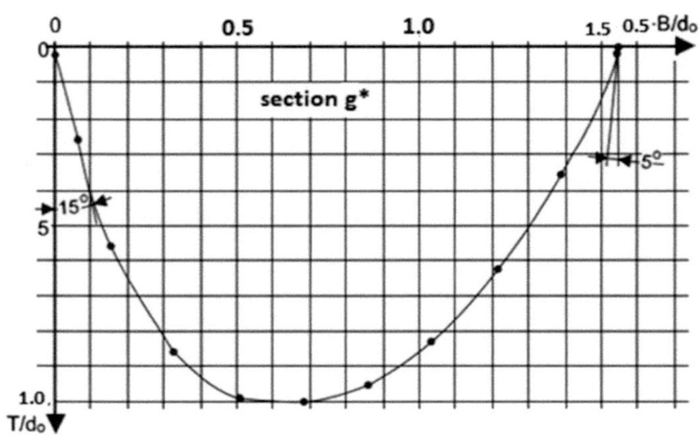

Fig. 10.25 Graph of normal cross section g*

$$\frac{T}{d_o} = \frac{0.012712 + 4.864891 \cdot \frac{B}{2 \cdot d_o} - 3.116694 \cdot \left(\frac{B}{2 \cdot d_o}\right)^2}{1 + 1.754592 \cdot \frac{B}{2 \cdot d_o} - 0.689013 \cdot \left(\frac{B}{2 \cdot d_o}\right)^2} \rightarrow R^2 = 0.9998$$

10.6.8 Graph of Sections A* to G*—Fig. 10.26

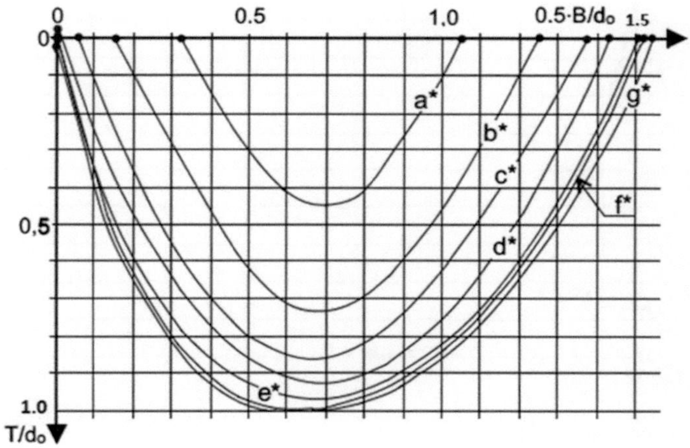

Fig. 10.26 Graph of the normal transversal sections from a* to g*

10.6.9 Section H*

- *Dimensionless Coordinates*—*Table 10.20.*
- *Graph*—Fig. 10.27.
- *Experimental Equation and Variance.*

$$\frac{T}{d_o} = \frac{-0.005041 + 5.340854 \frac{B}{2 \cdot d_o} - 3.432701 \cdot \left(\frac{B}{2 \cdot d_o}\right)^2}{1 + 2.257413 \cdot \frac{B}{2 \cdot d_o} - 0.963853 \cdot \left(\frac{B}{2 \cdot d_o}\right)^2} \rightarrow R^2 = 0.9998$$

Table 10.20 Section g* in the longitudinal normal plane T/dox0.5·B/do

Num.	T/d_o	0.5·B/do	Num.	T/d_o	0.5·B/do
44	−0.0081	0.0000	106–106*	0.9220	0.8590
101–101*	0.2650	0.0610	107–107*	0.8100	1.0320
102–102*	0.5450	0.1530	108–108*	0.5980	1.2190
103–103*	0.8450	0.3260	109–109*	0.3440	1.3920
104–104*	0.9550	0.5060	19–18*	0.0000	1.5550
105–105*	0.9750	0.6790	–	–	–

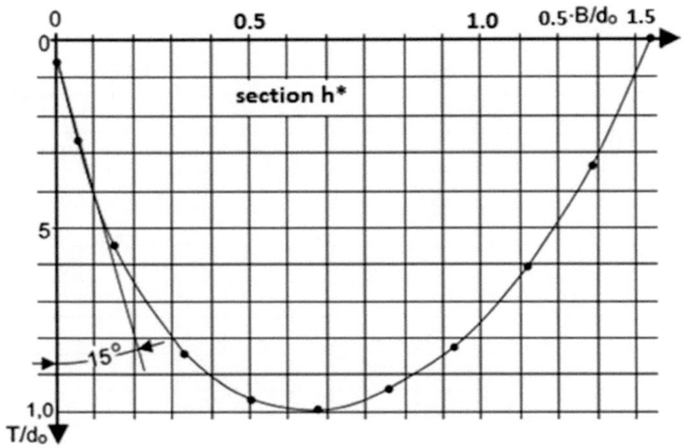

Fig. 10.27 Graph of normal cross section h*

10.6.10 Section I*

- *Dimensionless Coordinates*—Table 10.21.
- *Graph*—Fig. 10.28.
- *Experimental Equation and Variance*

$$\frac{T}{d_o} = \frac{-0.001605 + 4.941724 \cdot \frac{B}{2 \cdot d_o} - 3.234048 \cdot \left(\frac{B}{2 \cdot d_o}\right)^2}{1 + 1.685217 \cdot \frac{B}{2 \cdot d_o} - 0.134407 \cdot \left(\frac{B}{2 \cdot d_o}\right)^2 - 0.628294 \cdot \left(\frac{B}{2 \cdot d_o}\right)^3} \rightarrow R^2$$
$$= 0.9995$$

Table 10.21 Section i* in the longitudinal normal plane T/dox0.5·B/do

Num.	T/d_o	0.5·B/do	Num.	T/d_o	0.5·B/do
43	0.0028	0.0000	115–115*	−0.8700	0.8590
110–110*	−0.2650	0.0610	116–116*	−0.7600	1.0320
111–111*	−0.5330	0.1530	117–117*	−0.5600	1.2190
112–112*	−0.8300	0.3260	118–118*	−0.3290	1.3920
113–113*	−0.9150	0.5060	19–19*	0.0000	1.5410
114–114*	−0.9280	0.6790	–	–	–

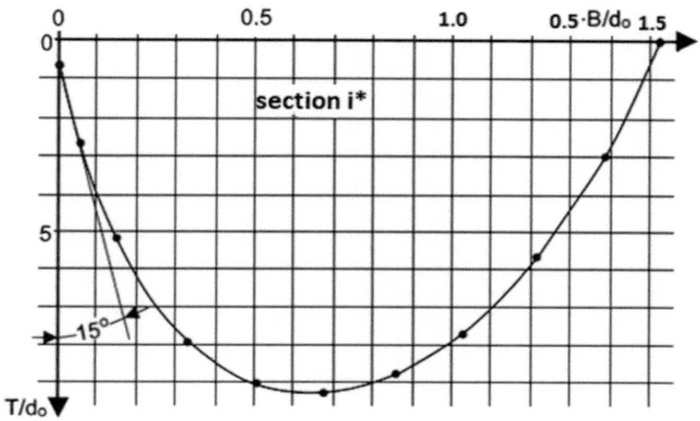

Fig. 10.28 Graph of normal transversal Sect. i*

*10.6.11 Section J**

- *Dimensionless Coordinates*—Table 10.22.
- *Graph*—Fig. 10.29.
- *Experimental Equations and Variances.*

Numbers 42; 42 $*$ − 123; 123 $* \ *$ → Interval $0 \leq \frac{B}{2 \cdot d_o} \leq 0.679$

$$\frac{T}{d_o} = 0.002389 + 4.600965 \cdot \frac{B}{2 \cdot d_o} - 8.279836 \cdot \left(\frac{B}{2 \cdot d_o}\right)^2 - 4.600965 \cdot \left(\frac{B}{2 \cdot d_o}\right)^3 \rightarrow R^2 = 0.9996$$

Numbers 123; 123 $*$ –21; 21 $* *$ → Interval $0.679 \leq \frac{B}{2 \cdot d_o} \leq 1.483$

Table 10.22 Section j* in the longitudinal normal plane T/dox0.5·B/do

Num.	T/d_o	0.5·B/do	Num.	T/d_o	0.5·B/do
42	−0.0022	0.0000	124–124*	0.8100	0.8590
119–119*	0.2650	0.0610	125–125*	0.6890	1.0320
120–120*	0.5200	0.1530	126–126*	0.5000	1.2190
121–121*	0.7950	0.3260	127–127*	0.2300	1.3920
122–122*	0.8620	0.5060	20–20*	0.0000	1.4830
123–123*	0.8710	0.6790	–	–	–

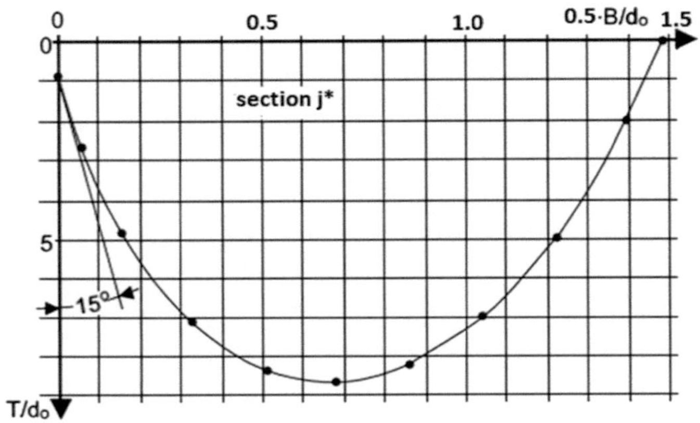

Fig. 10.29 Graph of normal cross section j*

$$\frac{T}{d_o} = -3.399285 + 17.793369 \cdot \frac{B}{2 \cdot d_o} - 26.767078 \cdot \left(\frac{B}{2 \cdot d_o}\right)^2 + 17.393836 \cdot \left(\frac{B}{2 \cdot d_o}\right)^3 - 4.310417 \cdot \left(\frac{B}{2 \cdot d_o}\right)^4 \rightarrow R^2 = 0.9999$$

10.6.12 Section K*

- *Dimensionless Coordinates*—Table 10.23.
- *Graph*—Fig. 10.30.
- *Experimental Equation and Variance*

$$\frac{T}{d_o} = \frac{0.103669 + 2.741143 \cdot \frac{B}{2 \cdot d_o} - 1.927216 \cdot \left(\frac{B}{2 \cdot d_o}\right)^2}{1 + 0.859704 \cdot \frac{B}{2 \cdot d_o} - 0.403251 \cdot \left(\frac{B}{2 \cdot d_o}\right)^2} \rightarrow R^2 = 0.9998$$

Table 10.23 Section k* in the longitudinal normal plane T/dox0.5·B/do

Num.	T/d_o	0.5·B/do	Num.	T/d_o	0.5·B/do
41	−0.0076	0.0000	133–133*	0.7220	0.8590
128–128*	0.2650	0.0610	134–134*	0.5900	1.0320
129–129*	0.4940	0.1530	135–135*	0.4000	1.2190
130–130*	0.7300	0.3260	136–136*	0.1240	1.3920
131–131*	0.7900	0.5060	21–21*	0.0000	1.4280
132–132*	0.8000	0.6790	–	–	–

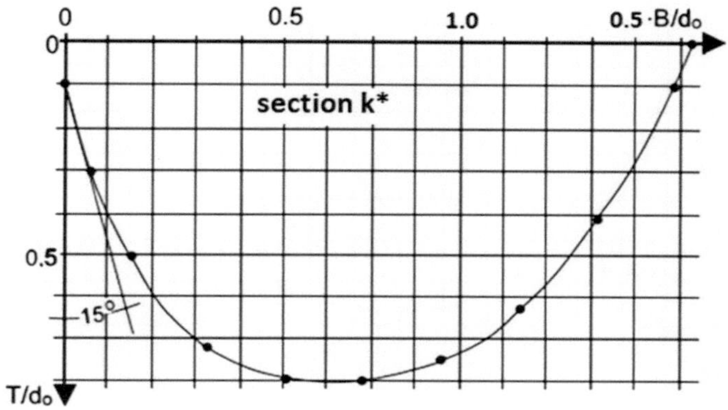

Fig. 10.30 Graph of normal cross section k*

10.6.13 *Section L**

- *Dimensionless Coordinates*—Table 10.24.
- *Graph*—Fig. 10.31.
- *Experimental Equations and Variances.*

Numbers 40 – 140; 140 $*$ → Interval $0 \le \frac{B}{2 \cdot d_o} \le 0.866$

$$\frac{T}{d_o} = \frac{0.012848 + 4.135827 \cdot \frac{B}{2 \cdot d_o}}{1 + 1.918487 \cdot \frac{B}{2 \cdot d_o} + 4.753380 \cdot \left(\frac{B}{2 \cdot d_o}\right)^2} \rightarrow R^2 = 0.9999$$

Numbers 140; 140 $*$ – 22; 22 $*$ → Interval $0.660 \le \frac{B}{2 \cdot d_o} \le 1.355$

Table 10.24 Section l* in the longitudinal normal plane T/dox0.5·B/do

Num.	T/d_o	0.5·B/do	Num.	T/d_o	0.5·B/do
40	0.0124	0.0000	141–141*	0.6750	0.6790
137–137*	0.2600	0.0610	142–142*	0.5950	0.8590
138–138*	0.4580	0.1530	143–143*	0.4580	1.0320
139–139*	0.6400	0.3260	144–144*	0.2650	1.2190
140–140*	0.6800	0.5060	22–22*	0.0000	1.3550

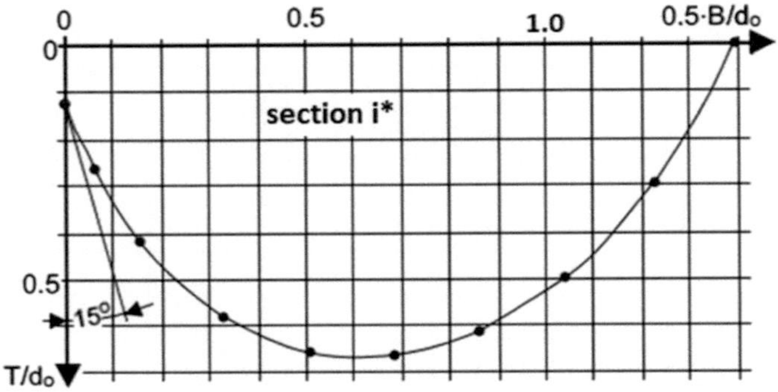

Fig. 10.31 Graph of normal cross section l*

$$\frac{T}{d_o} = 0.504597 + 1.050772 \cdot \frac{B}{2 \cdot d_o} - 2.071494 \cdot \left(\frac{B}{2 \cdot d_o}\right)^4 + 1.326114 \cdot \left(\frac{B}{2 \cdot d_o}\right)^6 - 0.321920 \cdot \left(\frac{B}{2 \cdot d_o}\right)^8 \rightarrow R^2 = 0.9998$$

10.6.14 Section M*

- *Dimensionless Coordinates*—Table 10.25.
- *Graph*—Fig. 10.32.
- *Experimental Equations and Variances.*

Numbers 39 – 34→ Interval $0 \le \frac{B}{2 \cdot d_o} \le 0.261$

$$\frac{T}{d_o} = 0.017408 + 0.290894 \cdot \frac{B}{2 \cdot d_o} - 0.867338 \cdot \left(\frac{B}{2 \cdot d_o}\right)^{0.5}$$

Numbers 31 – 23→ Interval $0.599 \le \frac{B}{2 \cdot d_o} \le 1.311$

Table 10.25 Section m* in the transverse plane T/dox0.5·B/do

Num.	T/d$_o$	0.5·B/do	Num.	T/d$_o$	0.5·B/do
39	0.0173	0.0000	31	0.5000	0.5990
145	0.2500	0.0610	147	0.4970	0.6790
146	0.4000	0.1530	148	0.4280	0.8590
34	0.4840	0.2160	149	0.2800	1.0320
–	–	–	150	0.0950	1.2190
–	–	–	23	0.0000	1.2750

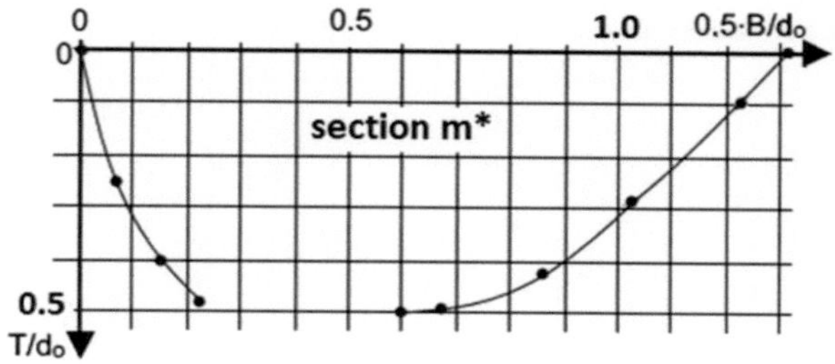

Fig. 10.32 Graph of normal cross section m*

$$\frac{T}{d_o} = -2,971161 + 6,512261 \cdot \left(\frac{B}{2 \cdot d_o}\right)^{-1} - 4,083773 \cdot \left(\frac{B}{2 \cdot d_o}\right)^{-2} + 0,855576 \cdot \left(\frac{B}{2 \cdot d_o}\right)^{-3} \to R^2 = 0,9999$$

*10.6.15 Section N**

- *Dimensionless Coordinates*—Table 10.26.
- *Graph*—Fig. 10.33.
- *Experimental Equations and Variances.*

Numbers 38 – 36→ Interval $0 \leq \frac{B}{2 \cdot d_o} \leq 0.093$

$$\frac{T}{d_o} = 0.022419 + 0.779327 \cdot \left(\frac{B}{2 \cdot d_o}\right)^{0.5} \to R^2 = 0.9999$$

Numbers 30 – 25→ Interval $0.559 \leq \frac{B}{2 \cdot d_o} \leq 1.097$

Table 10.26 Section n* in the transverse plane T/dox0.5·B/do

Num.	T/d_o	0.5·B/do	Num.	T/d_o	0.5·B/do
38	0.0224	0.0000	30	0.2900	0.5990
151	0.2150	0.0610	152	0.2800	0.6760
36	0.2600	0.0880	153	0.2000	0.8590
–	–	–	154	0.0600	1.0320
–	–	–	25	0.0000	1.0970

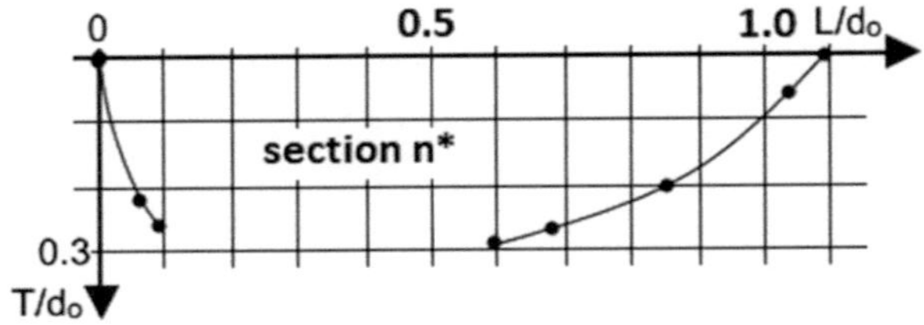

Fig. 10.33 Graph of normal transversal section n*

$$\frac{T}{d_o} = 0.214063 + 0.514433 \cdot \frac{B}{2 \cdot d_o} + 2.709009 \cdot \left(\frac{B}{2 \cdot d_o}\right)^2 - 0.956102 \cdot \left(\frac{B}{2 \cdot d_o}\right)^4 + 0.316496 \cdot \left(\frac{B}{2 \cdot d_o}\right)^6 \rightarrow R^2 = 0.9999$$

10.6.16 *Graph of Sections G* to N*—Fig. 10.34*

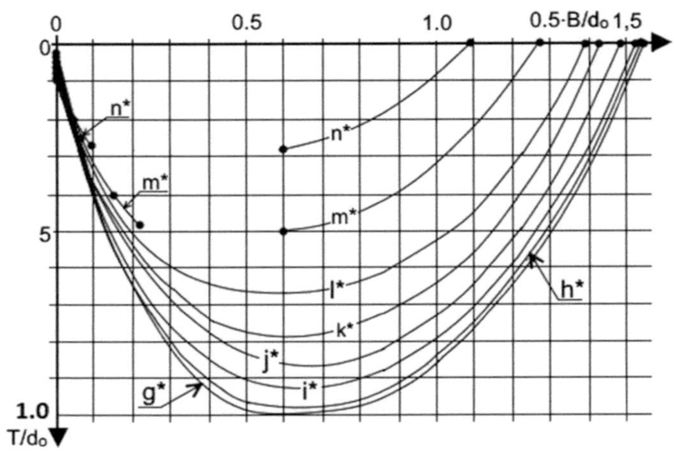

Fig. 10.34 Graph of the longitudinal normal sections from g* to n*

10.7 Reviews, Comments, and Recommendations

- The drawings for the set of curves representing the cross sections (Figs. 10.13, 10.18, 10.26, 10.34) is very important, since they allow one to verify any unforeseen intersections or discontinuities that must be corrected, along with and points that must have the same coordinates.
- Figure 10.35 shows, in detail and in scale, the outline of the shell cutaway in the vertical projection, and the magnitude is the variable at the position of the normal transversal section relative to the dividing edge (points 1; 47;..0.37).
- The angles of the flow paths, β_5, at the exit edge of the shells, which do not coincide with the corresponding angles at the exit of the cross sections, except for at the g^* section, must be corrected after structuring the shells when necessary to avoid impacts with the backs of the anterior shells (see Figs. 9.32).
- In principle, the values of the dimensionless coordinates of the points shown in Table 10.1 must be of the same order as those obtained via the equations, at least to the tenth place. For example, point 87 from longitudinal section g and corresponding value in cross section $f*$, with the value of its dimensionless coordinate from Table 10.1 as $T/d_o = 0.945$ results in:

Section g –

$$\frac{L}{d_o} = -0.379 \rightarrow \frac{T}{d_o} = \frac{0.999512 + 0.105460 \cdot \frac{L}{d_o} - 0.589885 \cdot \left(\frac{L}{d_o}\right)^2}{1 + 0.188244 \cdot \frac{L}{d_o} - 0.281315 \cdot \left(\frac{L}{d_o}\right)^2} \rightarrow R^2$$

$$= 0.9995 \frac{T}{d_o} = \frac{0.999512 + 0.105460 \cdot (-0.379) - 0.589885 \cdot (-0.379)^2}{1 + 0.188244 \cdot (-0.379) - 0.281315 \cdot (-0.379)^2}$$

$$= 0.985$$

Section f*-

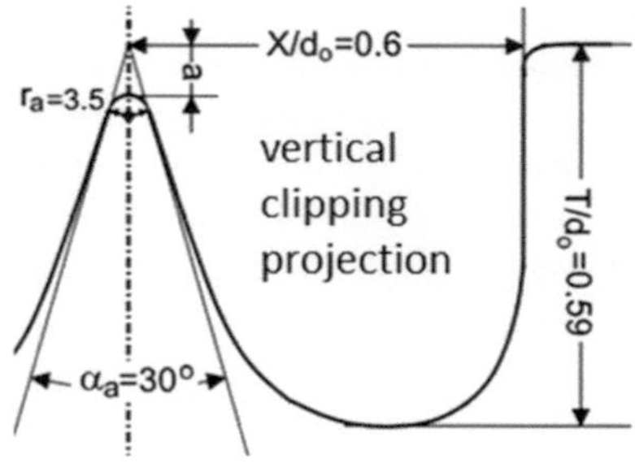

Fig. 10.35 Vertical projection of the shell cutaway, to scale

$$\frac{B}{2\cdot d_o}=0.859\rightarrow\frac{T}{d_o}=0.016547+4.434314\cdot\frac{B}{2\cdot d_o}-8.070254\cdot\left(\frac{B}{2\cdot d_o}\right)^2$$

$$+7.951069\cdot\left(\frac{B}{2\cdot d_o}\right)^3-4.409370\cdot\left(\frac{B}{2\cdot d_o}\right)^4+0.928988\cdot\left(\frac{B}{2\cdot d_o}\right)^5\rightarrow R^2$$

$$=0.9999\ \frac{T}{d_o}=0.016547+4.434314\cdot 0.859-8.070254\cdot 0.859^2+7.951069$$

$$\cdot 0.859^3-4.409370\cdot 0.859^4+0.928988\cdot 0.859^5=0.944$$

The results show what we already expected, i.e., the value obtained via the equation shows that the greater the value of R^2 the smaller the difference with respect to the value of the predicted coordinate in Table 10.1.

To analyze this, either obtain the value of the T coordinate of point 87, where $d_o = 150$ (mm) or $T_{co} = 0.945 \cdot 150 = 141.75$ (mm).

From the equation for section g we have: $T_g = 0.985 \cdot 150 = 147.75$ (mm), and a greater T_{co} value in $\Delta T = 147.75 - 141.75 = 6$ (mm), or 4.06 (%) greater. Therefore, one should make adjustments via machining.

From the equation for section $f*$ we have: $T_{f*} = 0.944 \cdot 150 = 141.6$ (mm), and a lesser T_{co} value in $\Delta T = 141.6 - 141.75 = -0.15$ (mm), or 0.106 (%) lesser. Therefore, one should make adjustments via machining.

Since two equations pass through each point, in principle, if the modeling of the inner surface of the shell is done analogically, to give values for making adjustments via machining, an equation must be chosen that, for the considered point, results in coordinate $\frac{T}{d_o}$ being smaller than the numerical result from the table, and therefore smaller than the equation with R^2 closer to unity, i.e., $R^2 \cong 0.9999$.

If the section control templates are manufactured using discrete digital values, with reference to the coordinate values given in Table 10.1, we recommend that the mold of the inner surface of the shells also be digital. If they are analog, always use the equations that result in machining adjustments, or values lower than those corresponding to the values in Table 10.1.

- Structural *RO* issues on a whole, and model tests in general imply a need for machining the back of the shells so that shocks can be avoided, given the hydrodynamic flow, particularly at the exit of the shells, as analyzed in Item 9.10, and Figs. 9.30, 9.31, and 9.32.
- At our request, Professor Edson da Costa Bortoni, from the Federal University of Itajuba (*UNIFEI*) carried out a study on the possibility of drawing the inner surface of a *PHIT* shell within the criteria corresponding to the studies that resulted in the coordinates shown in Table 10.1. Given the available programs, he used "Curve 3D – v. 4.0" with an input condition set for changing only the $\frac{T}{d_o}$ Table values. He obtained better results, which are shown in Table 10.27 where the new $\frac{T}{d_o}$ coordinates were transcribed in *italics*.

Table 10.27 Coordinates of the Inner Shell Surface

Num.	0.5·B/do	L/d_o	T/d_o	Num.	0.5·B/do	L/d_o	T/d_o
1	0.00000000	−0.54500000	−0.08349770	77	0.67900000	−0.37900000	0.97380900
2–2*	0.06100000	−0.71600000	0.00632100	78	0.85900000	−0.37900000	0.92314100
3–3*	0.15300000	−0.89100000	0.04114600	79	1.03200000	−0.37900000	0.79341900
4–4*	0.32600000	−1.08500000	0.01606800	80	1.21900000	−0.37900000	0.55981300
5–5*	0.50600000	−1.19300000	−0.01372320	81	1.39200000	−0.37900000	0.25990400
6–6*	0.67900000	−1.23500000	−0.02942660	82	0.06100000	−0.18500000	0.24696200
7–7*	0.85900000	−1.17500000	0.07233670	83	0.15300000	−0.18500000	0.51930300
8–8*	1.03200000	−1.09000000	0.04175000	84	0.32600000	−0.18500000	0.82765200
9–9*	1.06700000	−1.06900000	0.02304160	85	0.50600000	−0.18500000	0.96551200
10–10*	1.21900000	−0.93100000	−0.02457640	86	0.67900000	−0.18500000	0.99128700
11–11*	1.25200000	−0.89100000	−0.02838760	87–87*	0.85900000	−0.18500000	0.93972200
12–12*	1.37200000	−0.71500000	−0.02880700	88–88*	1.03200000	−0.18500000	0.82082700
13–13*	1.39200000	−0.67900000	−0.02516870	89–89*	1.21900000	−0.18500000	0.60941200
14–14*	1.45000000	−0.53000000	0.02141400	90–90*	1.39200000	−0.18500000	0.33298500
15–15*	1.49900000	−0.35900000	0.05517600	91–91*	0.06100000	0.00000000	0.26121600
16–16*	1.53600000	−0.18200000	0.05141900	92–92*	0.15300000	0.00000000	0.53841200
17–17*	1.56400000	0.00000000	−0.00099510	93–93*	0.32600000	0.00000000	0.84283100
18–18*	1.55500000	0.17300000	−0.02698050	94–94*	0.50600000	0.00000000	0.96958100
19–19*	1.52700000	0.35900000	−0.04101160	95–95*	0.67900000	0.00000000	0.98686500
20–20*	1.48300000	0.53800000	0.08223100	96–96*	0.85900000	0.00000000	0.93375900
21–21*	1.42800000	0.72500000	−0.02087070	97–97*	1.03200000	0.00000000	0.82098300
21*-21**	1.39200000	0.81100000	0.00220900	98–98*	1.21900000	0.00000000	0.62046300
22–22*	1.35500000	0.88500000	0.02745500	99–99*	1.39200000	0.00000000	0.34997500
23–23*	1.31100000	0.97700000	0.03735200	100–100*	1.54500000	0.00000000	0.03252100
24–24*	1.21800000	1.13500000	0.03100800	101–101*	0.06100000	0.17300000	0.26451600

(continued)

Table 10.27 (continued)

Num.	0.5·B/do	L/d$_o$	T/d$_o$	Num.	0.5·B/do	L/d$_o$	T/d$_o$
25–25*	1.0970000	1.2690000	−0.00771300	102–102*	0.1530000	0.17300000	0.54367800
26–26*	1.0150000	1.3180000	−0.02531620	103–103*	0.3260000	0.17300000	0.84207400
27–27*	0.8590000	1.3550000	−0.02161890	104–104*	0.5060000	0.17300000	0.95819800
28–28*	0.6790000	1.3720000	0.01804700	105–105*	0.6790000	0.17300000	0.96801400
29–29*	0.5990000	1.3750000	0.05453100	106–106*	0.8590000	0.17300000	0.91319500
30–30*	0.5990000	1.2490000	0.39341900	107–107*	1.0320000	0.17300000	0.80372800
31–31*	0.5990000	1.0700000	0.48942400	108–108*	1.2190000	0.17300000	0.60680700
32–32*	0.5060000	0.9900000	0.59584500	109–109*	1.3920000	0.17300000	0.33063300
33–33*	0.3260000	1.0030000	0.57724900	110–110*	0.0610000	0.35900000	0.26285600
34–34*	0.2160000	1.0700000	0.46892400	111–111*	0.1530000	0.35900000	0.53940000
35–35*	0.1530000	1.1450000	0.36457100	112–112*	0.3260000	0.35900000	0.82589200
36–36*	0.0930000	1.2490000	0.24354600	113–113*	0.5060000	0.35900000	0.92840200
36*-36**	0.0610000	1.3040000	0.18000400	114–114*	0.6790000	0.35900000	0.92988000
37	0.0000000	1.3750000	0.06006700	115–115*	0.8590000	0.35900000	0.87315400
38	0.0000000	1.2490000	0.04201700	116–116*	1.0320000	0.35900000	0.76589900
39	0.0000000	1.0700000	0.02717700	117–117*	1.2190000	0.35900000	0.56825300
40	0.0000000	0.8970000	0.01926900	118–118*	1.3920000	0.35900000	0.27722700
41	0.0000000	0.7250000	0.01474200	119–119*	0.0610000	0.53800000	0.25773700
42	0.0000000	0.5380000	0.01262800	120–120*	0.1530000	0.53800000	0.52518100
43	0.0000000	0.3590000	0.01324400	121–121*	0.3260000	0.53800000	0.79198600
44	0.0000000	0.1730000	0.01578500	122–122*	0.5060000	0.53800000	0.87726000
45	0.0000000	0.0000000	0.01701300	123–123*	0.6790000	0.53800000	0.86950500
46	0.0000000	−0.1850000	0.01011100	124–124*	0.8590000	0.53800000	0.81098300
47	0.0000000	−0.3790000	−0.02049060	125–125*	1.0320000	0.53800000	0.70564800
48–48*	0.5060000	−1.0740000	0.28546200	126–126*	1.2190000	0.53800000	0.50510900

49–49*	0.67900000	−1.07400000	0.37853100	127–127*	1.39200000	0.53800000	0.19439600
50–50*	0.85900000	−1.07400000	0.31469400	128–128*	0.06100000	0.72500000	0.24794300
51–51*	1.03200000	−1.07400000	0.08022300	129–129*	0.15300000	0.72500000	0.49681900
52–52*	0.32600000	−0.89100000	0.39028800	130–130*	0.32600000	0.72500000	0.73117600
53–53*	0.50600000	−0.89100000	0.59985500	131–131*	0.50600000	0.72500000	0.79184700
54–54*	0.67900000	−0.89100000	0.67418200	132–132*	0.67900000	0.72500000	0.77208100
55–55*	0.85900000	−0.89100000	0.61760100	133–133*	0.85900000	0.72500000	0.71170800
56–56*	1.03200000	−0.89100000	0.42386300	134–134*	1.03200000	0.72500000	0.60922100
56*-56**	1.21900000	−0.89100000	0.05370900	135–135*	1.21900000	0.72500000	0.40570500
57–57*	0.15300000	−0.71600000	0.26626500	136–136*	1.39200000	0.72500000	0.07204300
58–58*	0.32600000	−0.71600000	0.59681500	137–137*	0.06100000	0.89700000	0.23357000
59–59*	0.50600000	−0.71600000	0.78140100	138–138*	0.15300000	0.89700000	0.45523300
60–60*	0.67900000	−0.71600000	0.83911000	139–139*	0.32600000	0.89700000	0.64605000
61–61*	0.85900000	−0.71600000	0.71170800	140–140*	0.50600000	0.89700000	0.67603600
62–62*	1.03200000	−0.71600000	0.62316600	141–141*	0.67900000	0.89700000	0.64201600
63–63*	1.21900000	−0.71600000	0.31261200	142–142*	0.85900000	0.89700000	0.57983300
64–64*	0.06100000	−0.54500000	0.13705300	143–143*	1.03200000	0.89700000	0.48155600
65–65*	0.15300000	−0.54500000	0.39907700	144–144*	1.21900000	0.89700000	0.27728300
66–66*	0.32600000	−0.54500000	0.71830300	145–145*	0.06100000	1.07000000	0.21337000
67–67*	0.50600000	−0.54500000	0.88393400	146–146*	0.15300000	1.07000000	0.39593700
68–68*	0.67900000	−0.54500000	0.93033700	147–147*	0.67900000	1.07000000	0.46341400
69–69*	0.85900000	−0.54500000	0.87942400	148–148*	0.85900000	1.07000000	0.39908100
70–70*	1.03200000	−0.54500000	0.73551400	149–149*	1.03200000	1.07000000	0.30726400
71–71*	1.21900000	−0.54500000	0.47169000	150–150*	1.21900000	1.07000000	0.10589400
72–72*	1.39200000	−0.54500000	0.13442300	151–151*	0.06100000	1.24900000	0.18801100
73–73*	0.06100000	−0.37900000	0.20744300	152–152*	0.67900000	1.24900000	0.22096400
74–74*	0.15300000	−0.37900000	0.47375200	153–153*	0.85900000	1.24900000	0.15371200

(continued)

Table 10.27 (continued)

Num.	0.5·B/do	L/d_o	T/d_o	Num.	0.5·B/do	L/d_o	T/d_o
75–75*	0.32600000	−0.37900000	0.78653800	154–154*	1.03200000	1.24900000	0.07154100
76–76*	0.50600000	−0.37900000	0.93778700	–	–	–	–

Note: The coordinates of column Num. On the right with 1 or 2 asterisks have reversed signs ie: Those with more (+); minus (−) eas minus (−); plus (+). Except the coordinates of the dividing edge of the shell

Table 10.28 Constants of the Chebyshev Eq. 10.5

a	−0.0232033100	h	−22950966.0000000000	o	0.0768856400
b	−0.3537865600	i	−0.2667354600	p	0.0095562740
c	−22903283.0000000000	j	−0.1311695400	q	0.0581457240
d	0.5628104320	k	0.0475549920	r	−0.0729220300
e	−0.0655140400	l	−0.0245130600	s	−0.0747218600
f	0.5820717340	m	0.9876574710	t	−0.0082757200
g	−9.6061.10–5	n	0.0134681080	u	−0.0148746600

The values of this coordinate from number 2 to 29 in the table, as they belong to the limit line of the shell surface, must be equal to zero, which means that they have to be corrected after modeling. The equation for the inner surface of the symmetrical half of the chosen shell, generated in 3*D*, was the Chebyshev Series, bivariable—X; Y in the order of 5—$R^2 = 0.9937696638$, transcribed below:

$$\begin{aligned} z = {} & a + bT1(x') + cT1(y') + dT2(x') + eT1(x')T1(y') + fT2(y') \\ & + gT3(x') + hT2(x')T1(y') + iT1(x')T2(y') + jT3(y') + kT4(x') \\ & + lT3(x')T1(y') + mT2(x')T2(y') + nT1(x')T3(y') + oT4(y') \\ & + pT5(x') + qT4(x')T1(y') + rT3(x')T2(y') + sT2(x')T3(y') \\ & + tT1(x')T4(y') + uT5(y') \end{aligned} \tag{10.4}$$

x' x (normalized between −1 and +1; $y'y$—(normalized between −1 and +1); $Tn(x') = \cos\,(n a\cos(x'))$

In this Equation (10.4), the coordinate values from Table 10.27 have the following correspondence:

- $$x = 0.5 \cdot B/d_o$$
- $$x = 0.5 \cdot B/d_o$$
- $$z = T/d_o$$

The values of the constants from Eq. (10.4) have been transcribed in Table 10.28.

Figure 10.36 show images of the shells with corrected limits after applying the Chebyshev Equation (Table Curve 3D) using the constants from Table 10.28 and the dimensionless coordinates from Table 10.27.

In these Figures, one should pay particular attention to the fact that the outline of the inner surface of the shell, in blue, is not part of the surface of the shell colored in purple.

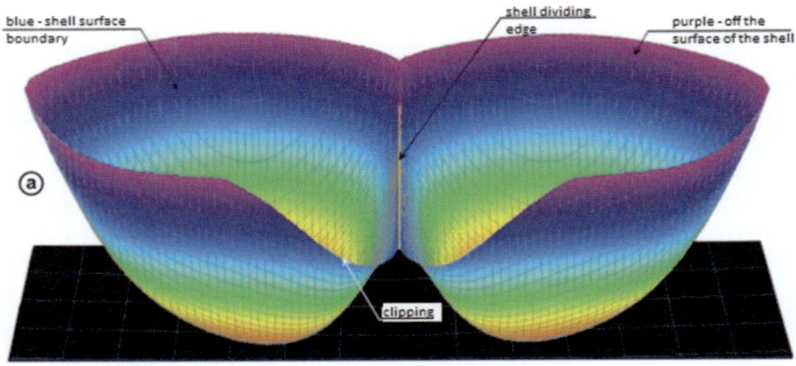

Fig. 10.36 (**a**) Perspective View. (**b**) Front View. (**c**) Blueprint

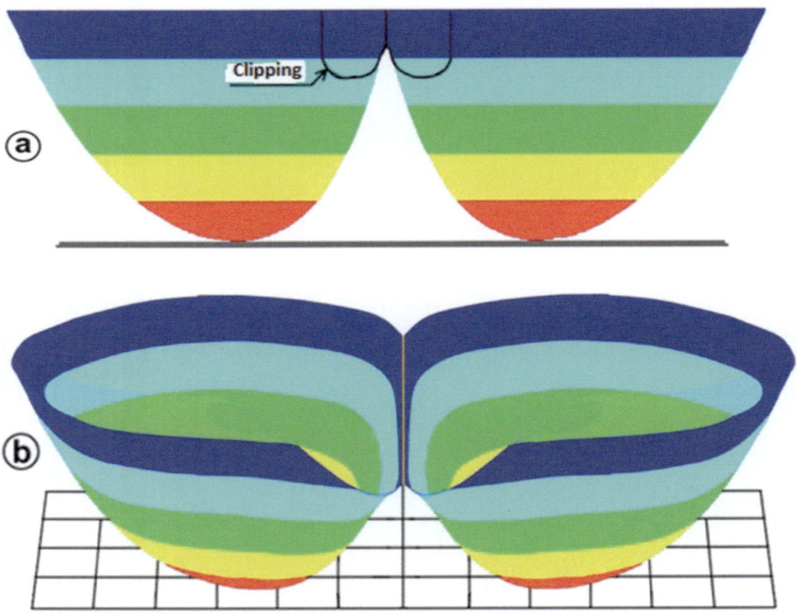

Fig. 10.37 Shell with levels or topographic horizontal lines. (**a**) Projection in the vertical view. (**b**) Perspective view

- Figure 10.37a, b, obtained using the "Table Curve 3D" software program, show the colored vertical projection and perspective of the inner surface of the shell and the lines separating them, which are called the levels or topographic lines.
- The full contour lines were drawn in relation to the $y = T$ direction using the computer program "datageos", with the dimensionless coordinates in their

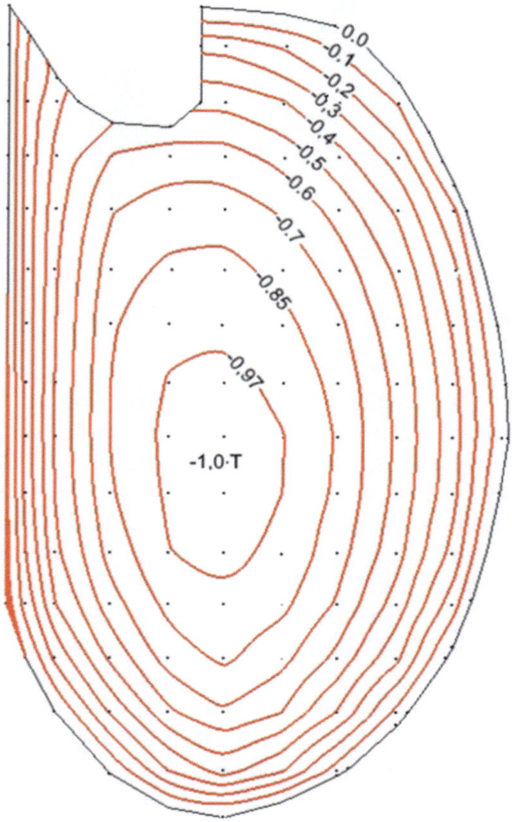

Fig. 10.38 Full contours in blueprint

respective maximum values in the direction $x \to x_{max} = (0.5 \cdot B/d_o)_{max} = 0.679$ and in the direction $y \to y_{max} = -(T/d_o)_{max} = -1.0$ (Table 10.1), in the blueprint of Fig. 10.38.

10.8 Modeling[1]

3D printing, also called prototyping, is the process by which physical objects are generated by layering materials based on a digital model. This technology is advancing rapidly and is becoming a trusted resource for mass manufacturing parts.

[1] Eng. Thiago Soares Correa.

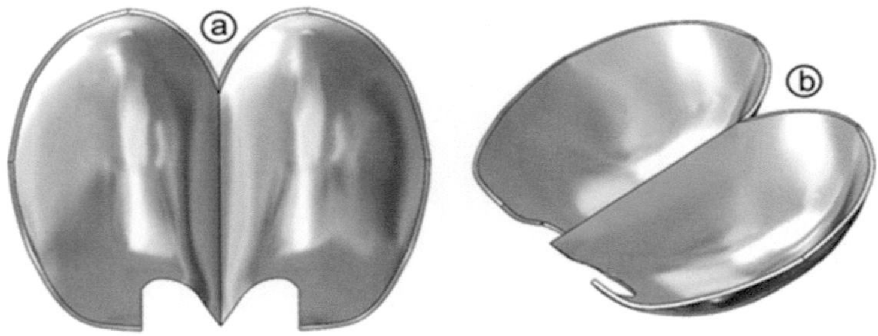

Fig. 10.39 Drawings generated by "SolidWorks 2016"

To manufacture of the 3D printer model to detail the hydrodynamic geometric outline of the inner surface of the shells, we have used Table 10.1, which has dimensionless values for the coordinates $0.5 \cdot B/d_o$, $L/d_o$and T/d_o.

Initially, a 3D drawing, like the ones in Fig. 10.39, was generated by "SolidWorks 2016", which is a three-dimensional modeling software program.

This will result is an STL format file that will be interpreted by the "INSIGHT 10.4" software program, where one can configure the position of the part on the printing table, to scale, and the print density. "INSIGHT 10.4" exports the information to the "Control Center 10.4" software program, which in addition to providing information on the amount of material that will be used, and the time taken to complete the part, divides the three-dimensional model into layers and sends the commands to perform the actual printing. The thermoplastic filament, once loaded into the 3D printer, is drawn to an extrusion head, where it will be heated at high temperatures to reach its lowest viscosity point, the so-called glass melting point, and is then deposited layer upon layer, and immediately after this deposition, it returns to a solid state, forming a 3D object.

The model studied here was built at the Product Innovation Laboratory (*PIL*) at UNIFEI, with the help of laboratory technician Luiz Fernando Rodrigues Alves. The 3D printer used was a "FORTUS 250mc" model from "Stratasys" (Fig. 10.40), which uses 1.5 mm ABS P430 polymer as the printing material.

This machine model uses a 3D printing process known as Fused Deposition Modeling (*FDM*), or here, ABS thermoplastic.

In this process, we were not concerned with the structural calculations, since the main objective is to detail the hydrodynamic geometric layout of the inner surface of the shells.

Figure 10.41 shows photographs of the result.

- It is fundamentally important to remember and comment that the studies developed in this chapter were theoretical, initially requiring structural complementation. Once this is done, the laboratory stage must be developed with the designed

Fig. 10.40 3D Printer—"FORTUS 250mc" from "Stratasys"

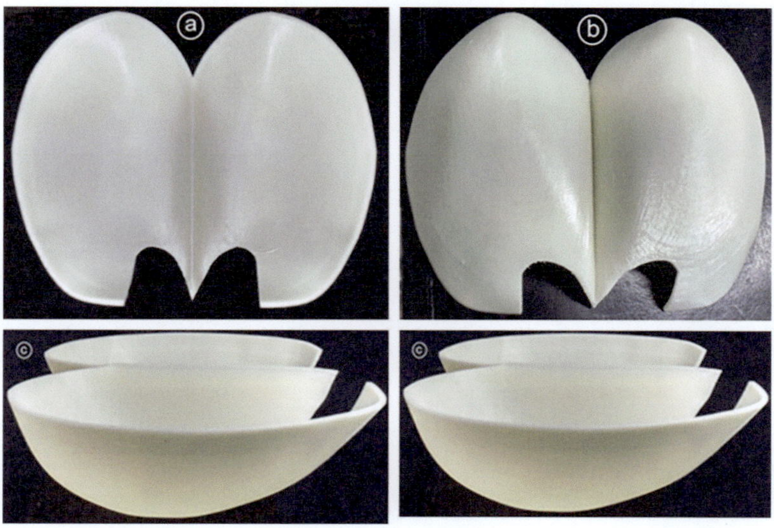

Fig. 10.41 (**a**, **b**) Front and back top views. (**c**, **d**) Side wiews

and manufactured model, which, once tested, will allow us, with this analysis, to obtain documented records, patents, and their respective publication. This procedure will give all contracting parties the opportunity to negotiate and develop the stages to make a competitive product available in the market.

Chapter 11
Fundamentals for the Preliminary Structural Calculation of a Pelton Hydraulic Impulse Turbine *PHIT*

The main objective of this chapter will be to establish the foundations that allow us to obtain equations, that preliminarily give results for the hydrodynamic calculations and in-scale drawings of the *HFP* of the *PHIT*, and to determine a first approximation of the thickness of each component composing the physical parts. Once the critical sections have been established, we will need to specify the materials that will be used to manufacture the corresponding parts, within the allowable stress levels. Subsequently, the Manufacturing Project for the *PHIT* model can be developed by *PT*, taking into account the limitations of the laboratories where it will be tested. The resulting report will list the elements for carrying out the Executive and Manufacturing Project so that the *PHIT* can ultimately be built. Furthermore, a form will be identified, analyzed, and presented for the established geometries, that will allow us to calculate the main forces necessary to keep both the *SCC* and the *QCC* open and closed.

11.1 Considerations and Hypotheses

- The preliminary structural calculation will be elaborated with reference to the calculations contained in Chaps. 8, 9, and 10, particularly with regards to the figures, symbols, conventions, graphics, and recommended values and *HFP*.
- In terms of components, the *PHIT*will observe the following division:
  - **The Transition Component (*TC*)** conducts the flow from the inlet section 0 of the *PHIT* to inlet Section 1$*$ in the control system. The *TC* is formed in series by tubes, bifurcations, injectors, and deflectors.
  - **The Slow Control Component (*SCC*)** conducts the flow from Section 1$*$ to section 2 at the output of the *IN*. This system must act after the Quick Control Component (*QCC*) is engaged, when it in fact exists, within a predefined time,

© The Author(s), under exclusive license to Springer Nature Switzerland AG 2025

Z. de Souza, G. L. Tiago Filho, *Pelton and Crossflow Impulse Hydraulic Turbines*,
https://doi.org/10.1007/978-3-031-87086-6_11

Table 11.1 Waterhammer fraction

$H_{\max}$	$h_s/H_{\max}$
mca	—
< 50	0.50
de 50 to 250	hs = √(14,254/$H_{\text{máx}}$) - 0,0341
>250	0.10

to limit the predicted water hammer. This time, also known as the needle displacement time, $t_a(s)$, is fixed by the *PT*, normally in the following range:

$$10 < t_{\text{CCL}} = t_a(s) < 40 \tag{11.1}$$

– **The Quick Control Component (*QCC*)**, also called the deflector (*DF*) (see Fig. 1.5 and subitem 8.11), maneuvers to totally or partially deflect the jet from the shells in a very short time, t_d (s). This time depends on the inertia of the rotating system which is linked to the transient or acceleration time of the *GG*, $t_g(s)$, which is the time measured in seconds, to take the *GG* from rotation $n = 0$ to the nominal rotation, $n = n_n$ (rpm), or the nominal moment acting on the axis. This connection is established in the following range:

$$0.25 \cdot t_g < t_d(s) < 0.75 \cdot t_g \tag{11.2}$$

These three components make up the *PHIT SA* (Admission System).

– **The Rotor (*RO*),** or the rotating component of the *PHIT*, transforms the kinetic energy of the flow into mechanical energy. This transformation occurs, initially, in the double shells, which are fixed or fused radially in a disk, centered with the axis of the *PHIT*, rotating the set at atmospheric pressure. This axis receives the flow energy in the *RO*, minus the ventilation losses, and transforms it into mechanical energy.
– **The Hydraulic Brake (*HB*)** is a cylindrical/conical component that acts on the back of the shells to reduce the rotation of the *RO* when the *PHIT* load is removed.
– **A Flywheel (*FW*)** is a cylindrical disk component, installed on the axis of the *PHIT*, when necessary, to complete the inertia of the *EG*, and to maintain the stability of the interconnected electrical system during load variations, with frequency variations lower than 4 (%).
– **Casing (*CAS*)** is a component that separates the flow of the *PHIT* from the external environment and the downstream level without drowning it (see Figs. 8.11 and 8.12).

• To structurally calculate and design the *SA*, we recommend taking:

$Q_1 = \frac{Q_{\max}}{z_j}$ (m^3/s), $H_1 = (1 + h_s) \cdot H_{\max}$ (mca), with h_s representing the water hammer fraction that can be preliminarily fixed, as shown in Table 11.1.

11.2 Admission System (*SA*)

11.2.1 Considerations and Hypotheses

Ultimately, the *SA* (see Figs. 8.1, 8.2a, 8.3a–c) comprises pieces with a circular cross section, and can, in a first approximation with equivalent modeling per unit of length, be structurally calculated from the cylindrical pieces, herein simply called tubes, subject to maximum internal flow pressure, $p = 0.981 \cdot H_1$ (N/cm^2).

11.2.2 Small Tubes

When the tube wall thickness is small relative to the internal diameter, d_i, the variation of stresses in the thickness is very small, and one can take only the tensile stresses as being uniformly distributed.

Observing this hypothesis, the tube wall thickness can be determined considering the principle of action and reaction, as shown in Fig. 11.1, where the equality between the forces, $p \cdot d_i \cdot L = 2 \cdot e \cdot \sigma_t \cdot L$, and $p = 0.981 \cdot H_1$, can determine the thickness, e_s of the small tubes as follows:

$$e_{\min} = \frac{p \cdot d_i}{2 \cdot \sigma_t} \rightarrow e = \frac{p \cdot d_i}{2 \cdot k_f \cdot \sigma_{t_{\text{ad}}}} + e_s = \frac{0.981 \cdot H_1 \cdot d_i}{2 \cdot k_f \cdot \sigma_{t_{\text{ad}}}} + e_s \rightarrow p\ \left(\text{N/cm}^2\right)$$

$$\sigma_t\ \left(\text{N/cm}^2\right)$$

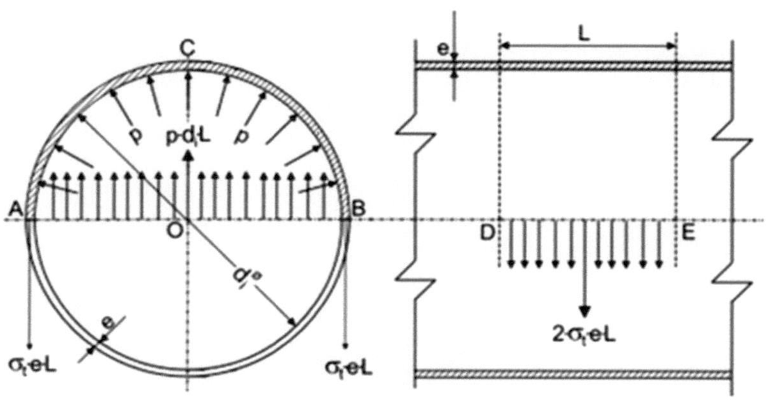

Fig. 11.1 Schematic for calculating the thickness of small tubes

$$H_1\ (\text{mca}) \tag{11.3}$$

In this expression, $k_f\,(1)$ is the efficiency coefficient that considers manufacturing defects, and the following values are indicated for tubes made using steel plates:

- $k_f = 1.0$ –seamless or welded tubes subjected to radiography and stress relief;
- $k_f = 0.9$ –soldiered tubes submitted to radiography without stress relief;
- $k_f = 0.8$ –welded tubes not subjected to radiography nor stress relief.

e_s (m) –is the supplementary thickness, which, in conduits subject to rust (oxidation) for example, can have values ranging from 1 to 3 (mm).

The minimum thickness for a steel pipe, $e_{\min}$(mm), needed to maintain its shape when empty, can be determined by:

$$e_{\min} = (d_i + 0.51)/400 \geq 6.35\ \ (\text{mm}) \tag{11.4}$$

If $e < e_{\min}$ then one must take $e_{\min}$, or keep e, and then fix external straps onto the tube.

11.2.3 Large Tubes

The schematics in Fig. 11.2 show cutaways of a thick steel tubes length $L = 1.0$ (m), subjected to internal pressure p (N/cm^2), with geometric conventions and radial and tangential stresses in an element of radial dimensions, dx, $d\alpha$.

The action of the internal pressure p reduces the thickness $e*$ resulting in, according to the theory of elasticity, forces in the radial and tangential directions, with the largest force occurring in the tangential direction, σ_t, which is decisive in predicting ruptures in the longitudinal direction, starting at the outer diameter d_e.

The application of the theory of elasticity, using the conventions in Fig. 11.1, allows us to express the stress σ_t for the steel tubes as:

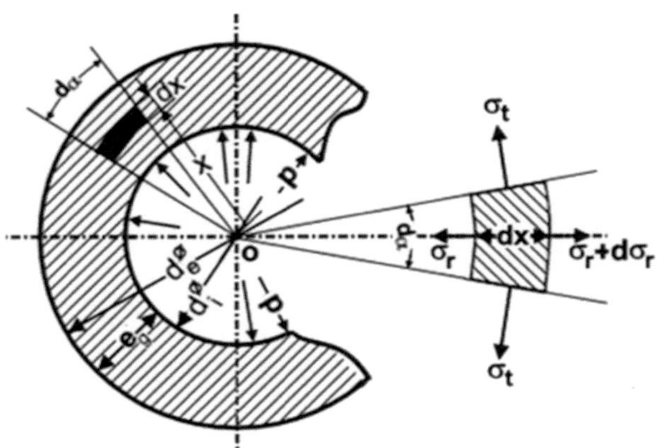

Fig. 11.2 Schematic for calculating thick tubes

$$\sigma_t = \frac{p}{d_e^2 - d_i^2} \cdot \left(0.7 \cdot d_i^2 + 1.3 \cdot d_e^2\right) \qquad (11.5)$$

Based on this expression, and defining the material that should be used to manufacture the tube, and defining the allowable tensile stress, $\sigma_{t_{\text{adm}}}$, and setting $d_e = d_i + 2 \cdot e_g$ algebraically, the thickness e_g per unit of length can be determined as follows:

$$\sigma_{t_{\text{adm}}} = \frac{p}{\left(d_i + 2 \cdot e_g\right)^2 - d_e^2} \cdot \left[0.7 \cdot d_i^2 + 1.3 \cdot \left(d_i + 2 \cdot e_g\right)^2\right]$$

Then we can obtain a second degree equation, which allows us to calculate e_g as:

$$e_g = \left(\sqrt{1 + \frac{2 \cdot p}{\sigma_{t_{\text{adm}}} - 1.3 \cdot p}} - 1\right) \cdot \frac{d_i}{2} \left(\sqrt{1 + \frac{1.962 \cdot H_1}{\sigma_{t_{\text{adm}}} - 1.275 \cdot H_1}} - 1\right) \cdot \frac{d_i}{2} \text{ (cm)} \qquad (11.6)$$

$$H_1 \text{ (mca)}$$

$$\sigma_{t_{\text{adm}}} \left(N/\text{cm}^{2.}\right)$$

$$d_i \text{ (cm)}$$

11.2.4 Setting the Expressions Together

Expression (11.4), for small tubes, considering $k_f = 1$ and $e_s = 0$, shows a linear relationship between the tube thickness and the internal diameter, once the allowable tensile stress of the steel in the tube is set, for example, $\sigma_{t_{\text{adm}}} = 11772 \ (N/\text{cm}^2)$, which then allows us to write:

$$\frac{e}{d_i} = \frac{0.981}{2 \cdot \sigma_{t_{\text{adm}}}} \cdot H_1 = \frac{0.981}{2 \cdot 11772} \cdot H_1 = 4.167 \cdot 10^{-5} \cdot H_1 \qquad (11.7)$$

Likewise, observing the same conditions for thick tubes, we have the following for Expression (11.6):

Table 11.2 Function values $e_g/d_i = f(H_1)$

H_1	$(e_g/d_i)\cdot 10^3$	H_1	$(e_g/d_i)\cdot 10^3$
m	-	m	-
0	0.00	100	41.95
10	0.42	250	105.94
20	0.83	500	215.62
30	12.53	1000	447.28
40	16.81	1500	697.60
50	21.98	2000	969.78

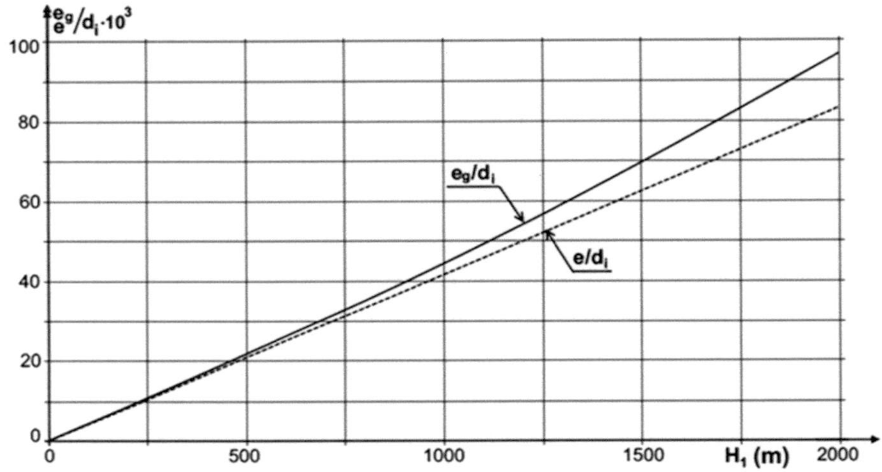

Fig. 11.3 Functions for comparing small and large tube thicknesses

$$\frac{e_g}{d_i} = \left[\left(1 + \frac{1.962 \cdot H_1}{\sigma_{t_{\text{adm}}} - 1.275 \cdot H}\right)^{0.5} - 1\right] \cdot 0.5 = \left[\left(1 + \frac{1.962 \cdot H_1}{11772 - 1.275 \cdot H_1}\right)^{0,5} - 1\right] \cdot 0.5 \tag{11.8}$$

Table 11.2 transcribes the values of $e_g/d_i = f(H_1)$ that generated the respective average functions plotted in Fig. 11.3

The functions show that any of the established equations can be used up to $H_1 < 500\ (m)$.

11.3 Slow Control Component

Figure 11.4 schematically represents the longitudinal section of the *SCC* with external command in open and closed positions, names its main parts, and indicates the forces acting on them. This is most commonly used in small and medium

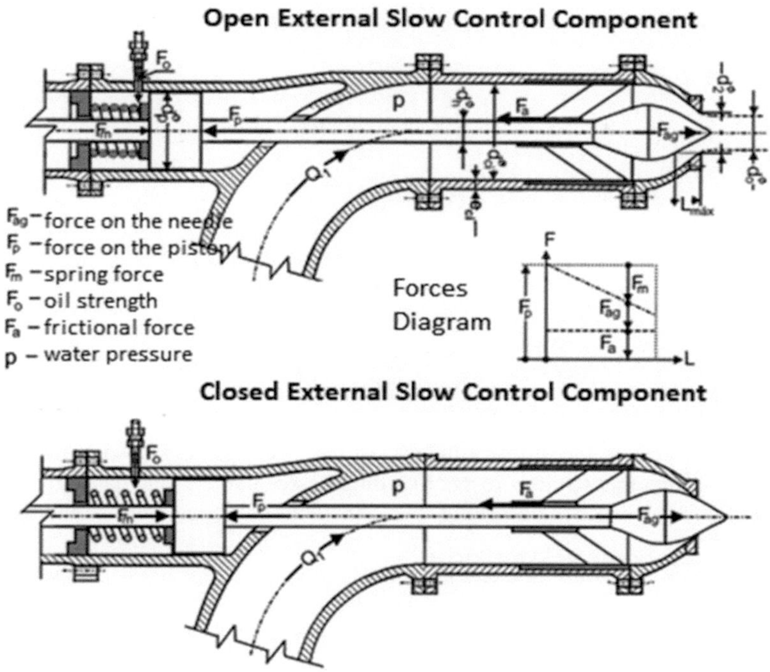

Fig. 11.4 Schematics of longitudinal cutaways of the *SCC* with open and closed external command

horizontal axis *PHIT*s, but also in certain cases on the vertical axis, as for example at the 2 *PHIT* in Devil Canyon in California, designed and manufactured by J.M. Voith, in Heidenheim, Germany, with:

- $$76\ (\text{MW})$$
- $$H_{\max} = 423.7\ (\text{m})$$
- $$Q = 22.65\left(\text{m}^3/\text{s}\right)$$
- $$n = 176.9(\text{rpm})$$

- $$n_q = 15.3$$
- $$z_r = 20 \text{ shells}$$
- $$z_j = 6 \text{ jets}$$

The same is true of Fig. 11.5, for the *SCC*, but here with internal command, which is currently used mostly in vertical axis *PHITs* with more than two jets, and medium to high power ratings.

In these two cases, the limits of the largest part, through which water flows, has a cylindrical geometry, so their respective thicknesses can be calculated by Expressions (11.3) and (11.6), observing the constants in items and subitems (11.1 and 11.2).

For example, the rod diameters, d_h (cm), which are subjected to simple tension, depending on the maximum resulting force, F_{r_a}(N) (calculated by Expression 11.12) acting on the axis, and the permissible tensile stress, $\sigma_{t_{\text{adm}}}(\text{N}/\text{cm}^2)$ of the material used when manufacturing the rods, as shown in the following expression:

$$d_h = \sqrt{\frac{4 \cdot F_{r_a}}{\pi \cdot \sigma_{t_{\text{adm}}}}} (\text{cm}) \rightarrow F_{r_a}\ (N) \rightarrow \sigma_{t_{\text{ad}}}\ \left(\text{N}/\text{cm}^2\right) \tag{11.9}$$

Evidently, there are several geometries that have been used by *PHIT* manufacturer regulation systems, which are traditionally called double (needle + *DF*) regulation systems, comprising a regulator and regulation mechanisms, first using mechanical regulators, and later oil-hydraulic regulators, then analog electronics, and now digital electronics.

Given the variety, we will show using a script how the resulting force F_r can be determined for the geometries, with minimal parts, e.g., those schematized in Figs. 11.4 and 11.5, comprising basically a piston/spring system, without or with oil compensators.

In Fig. 11.4 the force on the needle, F_{ag} (N) can be determined via the impulse law, which reaches its maximum value, $F_{\text{ag}_{\text{max}}}(N)$, when the *IN* is closed, and then decreases, approximately proportional to the needle stroke, L_{ag}, until reaching a minimum value, $F_{\text{ag}_{\text{min}}}(N)$, which depends on the geometry of the *IN* and the needle itself.

Thus, knowing the diameter of the output of the *IN*, d_2, of the needle shaft, d_h, and of the static pressure, p (N/cm^2), in front of the *IN*, we obtain:

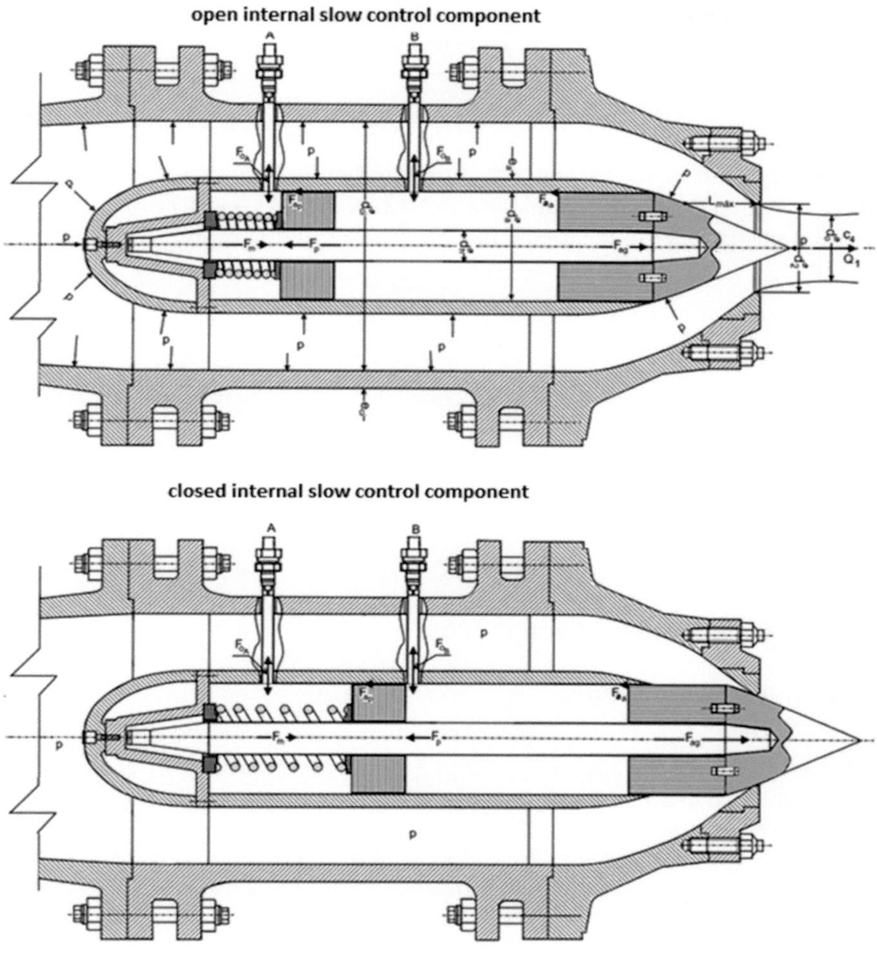

Fig. 11.5 Schematics of the longitudinal sections of the *SCC* with open and closed internal command

$$F_{\mathrm{ag}_{\max}} = p \cdot \frac{\pi}{4} \cdot \left(d_2^2 - d_h^2\right) \ (N) \tag{11.10}$$

To obtain small and constant resulting forces, make the spring force F_m (N) act in the closing direction, proportional to the stroke L_{ag} (cm) of the needle, and the force on the piston, F_p (N), in the opposite direction, calculated by:

$$F_p = p \cdot \frac{\pi}{4} \cdot \left(d_p^2 - d_h^2\right) \ (N) \tag{11.11}$$

This force must be such that the resulting force, F_{r_a} (N), always acts in the direction of the needle opening, to mitigate the water hammer:

$$F_{r_a} = F_{ag} + F_m - \left(F_p + F_a\right) \ (N) \tag{11.12}$$

In this expression, the friction force F_a (N) is the algebraic sum of all the friction forces acting on the entire control mechanism (see Figs. 11.4 and 11.5).

Thus, the diameter of the needle shaft d_h can be determined by:

$$d_h = \sqrt{\frac{4 \cdot F_{r_a}}{\pi \cdot \sigma_{t_{\mathrm{adm}}}}} \ (\mathrm{cm}) \tag{11.13}$$

The needle displacement work (fraction of the servomotor work capacity) is determined by:

$$T_{\mathrm{ag}} = F_{r_a} \cdot L_{\mathrm{ag}} \ (\mathrm{N} \cdot \mathrm{cm}) \tag{11.14}$$

11.4 Quick Control Component

11.4.1 Considerations

As seen in Item 8.5 in Chap. 8, the *DF* is a component that, once its geometry has been established to act on the jet, and once it has been attached to the body of the *IN*, and to the corresponding control system, establishes a form for preliminarily carrying out the structural calculations of the parts based on the *HFP* data, and the maximum working conditions, which are characterized by:

- $$H_1 \ (\mathrm{mca})$$

- $$Q_1 = \frac{Q_{\max}}{z_j} \ \left(\mathrm{m}^3/\mathrm{s}\right)$$

- $$c_{4_1} = \phi_{0;4} \cdot \sqrt{2 \cdot g \cdot H_1} \ (\mathrm{m/s})$$

- $$d_{o_1} = \sqrt{\frac{4 \cdot Q_1}{\pi \cdot c_{4_1}}} \ (m) \tag{11.15}$$

Once this is done, the *DF* must be modeled and tested under loads, Q_1^* (m^3/s), H^*(mca) which will give, for geometrically similar prototype:

The forces as:

$$k_{\text{esf}} = \frac{H_1}{H*} \cdot \left(\frac{d_{o_1}}{d_o^*}\right)^2 \tag{11.16}$$

The work as:

$$k_{\text{trb}} = \frac{H_1}{H*} \cdot \left(\frac{d_{o_1}}{d_o^*}\right)^3 \tag{11.17}$$

11.4.2 Head Thickness

To determine the thickness of the DF head, e_d, which will be taken as being equal to the thickness of the rods, e_h, we will reference the schematic in Fig. 11.6, with the critical longitudinal section of length, L_d, subjected to bending and shear, allowing us to develop the following equation:

Knowing the resulting force F_{r_d} (N) acting on the *DF* head, knowing the inner radius r_i (cm), and knowing the length L_d (cm), the bending stress σ_f (N/cm^2), can then be obtained as follows:

$$M_f = F_{r_d} \cdot r_i \ (\text{N} \cdot \text{cm}) \rightarrow W = \frac{e_d \cdot L_d^2}{6} \ (\text{cm}^3) \rightarrow \sigma_f = \frac{M_f}{W} = \frac{6 \cdot F_{r_d} \cdot r_i}{L_d^2 \cdot e_d} \ (\text{N/cm}^2)$$

Now, the shear stress can also be equated as a function of e_d as:

$$\tau_{\text{cis}} = \frac{F_{r_d}}{L_d \cdot e_d} \ (\text{N/cm}^2)$$

Combining these two tensions we get the maximum tension, $\tau_{\max}$ (N/cm^2), obtained using the following resistance criterion:

$$\tau_{\max} = \sqrt{(0.5 \cdot \sigma_f)^2 + \tau_{\text{cis}}} = \sqrt{\left(\frac{3 \cdot F_{r_d} \cdot r_i}{L_d^2 \cdot e_d}\right)^2 + \left(\frac{F_{r_d}}{L_d \cdot e_d}\right)^2} = \frac{F_{r_d}}{L_d \cdot e_d} \cdot \sqrt{\left(\frac{3 \cdot r_i}{L_d}\right)^2 + 1} \ (\text{N/cm}^2)$$

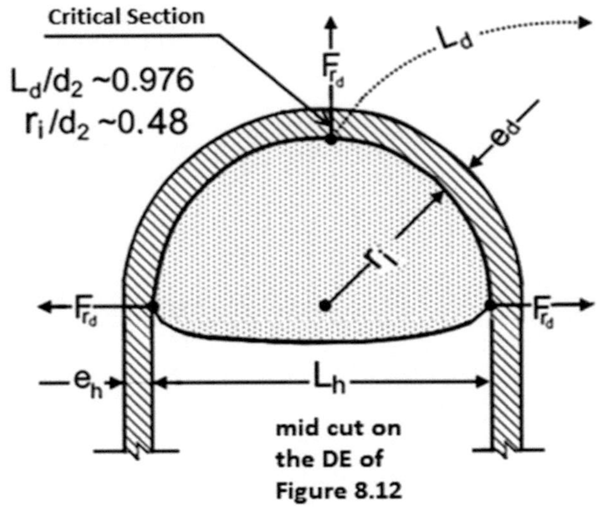

Fig. 11.6 Schematic section of the *DF* showing the critical structural section

With this tension, and considering the expression for the factor of safety, $f_s = \frac{0.6 \cdot \sigma_{\text{esc}}}{\tau_{\max}}$, and its range, $2 \le f_s \le 4$, the interval for $e_d = e_h$ be determined as:

$$e_d = e_h = \frac{F_{r_d} \cdot f_s}{0.6 \cdot \sigma_{\text{esc}} \cdot L_d} \cdot \sqrt{1 + \left(\frac{3 \cdot r_i}{L_d}\right)^2} \ (\text{cm}) \tag{11.18}$$

11.4.3 Mass

Considering the geometry of the *DF* drawn proportionately in Fig. 11.7, its weight or mass can be calculated with good approximation by subdividing it into several parts, as shown in the figure below:

- The head volume:

$$(aa; bb) \cdot L_d$$

$$V_{\text{ca}} = \frac{\pi}{2} \cdot (2 \cdot r_i + e_d) \cdot e_d \cdot L_d \ \left(\text{cm}^3\right)$$

- The volume of the two rods:

$$2 \cdot \text{Area of the Section } (a * b * cdgfa*) \cdot e_h$$

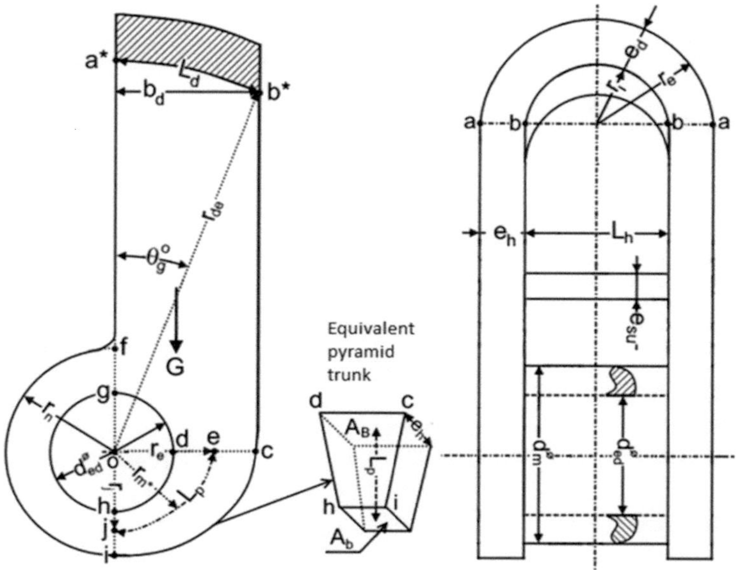

Fig. 11.7 Geometry and conventions for determining the *DF* mass

$$V_{h_1} = 2 \cdot [(ocb*o) + (ob*a*o) - (odgo)] \cdot e_h$$
$$= \left[\left(b_d \cdot sin\,\theta_g^o + \frac{\pi \cdot \theta_g^o}{90}\right) \cdot r_{de} - \frac{\pi \cdot d_{ed}^2}{8}\right] \cdot e_h \ (\text{cm}^3)$$

- 2· area of the section (*fi*; *gh*) · e_h

$$V_{h_2} = \pi \cdot \left[r_n^2 - \left(\frac{d_{ed}}{2}\right)^2\right] \cdot e_h \ (\text{cm}^3)$$

- 2· area of the section (*ic*; *dh*), with the corresponding volume calculated using the equivalent truncated pyramid, shown in Fig. 11.7.

$$V_{h_3} = \frac{2 \cdot h_p}{3} \cdot \left(A_B + A_b + \sqrt{A_B \cdot A_b}\right) \ (\text{cm}^3) \rightarrow$$

$$h_p = 0.5 \cdot \pi \cdot r_{m*} \rightarrow$$

$$r_{m*} = 0.5 \cdot (r_e + r_j) \rightarrow$$

$$r_e = 0.5 \cdot b_d + 0.25 \cdot d_{\text{ed}} \rightarrow$$

$$r_j = 0.5 \cdot r_n + 0.25 \cdot d_{\text{ed}} \rightarrow$$

$$A_B = (b_d - 0.5 \cdot d_{ed}) \cdot e_h \rightarrow$$

$$A_b = (r_n - 0.5 \cdot d_{\text{ed}}) \cdot e_h \left(\text{cm}^2\right)$$

- The volume of the plate between the rods:

$$V_{\text{pd}} = L_h \cdot b_d \cdot e_{\text{su}} \left(cm^3\right)$$

- The volume of the hollow shaft between the rods:

$$V_{ev} = \frac{\pi}{4} \cdot \left(d_m^2 - d_{\text{ed}}^2\right) \cdot L_h \left(\text{cm}^3\right)$$

- The *DF* volume:

$$V_{\text{DE}} = V_{\text{ca}} + V_{h_1} + V_{h_2} + V_{h_3} + V_{\text{pd}} + V_{\text{ev}} \left(\text{cm}^3\right) \quad (11.19)$$

- The *DF* mass:

$$G_{\text{DE}} = \rho_{\text{steel}} \cdot V_{\text{DE}}\ (N) \rightarrow \rho_{\text{steel}} \cong 7.698 \left(\text{N/cm}^3\right) \quad (11.20)$$

11.5 Rotor

11.5.1 Considerations and Hypotheses

Consider the part of the *RO* comprising the shells and discs that interconnect with the axis of the *PHIT*.

This interconnection can be made directly as long as the *RO* is manufactured as a single cast and machined part, which is fixed to the shaft, as shown in Fig. 8.3b, or in two parts, with the shells cast and machined individually, or together, which are then fixed to a disk, which, in turn, is then fixed to the shaft, for example, as shown in Fig. 11.8.

Particularly regarding the shaft, Fig. 11.8 is especially important, since it was drawn straight, with a constant diameter, and without any bearings (guide + strut or guide/stanchion). Its importance lies in the fact that I has a small diameter, d_e, in the cylindrical case, which has been referenced to the average constant ratios in the lower part of the figure.

These relations can be used in a first sketch, which, once made compatible, will allow for a first approximation, and, as we see, will allow us to obtain a first idea of the mass or weight G_{RO} of the *RO*, as long as it is homologous to another *PHIT RO* that is geometrically similar, and has the same specific rotation, n_{qA}. If the average

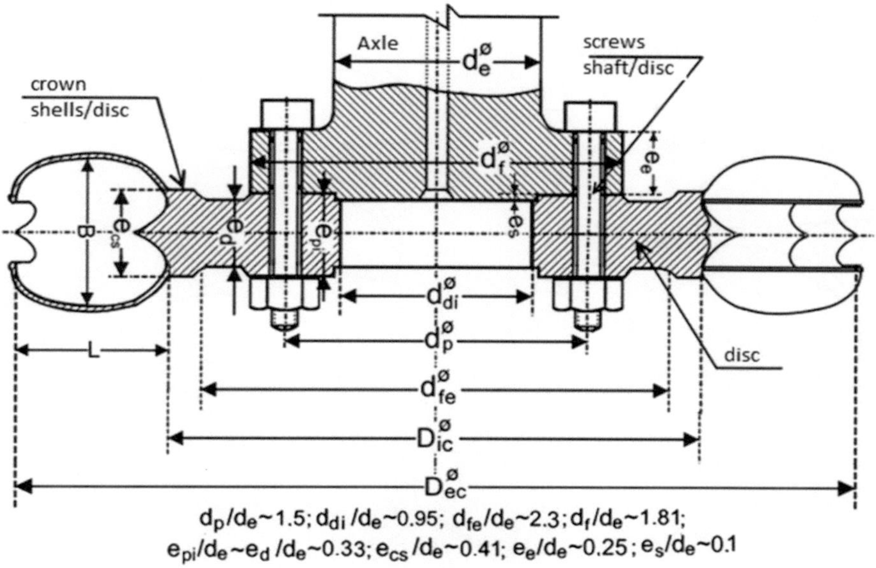

Fig. 11.8 Schematical of the *RO* section of the *PHIT* with shells and discs cast in one piece

specific masses, p_m, of the materials used in the manufacturing are different, multiply the result by the ratio between them.

In Fig. 11.9, showing a schematic of shells that were cast separately from the disk, and later fixed to it using screws, with the set coupled to the horizontal or vertical axis. In this case, one can equalize the forces between the shells using tensioning pins or radial wedges, as shown in the figure.

Figure 11.10 shows photographs of an old out-of-use *PHIT RO* Voith shell from the Henry Bordem *HPP*, showing the legs with the respective holes for individually attaching the shells to the disk, showing the critical section relative to the structural part, and showing the part that interconnects the shell with the disk, which is also part of the tension equalization system in the *RO* using pins or tensioning wedges.

The photograph in Fig. 11.11 shows, in sequence, two shells, highlighting the machining on the back and on the inner surface.

One must also observe the thickness of the shell, e_c, and the rib, e_n, which absorb the maximum force of the jet on the critical section of the shells (see item 7 of Chap. 9). As shown in the photographs, the rib thickness, e_n, is minimal at the edges of the shells.

Projecting for the structural design of the shells and their respective coupling to the *PHIT* axis are essential steps before manufacturing and testing a *RO* model/prototype.

Once the inner surface of the shells is known, e.g., the surface developed in Chap. 10, and using the critical sections established in Fig. 11.10 as a reference, the part of the shell between these two sections can be used to set the respective parts.

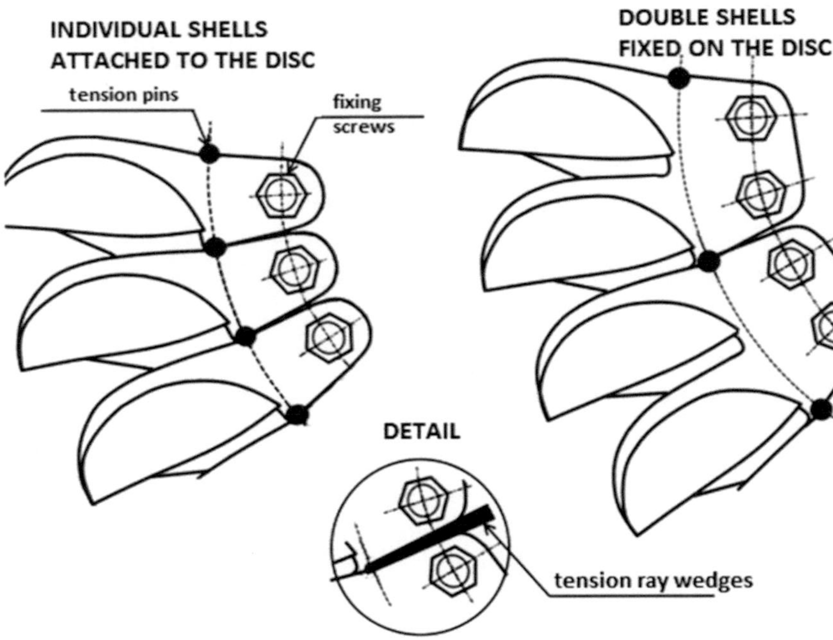

Fig. 11.9 Shells cast and machined individually, in or double for fixing on disk, and later coupling to the *PHIT RO*

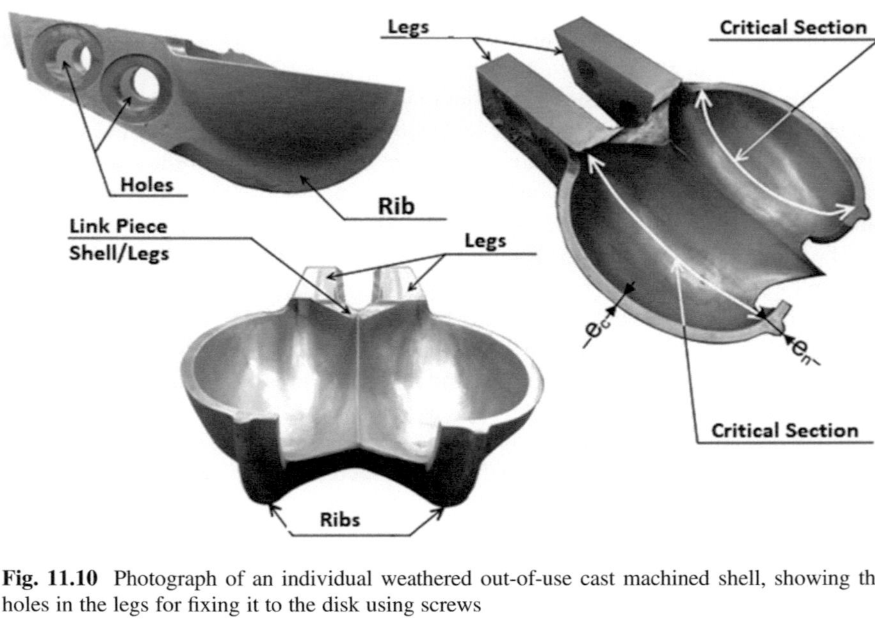

Fig. 11.10 Photograph of an individual weathered out-of-use cast machined shell, showing the holes in the legs for fixing it to the disk using screws

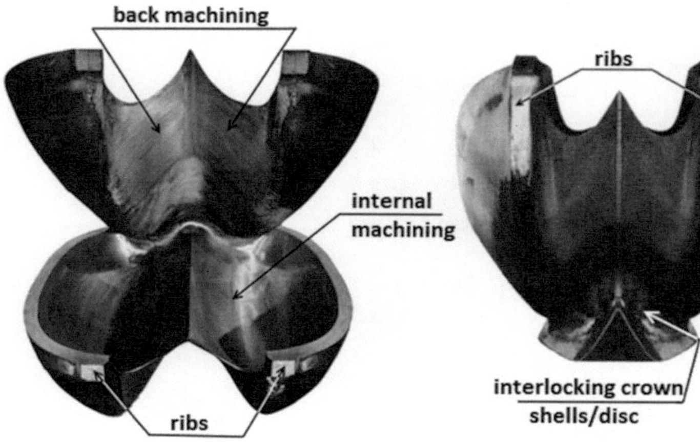

Fig. 11.11 Photograph of the sequence of two shells with the part names

11.5.2 Average Shell Thickness

The equation for preliminary obtaining of the average thickness of the shells will be made observing the following hypotheses:

- The flow is subjected to maximum conditions at the *IN*, with the shell stopped.

$$Q_1 = \frac{Q_{\max}}{z_j}\ (\mathrm{m}^3/\mathrm{s})$$

$$H_1 = (1 + h_s) \cdot H_{\max}\ (\mathrm{mca})$$

- There will be crimping in the critical sections established in the photographs of Fig. 11.10.
- The Inner surface of shell will be established and drawn as was done in Chap. 10 or 9 supplemented with cross sections and longitudinal sections.

Having established these conditions, Fig. 11.12 was started by drawing a stationary shell with its internal surface and normal flow set to the dividing edge, marking length L_n in the critical section, limited by the flow. In the absence of other indications, $L_n \approx 1.6 \cdot d_{0_1}$ was taken. Next, b; b was drawn, showing that the symmetry of the shell eliminates stresses arising from the flow between sections $S = (ac4a) = (a * c * 4a*)$, which allows for setting section $(a; a)$ and section $(a*; a*)$. Continuing, to the left of the figure, section $(a; a)$ was drawn with thickness e_c^* taken as being maximum.

Due to the drop in flow stresses between section $(a; a)$ and the outlet at section 5, and considering further machining to avoid shocks, thickness $e_{c_{\max}}^*$ is normally

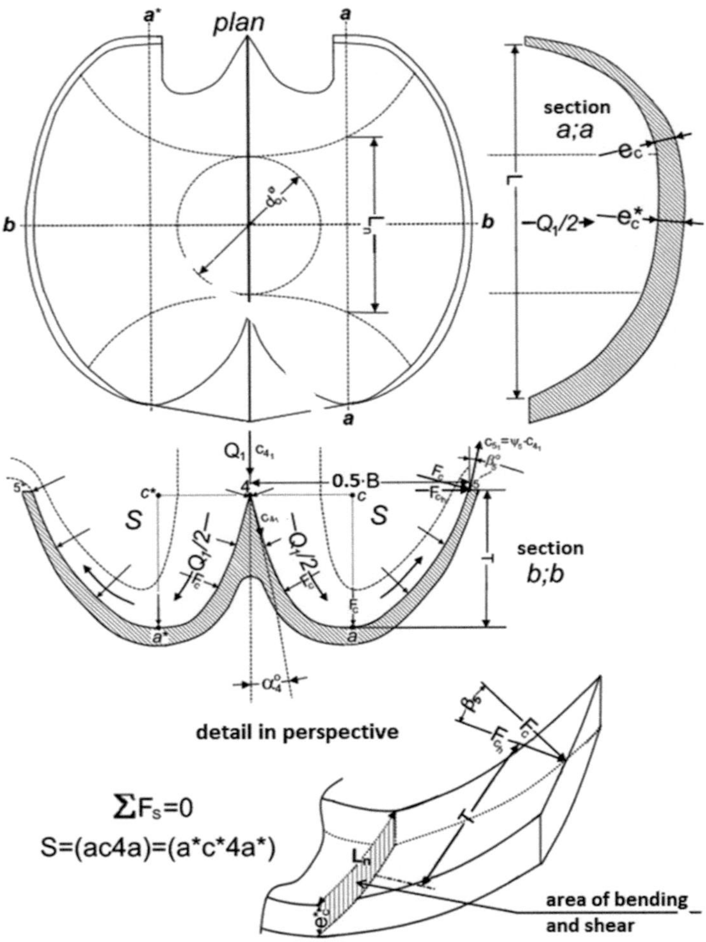

Fig. 11.12 Shell schematic for analytical studies

reduced to one third its size, giving rise to the ribs shown in Figs. 11.10 and 11.11, which substantially reduce the weight of the shells, particularly large-sized shells subjected to high heads and flows.

Referencing Fig. 11.12, the crimping section, $S_e = e_c^* \cdot L_n$ will be subjected to bending and shear stresses, which allow us to combine the stresses as:

$$\tau_{\max} = \sqrt{\left(\frac{\sigma_f}{2}\right)^2 + \tau_{\text{cis}}^2} \tag{11.21}$$

The bending stress, σ_f is obtained from the relationship between the bending moment and the resistance moment:

$$\sigma_f = \frac{M_f}{W}$$

i.e.,

$$M_f = F_{c_h} \cdot T \rightarrow W = \frac{e_c^* \cdot L_n^2}{6} \rightarrow \sigma_f = \frac{6 \cdot F_{c_h} \cdot T}{e_c^* \cdot L_n^2} \ (\mathrm{N/cm^2})$$

The shear stress, thus, results in:

$$\tau_{\text{cis}} = \frac{F_{c_h}}{e_c^* \cdot L_n}$$

Substituting (11.19) in the expressions for $\sigma_f$and τ_{cis}, one can obtain the maximum stress as:

$$\tau_{\max} = \sqrt{\left(\frac{3 \cdot F_{c_h} \cdot T}{e_c^* \cdot L_n^2}\right)^2 + \left(\frac{F_{c_h}}{e_c^* \cdot L_n}\right)^2} = \frac{F_{c_b}}{e_c^* \cdot L_n} \cdot \sqrt{\left(\frac{3 \cdot T}{L_n}\right)^2 + 1}$$

Combining this expression with the expression for the factor of safety, $2 < f_s = \frac{0.6 \cdot \sigma_{\text{esc}}}{\tau_{\max}} < 4$, the thickness e_c^* can be determined as:

$$e_c^* = \frac{F_{c_h} \cdot f_s \cdot \sqrt{1 + \left(\frac{3 \cdot T}{L_n}\right)^2}}{0.6 \cdot \sigma_{\text{esc}} \cdot L_n} \tag{11.22}$$

In this expression,

$$F_{c_h} = F_c \cdot cos\beta_5 \rightarrow F_c = 0.5 \cdot \rho \cdot Q_1 \cdot c_{4_1} \cdot (cos\alpha_4 + \psi_5 \cdot cos\beta_5) \ (N) \rightarrow L_n \approx 1.26 \cdot d_{o_1}.$$

11.5.3 The PHIT *Shaft*

In preliminary design studies, the shaft diameter can be calculated by the following expressions:

The horizontal *PHIT* axis:

$$d_e = 130 \cdot \sqrt[4]{\frac{P_e}{n}} \ (\mathrm{mm}) \rightarrow P_e \ (\mathrm{kW}) \rightarrow n \ (\mathrm{rpm}) \tag{11.23}$$

The vertical *PHIT* axis:

$$d_e = 118 \cdot \sqrt[3]{\frac{P_e}{n}}\ (\text{mm}) \rightarrow P_e\ (\text{kW}) \rightarrow n\ (\text{rpm}) \tag{11.24}$$

11.6 The Mass of the *PHIT RO*

11.6.1 Considerations

The masses of the *RO*, whether of the model or the prototype, must be obtained to scale, with unexpired verification certificates, whenever possible. As of early 2020, very little reliable data exists on the weights of *PHIT* components in the public domain.

One exception is the constant from Pages 01 to 64 of:

Turbomachines Hidraulique, Henry P., Presses Polytechniques et Universitaires Romandes—Lausanne—ISBN 2–88–74-209-9—1992

… where one can find 9 *RO* masses for large *PHIT*s, from which the segmented line in Fig. 11.13 was drawn.

Then, in this graph, the limits of the function, $G_{RO} = f\left(\frac{z_i^{0,5} \cdot D}{z_c \cdot B} \cdot n_{qA_1}\right)$ were drawn.

where:

- $G_{RO}$the mass of *RO*
- z_j the number of jets
- z_c the number of shells
- D (m) the reference or pitch diameter
- $n_{qA_1} = 3 \cdot n \cdot \frac{(Q/z_j)^{0.5}}{H^{0.73}}$ the specific rotation

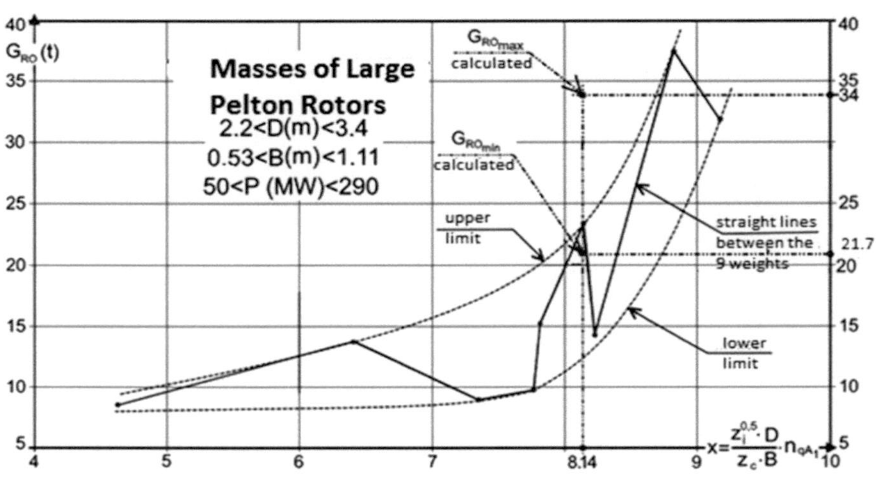

Fig. 11.13 Graph of *PHIT RO* masses referencing the 9 points from Pierre Henry's book

- n (rpm) the rotor rotation
- Q (m^3/s) the design flow
- H (m) the project head in meters of water column

We should note that the use of the letter G and the word weight stem from an old engineering habit originally of using the Technical System of Units (*TSU*) (weight; length; second). However, with the universalization of the International System of Units (*SI*) (mass; length; second), numerically, weight and mass have the same values, but different units: weight is kgf and mass is kg

The final geometry of the *PHIT RO*, particularly the inner surface of the shells, is not unique, and is ultimately the industrial property of each manufacturer, which has been developed over years by their design and research teams, as has been studied and commented on in the chapters of this book.

These particularities cause immense difficulties in developing analytical formulations for obtaining the G_{RO}, even if executive and manufacturing projects, test results, and the masses of geometrically similar analogous models with the same specific rotation are available.

The following should be taken as more of a methodical reasoning, than as a method for calculating the G_{RO} analytically.

Therefore, this study will separately address the shell mass, and its assembly, forming the shell crown mass, and then deal with the system (disks) that will interconnect them to the axis of the respective *PHIT*.

11.6.2 Shell and Shell Crown Mass

We will propose obtaining the mass of the shell, G_c analytically in first approximation by starting with an equation of a parabola $y^2 = 2 \cdot x$ and the corresponding volume of the paraboloid of revolution $V_p = \frac{\pi}{2} \cdot R^2 \cdot L$, from the application of the Pappus–Guldin theorem, Fig. 11.14a, b. This figure was drawn taking the ray R of the base of the paraboloid of revolution in the horizontal direction x and its maximum height length L (cm) from the inner surface of the Pelton shell in the vertical direction y as orthogonal coordinates (see symbology in Fig. 9.4).

Then, the upper part of this paraboloid of revolution was separated and shown in Fig. 11.14c, which was later rotated in Fig. 11.14d, with the volume of the upper half b of the paraboloid of revolution shown in Fig. 11.14a, b.

Figure 11.14e shows the result of joining parts $e = a + d$, which, within the constraints, forms half the volume of the inner surface of the Pelton shell, equivalent to the paraboloid of revolution in Fig. 11.14b.

Figure 11.15, contains the projections of half the inner surface of the shell to scale, in solid lines, proposed and studied in Chap. 10, superimposed on the corresponding dashed lines of the joint of the halves of the paraboloid of revolution (Fig. 11.14e).

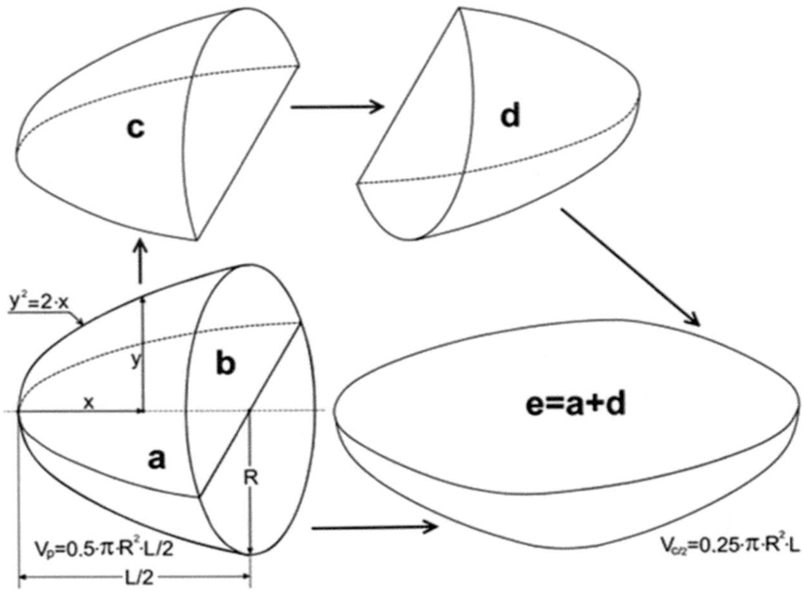

Fig. 11.14 Paraboloid of revolution and schematic to obtain half the volume of the equivalent Pelton shell

Referencing these figures, we will calculate the volume of a Pelton V_c (cm^3) shell, and later, after knowing z_c and the average mass of the material that will be used in the manufacturing, ρ_m (kg/cm^3), with a first idea of the mass of the G_c (kg) shell and the corresponding G_{cc} (kg) shell crown.

Considering the conventions in Figs. 11.14 and 11.15, and Fig. 9.4, the volumes and weights can be obtained as follows:

- The average radius of equivalent paraboloid of revolution:

$$R_m = 0.5 \cdot (T + 0.25 \cdot B)\ (\text{cm}) \tag{11.25}$$

- The equivalent internal volume of the half Pelton shell:

$$V_{in_{c/2}} = 0.25 \cdot \pi \cdot R_m^2 \cdot L\ \left(\text{cm}^3\right) \tag{11.26}$$

- The equivalent external volume of the half Pelton shell:

$$V_{ex_{c/2}} = 0.25 \cdot \pi \cdot (R_m + e_c)^2 \cdot L \tag{11.27}$$

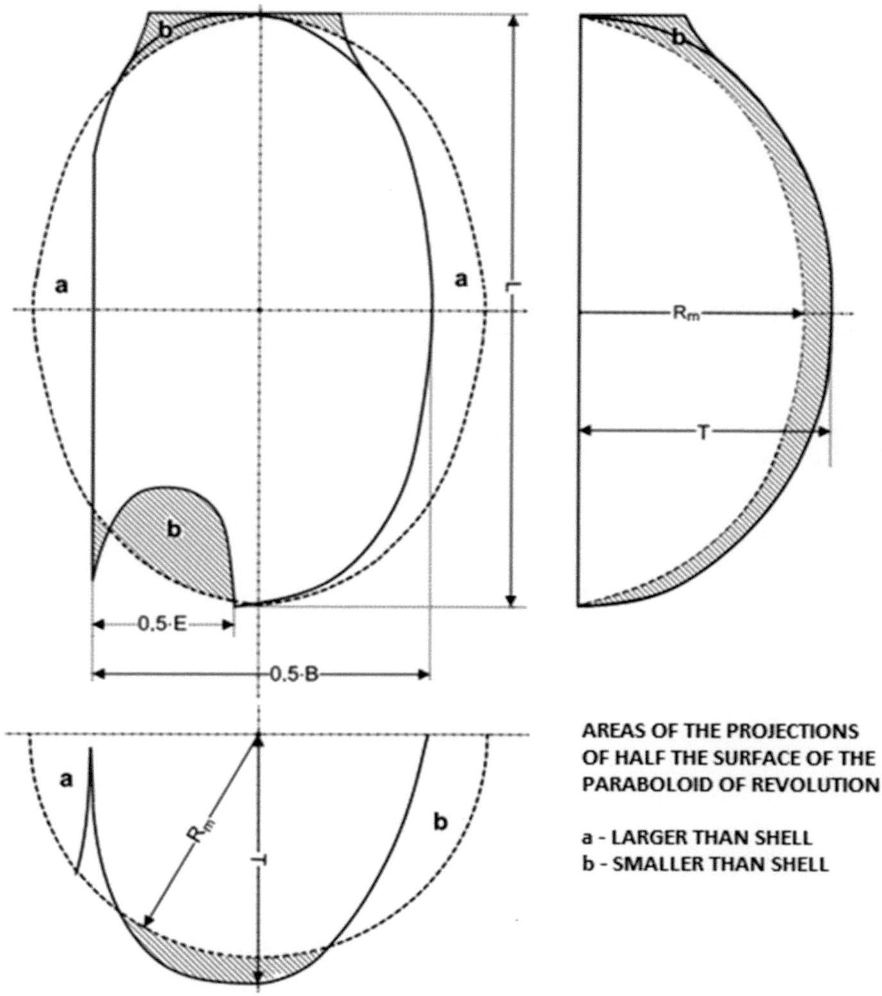

Fig. 11.15 Projections, at the same scale, of half of the inner surfaces of the Pelton shells and the corresponding paraboloid of revolution

- The equivalent volume of the Pelton shell:

$$V_c = 2 \cdot \left(V_{ex_{c/2}} - V_{\mathrm{in}_{c/2}} \right) \tag{11.28}$$

- The equivalent masses of the shell and crown of the Pelton shells:

$$G_c = \rho_m \cdot V_c \rightarrow G_{cc} = z_c \cdot G_c \cdot (kg) \tag{11.29}$$

11.6.3 The Masses of the Circular Crowns and Discs

If the *RO* will be cast in one piece, or in more than one piece, the shaft diameter, d_e, of the *PHIT* determines the available circular space d_d, between d_{i_c} (diameter of the inner surface of the shell crown) $d_d = d_{i_c} - d_e$ for where the interconnection can be designed.

If the shells are cast and later fixed to the disk (Figs. 11.8, 11.9, and 11.10), this circular space, will be in part occupied by the legs of the shells, and in part by the disc that must contain the drill holes so that it can be attached by screws and with stress distributors.

For the equation we will use an example for fixing the *RO* to the constant axis in the vertical projection of Fig. 11.8, noting that the equation must be adapted and/or made for each specific study and/or project.

In Fig. 11.8, the project for a one piece casting of the shell crown and a disk with holes for fixing it to the *PHIT* axis, results in the following equations:

- The volume of the crown shells/disc:

$$V_{c/d} = \frac{\pi}{4} \cdot \left(D_{ic}^2 - d_{fe}^2\right) \cdot e_{cs}\ (\mathrm{cm}^3) \tag{11.30}$$

- The disk volume:

$$V_{fe/f} = \frac{\pi}{4} \cdot \left(d_{fe}^2 - d_f^2\right) \cdot e_d\ (\mathrm{cm}^3) \tag{11.31}$$

- The disk/axis volume:

$$V_{d/e} = \frac{\pi}{4} \cdot \left(d_f^2 - d_{di}^2\right) \cdot e_{pi}\ (\mathrm{cm}^3) \tag{11.32}$$

- The total disk volume:

$$V_d = V_{c/d} + V_{fe/f} + V_{d/e}\ (\mathrm{cm}^3) \tag{11.33}$$

- The mass of the discs:

$$G_d = \rho_m \cdot V_d\ (\mathrm{kg}) \tag{11.34}$$

11.6.4 The Mass of the PHIT RO Using the Analytical Method

$$G_{RO} = G_{cc} + G_d\ (\mathrm{kg}) \tag{11.35}$$

11.6.5 Considerations

The analytical method presents errors compared to the weighing method, particularly when determining the volumes of the shells, the crowns, and the interconnection with the disks.

The biggest error is concentrated in the shell, since the crown or disk geometry is practically defined. Therefore, the areas and thicknesses can be obtained with maximum errors specified in the graphical software program used for the drawings, e.g., AutoCAD.

The volume of the shell in this proposed analytical method is equivalent to half of an equation of a paraboloid of revolution with volume compensation (see Fig. 11.15). Therefore, unpredictable errors values may be present in this equivalence, as will be shown in the analysis presented below:

- The average radius calculation, $R_m = 0.5 \cdot (T + 0.25 \cdot B)$, will have a decreasing error when T is closer to $0.25 \cdot B$, or at the limit of equality (Fig. 11.15).
- The R_m error increases the corresponding error in the formula for calculating $V_{\text{in}_{c/2}}$ (Formula 11.26), since it is squared, and in $V_{ex_{c/2}}$ (Formula 11.27) the error is even more, because before being squared the average shell thickness, e_c, has to be added.
- This average thickness, e_c, also contains errors in the shells with little variable thickness, but mainly in shells containing ribs, as shown in Figs. 11.10 and 11.11.

This means that the analytical calculation results of G_{RO}, using Formula 11.35, must be multiplied by a factor of $f_r = \frac{G_{RO}}{G_{RO_{pe}}}$, where $G_{RO_{pe}}$ (kg) is the mass of the *RO* resulting from weighing either the model or the corresponding prototype on a scale with an unexpired verification certificate. The range of variation from f_r to a homologous *PHIT RO*, for example, as a function of n_{qA} or d_o/D, is not yet available, but it can be done depending mainly on the permission that researchers have in accessing manufacturing, designer, laboratory, or other private information.

11.6.6 Applying the Analytical Method for a SHP

To first test the method proposed here, mainly with regard to the considerations contained in the previous subitem, with emphasis on variables R_m; e_c and f_r, we will use a micro *RO* Pelton that has been worn by use, which was available at the *LHPCH* at *UNIFEI*, with photographs in Fig. 11.16, and which has the following characteristics:

- Geometry measured with a caliper (in millimeters):

Fig. 11.16 Rotor used in the micro *PHIT*

$$d_{de} = 3.9$$

$$d_e = 2.1$$

$$e_c = 0.35$$

$$e_{cs} \cong e_d = 3.8$$

$$B = 6.8$$

$$T = 2.1$$

$$L = 5.1$$

$$D = 15.1\ (\text{cm})$$

$$z_c = 8\ \text{shells}$$

We took the steel mass (weight) of the crown as interconnected with the shells in the disk as compensating for the sum of the masses of the recesses existing in the disk $D_{i_c} \cong d_{\mathrm{de}}$.

- The mass was obtained on a scale, measured in grams $G^*_{RO_{pe}} = 874$ (gr). Considering wear from use around 10 (%), gave a reference mass at $G_{\mathrm{RO_{pe}}} = 1.1 \cdot 874 \cong 962$ (gr). The specific mass of the *RO* material was $\rho_{\mathrm{steel}} = 7698 \cdot 10^{-6}$ (kg/cm^3).

  Calculations:

1. The average radius of the equivalent paraboloid of revolution:

$$R_m = 0.5 \cdot (T + 0.25 \cdot B) = 0.5 \cdot (2.1 + 0.25 \cdot 6.8) = 1.9 \text{ (cm)}$$

2. The equivalent internal volume of the half Pelton shell:

$$V_{in_{c/2}} = 0.25 \cdot \pi \cdot R_m^2 \cdot L = 0.25 \cdot \pi \cdot 1.9^2 \cdot 5.1 = 14{,}460 \left(\text{cm}^3\right)$$

3. The equivalent external volume of the half Pelton shell:

$$V_{ex_{c/2}} = 0.25 \cdot \pi \cdot (R_m + e_c)^2 \cdot L = 0.25 \cdot \pi \cdot (1.9 + 0.35)^2 \cdot 5.1 = 20.278 \left(\text{cm}^3\right)$$

4. The equivalent volume of the Pelton shell:

$$V_c = 2 \cdot \left(V_{ex_{c/2}} - V_{in_{c/2}}\right) = 2 \cdot (20.278 - 14.46) = 11.636 \left(\text{cm}^3\right)$$

5. The equivalent mass of the Pelton shell:

$$G_c = \rho_m \cdot V_c = 7698 \cdot 10^{-6} \cdot 11.636 = 0.0896 \text{ (kg)} = 89.6 \text{ (gr)}$$

6. The equivalent mass of the shell crown:

$$G_{cc} = z_c \cdot G_c = 8 \cdot 0.0896 = 0.717 \text{ (kg)} = 717 \text{ (gr)}$$

7. The volume of the disc:

$$V_d = 0.25 \cdot \pi \cdot \left(d_{de}^2 - d_e^2\right) \cdot e_d = 0.25 \cdot \pi \cdot \left(3.9^2 - 2.1^2\right) \cdot 3.8 = 32.233\ \left(\text{cm}^3\right)$$

8. The mass of the disc:

$$G_d = \rho_m \cdot V_d = 7698 \cdot 10^{-6} \cdot 32.233 = 0.248\ (\text{kg}) = 248\ (\text{gr})$$

9. The mass of the rotor:

$$G_{RO} = G_{cc} + G_d = 0.717 + 0.248 = 0.965\ (\text{kg}) = 965\ (\text{gr})$$

10. The factor $f_r = \frac{G_{RO}}{G_{\text{ROpe}}} = \frac{965}{962} = 1.003\ (\%)$, i.e., the calculated rotor mass was 0.3 (%) greater than what was weighed, considering an estimated wear of 10 (%).

    - Both the model and the prototype *PHIT RO* undergo corrections after manufacturing to avoid shocks in the *HFP* , which result in energy losses (see photographs in Fig. 9.31) and to correct, or at least limit, vibrations in the operational field from exceeding the normalized limits.
    - This means correcting the geometry of the Executive Project and weighing the components after the corrections are made.
    - Thus, the measurements are:

        - *B.*
        - *T.*
        - *L*
        - D_i
        - d_{ec}
        - d_p
        - d_i
        - d_e
        - e_c
        - e_{cs}
        - e_d

      and others necessary measurements for obtaining the *RO* mass using analytical method must be obtained with their respective errors (accuracy/precision) directly on the shells, disks, and *RO* axis after the corrections have been made.

- To have a first idea of the influence of errors when obtaining these measurements which were used in the application in the previous subitem, we will assume that the measurement $e_c = 0.4$ (cm) now contains an error of ± 10 (%). Thus, the script

Table 11.3 Results

e_{c1}	$G_{RO\ 1}$	f_{r1}	e_{c2}	$G_{RO\ 2}$	f_{r2}
cm	gr	—	cm	gr	-
0.32	887.00	0.92	0.39	1043.00	1.08
Smaller mass by 8 (%)			Greater mass by 8 (%)		

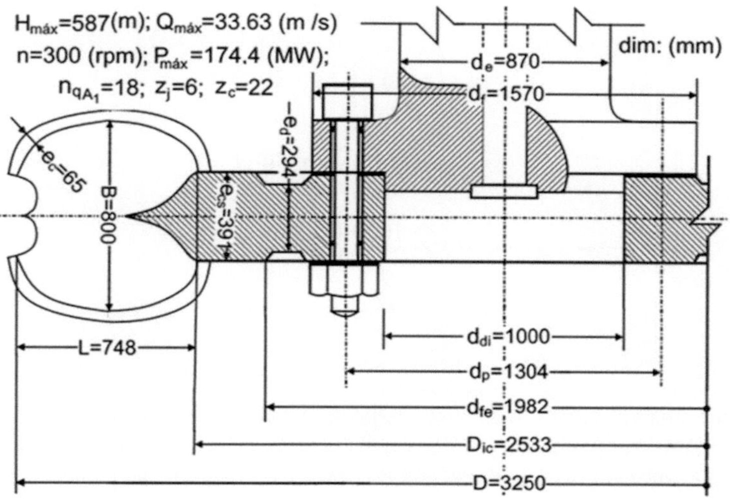

Fig. 11.17 Longitudinal section in the *PHIT* to calculate the *RO* weight using the analytical method

will be applied to $e_{c_1} = 0.35 \cdot 0.9 = 0.315$ (cm) and $e_{c_2} = 0.35 \cdot 1.1 = 0.385$ (cm), with results transcribed in Table 11.3.

- These results indicate that the error in the calculation of the mass of the *RO* is **smaller** than the error made in the measurement of the independent variable that has the **greatest** influence on determining the mass using the analytical method proposed here.

11.6.7 Application of the Analytical Method for LHP

11.6.7.1 Input Data

For this application we will use the characteristics contained in Fig. 11.17 of the drawings obtained in the bibliography with nonconstant dimensions obtained via proportion and complemented by others as:

- $$\phi_{0;4} = 0.97$$
- $$T = 33 \text{ (cm)}$$

- $$\alpha_4 = 15^o$$
- $$\beta_5 = 5^o$$
- $$\psi_5 = 0.98$$
- $$h_s = 0.1$$
- $$g = 9.81\ \left(\mathrm{m/s^2}\right)$$
- $$\sigma_{\mathrm{esc}} = 14623\left(\mathrm{N/cm^2}\right)$$
- $$\rho = 1000\ \left(\mathrm{kg/m^3}\right)$$
- $$\rho_m = 7698 \cdot 10^{-6}\left(\mathrm{kg/m^3}\right)$$

All the characteristics with their respective values and units contained in the figure and the complementary figures must be stored in the computer system to be used in the software program that will be developed from the sequence script.

11.6.7.2 Script Form and Calculations

Calculate:

1. The maximum Head:

$$H_1 = (1 + h_s) \cdot H_{\max} = (1 + 0.1) \cdot 587 = 645.7\ (\mathrm{m})$$

2. The average absolute velocity of the flow at the entrance of the shells:

$$c_{4_1} = \phi_{0;4} \cdot \sqrt{2 \cdot g \cdot H_1} = 0.97 \cdot \sqrt{2 \cdot 9.81 \cdot 645.7} = 109.178\ (\mathrm{m/s})$$

3. The maximum flow through *IN*:

$$Q_1 = Q_{\max}/z_j = 33.63/6 = 5.605\ \left(\mathrm{m^3/s}\right)$$

4. The diameter of the flow Q_1:

$$d_{o_1} = \sqrt{\frac{4 \cdot Q_1}{\pi \cdot c_{4_1}}} = \sqrt{\frac{4 \cdot 5.605}{\pi \cdot 109.178}} = 0.256\ (m) = 25.6\ (\text{cm})$$

5. The average length of the jet flowing through the shell:

$$L_n = 1.26 \cdot d_{o_1} = 1.26 \cdot 25.6 = 32.3\ (\text{cm})$$

6. The horizontal projection of the half-jet force at the exit of each shell half:

$$F_{c_h} = 0.5 \cdot \rho \cdot Q_1 \cdot c_{4_1} \cdot (cos\,\alpha_4 + \psi_5 \cdot cos\,\beta_5) \cdot cos\,\beta_5 = = 0.5 \cdot 1000 \cdot 5.605 \cdot 109.178 \cdot (cos\,15 + 0.98 \cdot cos\,5) \cdot cos\,5 = 591995.190\ (\text{N})$$

7. The shell thickness:

$$e_x^* = \frac{F_{c_h} \cdot f_s \cdot \sqrt{1 + \left(\frac{3 \cdot T}{L_n}\right)^2}}{0.6 \cdot \sigma_{\text{esc}} \cdot L_n} = \frac{591995 \cdot f_s \cdot \sqrt{1 + \left(\frac{3 \cdot 33}{32.3}\right)^2}}{0.6 \cdot 24623 \cdot 32.3} = 4 \cdot f_s$$

so:

$$e^*_{c_{\min}} = 2 \cdot f_c = 2 \cdot 4 = 8\ (\text{cm})$$

and

$$e^*_{c_{\max}} = 4 \cdot f_c = 4 \cdot 4 = 16\ (\text{cm})$$

Store and continue with both values.

8. The average radius of the equivalent paraboloid of revolution:

$$R_m = 0.5 \cdot (T + 0.25 \cdot B) = R_m = 0.5 \cdot (T + 0.25 \cdot B) = 0.5 \cdot (33 + 0.25 \cdot 80) = 26.5\ (\text{cm})$$

9. The equivalent internal volume of a half Pelton shell:

$$V_{\text{in}_{c/2}} = 0.25 \cdot \pi \cdot R_m^2 \cdot L = 0.25 \cdot \pi \cdot 26.5^2 \cdot 74.8 = 41255.630\ \left(\text{cm}^3\right)$$

10. The equivalent external volumes of the half Pelton shell:

$$V\left(R_m + e^*_{e_{\min}}\right)^2_{ex_{c/2\min}} = 0.25 \cdot \pi \cdot (26.5 + 8)^2 \cdot 74.8 = 69924.548\ \left(\text{cm}^3\right)$$

$$V_{ex_{c/2\max}} = 0.25 \cdot \pi \cdot \left(R_m + e^*_{e\max}\right)^2 \cdot L = 0.25 \cdot \pi \cdot (26.5 + 16)^2 \cdot 74.8$$
$$= 106113.182 \left(\text{cm}^3\right)$$

11. The equivalent volumes of the Pelton shell:

$$V_{c\min} = 2 \cdot \left(V_{ex_{c/2\min}} - V_{\text{in}_{c/2}}\right) = 2 \cdot (69924.548 - 41255.639) = 57337.818 \left(\text{cm}^3\right)$$

$$V_{c\max} = 2 \cdot \left(V_{ex_{c/2\max}} - V_{\text{in}_{c/2}}\right) = 2 \cdot (106113.182 - 41255.639)$$
$$= 129715.086 \left(\text{cm}^3\right)$$

12. The equivalent masses of the Pelton shells:

$$G_{c\min} = \rho_m \cdot V_{c\min} = 7698 \cdot 10^{-6} \cdot 57337.818 = 441.386 \text{ (kg)}$$
$$G_{c\max} = \rho_m \cdot V_{c\max} = 7698 \cdot 10^{-6} \cdot 129715.086 = 998.547 \text{ (kg)}$$

13. The equivalent masses of the shell crowns:

$$G_{cc\min} = z_c \cdot G_{c\min} = 22 \cdot 441.386 = 9710.492 \text{ (kg)} \cong 9.710 \ (t)$$
$$G_{cc\max} = z_c \cdot G_{c\max} = 22 \cdot 998.547 = 21968.034 \text{ (kg)} \cong 21.968 \ (t)$$

14. The mass of the diameter of the crown, D_{ic}, d_{fe}:

$$G_{d_1} = \rho_m \cdot 0.25 \cdot \pi \cdot \left(D_{ic}^2 - d_{fe}^2\right) \cdot e_{cs} = 7698 \cdot 10^{-6} \cdot 0.25 \cdot \pi \cdot \left(253.3^2 - 198.2^2\right)$$
$$\cdot 39.1 = 5881.037 \text{ (kg)} \cong 5.881 \ (t)$$

15. The mass of the diameter of the crown, d_{fe}, d_f (cm):

$$G_{d_2} = \rho_m \cdot 0.25 \cdot \pi \cdot \left[\left(d_{fe}^2 - d_f^2\right]\right. \cdot e_d = 7698 \cdot 10^{-6} \cdot 0.25 \cdot \pi \cdot \left(198.2^2 - 157^2\right)$$
$$\cdot 29.4 = 2691.269 \text{ (kg)} \cong 2.691 \ (t)$$

16. The mass of the diameter of the crown, d_f, d_{di} (cm):

$$G_{d_3} = \rho_m \cdot 0.25 \cdot \pi \cdot \left(d_f^2 - d_{di}^2\right) \cdot e_{cs} = 7698 \cdot 10^{-6} \cdot 0.25 \cdot \pi \cdot \left(157^2 - 100^2\right) \cdot 39.1$$
$$= 3463.000 \text{ (kg)} = 3.463 \ (t)$$

17. The mass of the sum of the disk crowns:

$$G_{cd} = G_{d_1} + G_{d_2} + G_{d_3} = 5881.037 + 2691.269 + 3463 = 12035.306\ (\text{kg}) \cong 12.035\ (t)$$

18. The masses of the *RO*:

$$G_{RO_{\min}} = G_{cc_{\min}} + G_{cd} \cong 9.710 + 12.035 \cong 21.745\ (t)$$
$$G_{RO_{\max}} = G_{cc_{\max}} + G_{cd} \cong 21.968 + 12.035 = 34.003\ (t)$$

19. The horizontal coordinate:

$$x = \frac{z_j^{0,5} \cdot D}{z_c \cdot B} \cdot n_{qA_1} = \frac{6^{0,5} \cdot 3.25}{22 \cdot 0.8} \cdot 18 = 8.142$$

20. Using the value of x, remove the following from the graph in Fig. 11.13:

$$G^*_{RO_{\min}} = 13.0\ (t) \rightarrow G^*_{RO_{\max}} = 23.5\ (t) \rightarrow$$

21. Considering the limit values obtained from the graph in Fig. 11.13, along with those weighed on certified scales, determine the respective factors:

$$f_{r_{\min}} = \frac{G_{RO\min}}{G^*_{RO_{\min}}} = \frac{21{,}745}{13} = 1,67$$

$$f_{r_{\max}} = \frac{G_{RO\max}}{G^*_{RO_{\max}}} = \frac{34.003}{23.5} = 1.46$$

11.6.8 Comments on the Analytical Method for LHP

- The first three comments made for *SHP* are equally valid for *LHP*.
- The mass range resulting from the application of the analytical method, $1.25 \leq f_r \leq 1.48$, transcribed in the graph in Fig. 11.13, shows results, i.e., $G_{RO_{\min}} = 21.745\ (t)$ which is within the range.
- Probably when function $G_{RO} = f(x)$ is plotted with less uncertainty and with more points obtained by direct weighing, it is possible to substantially reduce the range of f_r.
- Therefore, the *RO* mass, obtained using the analytical method, must also have its characteristics taken from the details of the final *RO* project, i.e., the manufacturing project compatible with the analogous tested model.

  Research, particularly among manufacturers and designers, must be carried out to obtain a graph with average curves with uncertainty limits that cover the entire range of *PHIT*s, from micro to high power *PHIT*s, i.e., across all analogous *PHIT* ranges.

11.6.9 The Mixed Method

The mixed method determines a first approximation of the mass of the prototype G_{RO_p} (N) as long as the following are available:

- Executive projects of the *RO* of the prototype and the homologous geometrically similar model with the same specific rotation.
  Homologous manufactured *RO* model.
- Average specific mass of the material, ρ_m (N/cm^3) used to build the model, taken as being equal to the average of the material used to build the prototype $\rho_m = \rho_p$ (N/cm^3).

Then, the mass of the prototype *RO* G_{RO_p} (N) of the *PHIT* can be obtained as a first approximation by observing the following steps:

Weigh the mass of the model rotor on a scale with a verification certificate and corresponding accuracy/precision, G_{RO_m} $(N)L$;

1. Calculate the volume of the *RO* model, $V_{RO_m} = \frac{G_{ROpe}}{\rho_m}$ (cm^3);
2. Using the analytical method, calculate the volume of the *RO* model, $V^*_{RO_m}$ (cm^3) using characteristics taken from the Executive Project corresponding to the manufactured project;
3. Calculate the relationship between the calculated volumes, $k_m = \frac{V^*_{ROm}}{V_{ROm}}$;
4. Using the analytical method calculate the volume of the *RO* prototype $V^*_{RO_p}$ (cm^3) using characteristics taken from the Executive Project corresponding to the manufactured project;
5. Calculate the volume of the *RO* prototype, $V_{RO_p} = \frac{G^*_{ROp}}{k_m}$ (cm^3);
6. Calculate the mass of the *RO* prototype, $G_{RO_p} = \rho_m \cdot V_{RO_p}$ (N).

11.7 The Hydraulic Brake

There is a certain time, *t* (*s*), between load removals in any turbomachine rotating at *n* (*rpm*), until it stops, which basically depends on the interconnections between the inertia of the rotating system in the air, and the friction of the system with the *GG* fixing system.

This interconnection is made via a bearing system, which is either horizontal or vertical, depending on the positioning of the rotating system axis, normally designed

Fig. 11.18 Hydraulic brake diagram for a *PHIT*

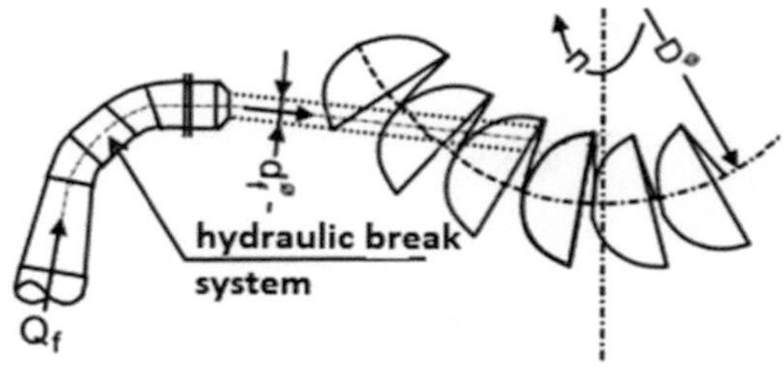

to operate with constant rotation n (rpm). Outside this rotation, wear increases, and the service life decreases, because the friction varies with n.

This means that the time t_p (s), between load removals and stopping the *GG*, is as short as minimum possible, which can be achieved by using mechanical or hydraulic brakes.

For *PHIT*s, with *RO* rotating in the air, part of the hydraulic energy must be available after the load is removed to decelerate it to overcome friction, stopping the system from rotating in the time specified by the *PT*, as $t(s)$.

This can be done using hydraulic energy (head) as shown in the diagram in Fig. 11.18, known as a hydraulic brake.

This system begins in most cases at the end of the penstock, normally before the valve that interconnects it with the *PHIT* inlet, which is not shown in the diagram of Fig. 11.18, however we can see a typical example in the schematic section from Fig. 1.2 in Chap. 1.

In this section, a singular (tube) is welded to a valve or register to withdrawal the flow Q_f (m^3/s).

Once this is done, one can design the hydraulic brake system and its calculate its structural, since d_f and Q_f are known using Expression (11.4) or (11.6) in the tube part, and the constant in subitem 8.4.2 in Chap. 8 for the *IN*.

Evidently, each manufacturer of each type of *PP* with a *PHIT*, already has indications that allow the *ET* to design and install a Hydraulic Brake *PHIT* system.

To start this study, one part from the premise that when the *DE* removes the hydrodynamic load from the *PHIT*, diverting the flow, the tangential velocity of the *RO* is u (m/s), and must be quickly reduced to zero. The average speed of the shock flow coming from the hydraulic brake system must not exceed u (m/s), therefore, $c_f = u$ (m/s).

Adopting $d_f \leq 0.5 \cdot d_o$ (m), the flow that must go through the *IN* of the hydraulic brake can be calculated as:

$$Q_f = u \cdot \frac{\pi \cdot d_f^4}{4} \quad (\text{m/s}) \tag{11.36}$$

With these and the *SA* characteristics, one can design and preliminarily calculate the structure of the *HFP* of the hydraulic brake system.

11.8 The Flywheel

The double regulation criterion in the *PHIT* is that the penstock hydraulic time,[1] t_h, must be in the following range:

$$0.6 < t_h = \frac{L_{cf} \cdot c_{0;\,1}}{g \cdot H_{cf}}\ (s) < 1.0 \tag{11.37}$$

where:

- L_{cf}—is the penstock length;
- $c_{0;\,1}$ is the average flow velocity in the inlet tube of the *PHIT*;
- g (m/s^2)—is the local acceleration of gravity;
- H_{cf}—is the vertical distance between the upper water level in the penstock (at the water intake or in the balance chimney) and at the exit of the water in the shells.

When t_h is within the range, or is less than 0.6 (s), the adjustment is simple, just move the needle so it does not need the *DF*. When $t_h > 1.0$ (s) the adjustment is double, initially activating the *DF* and then the needle. In this case, the time it takes the *DE* to completely deflect the jet $t_d\,(s)$ is linked to the inertia time of the *GG*, $t_d\,(s)$, currently in the range $t_g = \frac{0.6 < t_d < 1}{0.25 \text{ to } 0.75}$, thus obtaining $t_d\,(s)$ for the *GG* having minimal inertia, $t_d = 0.25\ (s)$, and a maximum $t_d = 0.74 \cdot t_g\ (s)$.

Only a *FW* is needed with inertia GD^2_{VO} when the sum of the inertias of the *GE*, GD^2_{GE}, and *HT*, GD^2_{TH}, is less than the inertia of the rotating system that connects the axis of the *HT* to that of the *GE*, i.e., the inertia of the *GG*, GD^2_{GG}, thus:

$$GD^2_{GE} + GD^2_{TH} < GD^2_{GG} \tag{11.38}$$

The following first approximation inertia expressions from (11.38) can be determined for an *EG*, *HT* and *GG* as:

$$GD^2_{GE} = 1608252.9 \cdot \frac{P_{el}^{1.333}}{n^{2.15}} \rightarrow GD^2_{TH} \cong 0.125 \cdot GD^2_{GE} \rightarrow GD^2_{GG}$$
$$= 3578256.4 \cdot \frac{P_{el}}{n^2} t_g \tag{11.39}$$

Substituting (11.39) into (11.38), we have:

[1] Centrais Hidrelétricas, Souza, Z./outros—3ª edição—pgs. 230 a 234—Ed. Interciência—RJ—BR—2018.

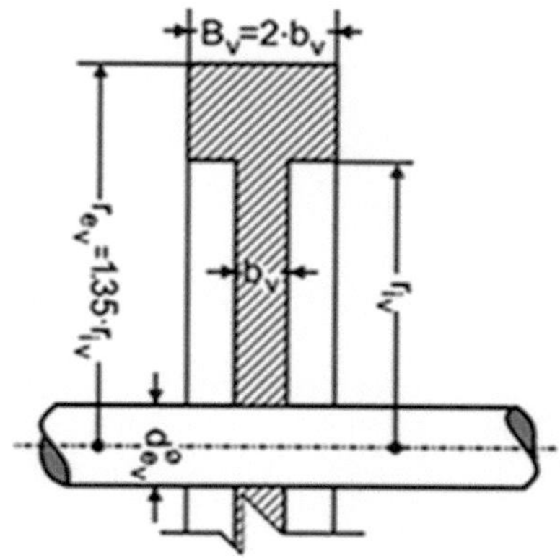

Fig. 11.19 Steel flywheel geometry and characteristics

$$1.125 \cdot 1608252.9 \cdot \frac{P_{el}^{1.333}}{n^{2.15}} < 3578256.4 \cdot \frac{P_{el}}{n^2} \cdot t_g \rightarrow t_g > \frac{1.125 \cdot 1608252.9 \cdot P_{el}^{0.333}}{3578256.4 \cdot n^{0.15}}$$

or

$$t_g > 0.5056 \cdot \frac{P_{el}^{0.333}}{n^{0.15}}\ (s) \tag{11.40}$$

where:

P_{el}—is the rated electrical power on the *EG* output terminals for direct coupling or mechanical power on the *PHIT* shaft for indirect coupling;

t_g (*s*)—is the transient time of the GG;

n (rpm)—is the rated speed of the *PHIT* for indirect coupling, or of the *GG* for direct coupling.

Having performed the calculations, a *FW* is needed when Expression (11.38) is not equal, e.g., and so GD_{VO}^2 can be by the following expression:

$$GD_{VO}^2 = GD_{GG}^2 - \left(GD_{GE}^2 + GD_{TH}^2\right) = GD_{GG}^2 - 1.125 \cdot GD_{GE}^2\ \left(\mathrm{N \cdot m^2}\right) \tag{11.41}$$

Here, considering the geometry and characteristics shown in Fig. 11.19, the *FW* can be sized as follows:

$$r_{e_v} = 0.156 \cdot \sqrt[5]{GD_{VO}^2}\,(\mathrm{m}) \rightarrow r_{i_v} = 0.74 \cdot r_{e_v} \rightarrow B_v = r_{e_v} - r_{i_v} \rightarrow b_v = 0.5 \cdot B_v \tag{11.41}$$

11.9 Shell/Hull

The structural calculation of the *CAS* is conducted by the *PT*, with studies to define the Engine Room project, particularly for each of the *GGs*, since they depend on interconnections with the penstock and the Escape Channel.

Chapter 12
An Algorithm for Calculating and Designing a Pelton Hydraulic Impulse Turbine (*PHIT*)

In this chapter we will develop an algorithm to hydrologically and structurally calculate a *PHIT* so it can be preliminarily designed. To do this, the file content, bibliography, Basic HPP Project, and the Authorization or Concession Agreements from ANEEL must all be available to the project team, not to mention the studies developed in the previous chapters. Using this documentation, the project team will determine the input data, which must be stored in the computer's memory banks so that each result can be obtained as intended by the designers and programmers. The algorithm will detail the applicable steps and which any established alternatives to maintain continuity, so the results give possible or best alternatives within the corresponding hypotheses. Once the algorithm is validated, it can be used both in *PHIT* models and in prototypes, particularly as a foundation for conducting the Basic, Executive, and Manufacturing Projects, along with the test programs for the laboratory, factory, or site.

12.1 Recommendations, Considerations, and Assumptions

- In general, after characterizing each step of the algorithm, observations can be made for the consideration of the *PT* and the programmers, which must be included in the corresponding software user's manual.
- The Input Data will be subdivided into two groups:
  - Determinants that cannot be changed, since they are included in the contract clauses between the parties, or have been established by the *PT* and agreed upon with the parties. To obtain the latter, the *PT* can use files, projects, and characteristics related to the site where the *HPP* will be implemented, among others. These determinative data are basically:

    The flows in (m^3/s);
    The design or nominal flow $Q = Q_n$;

© The Author(s), under exclusive license to Springer Nature Switzerland AG 2025

Z. de Souza, G. L. Tiago Filho, *Pelton and Crossflow Impulse Hydraulic Turbines*,
https://doi.org/10.1007/978-3-031-87086-6_12

The maximum flow Q_{max};
The minimum flow Q_{min};
The head in (mca);
The net, nominal, or available Head $H = H_0$;
The maximum Head H_{max};
The minimum Head H_{min};
The loads and yields, and the range of uncertainties obtained and observed for the model and from on-site tests;
The frequency in (Hz) of the consumer f_r;
The average specific mass in (Kg/m^3) of air ρ_{air} and water ρ;
The average accelerations of gravity in (m/s^2), at the site g_s, and for the universal gravity $g = 9.81$.
The materials and respective voltages that will be used to manufacture the components and their corresponding parts.
All these data must be stored in the memory banks of the computer system that will be used.

- There are data set by the *PT* that can be changed e.g., the current limits obtained in the archives, the constants in the bibliography, the constants at the end of this book, or constants from other sources. These data must be analyzed and detailed with the programmers, since they may need to be replaced in the memory banks of the computer system, and subsequently be used in the corresponding steps of the software program.
- We also recommend that scientific material, a laboratory, and particularly the contracting parties, be available to the *PT*, so that they can establish the *PHIT* model for study, design, and manufacture, e.g., the general arrangement and the main geometric and proportional characteristics of each of the components (*CA*; *CCL*; *CCR*; *RO*; *FH*; *VO*; *CAR*). For the *RO*, the shells must be detailed, particularly the inner surface, and the *PT* must decide if the inductive method prescribed in Chap. 10 will be used or not.

We recommend adopting $P_e \leq 1000$ (kW) & $n_T < 600$ (rpm) for indirect coupling with a gear ratio *EG*/*PHIT* at $r_{\text{ts}} \leq 5$. For $n_T \geq 600$ (rpm) and greater P_e, we recommend direct coupling, $n_{\text{TH}} = n_{\text{GE}} = n$, either on the horizontal or vertical axis.

In the steps that need to-scale plotting removing characteristic values to proceed with the calculations in the following steps, one can use a computational routine in a graphical software program like AutoCAD, and these results must be stored and used both in the early and later steps of the computer program.

The symbology, data, and characteristics determined as final predicted values, obtained in the script steps, must be stored in the memory banks of the computing system, thereby permitting analysis, calculation continuity, and output into tables and graphs, at the discretion of the users.

The computer software program must be able to perform recalculations from step 1, at the end of the hydrodynamic, geometric, and structural part, modifying the input data set by the *PT* and Tabling the resulting alternatives, so that the *PT* can choose the alternative(s) for continuing with the calculations, and make decisions from the specified step, with all of the new data, or only part of the new data.

The quantities and conventions have been described in the previous chapters, and were synthetically named at the beginning of this publication.

The software resulting from the algorithm for the horizontal axis of the *PHIT* , when applicable, should be able to account for 1 or 2 jets, and horizontal or vertical axes for 4, 5, or 6 jets (see Sect. 8.1).

If the internal diameter, D_{cf}, of the penstock is not available at the valve outlet at the engine room inlet, one can determine this using the Bondschu Formula, $d_{\mathrm{cf}} = 1.27 \cdot \frac{Q^{0.4286}}{(1.3 \cdot H_{\max})^{0.1429}}$ (m):

(Manual de Inventário das Bacias Hidrográficas—Eletrobrás—RJ—BR—1997)

… or preliminarily use the value obtained as being equal to the inner diameter, d_0, at the entrance of the cross flow section 0 in the *PHIT*.

The average flow velocity in the cylinder/needle crown, should be used as $c_{\mathrm{ca}} \leq c_t = \mathrm{const}$ whenever possible, or as normal, $c_{ca} > c_t$, which reduces volume, but increases energy losses.

The *PT* must specify, together with manufacturers, the minimum external dimensions of the *CS*, i.e.,

- $$d_a$$
- $$L_{\mathrm{cii}}$$
- $$L_{\mathrm{og}}$$
- $$L_{\mathrm{in}}$$

In the absence of these dimensions, the preliminary studies should use the relationships shown in Fig. 8.8.

The average absolute velocity of the flow in the pipes of the Admissions System (*SA*) c_t (m/s) between cross section 0 and cross section 1∗, will be taken as being constant, and determined by referencing the flow $Q = Q_n$ (m^3/s).

The intervals of the main relationships between the geometric and dynamic quantities are currently in the public domain, and are as follows:

- $$40 \leq \theta^{o}_{co} \leq 60$$
- $$60 \leq \theta^{o}_{in} \leq 90$$

- $$35 \leq \theta_d^o \leq 55$$
- $$4 \leq \beta_5^0 \leq 7$$
- $$0.97 \leq \eta_f \leq 0.99$$
- $$0.97 \leq \psi_5 \leq 0.99$$
- $$9 \leq D/d_o \leq 16$$
- $$1.7 \leq d_{ci}/d_a \leq 2$$
- $$1.0 \leq d_{ci}/d_t \leq 1.1$$
- $$1.2 \leq d_2/d_o = \leq 2.5$$
- $$\varepsilon - \text{table 8.3}$$
- $$k_v - \text{error at 50 } (\%)$$
- $$0.95 \leq \eta_m \leq 0.99$$
- $$0.87 \leq \eta_t \leq 0.92$$
- $$0.96 \leq \eta_e \leq 0.99$$

In principle, the smallest values are for small *HPP*s, with each *GG* not exceeding P_e~1.0 (MW).

We will use the geometries, specifications, and relationships contained in the previous chapters of this book, for example, the preliminary structural calculation, as established in Chap. 11, to validate the algorithm.

The analytical method (Fig. 9.10) and the script from subitem 9.7.2.2 will be used to determine the theoretical step, p_{c*}.

Some uncertainties occur within the hypotheses and forms used in the preliminary calculation of the energy losses in the *SA*. Therefore, we recommend using a safety factor when calculating the specific works, $1.0 \leq f_{se} \leq 1.1$, or in other words $Y_{i_{p_j}} = f_{se} \cdot Y_{i_{p_j\,\text{calculated}}}$.

12.2 Algorithm and Validation

12.2.1 Determining the Input Data and Fixed Values for Validation

- $$Q = Q_n = 14.516\ \left(\text{m}^3/\text{s}\right)$$
- $$Q_{\max} = 15.020\ \left(\text{m}^3/\text{s}\right)$$
- $$Q_{\min} = 13.790\ \left(\text{m}^3/\text{s}\right)$$
- $$H = H_0 = 630\ (\text{mca})$$
- $$H_{\min} = 627\ (\text{mca})$$
- $$H_{\max} = 640\ (\text{mca})$$
- $$fr = 60\ (\text{Hz})$$
- $$g = 9.81\ \left(\text{m/s}^2\right)$$
- $$\rho = 1000\ \left(\text{kg/m}^3\right)$$
- $$\rho_{ar} = 1.32\ \left(\text{kg/m}^3\right)$$

- $$g_s \cong g = 9.81\ \left(\mathrm{m/s^2}\right)$$
- $$\nu_a = 10^{-6}\ \left(\mathrm{m^2/s}\right)$$

Table with the values of the dimensionless coordinates of the reference shell, and the constants in Table 10.1 in the validation.

12.2.2 *Input Data Preliminarily Set by the* PT *and Values for Validation*

In the validation process, the constants in Chapters 8 and 9 will be used as geometric models, along with their proportions, where applicable, except for items referring to the inner surface of the shells, for which we will use the constant in Chap. 10, as well as for the mechanical calculations in Chap. 11.

- $$\boldsymbol{\phi_{0;4} = 0.98}$$
- $$\boldsymbol{d_{\text{in}} = d_2 = 1.23 \cdot d_o}$$
- $$\boldsymbol{\varepsilon_t = \varepsilon_{ca} = 0.02\ (\text{mm})}$$
- $$\boldsymbol{\psi_5 = 0.98}$$
- $$\boldsymbol{r_{v_1} = 0.48}$$
- $$\boldsymbol{r_{v_{21}} = 0.485}$$
- $$\boldsymbol{r_{v_3} = 0.49}$$
- $$\boldsymbol{h_s = 0.1}$$

- $$\mu_{ad} = 0.01$$
- $$\eta_{f_{2;4}} = \eta_{f_{4;5}} = 0.995$$
- $$\beta_c = 10^o$$
- $$\eta_m = \eta_{el} = 0.98$$
- $$f_{se} = 1.1$$
- $$\beta_5 = 5^o$$
- $$d_a/d_{ci} = 0.54$$
- $$d_{\text{in}} = d_2 = 1.22 \cdot d_o$$
- $$d_{ci}/d_{tb} = 1.1$$
- $$L_{dc}/d_a = 0.64$$
- $$L_{cii}/d_o = 4.83$$
- $$L_{df}/d_o = 1.15$$
- $$\theta_{co} = 45^o$$
- $$\theta_{\text{in}} = 75^o$$
- $$\delta_b = 40^o$$

- $$e_d = e_h = e_{su}$$
- $$(L/d_o)_{fix} = 2.9$$
- $$(T/d_o)_{fix} = 1.13$$
- $$(B/d_o)_{fix} = 3$$
- $$d_f = 0,5 \cdot d_o$$
- $$\bar{E}e = d_{t_1}$$
- $$\lambda_r = 0.02$$

The values taken from the shell used here from Expression (10.1), are:

- $$(L/d_o)_{\text{con}} = 2,61$$
- $$(0.5 \cdot B/d_o)_{\text{con}} = 1.564$$
- $$(T/d_o) = 1.0$$

To facilitate manual calculations in the validation process, a single material will be used for all parts with the following tensions and specific mass:

$$\sigma_{esc} = 24623\ \left(\text{N/cm}^2\right)$$

$$\sigma_{f_{\text{adm}}} = 14470\ \left(\text{N/cm}^2\right)$$

$$\boldsymbol{\sigma_{t_{\text{adm}}} = 16744\ \left(\text{N/cm}^2\right)}$$

$$\boldsymbol{\tau_{cis_{\text{adm}}} = 9565\ \left(\text{N/cm}^2\right)}$$

$$\boldsymbol{\tau_{tor_{\text{adm}}} = 2500\ \left(\text{N/cm}^2\right)}$$

$$\boldsymbol{\tau_{tor_{\text{adm}}} = 2500\ \left(\text{N/cm}^2\right)}$$

$$\boldsymbol{\rho_m = 7698 \cdot 10^{-6}\ \left(\text{kg/cm}^3\right)}$$

Also, the symbols, geometries, and relationships from Figs. 8.1, 8.2, 8.3, 8.11, 8.12, 8.13, 9.15, 11.6, 11.7, 11.8, and 11.12 will be used.

12.2.3 Calculations and Decisions

Calculate:

1. The nominal internal specific work of the flow at inlet 0:

$$Y_{i_0} = g \cdot H_0$$

$$Y_{i_0} = 9.81 \cdot 630 = 6180.3\ (\text{J/kg})$$

2. The average absolute velocity of the flow at inlet 0:

$$c_0 = \sqrt{2 \cdot Y_{i_0}} = \sqrt{2 \cdot 6180.3} = 111.178\ (\text{m/s})$$

3. The predicted average absolute velocity of the flow at the entrance of shell 4,

$$c_4 = \phi_{0;4} \cdot c_0 = 0.98 \cdot 111.178 = 108.954\ (\text{m/s})$$

Table 12.1 Number of zj

1	2	3
Z_{zj}	Q_{1j}	$d_{o}j$
–	m³/s	m
H–1	14.52	0.41
H–2	7.26	0.29
V–4	3.63	0.21
V–5	2.90	0.18
V–6	2.42	0.17

4. Format Table 12.1, placing the horizontal axis H or vertical axis V and number of jets in column 1:

$$Q_{1_j} = Q_j / z_{z_j} = 14.516 / z_j \ (\mathrm{m}^3/\mathrm{s})$$

5. Place the flows in column 2:

$$Q_{1_j} = Q_j / z_{z_j} = 14.516 / z_j \ (\mathrm{m}^3/\mathrm{s})$$

6. Place the minimum jet diameters:

$$d_{o_j} = \sqrt{\frac{4 \cdot Q_{1_j}}{\pi \cdot c_4}} = \sqrt{\frac{4 \cdot Q_{1_j}}{\pi \cdot 108.954}} = 0.1081019 \cdot \sqrt{Q_{1_i}}$$

… in column 3, storing and continuing with Table results 12.1 with $d_o \leq 0{,}250$ (m), which covers:

$$z_j = V - 4 \rightarrow V - 5 \rightarrow V - 6 \rightarrow Q_{1_j} = 3.629$$

$$2.903$$

$$2.419 \ (\mathrm{m}^3/\mathrm{s})$$

To continue the validation, stored $z_j = 4$ and vertical axis $Q_1 = 3.629$, $d_0 = 0.206\ (\text{m}) = 206\ (\text{mm})$.

7. The synchronous rotation of the *EG*, n_G, is the same as the *PHIT*, n_T, or $n_G = n_T = n$, thus observe the higher yield range for n_{qA_1} (see Fig. 9.3):

For:

$$n_{qA_{1\,\text{min}}} = 14 \rightarrow n^*_{\text{min}} = \frac{H^{0.75} \cdot n_{qA_{1\,\text{min}}}}{3 \cdot Q_{j4}^{0,5}} = \frac{630^{0.75} \cdot 14}{3 \cdot 3.629^{0.5}} = 308.048 \rightarrow$$

Synchronous:

$$n_{\text{min}} = 360\ (\text{rpm})$$

$$z_{pp_{\text{max}}} = \frac{3600}{n_{\text{min}}\,\frac{3600}{360}} n_{qA_{1\,\text{min}}} = 3 \cdot n_{\text{min}} \cdot \frac{Q_{j4}^{0.5}}{H^{0.75}} = 3 \cdot 360 \cdot \frac{3.629^{0.5}}{630^{0.75}} = 16.361$$

For:

$$n_{qA_{1\,\text{max}}} = 20 \rightarrow n^*_{\text{max}} = \frac{H_0^{0.75} \cdot n_{qA_{1\,\text{max}}}}{3 \cdot Q_{j4}^{0.5}} = \frac{630^{0.75} \cdot 20}{3 \cdot 3.629^{0.5}} = 440.069 \rightarrow$$

Synchronous:

$$n_{\text{max}} = 450\ (\text{rpm})$$

$$z_{pp_{\text{min}}} = \frac{3600}{n_{\text{max}}} = \frac{3600}{450} = 8\ pp \rightarrow n_{qA_{1\,\text{max}}} = 3 \cdot n_{\text{max}} \cdot \frac{Q_{j4}^{0.5}}{H^{0.75}} = 3 \cdot 450\ \frac{3.629^{0.5}}{630^{0.75}}$$
$$= 20.51$$

8. Format Table 12.2, placing the synchronous rotations between those established in the previous step and the intermediate ones in column 1, in this case $n = 400\ (\text{rpm})$.

Table 12.2 Setting characteristics

1	2	3	4	5	6
n_i	z_{ppi}	$n_{qA\ 1i}$	r_{vi}	D_i	u_i
rpm	p.poles	–	-	m	m/s
360	10.00	16.36	0.48	2.78	52.30
400	9.00	18.18	0.49	2.52	52.84
450	8.00	20.51	0.49	2.26	53.39

9. Set the number of *EG* pole pairs in column 2:

$$z_{pp_i} = \frac{f_r}{n_i} = \frac{3600}{n_i}$$

10. Set the specific rotations in column 3:

$$n_{qA_{1_i}} = 3 \cdot n_i \cdot \frac{Q_{i4}^{0.5}}{H^{0.75}} = 3 \cdot n_i \cdot \frac{3.629^{0.5}}{630^{0.75}} = 0.3143 \cdot n_i$$

11. Set the three velocity ratios for the data in column 4:

$$r_{v_i} = 0.480$$

$$0.485$$

$$0.490$$

12. Transcribe the values of the calculated primitive diameters in column 5:

$$D_i = 473.36 \cdot \phi_{0;4}^{1.5} \cdot d_o \cdot \left(\frac{r_{v_i}}{n_{qA_1}}\right)_i = 473.36 \cdot 0.98^{1.5} \cdot 0.206 \cdot \left(\frac{r_{v_i}}{n_{qA_1}}\right)_i$$

$$= 94.60147 \cdot \left(\frac{r_{v_i}}{n_{qA_1}}\right)_i \ (\mathrm{m})$$

13. Set the tangential velocities D_i, $u_i = \frac{\pi}{60} \cdot (D \cdot n)_i$ (m/s) in column 6:

In principle, the computer software program should continue the script for each r_{v_i}, but in the end this choice is made by the *PT*. For validation, the values in *italics* in the second line should be $n = 400$(rpm), $z_{pp} = 9$ pairs of poles in the *EG*:

$$n_{qA_1} = 18.179$$

$$r_v = 0.485$$

$$D = 2.523 \text{ (mm)}$$

$$U = 52.844 \text{ (m/s)}$$

14. The coefficients of similarity between the fixed dimensions of the inner surface of the shells and those of the inner surface of the standard shell, are:

$$f_B = \frac{(0.5 \cdot B/d_o)_{\text{fix}}}{(0.5 \cdot B/d_o)_{\text{con}}} = \frac{0.5 \cdot 3}{1.564} = 0.959$$

$$f_L = \frac{(L/d_o)_{\text{fix}}}{(L/d_o)_{\text{con}}} = \frac{2.9}{2.61} = 1.111$$

$$f_T = \frac{(T/d_o)_{\text{fix}}}{(T/d_o)_{\text{con}}} = \frac{1.13}{1} = 1.13$$

15. Multiply each value of the scale coefficients calculated in the previous step by $d_0 = 206$ (mm) and then by the respective coordinates of each of the three columns in Table 10.1, using expression:

$$(0.5.B) = f_B \cdot d_o \cdot (0,5 \cdot B)_{\text{Tab}} = 0.959 \cdot 206 \cdot (0.5 \cdot B)_{\text{Tab}} = 197.554 \cdot (0.5 \cdot B)_{\text{Tab}.j} \text{ (mm)}$$

$$L_j = f_L \cdot d_o \cdot L_{\text{Tab}.j} = 1.111 \cdot 206 \cdot L_{\text{Tab}.j} = 228.866 \cdot L_{\text{Tab}.j} \text{ (mm)}$$

$$T_j = f_T \cdot d_o \cdot T_{\text{Tab}.j} = 1.13 \cdot 206 \cdot T_{\text{Tab}.j} = 232.78 \cdot T_{\text{Tab}.j} \text{ (mm)}$$

16. Format and fill in Table 12.3 with the values resulting from the multiplication of the coordinates obtained in the previous step by the respective coordinates of the shell taken as a reference, in the validation of Table 10.1;
17. Format and fill in Table 12.4 with the values taken from Table 12.3 corresponding to coordinates L and $0.5 \cdot B$ of the line that delimits the flow exit from the inner surface of the shells in the right part between the points 1 and 371.
18. Choose a graphical design computer program e.g., AutoCAD, setting a scale in millimeters, then draw the line that delimits the flow exit of the internal surface of the shells in the right part of Fig. 12.1, between points 1 and 29, making the changes, where it fits, obtaining the best equations using computer programs like "Table Curve 2D," for example.

- Equation between points 7 to 17

$$L = -1765.6974 + 50.92753 \cdot (0.5 \cdot B) - 0.62069399 \cdot (0.5 \cdot B)^2 + 0.0035174529 \cdot (0.5 \cdot B)^3 - 9.4655796 \cdot 10^{-6} \cdot (0.5 \cdot B)^4 + 9.8718879 \cdot 10^{-9} \cdot (0.5 \cdot B)^5 \rightarrow R^2 = 0.99972107$$

- Equation between points 17 to 22

$$L = \frac{61.910367 - 0.38704295 \cdot (0.5 \cdot B) + 0.0006041567 \cdot (0.5 \cdot B)^2}{1 - 0.0098170065 \cdot (0.5 \cdot B) + 3.2725438 \cdot 10^{-5} \cdot (0.5 \cdot B)^2 - 3.6974029 \cdot 10^{-8} \cdot (0.5 \cdot B)^3} \rightarrow R^2 = 0.9999594$$

Table 12.3 Surface coordinates

Inner Shell in (mm)				Inner Shell in (mm)			
Num.	0.5·B	*L*	*T*	Num.	0.5·B	L	T
1	0.00	−124.73	−6.63	77	134.14	−86.74	226.96
2-2*	12.05	−153.87	0.00	78	169.70	−86.74	215.09
3-3*	30.23	−203.92	0.00	79	203.88	−86.74	181.57
4-4*	64.40	−248.32	0.00	80	240.82	−86.74	130.36
5-5*	99.96	−273.04	0.00	81	275.00	−42.34	65.18
6-6*	134.13	−280.36	0.00	82	12.05	−42.34	58.89
7-7*	169.70	−268.92	0.00	83	30.23	−42.34	125.47
8-8*	203.88	−249.46	0.00	84	64.40	−42.34	193.21
9-9*	210.79	−244.66	0.00	85	99.96	−42.34	224.63
10-10*	240.82	−213.07	0.00	86	134.14	−42.34	231.62
11-11*	247.34	−203.92	0.00	87-87*	169.70	−42.34	219.98
12-12*	271.04	−169.64	0.00	88-88*	203.88	−42.34	190.88
13-13*	275.00	−155.40	0.00	89-89*	240.82	−42.34	139.20
14-14*	286.64	−121.30	0.00	90-90*	275.00	−42.34	80.54
15-15*	296.13	−82.16	0.00	91-91*	12.05	0.00	61.69
16-16*	303.44	−41.65	0.00	92-92*	30.23	0.00	128.03
17-17*	308.97	0.00	0.00	93-93*	64.40	0.00	197.86
18-18*	307.20	39.59	0.00	94-94*	99.96	0.00	226.96
19-19*	301.66	82.16	0.00	95-95*	134.14	0.00	232.78
20-20*	292.97	123.12	0.00	96-96*	169.70	0.00	221.14
21-21*	282.11	165.93	0.00	97-97*	203.88	0.00	193.21
21*-21**	274.98	185.61	0.00	98-98*	240.82	0.00	142.00
22-22*	267.69	202.55	0.00	99-99*	275.00	0.00	83.80
23-23*	258.99	223.60	0.00	100-100*	305.22	0.00	9.31
24-24*	240.62	259.76	0.00	101-101*	12.05	39.59	61.69
25-25*	216.62	290.43	0.00	102-102*	30.23	39.59	126.87
26-26*	200.51	301.65	0.00	103-103*	64.40	39.59	196.70
27-27*	169.70	310.00	0.00	104-104*	99.96	39.59	222.30
28-28*	134.14	312.50	0.00	105-105*	134.14	39.59	226.96
29-29*	118.33	313.50	0.00	106-106*	169.70	39.59	214.62
30-30*	118.33	285.85	67.51	107-107*	203.88	39.59	188.55
31-31*	118.33	277.59	116.39	108-108*	240.81	39.59	139.20
32-32*	99.96	226.58	135.01	109-109*	275.00	39.59	90.08
33-33*	64.40	229.55	130.36	110-110*	12.05	82.16	61.69
34-34*	42.67	244.89	112.67	111-111*	30.23	82.16	124.07
35-35*	30.22	262.64	90.09	112-112*	64.40	82.16	193.21
36-36*	17.38	285.85	65.18	113-113*	99.96	82.16	212.99
36*-36**	12.05	295.24	46.79	114-114*	134.14	82.16	216.02
37	0.00	314.69	6052.00	115-115*	169.70	82.16	202.52
38	0.00	285.85	5214.00	116-116*	203.88	82.16	176.91
39	0.00	244.89	4027.00	117-117*	240.82	82.16	130.36

(continued)

Table 12.3 (continued)

Inner Shell in (mm)				Inner Shell in (mm)			
Num.	0.5·B	*L*	*T*	Num.	0.5·B	L	T
40	0.00	205.29	2886.00	118-118*	275.00	82.16	68.67
41	0.00	165.93	1769.00	119-119*	12.05	123.13	61.69
42	0.00	123.13	0.51	120-120*	30.23	123.13	121.05
43	0.00	82.16	−0.65	121-121*	64.40	123.13	185.06
44	0.00	39.59	−1886.00	122-122*	99.96	123.13	200.66
45	0.00	0.00	−3026.00	123-123*	134.14	123.13	202.75
46	0.00	−42.34	−4260.00	124-124*	169.70	123.13	188.55
47	0.00	−86.74	−5540.00	125-125*	203.88	123.13	160.39
48-48*	99.96	−245.80	74.26	126-126*	240.82	123.13	116.39
49-49*	134.14	−245.80	108.94	127-127*	275.00	123.13	53.54
50-50*	169.70	−245.80	76.82	128-128*	12.05	165.93	61.69
51-51*	203.87	−245.80	12.34	129-129*	30.23	165.93	114.99
52-52*	64.40	−203.92	77.50	130-130*	64.40	165.93	169.93
53-53*	99.96	−203.92	142.46	131-131*	99.96	165.93	183.90
54-54*	134.14	−203.92	168.77	132-132*	134.14	165.93	186.22
55-55*	169.70	−203.92	146.65	133-133*	169.70	165.93	168.07
56-56*	203.88	−203.92	95.44	134-134*	203.88	165.93	137.34
56*-56**	240.82	−203.92	16.29	135-135*	240.82	165.93	93.11
57-57*	30.23	−163.87	55.87	136-136*	275.00	165.93	28.86
58-58*	44.40	−163.87	128.03	137-137*	12.05	205.29	60.52
59-59*	99.96	−163.87	183.20	138-138*	30.23	205.29	106.61
60-60*	134.14	−163.87	200.42	139-139*	64.40	205.29	148.98
61-61*	169.70	−163.87	179.24	140-140*	99.96	205.29	158.29
62-62*	203.88	−163.87	139.67	141-141*	134.14	205.29	157.13
63-63*	240.82	−163.87	77.52	142-142*	169.70	205.29	138.50
64-64*	12.05	−124.73	36.08	143-143*	203.88	205.29	106.61
65-65*	64.40	−124.73	93.11	144-144*	240.82	205.29	61.69
66-66*	64.40	−124.73	160.62	145-145*	12.05	244.89	58.20
67-67*	99.96	−124.73	202.98	146-146*	30.23	244.89	38.11
68-68*	134.14	−124.73	216.95	147-147*	134.14	244.89	115.69
69-69*	169.70	−124.73	201.82	148-148*	169.70	244.89	99.63
70-70*	203.88	−124.73	167.60	149-149*	203.88	244.89	65.18
71-71*	240.82	−124.73	110.57	150-150*	240.82	244.89	22.11
72-72*	275.00	124.73	34.92	151-151*	12.05	285.85	50.05
73-73*	12.05	−86.74	49.35	152-152*	134.14	285.85	65.18
74-74*	30.23	−86.74	116.86	153-153*	169.70	285.85	46.56
75-75*	64.40	−86.74	181.57	154-154*	203.88	285.85	18.62
76-76*	99.96	−86.74	219.51	-	-	-	-

Table 12.4 Coordinates of the limits of the shell in plan

Num.	*L* (mm)	0.5·B (mm)	Num.	*L* (mm)	0.5·B (mm)
1	−124.73	0.00	17	0.00	308.97
2	−163.87	12.05	18	39.59	307.20
3	−203.92	30.23	19	82.16	301.66
4	−248.32	64.40	20	123.12	292.97
5	−273.04	99.96	21	165.93	282.11
6	−280.36	134.13	21*	185.61	274.98
7	−268.92	169.70	22	202.55	267.69
9	−244.66	210.79	23	223.60	258.99
10	−213.07	240.82	24	259.76	240.62
11	−203.92	247.34	25	290.43	216.62
12	−169.64	271.04	26	301.65	200.51
13	−155.40	275.00	27	305.54	169.70
14	−121.30	286.64	28	314.00	134.14
15	−82.16	296.13	29	314.69	118.33
16	−41.65	303.44	–	–	–
Coordinates of shell boundaries in plan changed					
2	–	153.87	29	–	313.50
27	–	310.00	31	–	277.59
28	–	312.50	35	–	262.64

- Equation between points 0 to 7

$$L = -124.80166 - 3.7507776 \cdot (0.5 \cdot B) + 0.048441114 \cdot (0.5 \cdot B)^2 - 0.00042964435 \cdot (0.5 \cdot B)^3 + 2.1163516 \cdot 10^{-6} \cdot (0.5 \cdot B)^4 - 3.965497 \cdot 10^{-9} \cdot (0.5 \cdot B)^5 \rightarrow R^2 = 0.99999515$$

- Equation between points 22 to 27

$$L = 1194.1991 - 17.725233 \cdot (0.5 \cdot B) + 0.1282972 \cdot (0.5 \cdot B)^2 - 0.00038753819 \cdot (0.5 \cdot B)^3 + 3.8823055 \cdot 10^{-7} \cdot (0.5 \cdot B)^4 \rightarrow R^2 = 0.99999999$$

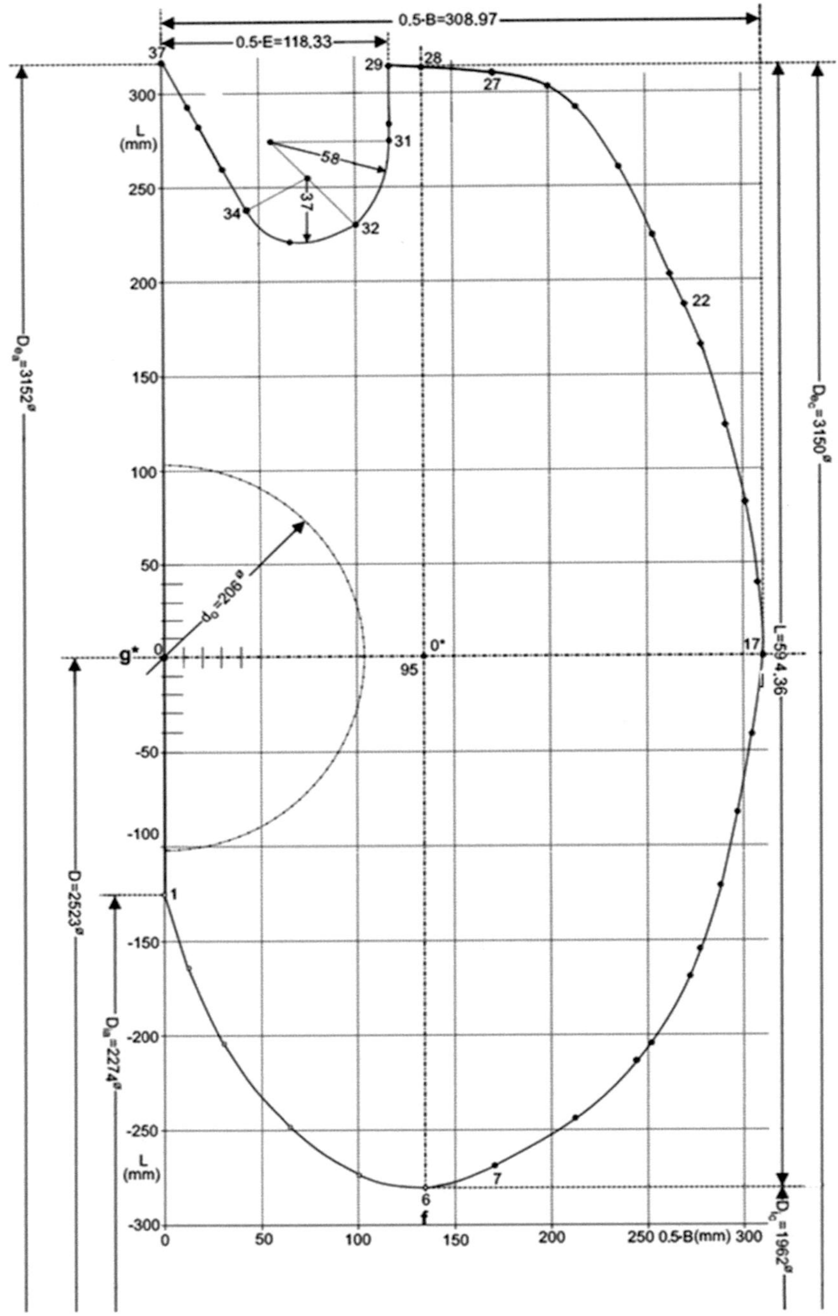

Fig. 12.1 Graphs and equations of the limit curve on the right of the output line of the inner surface of shells, in scale

19. Complement the right half of the cutaway by drawing agreements between points 27 and 29, circles between 31 and 34, and a straight segment between 34 and 37, with the line segment $\overline{37/1}$ in the projection, of the dividing edge of the shell.
20. Transcribe the respective values of the altered coordinates obtained in the graphic drawings in Tables 12.3 and 12.4, in *italics*, and make a drawing with a center at 0, $d_0 = 206$ (mm).
21. Calculate and mark the following in Fig. 12.1, with approximations in millimeters:

$$L = L_6 + L_{29} = 280.36 + 313.5 = 593.86 \text{ (mm)} \rightarrow$$

$$0.5 \cdot B = (0.5 \cdot B)_{17} = 308.97 \text{ (mm)} \rightarrow$$

$$0.5 \cdot E = (0.5 \cdot E)_{29} = 118.33 \text{ (mm)} \rightarrow$$

$$T = T_{95} = 232.78 \text{ (mm)} \rightarrow$$

$$D_{i_a} = D - 2 \cdot L_1 = 2523 - 2 \cdot 124.73 = 2273.54 \text{ (mm)} \rightarrow$$

$$D_{e_a} = D + 2 \cdot L_{37} = 2523 + 2 \cdot 314.69 = 3152.38 \text{ (mm)} \rightarrow$$

$$D_{e_c} = D + 2 \cdot L_{29} = 2523 + 2 \cdot 313.5 = 3150 \text{ (mm)} \rightarrow$$

$$D_{i_c} = D - 2 \cdot L_6 = 2523 - 2 \cdot 280.36 = 1962.28 \text{ (mm)}$$

To continue the validation, we will use and store:

$$L = 594.36 \text{ (mm)}$$

$$B = 2 \cdot 308.97 = 617.94 \text{ (mm)}$$

$$T = 232.78 \text{ (mm)}$$

$$E = 2 \cdot 118.33 = 236.66 \text{ (mm)}$$

$$D_{i_a} \cong 2274 \text{ (mm)}$$

$$D_{e_a} \cong 3152 \text{ (mm)}$$

$$D_{e_c} \cong 3150 \text{ (mm)}$$

$$D_{i_c} \cong 1962 \text{ (mm)}$$

The drawing in the curved part of half of the cutaway was drawn with 2 arcs of circumferences.

22. Choose and mark on the blueprint (Fig. 12.1) longitudinal section f and transversal section $g*$ that will be the basis for drawing the respective internal surfaces of the shell and subsequent templates, here referencing Fig. 10.4.
23. Format and fill in Tables 12.5 and 12.6 with the respective numbers and values of the right longitudinal section f and right transversal section $g*$ of the inner surface of the shell.
24. With the values in the table, draw and generate the respective average curves, in Figs. 12.2 and 12.3:

Table 12.5 Coordinates of the right longitudinal section

f with 0.5·B=Num. 95=134.1 (mm)=const.					
Num.	*L* (mm)	*T*(mm)	Num.	*L* (mm)	*T*(mm)
0	–280.36	0.00	105	39.59	–226.96
49	–245.80	–108.84	114	82.16	–216.02
54	–203.92	–168.77	123	123.13	–202.75
60	–163.87	–200.42	132	165.93	–186.22
68	–124.73	–216.95	141	205.29	–157.13
77	–86.74	–226.96	147	244.89	–115.69
86	–42.34	–231.62	152	285.85	–65.18
95	0.00	–232.78	28	314.00	0.00

Table 12.6 Coordinates of the right cross section

*g**with L=Num. 95=0 (mm)=const.					
Num.	0.5·B (mm)	*T*(mm)	Num.	0.5·B (mm)	*T*(mm)
45	0.00	3026.00	96	169.70	–221.14
91	12.05	–61.69	97	203.88	–193.21
92	30.23	–128.03	98	240.82	–142.00
93	64.40	–197.86	99	275.00	–83.80
94	99.96	–226.96	100	305.22	–9.31
95	134.14	–232.78	17	308.97	0.00

$$T = -232.70939 + 0.072468619 \cdot L + 0.0020652046 \cdot L^2 - 1.16171228 \cdot 10^{-7} \cdot L^3 - 7.43689993 \cdot 10^{-8} \cdot L^4 - 8.4500702 \cdot 10^{-11} \cdot L^5 + 2.8076001 \cdot 10^{-12} \cdot L^6 + 1.2847829 \cdot 10^{-15} \cdot L^7 - 2.0458103 \cdot 10^{-20} \cdot L^8 + 1.9754011 \cdot 10^{-22} \cdot L^9 \rightarrow R^2 = 0.99997827$$

$$T = \frac{3.02642554 - 8.2606001 \cdot (0.5 \cdot B)^{0.5} - 0.72621275 \cdot (0.5 \cdot B) + 0.067477427 \cdot (0.5 \cdot B)^{1.5}}{1 - 0.22822487 \cdot (0.5 \cdot B)^{0.5} + 0.031260392 \cdot (0.5 \cdot B) - 0.001880767 \cdot (0.5 \cdot B)^{1.5} + 0.067477427 \cdot (0.5 \cdot B)^{1.5}}$$

$$\longrightarrow \quad R^2 = 0.99999787$$

25. With the drawing of the right half of the inner surface of the shell in Fig. 12.1 transposed to Fig. 12.4, obtain the blueprint of the inner surface of the shell (Fig. 12.4a) rotating the transposed Fig. 180 (degrees) around the dividing line.
26. Obtain the longitudinal section cutaways of the inner surface of the shell, transposing the corresponding longitudinal sections on the right, as was done in the section *AA*, in this validation only with section *f* (Fig. 12.4b).
27. Obtain the cross sections of the inner surface of the shell, rotating the corresponding sections by 180 (degrees) on the right around the vertical line that passes through the projection of the dividing line, as was done here only for section $g*$ (Fig. 12.4c) section *B*, resulting in $\alpha_a = 20^o$ for this shell after measurements.

Calculate:

28. The tangential velocity in D_{e_a} of the *RO*:

$$u_{e_a} = \frac{\pi \cdot D_{e_a} \cdot n}{60} = \frac{\pi \cdot 3.152 \cdot 400}{60} = 66.015 \text{ (m/s)}$$

29. The angle:

$$\theta_1 = 2 \cdot \arcsin \frac{a\bar{a}_4}{D_{e_a}} = 2 \cdot \arcsin \frac{1.578}{3.152} = 60.084^o$$

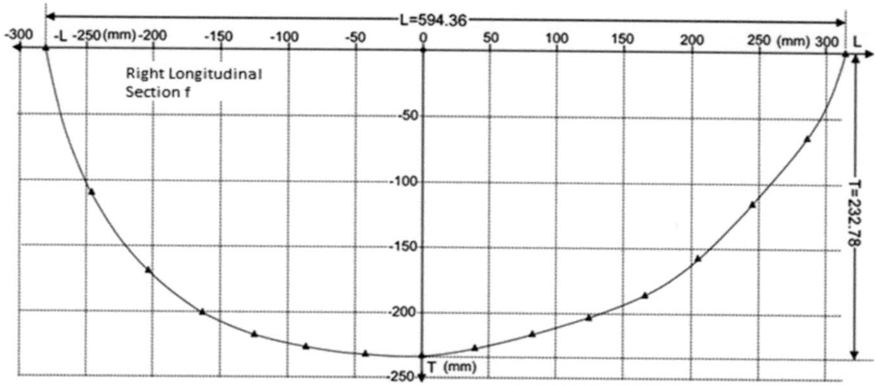

Fig. 12.2 Graph and equation of function $T = f(L)$, of the longitudinal section of the right part of the inner surface of the validated *PHIT RO* shell.

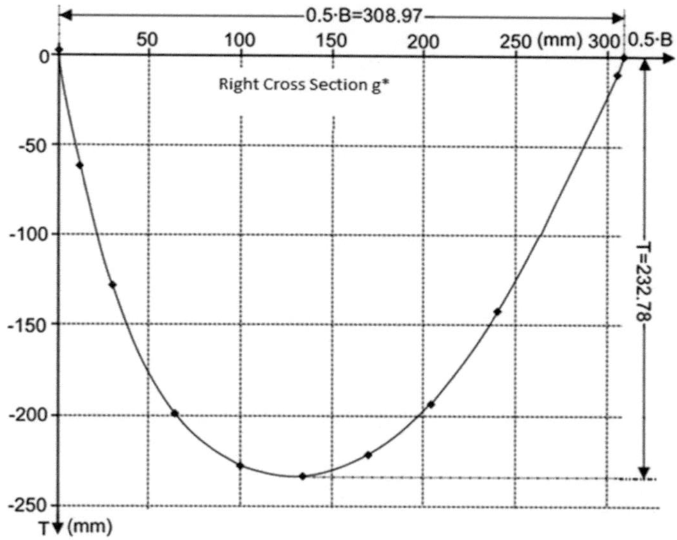

Fig. 12.3 Graph and equation of the function $T = f(0.5 \cdot B)$, of the cross section of the right part of the inner surface of the validation *THP RO* shell.

30. The arc:

$$\mathrm{arc}(aa_2) = \frac{u_{e_a}}{c_4} \cdot a\bar{a}_1 = \frac{66.023}{108.954} \cdot 0.789 = 0.478 \text{ (m)}$$

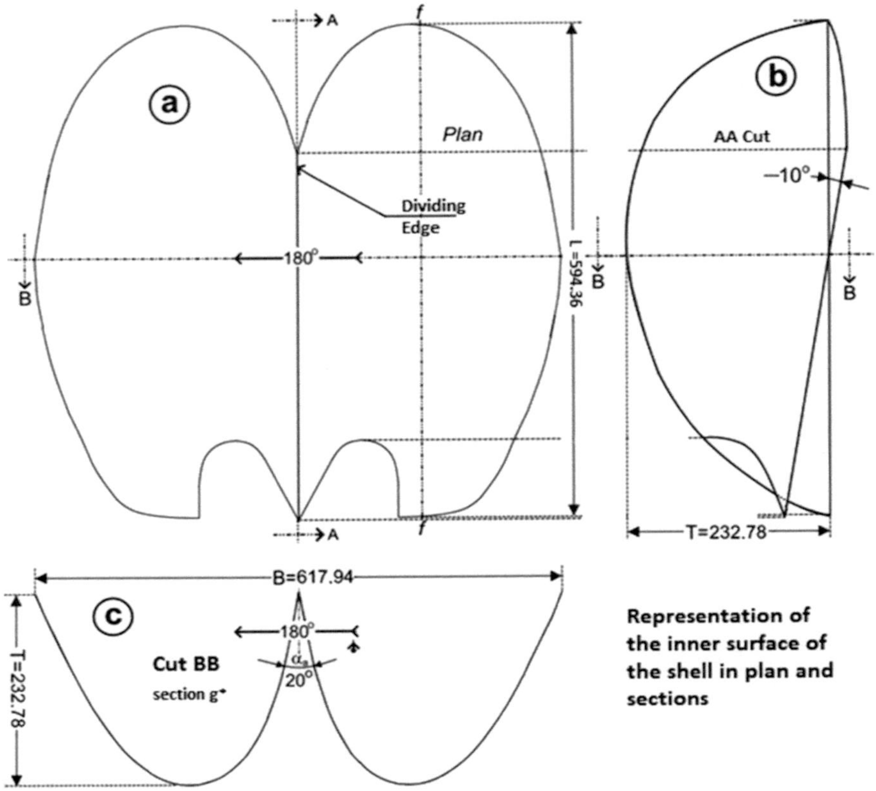

Fig. 12.4 Projections of the lines that delimit the blueprint and longitudinal and transversal section *f* and section $g*$

25.The angle:

$$\theta_2 = \frac{2 \cdot \mathrm{arc}(aa_2)}{D_{e_a}} = \frac{2 \cdot 0.478}{3.152} = 0.303299\ (\mathrm{rd}) = 0.303299 \cdot \frac{180}{\pi} = 17.378^o$$

29. The theoretical step between shells in degrees:

$$\alpha_{c*} = \theta_1 - 2 \cdot \theta_2 = 60.084 - 2 \cdot 17.378 = 25.328^o$$

30. The theoretical number of shells:

$$z_{c*} = \frac{360}{\alpha_{c*}} = \frac{360}{25.328} = 14.214 \text{ shells}$$

31. The length of the circumference, D_{e_a} :

$$C_{D_{e_a}} = \pi \cdot D_{e_a} = \pi \cdot 3.152 = 9.902 \text{ (m)}$$

32. The theoretical pitch between shells in meters:

$$p_{c*} = \frac{C_{D_{ea}}}{z_{c*}} = \frac{9.902}{14.214} = 0.697 \text{ (m)}$$

38. The limits of the actual step between the shells:

$$p_{c_{\min}} > r_{p_{c_{\min}}} \cdot p_{c*} > 0.7 \cdot 0.697 > 0.488 \text{ (m)} \rightarrow$$

$$p_{c_{\max}} < r_{p_{c\max}} \cdot p_{c*} < 0.8 \cdot 0.697 < 0.558 \text{ (m)}$$

39. The number of corresponding shells:

$$z_{c_{\max}} < \frac{C_{D_{ea}}}{p_{c_{\min}}} < \frac{9.902}{0.488} < 20.291$$

or

$$z_{c_{\max}} = 20 \text{ shells}$$

and

$$z_{c_{\min}} > \frac{C_{D_{ea}}}{p_{c_{\max}}} > \frac{9.902}{0.558} > 17.746.$$

or

$$z_{c_{\min}} = 18 \text{ shells}$$

40. The real steps between the shells, and we will continue with:

$$p_{c_{\max}} = \frac{C_{D_{ea}}}{z_{c_{\min}}} = \frac{9.902}{18} \cong 0.550\ (\mathrm{m}) \rightarrow$$

$$\alpha_{c_{\max}} = \frac{360}{z_{c_{\min}}} = \frac{360}{18} = 20^{o}$$

and

$$p_{c_{\min}} = \frac{C_{D_{ea}}}{z_{c_{\max}}} = \frac{9.902}{20} \cong 0.495\ (\mathrm{m}) \rightarrow \alpha_{c_{\min}} = \frac{360}{z_{c_{\max}}} = \frac{360}{20} = 18^{o}$$

The validation will continue with:

$$z_c = 20 \text{ shells}$$

$$\alpha_c = 18^{o}$$

$$p_c = 0.495\ (\mathrm{m})$$

41. The force acting on each half of the shell in the horizontal direction:

$$\begin{aligned} F_{c_h} &= 0,5 \cdot \rho \cdot Q_1 \cdot c_{4_1} \cdot (\cos \alpha_a/2 + \psi_5 \cdot \cos \beta_5) \cdot \cos \beta_5 \\ &= 0.5 \cdot 1000 \cdot 3.755 \cdot 115.176 \cdot (\cos 20/2 + 0.98 \cdot \cos 5) \cdot \cos 5 \\ &= 422455.681\ (\mathrm{N}) \end{aligned}$$

42. The length of the flow projection:

$$L_n = 1.26 \cdot d_{o_1} = 1.26 \cdot 20.4 = 25.704\ (\mathrm{cm})$$

43. The maximum shell thickness in the bezel or rib:

$$e_c^* = \frac{F_{C_h} \cdot f_s \cdot \sqrt{1 + \left(\frac{3 \cdot T}{L_n}\right)^2}}{0.6 \cdot \sigma_{esc} \cdot L_n} = \frac{422455.681 \cdot f_s \cdot \sqrt{1 + \left(\frac{3 \cdot 23.3}{25.704}\right)^2}}{0.6 \cdot 24623 \cdot 25.704} = 3.223 \cdot f_s$$

... with this we have:

$$e_{c_{\min}}^* = 3.223 \cdot f_{s_{\min}} = 3.223 \cdot 2 = 6.446 \text{ (cm)} \geq 64 \text{ (mm)}$$

and

$$e_{c_{\max}}^* = 3.222 \cdot f_{s_{\max}} = 3.223 \cdot 4 = 12.894 \text{ (cm)} \leq 129 \text{ (mm)}$$

Store these values for use in the shell design.

44. Establish a step-by-step route in sequence for drawing the position of the dividing edge of the shells, the circumference where their directions will be tangent, and the circumference where the direction of the projections of their exit lines will also be tangent (Table 12.7).

44.1. Choose a graphical design computer program, e.g., AutoCAD, then set the scale in millimeters, marking point *O* on the worksheet, and draw two orthogonal axes, thereby starting Fig. 12.5 in scale.

Table 12.7 Input data

H	H_{max}	H_{min}	Q	Q_{max}	Q_{min}	ρ	ρ_{ar}	ρ_m		
mca			m³/s			kg/m³				
630.00	640.00	627.00	14.52	15.02	13.75	1000.00	1.32	$7698 \cdot 10^{-6}$		
f_r	g	ε_t	λ_r	h_s	$\eta_{f\,2;4} = \eta_{f\,4;5}$	$\eta m = \eta el$	$\Phi_{0;5}$	Ψ_5	μ_{ad}	
Hz	m/s²	mm	–							
60.00	9.81	0.02	0.02	0.10	1.00	0.98	0.98	0.98	0.01	
β_c	β_5	θ_{co}	θ_{in}	δ_b	f_{se}	σ_{esc}	σf_{adm}	σt_{adm}	τcis_{adm}	$\tau tor\ adm$
degrees					–	N/cm²				
10	5	45	75	40	1.10	24623.00	14470.00	16744.00	9565.00	2500.00
Basic Geometries: Figs. 8.3a, 8.8, 8.12, 8.13, 8.15, 10.3, 10.4, 11.5, 11.6, 11.7, 11.8, 11.12, 11.15, 11.8 and 11.19, Table 10.1										

44.2. With the center at O, draw the circles:

$$D = 2523 \text{ (mm)}$$

$$D_{e_c} = 3152 \text{ (mm)}$$

$$D_{i_c} = 1962 \text{(mm)}$$

$$D_{e_a} = 3152 \text{ (mm)}$$

$$D_{i_a} = 2274 \text{ (mm)}$$

44.3. On the vertical axis, mark point B_1 meeting circumference D, drawing the jet axis through it.
44.4. From the jet axis, point B_1, draw horizontal line distant from $0.5 \cdot d_o = 0.5 \cdot 206 = 103$ (mm), obtaining the limits of the jet on the drawing, and points, B_2, B, and point A at point that meets the lower limit of the jet with circumference D_{e_a}.
44.5 Choose $D_{e_a}$and mark point A_1 connecting it with O
44.6. From A_1, draw a horizontal line that meets point A_3
44.7. From A_1, draw the *RAT* in contact with the inner surface of the shell, then with a circle that passes through points $A_1B_2A_3$and center at O_1, which was obtained by joining A_1 with B_2, determine midpoint M, and from this, draw a normal until finding the vertical axis of the coordinates, thus obtain the radius of the circle $r = \overline{O_1A_1}$.
44.8. Connect A_1 with O, marking the actual stored step to the right, in this case, $\alpha_c^o = 18^o \rightarrow p_c = 0.495 \text{ (m)} = 495 \text{ (mm)}$, thus obtaining point A_2 on the circumference D_{e_a}.
44.9. With a vertex at A_1, mark angle β_1 to the left of $\overline{AO}$, to start attempts for obtaining the dividing edge of the inner surface of the shells. Therefore, we recommend starting with $\beta_1 = \alpha_c$ in the validation as $\beta_1 = \alpha_c = 18^o$, an angle from which the script gave $\beta = 90^o$. If this does not occur, new attempts must be made with other angle values until this occurs, proceeding with the steps in sequence (Table 12.8)

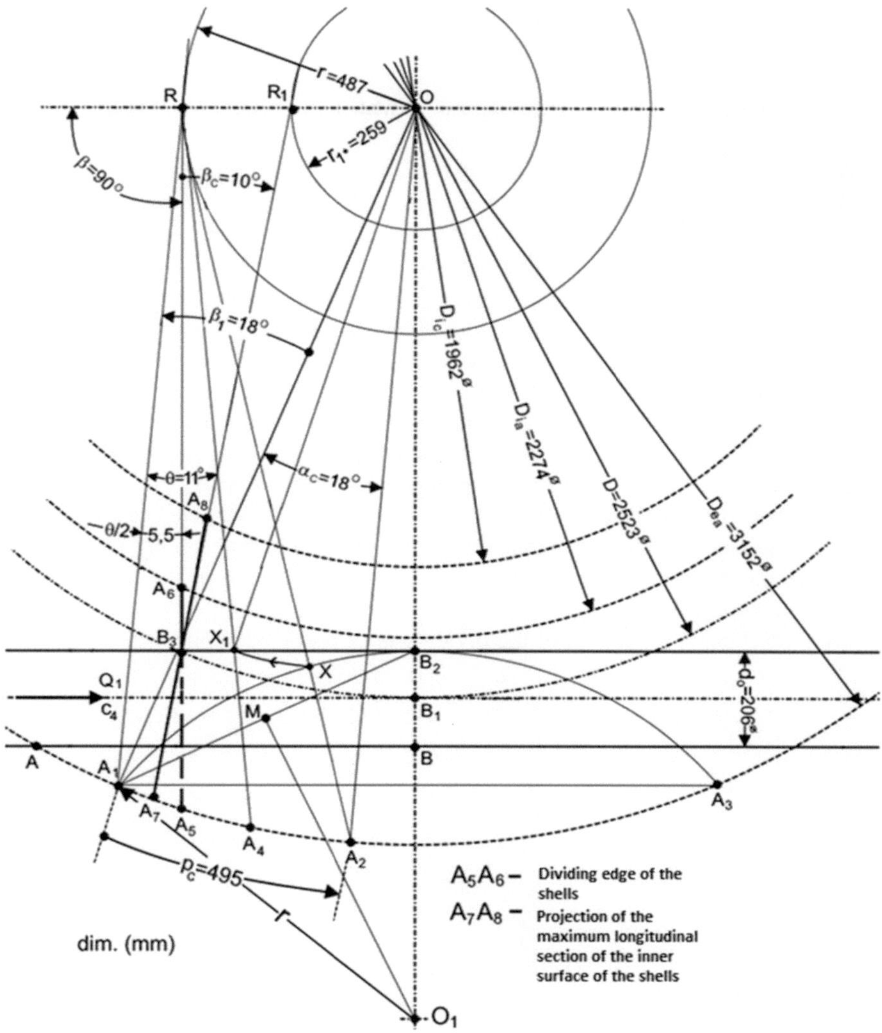

Fig. 12.5 Obtaining the dividing edge of the shells and the projection of the line of shell exit

44.10. The side of the marked angle from A_1 meets the axis passing through O at point R.

44.11. Measure radius r_1, in this case, $r_1 = 487$ (mm) on the scaled drawing.

44.12. With the center at O and radius r_1 draw a circumference.

44.13. From A_2 draw a tangent to the circle radius r_1 which meets the RAT at point X_1

44.14. Draw a circle arc that meets the projection of the upper limit of the jet at point X_1 whenever point X is not on the projection of the upper limit of the jet with center at O and radius OX.
44.15. Draw, a tangent to circle r_1 from X_1 which finds circle D_{e_a}at point A_4 in its extension.
44.16. Measure the angle θ formed by the tangents to the circle A_1 with vertices at A_4 and A_4, in this case, $\theta = 11^o$.
44.17. Draw the average line $\theta = 11^o$ containing the segment $\overline{RA_5}$ which meets the circles D_{i_a}, D, and $D_{e_a}$and respectively at points A_6, B_3, and A_5.
44.18. Measure the angle β that forms the average line drawn (segment $\overline{RA_5}$) with the horizontal axis passing through OR. If this angle is almost equal to $\beta = 90^o$, as was the case, continue, if not, return to step 44.3 changing the value of β_1 until this occurs, then continue.
44.19. From the average normal mark the angle fixed in the data $\beta_c = 10^o$, to the right, with a side that meets the horizontal axis OR at point R_1, then drawing a circle with center at O with the radius measured to scale, r_1, in this case $r_1 = 259$ (mm) with the center at O. In this circle radius r, all line projections of the maximum longitudinal sections of the shells will be tangent to it.
44.20. Mark the projection of the maximum section length of the shell longitudinal section, $L = \overline{A_7 A_8}$.
44.21. Calculate the length of the dividing edge and its components:

$$L_a = L_{A_5;A_6} = 0.5 \cdot (D_{e_a} - D_{i_a}) = 0.5 \cdot (3152 - 2264) = 439\ (\mathrm{m}) \rightarrow$$

$$L_{a_i} = L_{A_5;B_3} = 0.5 \cdot (D_{a_e} - D) = 0.5 \cdot (3152 - 2523) = 314.5\ (\mathrm{mm}) \rightarrow$$

$$L_{a_s} = L_{B_3;A_6} = 0.5 \cdot (D - D_{i_a}) = 0.5 \cdot (2523 - 1174) = 124.5\ (\mathrm{mm})$$

44.22 Calculate the length of the vertical projection of the inner surface of the shell:

$$L_{s_c} = 0.5 \cdot (D_{e_c} - D_{i_c}) = 0.5 \cdot (3150 - 1962) = 1188\ (\mathrm{mm})$$

45. The following shows the average internal diameter in the SA inlet tube, if not included in the stored data, if it is, skip this step:

Table 12.8 Hydrodynamic calculation

Y_{i0}	c_0	c_4	z_j	$Q=Q_1$	Q_2	Q_4	Q_6	Q_{4*}	Q_r	Q_f
J/kg	m/s		–	m^3/s						
6180.30	111.18	108.95	4.00	14.52	10.89	7.26	3.63	14.44	14.37	0.44

$n_T=n_G$	z_{pp}	$n_{qA\ 1}$	d_o	z_c	p_c	α_c	d_o	D	d_f	u	w_4	w_5
rpm	poles	–	mm	num.	mm	degrees	mm			m/s		
400.00	9.00	18.18	206.00	20.00	495.00	18	206.00	2,523.00	103.00	52.84	56.11	51.07

c_5	P_{i0}	$P_{i2}*$	P_{i4}	P_{ir}	P_e	P_m	P_{el}	$\eta_{i\ 0;2}*$	$\eta_{i\ 0;4}$	ηe	ηir
m/s	MW							%			
5.72	89.71	86.16	85.73	84.41	83.53	81.86	80.22	96.0	95.6	98.5	93.1

η_t	η_{GG}	H_1	$c_{4\ 1}$	Q_1*	d_{o1}	$c_{6;7}$	E_{ct}	H_{1*}	$e_{ci/a}$	$H1_{ci/a}$	L_a	L_{ai}	L_c
%		mca	m/s	m^3/s	mm	m/s	mca	m/s	mca		mm		
91.3	89.4	704.00	115.18	3.76	204.00	7.72	1.00	703.00	9.25	700.00	439.00	124.50	1188.00

$$d_{t_0} = d_{t_1} = 1.27 \cdot \frac{Q_{\max}^{0.4286}}{(1.3 \cdot H_{\max})^{0.1429}} = 1.27 \cdot \frac{15.02^{0.4286}}{(1.3 \cdot 640)^{0.1429}} = 1.552 \text{ (m)}$$

46. The average flow velocity, taken as being the same in the tube sections is:

$$c_t = \frac{4 \cdot Q}{\pi \cdot d_o^2} = \frac{4 \cdot 14.516}{\pi \cdot 1.552^2} = 7.673 \text{ (m/s)}$$

47. The flow in each tube section is:

$$Q_i = Q_{i-1} - Q_i / z_j \rightarrow$$

$$Q = Q_1 = 14.516 \left(\text{m}^3/\text{s}\right) \rightarrow$$

$$Q_2 = Q_3 = 14.516 - 14.516/4 = 10.887 \left(\text{m}^3/\text{s}\right) \rightarrow$$

$$Q_4 = Q_5 = 10.887 - 14.516/4 = 7.258 \left(\text{m}^3/\text{s}\right) \rightarrow$$

$$Q_6 = Q_7 = Q_8 = Q_{\text{in}} = Q/4 = 7.258 - 14.516/4 = 3.629 \left(\text{m}^3/\text{s}\right)$$

48. The internal diameter of each tube section is:

$$d_{t_i} = \sqrt{\frac{4 \cdot Q_i}{\pi \cdot c_0}} = \sqrt{\frac{4 \cdot Q_i}{\pi \cdot 7.673}} = 0.407354 \cdot \sqrt{Q_i} \text{ (m)}$$

$$d_{t_0} = d_{t_1} = \bar{E}e = 0.407354 \cdot \sqrt{14.516} = 1.552 \text{ (m)} \rightarrow$$

$$d_{t_2} = d_{t_3} = 0.407354 \cdot \sqrt{10.887} = 1.344 \text{ (m)} \rightarrow$$

$$d_{t_4} = d_{t_5} = 0.407354 \cdot \sqrt{7.258} = 1.097 \text{ (m)} \rightarrow$$

$$d_{t_6} = d_{t_8} = d_b = 0.407354 \cdot \sqrt{3.629} = 0.776 \text{ (m)}$$

49. The inner diameter of the *SCC* fixed cylinder is:

$$d_{ci} = 1.1 \cdot d_b = 1.1 \cdot 0.776 = 0.854 \text{ (m)}$$

50. The outer diameter of the *SCC* needle is:

$$d_a = 0.54 \cdot d_{ci} = 0.54 \cdot 0.854 = 0.461 \text{ (m)}$$

51. The diameter at the output of the *IN* is:

$$d_{in} = d_2 = 1.22 \cdot d_o = 1.22 \cdot 0.206 = 0.251 \text{ (m)}$$

52.The length of the *DI* is:

$$L_{di} = 4.072 \cdot (d_{ci} - d_b) = 4.072 \cdot (0.854 - 0.776) = 0.318 \text{ (m)}$$

53. The length of the cylindrical part of the needle of the *SCC* is:

$$L_{cii} = 4.83 \cdot d_o = 4.83 \cdot 0.206 = 0.995 \text{ (m)}$$

54. The complementary cylindrical length between the output of the *DI* and the input of the cylindrical part of the needle is:

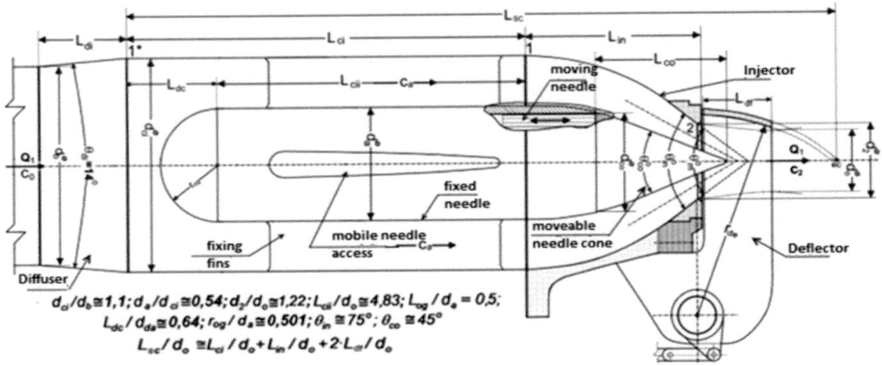

Fig. 12.6 Drawing of the *HFP* in the *SCC*

$$L_{dc} = 0.64 \cdot d_a = 0.64 \cdot 0.461 = 0.295 \text{ (m)}$$

55. The radius of the needle's ogive is:

$$r_{og} = 0.501 \cdot d_a = 0.5 \cdot 0.461 = 0.231 \text{ (m)}$$

56. The length of the cylindrical part of the *IN* is:

$$L_{ci} = L_{cii} + L_{dc} = 0.995 + 0.295 = 1.240 \text{ (m)}$$

57. The average flow velocity in the cylinder/needle crown of the *CS* is:

$$c_a = \frac{4 \cdot Q_{in}}{\pi \cdot \left(d_{ci}^2 - d_a^2\right)} = \frac{4 \cdot 3{,}629}{\pi \cdot \left(0.854^2 - 0.461^2\right)} = 8.941 \text{ (m/s)}$$

58. Choose a computer program for scaled drawings, e.g., AutoCAD with precision/accuracy to at least tenths of millimeters, and then proceed as follows in the program spreadsheet:

58.1. Fix any point on the worksheet (Fig. 12.6) drawing a horizontal axis from it to the right, thus starting the drawing of the *HFP* in the *SCC*.

58.2. From this point, using the horizontal axis as a reference, draw distant horizontal lines up and down, respectively from:

$$0.5 \cdot d_d = 0.5 \cdot 776 = 388\ (\text{mm}) \rightarrow$$

$$0.5 \cdot d_a = 0.5 \cdot 461 = 230.5\ (\text{mm}) \rightarrow$$

$$0.5 \cdot d_{ci} = 0.5 \cdot 854 = 427\ (\text{mm})$$

58.3. From the vertical line that passes through the point, set as the starting point, horizontally mark $L_{di} = 318$ (mm) and vertically draw at the right end:

$$L_{dc} = 295\ (\text{mm}) \rightarrow$$

$$L_{cii} = 945\ (\text{mm})$$

And above, horizontally as:

$$L_{ci} = 1240\ (\text{mm})$$

58.4. From the right end of the segment L_{ca} mark the angle $\theta_{co}/2 = 45/2 = 22.5^o$ obtaining, point F on the horizontal axis.

58.5 From point F, complete the cone that corresponds to the open *IN*.

58.6. Draw segment $0.5 \cdot d_2 = 0.5 \cdot 251 = 125.5$ (mm), from point F, vertically, up and down, the draw horizontally along the ends, obtaining point A and point A_1.

58.7. Starting at point A and point A_1 and the horizontal ones, marking angle $\theta_{\text{in}}/2 = 75/2 = 32.\ 5^o$ obtaining point B on the horizontal axis, and in the extension of the horizontals that delimit d_{ci} point C and point C_1.

Calculate:

58.8. The length of the *IN*:

$$L_{in} = \frac{0.5 \cdot d_a}{tg(0.5 \cdot \theta_{in})} = \frac{0.5 \cdot 461}{tg(0.5 \cdot 45)} \cong 556.5 = 557\ (mm)$$

58.9. The length of the *SCC*:

$$L_{\text{cci}} = L_{\text{ci}} + L_{\text{in}} = 1240 + 557 = 1797\ (\text{mm})$$

58.10. The needle stroke:

$$L_{\text{cu}} = \frac{0.5 \cdot d_2}{tg(0.5 \cdot \theta_{co})} = \frac{0.5 \cdot 251}{tg(0.5 \cdot 45)} = 303\ (\text{mm})$$

58.11. The length between the outlet section of the *SCC* and the inlet section, in the case of the deflector, *E*:

$$L_{\text{FE}} = L_{\text{ccl}} + L_{\text{di}} = 1797 + 318 = 2115\ (\text{mm})$$

58.12. The length of the horizontal projection of the *DE* head is $L_{df} = 1.5 \cdot d_o = 1.15 \cdot 206 = 237$ (mm), and mark point F $L_{ea} = 2 \cdot L_{df} = 2 \cdot 237 = 474$ (mm) to scale on the flow axis obtaining point ea, external to the dividing edge of the shell and the axis of the flow.

58.13. The total length of the jet shaft:

$$L_{\text{eaE}} = L_{\text{EF}} + L_{\text{ea}} = 2115 + 474 = 2589\ (\text{mm})$$

59. Choose a computer program for making the scale drawings, e.g., AutoCAD with precision/accuracy to at least tenths of millimeters, and then proceed as follows in the program spreadsheet to obtain the *HFP* in the *SA* of *PHIT* and later the calculation of the corresponding energy losses:

59.1. Fix point *O* in the worksheet (Fig. 12.7) drawing the circumferences of diameters, $D = 2523$ (mm) and $D_{e_a} = 3152$ (mm) to start the drawing of the *HFP* in the *SA* of the *PHIT*.

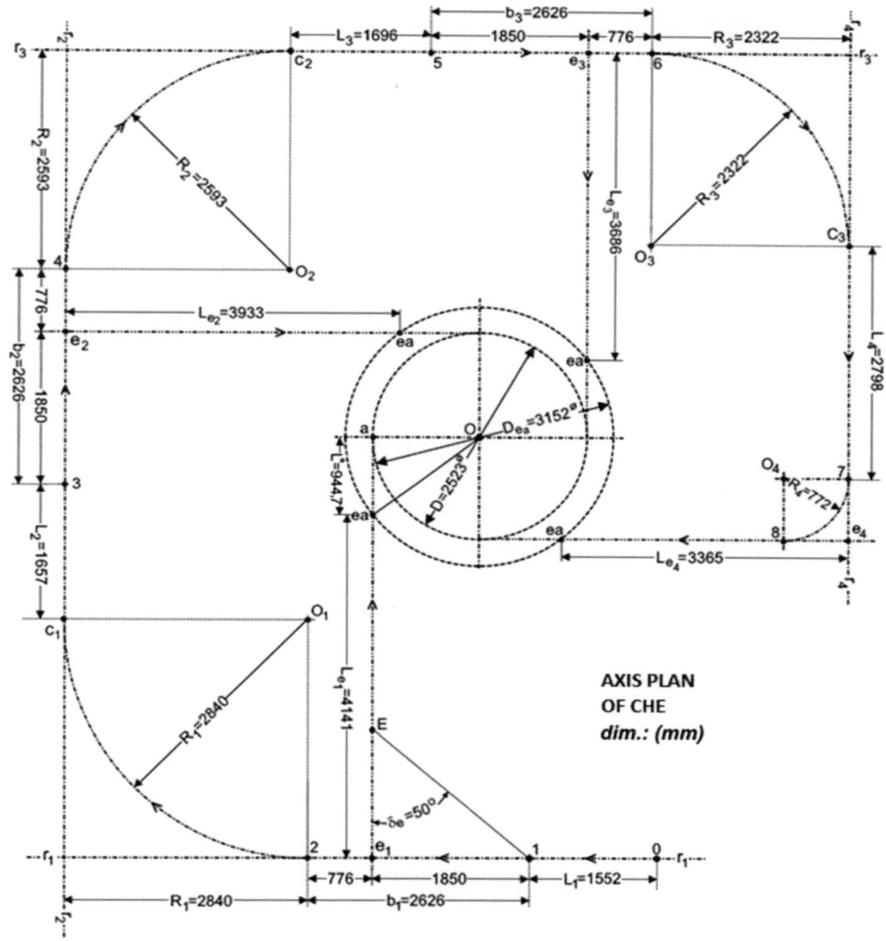

Fig. 12.7 To-scale blueprint, with the dimensions of the *HFP* axis of the flow in the *SA* of the *PHIT*

59.2. By tracing O two orthogonal axes and at the meeting of D, normal axes which meet D_{e_a} at point ea:

59.3. Calculate the lengths L_{e_j} between point ea, external to the dividing edges of the shells and the axis of flow, points e_j, $L_{e_j} = L_{eaE} + (\bar{E}e = d_t)_i = 2589 + d_{t_i}$ marking these distances from point ea on the normal axes, obtaining points e_1, e_2, e_3, e_4 on the axes of the flow, as:

$$L_{e_1} = 2589 + d_{t_1} = 2589 + 1552 = 4141\ (\text{mm}) \rightarrow$$

$$L_{e_1} = 2589 + d_{t_1} = 2589 + 1552 = 4141\ (\text{mm}) \rightarrow$$

$$L_{e_3} = 2589 + d_{t_4} = 2589 + 1097 = 3686\ (\text{mm}) \rightarrow$$

$$L_{e_4} = 2589 + d_{t_6} = 2589 + 776 = 3365\ (\text{mm})$$

59.4. Draw each point lines, r_1, r_2, r_3, r_4 normal to each L_e through e_j, which will form part of the directions of the flow axis.

59.5. From e_1 vertically mark $\bar{E} = d_{t_1} = 1552$ (mm), and with vertex at E, mark angle $\delta_e = 50^o$, a complement of angle $\delta_b = 40^o$, fixed for the bifurcation, obtaining point 1 on the axis.

Calculate

59.6. Mark segments:

$$\overline{e_1 1} = \overline{e_2 3} = \overline{e_3 5} = \overline{Ee_1} \cdot tg\, \delta_e = 1552 \cdot tg\, 50 = 1850\ (\text{mm})$$

… from points e_1, e_2, e_3 obtaining point 1, point 3, and point 5 on the axis.

59.7. Mark segments:

$$\overline{e_1 2} = \overline{e_2 4} = \overline{e_6 6} = 0.5 \cdot d_{t_1} = 0.5 \cdot 1552 = 776\ (\text{mm})$$

… from points e_1, e_2, e_3 obtaining point 2, point 4, and point 6 on the axis.

59.8. The length of the bifurcations:

$$b_1 = b_2 = b_3 = \overline{e_1 1} + \overline{e_1 2} = 1850 + 776 = 2626\ (\text{mm})$$

59.9. The side of triangle (O, e, aa):

$$L* = 0.5 \cdot \sqrt{D_{e_a}^2 - D^2} = 0.5 \cdot \sqrt{3152^2 - 2523^2} = 944.7\ (\text{mm})$$

59.10. At distance R_1(mm) from the normal line a to the axis (line r_1) passing through 2 and the line r_2:

$$R_1 = L* + L_{e_2} - \left(\overline{e_1 2} + 0.5 \cdot D\right) = 944.7 + 3933 - (776 + 0.5 \cdot 2523) = 2840\ (\text{mm})$$

59.11. Mark $R_1 = 2840$ (mm) normally to r_1, starting at point 2, obtaining center O_1, then draw the arc of circle $2c_1$ which is tangent to the line r_2 at c_1.

59.12. Calculate the straight length of the shaft:

$$L_2 = L_{e_1} + L* + 0.5 \cdot D - \left(R_1 + \overline{e_2 3}\right) = 4141 + 944.7 + 0.5 \cdot 2523 - (2840 + 1850) = 1657\ (\text{mm})$$

59.13. Calculate the radius:

$$R_2 = L* + L_{e_3} - \left(\overline{e_2 \cdot 3} + 0.5 \cdot D\right) = L* + L_{e_3} - \left(\overline{e_2 \cdot 3} + 0.5 \cdot D\right) = 944.7 + 3686 - (776 + 0.5 \cdot 2523) = 2593 \text{ (mm)}$$

59.14. Mark $R_2 = 2593$ (mm) normally to r_2 from point 4, obtaining center O_2, then draw the arc of circle $4c_2$ which is tangent to line r_3 at c_2.

59.15. Calculate the straight length of the shaft:

$$L_3 = L_{e_2} + L* + 0.5 \cdot D - \left(R_2 + \overline{e_3 5}\right) = = 3933 + 944.7 + 0.5 \cdot 2523 - (2593 + 1850) = 1696 \text{ (mm)}$$

59.16. The radius:

$$R_3 = L* + L_{e_4} - \left(\overline{e_3 \cdot 6} + 0.5 \cdot D\right) = 944.7 + 3365 - (776 + 0.5 \cdot 2523) = 2322 \text{ (mm)}$$

... then drawing the arc of circle $6c_3$ which is tangent to line r_4 at c_3.

59.17. Mark the radius $R_{4_{\min}}$ as $R_4 = d_{t_8} = 772$ (mm), drawing the arc of circle 87 with the center in O_4, which will be tangent to r_4 at point 7.

59.18. Mark length on the drawing:

$$L_4 = \overline{c_3 7} = L_{e_3} + L* + 0.5 \cdot D - (R_3 + R_4) = 3686 + 944.7 + 0.5 \cdot 2523 - (2322 + 772) = 2789 \text{ (mm)}$$

59.19. Draw to the right from point 1 on the flow axis at least distance, $L_1 = d_{t_1} = 1552$ (mm) to obtain point 0, and draw the flow axis, which will be a reference for starting (Fig. 12.8) corresponding to the *HFP* in the *SA* of the *PHIT*, which will be complemented by marking the internal dimensions of the components already established in scale, in of the parts are cast. In Fig. 12.9, the *HFP* design was drawn for parts that are manufactured in welded plates, which is more likely, given the sizes and weights.

60. The Reynolds number of the flow in the tube d_{t_1}:

$$N_{Re_{t_1}} = \frac{c_t \cdot d_{t_1}}{\nu} = 10^6 \cdot c_t \cdot d_{t_1} = 10^6 \cdot 7.673 \cdot 1.552 = 11.908496 \cdot 10^6$$

61. The energy loss coefficient of the flow in the tube d_{t_1}:

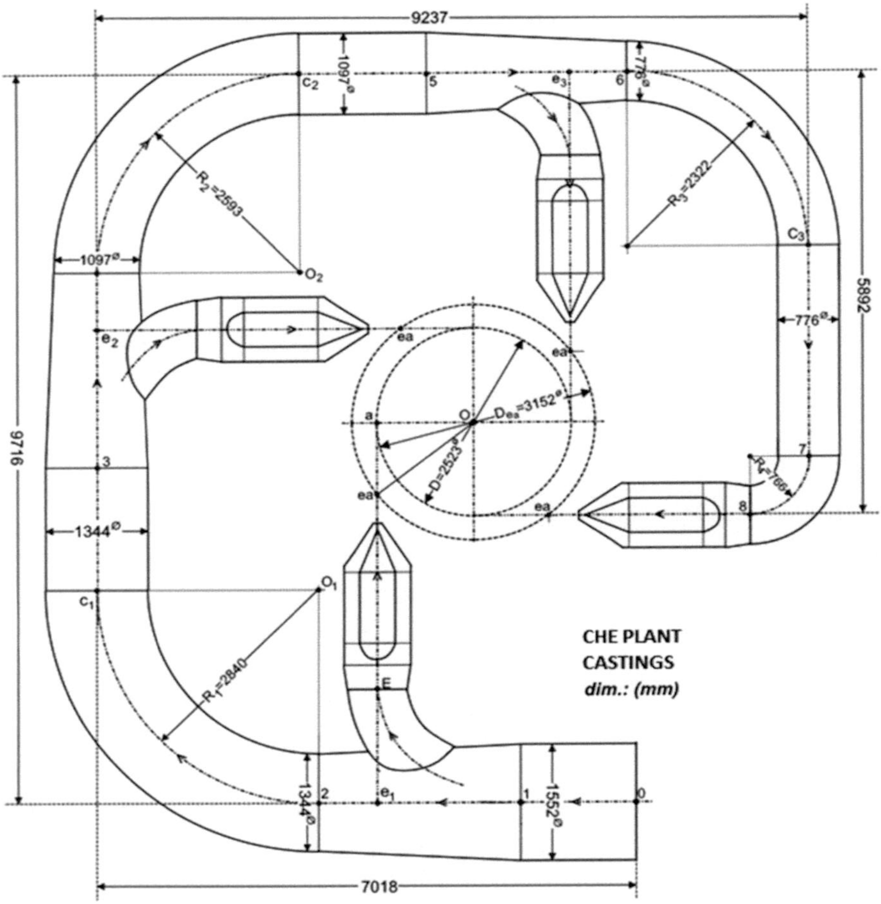

Fig. 12.8 To-scale blueprint, with the dimensions of the *HFP* in the *SA* of a *PHIT* with cast parts

$$\lambda_{t_1} = \frac{1.325}{\left[\ln\left(\frac{\varepsilon_t}{3.7 \cdot d_{t_1}} + \frac{5.74}{N_{Re_{t_1}}^{0.9}}\right)\right]^2} = \frac{1.325}{\left[\ln\left(\frac{0.02}{3.7 \cdot 1552} + \frac{5.74}{11908496^{0.9}}\right)\right]^2} = 0.00915008$$

62. The specific internal work of the flow lost in the tube d_{t_1}:

$$Y_{i_{p_{t_1}}} = f_{se} \cdot \lambda_{t_1} \cdot \frac{c_t^2}{2} \cdot \frac{L_1}{d_{t_1}} = 1.1 \cdot 0.00915008 \cdot \frac{14.516^2}{2} \cdot \frac{1.552}{1.552} = 1.060 \text{ (J/kg)}$$

63. The internal power of the flow lost in the tube d_{t_1}:

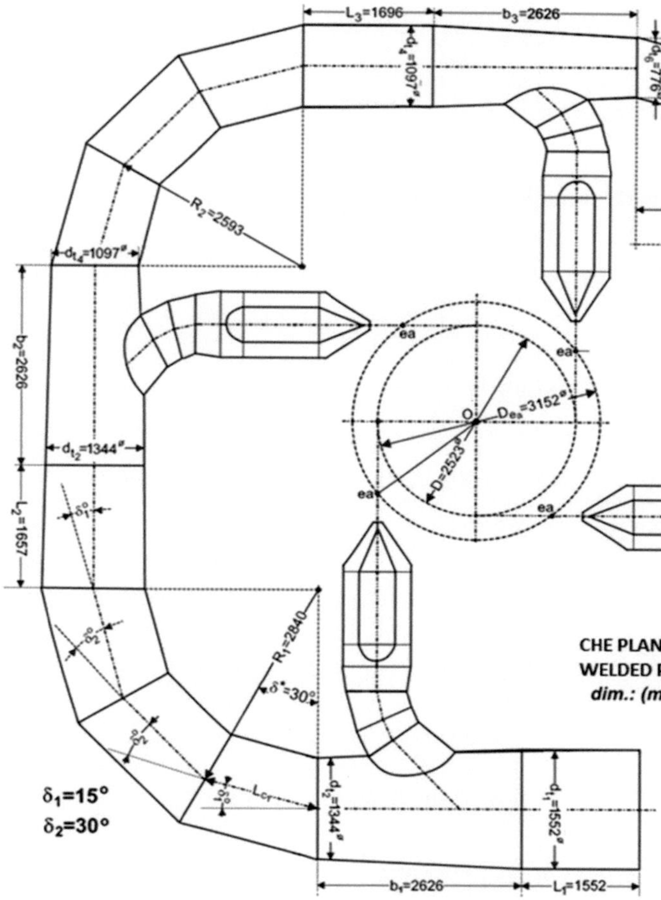

Fig. 12.9 To scale blueprint with the dimensions of the *HFP* in the *SA* of the *PHIT* with welded sheet metal parts

$$P_{i_{p_{t_1}}} = Q_1 \cdot Y_{i_{p_{t_1}}} = 14.516 \cdot 1.060 = 15.387 \text{ (kW)}$$

64. The Reynolds number of the flow in the tube d_{t_2}:

$$N_{Re_{t_2}} = 10^6 \cdot c_t \cdot d_{t_2} = 10^6 \cdot 7.673 \cdot 1.344 = 10.312512 \cdot 10^6$$

65. The energy loss coefficient of the flow in the tube d_{t_2}:

$$\lambda_{t_2} = \frac{1.325}{\left[\ln\left(\frac{\varepsilon_t}{3.7 \cdot d_{t_2}} + \frac{5.74}{N_{Re\,t_2}^{0.9}}\right)\right]^2} = \frac{1.325}{\left[ln\left(\frac{0.02}{3.7 \cdot 1344} + \frac{5.74}{10312512^{0.9}}\right)\right]^2} = 0.00936355$$

66. The equivalent length of the R_1 bend and the tube L_2:

$$L_{2_e} = \text{curve } R_1 + L_2 = 6 \cdot R_1 \cdot \ sin\, 0.5 \cdot \delta * + L_2$$

67. The specific internal work of the flow lost in the tube d_{t_2}:

$$Y_{i_{p_{t_2}}} = f_{se} \cdot \lambda_{t_2} \cdot \frac{c_t^2}{2} \cdot \frac{L_{2_e}}{d_{t_2}} = 1.1 \cdot 0.00936355 \cdot \frac{7.673^2}{2} \cdot \frac{6.067}{1.344} = 1.369\ (\text{J/kg})$$

68. The internal power of the flow lost in the tube d_{t_2}:

$$P_{i_{p_{t_2}}} = Q_2 \cdot Y_{i_{p_{t_2}}} = 10.886 \cdot 1.369 = 14.900\ (\text{kW})$$

69. The Reynolds number of the flow in the tube d_{t_4}:

$$N_{Re\,t_4} = 10^6 \cdot c_t \cdot d_{t_4} = 10^6 \cdot 7.673 \cdot 1.097 = 8.417281 \cdot 10^6$$

70. The energy loss coefficient of the flow in the tube d_{t_4}:

$$\lambda_{t_4} = \frac{1.325}{\left[\ln\left(\frac{\varepsilon_t}{3.7 \cdot d_{t_4}} + \frac{5.74}{N_{Re\,t_4}^{0.9}}\right)\right]^2} =$$

71. $= \frac{1.325}{\left[\ln\left(\frac{0.02}{3.7 \cdot 1097} + \frac{5.74}{8417281^{0.9}}\right)\right]^2} = 0.00967791$

72. The equivalent length of the R_2 bend and the tube L_3:

$$L_{3_e} = \text{curve } R_2 + L_3 = 6 \cdot R_2 \cdot sin(\delta * /2) + L_3 = 1.696 + 6 \cdot 2.593 \cdot sin\ 15$$
$$= 5.723\ (\text{m})$$

73. The specific work internal to the flow lost in the tube d_{t_4}:

$$Y_{i_{p_{t4}}} = f_{se} \cdot \lambda_{t4} \cdot \frac{c_t^2}{2} \cdot \frac{L_{3_e}}{d_{t4}} = 1.1 \cdot 0.00967791 \cdot \frac{7.673^2}{2} \cdot \frac{5.723}{1.097} = 1.635\ (\text{J/kg})$$

74. The internal power of the flow lost in the tube d_{t_4}:

$$P_{i_{pt_4}} = Q_4 \cdot Y_{i_{pt_4}} = 7.258 \cdot 1.635 = 11.863\ (\text{kW})$$

75. The Reynolds number of the flow in the tube d_{t_6}:

$$N_{Re_{t_6}} = 10^6 \cdot c_t \cdot d_{t_6} = 10^6 \cdot 7.673 \cdot 0.776 = 5.954248 \cdot 10^6$$

76. The energy loss coefficient of the flow in the tube d_{t_6}:

$$\lambda_{t6} = \frac{1.325}{\left[ln\left(\frac{\varepsilon_t}{3.7 \cdot d_{t_6}} + \frac{5.74}{N_{Re_{t6}}^{0.9}}\right)\right]^2} = \frac{1.325}{\left[ln\left(\frac{0.02}{3.7 \cdot 776} + \frac{5.74}{5954284^{0.9}}\right)\right]^2} = 0.0102519223$$

77. The equivalent length of bend R_3, pipe L_4 and bend R_4,:

$$L_{4_e} = curve\ R_3 + L_4 + curve\ R_4 = 6 \cdot R_3 \cdot sin(\delta * /2) + L_4 + 6 \cdot R_4$$
$$\cdot sin(\delta * /2) = 6 \cdot 2.322 \cdot sin\ 15 + 2.798 + 6 \cdot 0.776 \cdot sin\ 15 = 7.609\ (\text{m})$$

78. The specific internal work of the flow lost in the tube d_{t_6}:

$$Y_{i_{p_{t6}}} = f_{se} \cdot \lambda_{t6} \cdot \frac{c_t^2}{2} \cdot \frac{L_{4_e}}{d_{t6}} = 1.1 \cdot 0.0102519223 \cdot \frac{7.673^2}{2} \cdot \frac{7.609}{0.776} = 3.255 \text{ (J/kg)}$$

79. The internal power of the flow lost in the tube d_{t_6}:

$$P_{i_{p_{t6}}} = Q_6 \cdot Y_{i_{p_{t6}}} = 3.629 \cdot 3.255 = 11.813 \text{ (kW)}$$

80. The sum of the specific internal works lost in the tubes without considering deflections losses:

$$Y_{i_{p_t}} = Y_{i_{p_{t1}}} + Y_{i_{p_{t2}}} + Y_{i_{p_{t4}}} + Y_{i_{p_{t6}}} = 1.060 + 1.396 + 1.635 + 3.255$$
$$= 7.346 \text{(J/kg)}$$

81. The sum of the internal powers lost in the tubes without considering deflections losses:

$$P_{i_{p_t}} = P_{i_{p_{t1}}} + P_{i_{p_{t2}}} + P_{i_{p_{t4}}} + P_{i_{p_{t6}}} = 15.387 + 14.900 + 11.863 + 11.813$$
$$= 53.963 \text{(kW)}$$

82. The coefficient of internal energy loss due to flow deflections in the curves of each deflection:

$$\lambda_{d_j} = 0.9457 \cdot sin^2(\delta_j/2) + 2.05 \cdot sin^4(\delta_j/2).$$

In this case, each 90° curve has 4 deflections, 2 at 15° and 2 at 30°, which results in:

$$\lambda_{d_j} = \lambda_{d_1} = \lambda_{d_2} = \lambda_{d_3} = \lambda_{d_4} = \lambda_d$$
$$= 2 \cdot \Big[0.9457 \cdot sin^2(15/2) + 2.05 \cdot sin^4(15/2) + 0.9457 \cdot sin^2(30/2)$$
$$+ 2.05 \cdot sin^4(30/2)\Big]$$
$$= 2 \cdot 0.0892558507 = 0.178512$$

83. The specific internal work lost by deviaiton:

$$Y_{i_{p_{d_1}}} = Y_{i_{p_{d_2}}} = Y_{i_{p_{d_3}}} = Y_{i_{p_{d_4}}} = Y_{i_{p_{dc}}}$$

or

$$Y_{i_{p_{d_1}}} = f_{se} \cdot \lambda_d \cdot \frac{c_t^2}{2} = 1.1 \cdot 0.178512 \cdot \frac{7.673^2}{2} = 5.780\ (\mathrm{J/kg})$$

84. The specific internal work lost by deflection, in the 4 curves:

$$Y_{i_{p_{4d}}} = 4 \cdot Y_{i_{p_{dc}}} = 4 \cdot 5.78 = 23.122\ (\mathrm{J/kg})$$

85. The internal power of the flow lost by deflection, in the 4 curves:

$$P_{i_{p_{4d}}} = (Q_2 + Q_4 + 2 \cdot Q_6) \cdot Y_{i_{p_{dc}}} = (10.887 + 7.258 + 2 \cdot 3.629) \cdot 5.78$$
$$= 146.829\ (\mathrm{kW})$$

86. The specific internal work of the flow lost at each bifurcation in the straight section:

$$Y_{i_{p_{br}}} = Y_{i_{p_{b_1}}} = Y_{i_{p_{b_2}}} = Y_{i_{p_{b3}}} = f_{se} \cdot \lambda_r \cdot \frac{c_t^2}{2} = 1.1 \cdot 0.02 \cdot \frac{7.673^2}{2} = 0.648\ (\mathrm{J/kg})$$

87. The specific internal work of the flow lost in the 3 bifurcations in the direction of the stretch with the greatest energy loss:

$$Y_{i_{p_{3br}}} = 3 \cdot Y_{i_{p_{br}}} = 3 \cdot 0.650 = 1.950\ (\mathrm{J/kg})$$

88. The internal flow power lost in the 3 bifurcations in the straight sections, considering the inflows:

$$P_{i_{p_{3br}}} = (Q_1 + Q_2 + Q_4) \cdot Y_{i_{p_r}} = (14.516 + 10.887 + 7.258) \cdot 0.648 =$$
$$= 21.164\ (\mathrm{kW})$$

89. The average diameter of the injection system inlet *DF*:

$$d_{di_m} = 0.5 \cdot (d_{tb} + d_{ci}) = 0.5 \cdot (0.776 + 0.854) = 0.815 \text{ (m)}$$

90. The Reynolds number of the flow at the *DF* of the injection system inlet:

$$N_{Re\,di} \cong 1.26 \cdot 10^6 \cdot \frac{Q_6}{d_{di_m}} = 1.26 \cdot 10^6 \cdot \frac{3.629}{0.815} = 5.610478 \cdot 10^6$$

91. The energy loss coefficient of the flow in the *DF* of the injection system inlet:

$$\lambda_{di} = \frac{1.325}{\left[ln\left(\frac{\varepsilon_t}{3.7 \cdot d_{di_m}} + \frac{5.74}{N_{Re\,di}^{0.9}}\right)\right]} = \frac{1.325}{\left[ln\left(\frac{0.02}{3.7 \cdot 815} + \frac{5.74}{5610478^{0.9}}\right)\right]^2} = 0.010232931$$

92. The specific internal work of the flow lost in the *DF* of the injection system inlet

$$Y_{i_{P_{di}}} = f_{se} \cdot 0.8106 \cdot \lambda_{di} \cdot Q_6^2 \cdot L_{di} \cdot d_{di_m}^{-5} = 1.1 \cdot 0.8106 \cdot 0.010232931 \cdot 3.629^2 \cdot 0.318 \cdot 0.815^{-5} = 0.106 \text{ (J/kg)}$$

93. The internal flow power lost at the injection system input *DF*:

$$P_{i_{P_{di}}} = Q_6 \cdot Y_{i_{P_{di}}} = 3.629 \cdot 0.106 = 0.385 \text{ (kW)}$$

94. The area of the transversal section of the stretch of the cylinder/needle of the *SCC*:

$$A_{ci/a} = \frac{\pi}{4} \cdot \left(d_{ci}^2 - d_a^2\right) = \frac{\pi}{4} \cdot \left(0.584^2 - 0.461^{2)} = 0.40588978 \left(\text{m}^2\right)$$

95. The hydraulic diameter of the cylinder/needle section of the *SCC*:

$$d_h = d_{ci} - d_a = 0{,}854 - 0{,}461 = 0.393 \text{ (m)}$$

96. The flow in the cylinder/needle section of the *SCC*:

$$N_{Re_{ci;a}} \cong 1.26 \cdot 10^6 \cdot \frac{Q_8}{d_h} = N_{Re_{ci;a}} \cong 1.26 \cdot 10^6 \cdot \frac{3.629}{0.393} = 11.634962$$

97. The coefficient of loss of internal energy of the flow in the cylinder/needle section:

$$\lambda_{ci;a} = \frac{1.325}{\left[ln\left(\frac{\varepsilon_t}{3.7 \cdot d_h} + \frac{5.74}{N^{0.9}_{Re_{ci;a}}}\right)\right]} = \frac{1.325}{\left[ln\left(\frac{0.02}{3.7 \cdot 393} + \frac{5.74}{11634962^{0.9}}\right)\right]^2} = 0.01089779$$

98. The specific internal work of the flow lost in the cylinder/needle section:

$$Y_{i_{Pc;a}} = f_{se} \cdot 0.8106 \cdot \lambda_{ci;a} \cdot Q_8^2 \cdot L_{ci} \cdot d_h^{-5} = 1.1 \cdot 0.8106 \cdot 0.01089779 \cdot 3.629^2 \cdot 1.24 \cdot 0.393^{-5} = 16.927\ (\text{J/kg})$$

99. The internal specific power of the flow lost in the cylinder/needle section:

$$P_{i_{Pci;a}} = Q_8 \cdot Y_{i_{Pci;a}} = 3.629 \cdot 16.927 = 61.427\ (\text{kW})$$

100. The mean angle of the *IN*:

$$\theta_{m_{in}} = 0.5 \cdot (\theta_{in} + \theta_{co}) = 0.5 \cdot (75 + 45) = 60^o$$

101. The coefficient of loss of the internal energy of the flow in each *IN*:

$$\lambda_{in} = 0.0025 \cdot \left[1 - \left(\frac{d_2}{d_{ci}}\right)^2\right] \cdot cot\,g(0.5 \cdot \theta_{m_{in}}) = 0.0025 \cdot \left[1 - \left(\frac{0.251}{0.854}\right)^2\right] \cdot cot\,g(0.5 \cdot 60) = 0.001318692$$

102. The specific internal work of the flow lost in each *IN*:

$$Y_{i_{Pin}} = f_{se} \cdot \lambda_{in} \cdot \frac{c_a^2}{2} = 1.1 \cdot 0.001318692 \cdot \frac{9.941^2}{2} = 0.072\ (\text{J/kg})$$

103. The internal specific power of the flow lost in each *IN*:

$$P_{i_{Pin}} = Q_8 \cdot Y_{i_{Pin}} = 3.629 \cdot 0.072 = 0.261 \text{ (kW)}$$

104. The sum of the specific internal works of the flow in the *SA* with the highest loss in the *PHIT*:

$$Y_{i_{PSA}} = Y_{i_{Pt}} + Y_{i_{P4d}} + Y_{i_{P3_{br}}} + Y_{i_{Pdi}} + Y_{i_{Pc;a}} + Y_{i_{Pin}} = 7.346 + 23.122 + 1.95$$
$$+ 0.106 + 16.927 + 0.072 = 6.666 + 22.315 + 1.950 + 0.106 + 16.927$$
$$+ 0.072 = 49.523 \text{ (J/kg)}$$

105. The sum of the internal powers of the flow in the *SA* with the highest loss in the *PHIT*:

$$P_{i_{PSA}} = P_{i_{Pt}} + P_{i_{P4d}} + P_{i_{P3_{br}}} + P_{i_{Pdi}} + P_{i_{Pc;a}} + P_{i_{Pin}} = = 53.963 + 146.829$$
$$+ 21.164 + 0.385 + 61.427 + 0.261 = 284.029 \text{ (kW)}$$

106. The specific internal flow work available at the inlet section 0 of the *PHIT*,

$$Y_{i_0} = g \cdot H_0 = 9.81 \cdot 630 = 6180.300 \text{ (J/kg)}$$

107. The specific internal work of the flow available at the exit of the *SA* with the highest loss in the *PHIT* which is taken as being equal to the entrance of the outer end of the edge of the shells:

$$Y_{i_{2*}} = Y_{i_{SA}} = Y_{i_4} = Y_{i_0} - Y_{i_{PSA}} = 6180.3 - 49.523 = 6130.777 \text{ (J/kg)}$$

108. The internal power available from the flow at inlet section 0 of the *PHIT*:

$$P_{i_0} = Q \cdot Y_{i_0} = 14.516 \cdot 6180.3 = 89713.235 \text{ (kW)}$$

109. The available internal power of the flow at the inlet section 0 of the *PHIT*, per jet,

$$P_{i_{0/z_j}} = \frac{P_{i_0}}{z_j} = \frac{89713.235}{4} = 22428.309 \text{ (kW)}$$

110. The internal flow power available at the exit of the highest loss *SA* in the *PHIT*,

$$P_{i_{2*}} = P_{i_{SA}} = P_{i_{0/z_j}} - P_{i_{P_{SA}}} = 22428.309 - 284.029 = 22144.280 \text{ (kW)}$$

111. The average velocity of the flow in the exit section of the *SA* in the section with greater losses in the *PHIT*, which is taken as being equal to the entrance of the external end of the edge of the shells:

$$c_{SA} = c_{2*} = c_{4*} = \sqrt{2 \cdot Y_{i_{2*}}} = \sqrt{2 \cdot 6130.777} = 110.732 \text{ (m/s)}$$

112. The internal loss coefficient of the average flow velocity calculated in the highest loss *SA* in the *PHIT*:

$$\phi_{0;4*} = \frac{c_{4*}}{c_0} = \frac{110.732}{111.179} = 0.9960$$

Continue if:

$$0.999 \cdot \phi_{0;4} \leq \phi_{0;4*} \leq 1.001 \cdot \phi_{0;4}$$

If:

$$\phi_{0;4*}$$

… is out of range, print the data calculated up to step 112, including the following constant, and continue after choosing one of the propositions in sequence.

112.1. Continue with:

$$\phi_{0;4*} = \phi_{0;4}$$

… in the validation:

$$\phi_{0;4*} = \phi_{0;4} = 0,98$$

… with the hypothesis that the kWh generated can be increased, and the corresponding energy generated can be increased, then facturing whenever there is water availability greater than Q (m^3/s) in:

$$Q* = \frac{c_{4*}}{c_4} \cdot Q\ \left(\mathrm{m}^3/\mathrm{s}\right).$$

Or here as:

$$Q* = \frac{110.732}{108.954} \cdot Q = 1.01632 \cdot 14.516 = 14.753\ \left(\mathrm{m}^3/\mathrm{s}\right)$$

112.2. Continue with $\phi_{0;\,4}$fixed in the data from step 3, reduce diameter $d_{t0;\,1}$ in step 45, recalculate it until in step 111.2 and obtain $c_{4*} \cong c_4$, then continue. This reduces the fixed costs that affect the *SA*.

112.3. With $\phi_{0;\,4*}$ obtained in step 112 which is different (higher or lower) than $\phi_{0;\,4}$, set it in the data, return to step 3, recalculate it until reaching $0.999 \cdot \varphi_{0;\,4} \leq \varphi_{0;\,4*} \leq 1.001 \cdot \varphi_{0;\,4}$, then continue. Here, hypothesis 112.1 will be adopted to continue.

113. Recalculate the specific internal work of the flow available at the exit of the *IN* in the section with the highest energy loss, which is equal to the internal work at the entrance of the shells:

$$Y_{i_{2*}} = Y_{i_4} = \frac{c_4^2}{2} = \frac{108.954^2}{2} = 5935.487\ (\mathrm{J/kg})$$

114. Recalculate the specific internal work lost from the flow available at the exit of the *IN* in the section with the highest energy loss that is equal to the internal one at the entrance of the shells:

$$Y_{i_{P0;\,2*}} = Y_{i_{P0;\,4}} = Y_{i_0} - Y_{i_4} = 6180.3 - 5935.487 = 244.813\ (\mathrm{J/kg})$$

115. The internal flow power lost between sections 0 and 2* of the 4 *IN*:

$$P_{i_{P0;\,2*}} = Q \cdot Y_{i_{P0;\,2*}} = 14.516 \cdot 244.813 = 3553.706\ (\mathrm{kW})$$

116. The internal flow power available at the entrance to the *PHIT*:

$$P_{i_0} = Q \cdot Y_{i_0} = 14.516 \cdot 6180.3 = 89713.235\ (\mathrm{kW})$$

117. The internal power of the flow available at the output of the 4 *IN*:

$$P_{i_{2*}} = P_{i_0} - P_{i_{p0;2*}} = 89713.235 - 3553.706 = 86159.529 \text{ (kW)}$$

118. The *SA* yield calculated, considering the 4 *IN*:

$$\eta_{i_{0;2*}} = \frac{P_{i_{2*}}}{P_{i_0}} = \frac{86159.529}{89713.235} = 0.9604 = 96.04\ (\%)$$

119. The flow run available at the entrance of the shells:

$$Q_{4*} = \eta_{v_{2;4}} \cdot Q = 0.995 \cdot 14.516 = 14.443\ \left(\text{m}^3/\text{s}\right)$$

120. The internal flow power available at the entrance to edge 4 of the shells:

$$P_{i_4} = Q_4 \cdot Y_{i_4} = 14.443 \cdot 5935.487 = 85726.239 \text{ (kW)}$$

121. The internal flow power lost to leakage in the injector-shell free flow:

$$P_{i_{p2*;4}} = P_{i_{2*}} - P_{i_4} = 86159.529 - 85726.239 = 4333.290 \text{ (kW)}$$

122. The internal yield of the flow from this inlet section 0 from the *PHIT* to the entrance at edge 4 of the shells of the *RO*:

$$\eta_{i_{0;4}} = \frac{P_{i_4}}{P_{i_0}} = \frac{85726.239}{89713.235} = 0.9556 = 95.56\ (\%)$$

123. The internal flow yield from the *IN* outlet to the *RO* shell inlet:

$$\eta_{i_{2*;4}} = \frac{P_{i_4}}{P_{i_{2*}}} = \frac{85726.239}{86159.529} = 0.9950 = 99.50\ (\%)$$

124. The relative average velocity of the flow at the entrance of the shells:

$$w_4 = c_4 - u = 108.954 - 52.844 = 56.110\ (\text{m/s})$$

125. The specific internal work of the flow lost in the *RO*:

$$Y_{i_{p_r}} = 0.5 \cdot \left(1 - \psi_5^2\right) \cdot w_4^2 = 0.5 \cdot \left(1 - 0.98^2\right) \cdot 56.110^2 = 62.337 \text{ (J/kg)}$$

126. The specific internal work of the *RO* flow:

$$Y_{i_r} = Y_{i_4} - Y_{i_{p_r}} = 5935.487 - 62.337 = 5873.150 \text{ (J/kg)}$$

127. The flow available for use by the *RO* shells:

$$Q_r = \eta_{v4;5} \cdot Q_{4*} = 0.995 \cdot 14.443 = 14.371 \text{ (m/s)}$$

128. The internal power lost from the flow in the *RO*:

$$P_{i_{p_r}} = Q_r \cdot Y_{i_{p_r}} = 14.371 \cdot 62.337 = 84406.114 \text{ (kW)}$$

129. The internal power available in the *RO*:

$$P_{i_r} = Q_r \cdot Y_{i_r} = 14.371 \cdot 5873.364 = 84406.114 \text{ (kW)}$$

130. The internal yield of the *RO*:

$$\eta_{i_r} = \frac{P_{i_r}}{P_{i_4}} = \frac{84406.114}{85726.239} = 0.9846 = 98.46 \text{ (\%)}$$

131. The internal yield of the *PHIT* from 0, to the exit edge 5 of the shells:

$$\eta_{i0;5} = \frac{P_{i_r}}{P_{i_0}} = \frac{84406.114}{89713.235} = 0.9408 = 94.08 \text{ (\%)}$$

132. The relative average speed at the exit of the shells:

$$w_5 = \psi_5 \cdot w_4 = 0.98 \cdot 52.112 = 51.070 \text{ (m/s)}$$

133. The average absolute velocity at the exit of the shells:

$$c_5 = \sqrt{u^2 + w_5^2 - 2 \cdot u \cdot w_5 \cdot cos\,\beta_5} =$$
$$= \sqrt{52.844^2 + 51.070^2 - 2 \cdot 52.844 \cdot 51.070 \cdot cos\,6} = 5.720\ (\mathrm{m/s})$$

134. The specific internal work not used by *PHIT* in exiting the shells:

$$Y_{i_5} = 0.5 \cdot c_5^2 = 0.5 \cdot 5.720^2 = 16.357\ (\mathrm{J/kg})$$

135. The internal power not used by the *PHIT* at the exit of the shells:

$$P_{i_{p5}} = Q \cdot Y_{i_5} = 14.516 \cdot 16.357 = 237.438\ (\mathrm{kW})$$

136. The relative energy loss at the *RO* output:

$$\eta_{i_5} = \frac{P_{i_{p5}}}{P_{i_0}} = \frac{237.438}{89713.235} = 0.0026 = 0.26\ (\%)$$

137. Ventilation power losses using IEC/*IEC* 41 Formula for vertical axis *PHIT ROs* with housings, noting that, according to the CEI/*IEC* 41 Standard, this loss can be ±50 (%).

$$P_{p_{ve}} = 4.62607 \cdot 10^{-3} \cdot n^3 \cdot D_{e_c}^5 = 4.62607 \cdot 10^{-3} \cdot \left(\frac{400}{60}\right)^3 \cdot 3.150^5$$
$$= 425.100\ (\mathrm{kW})$$

Considering the maximum error:

$$P_{p_{ve}} = 1.5 \cdot 425 = 637.500\ (\mathrm{kW})$$

138. The energy loss from output 5 of the *RO* shells to the *PHIT* rotating system:

$$P_{p_{5;sg}} = P_{i_{p5}} + P_{p_{ve}} = 237.438 + 637.500 = 874.938\ (\mathrm{kW})$$

139. The power of *PHIT* delivered to the rotating system:

$$P_{sg} = P_e = P_{i_r} - P_{p_{5;sg}} = 84406.114 - 874.983 = 83531.176\ (\text{kW})$$

140. The relative energy loss from the exit of the *RO* shells to the rotating system of the *PHIT*,

$$\eta_{5;sg} = \frac{P_{sg}}{P_{i_r}} = \frac{83531.176}{84406.114} = 0.9896 = 98.96\ (\%)$$

141. The yield of the *PHIT* from 0 to the rotating system.

$$\eta_{i0;sg} = \eta_e = \frac{P_e}{P_{i_0}} = \frac{83531.176}{89713.235} = 0.9311 = 93.11\ (\%)$$

142. The mechanical power in the axis of the *PHIT* including its rotating system:

$$P_m = P_e \cdot \eta_m, P_m = 83531.176 \cdot 0.98 = 81860.552\ (\text{kW})$$

143. The total yield of the *PHIT*, including the rotation system:

$$\eta_t = \frac{P_e}{P_{i_0}} \cdot \frac{P_m}{P_e} = \eta_e \cdot \eta_m = 0.9311 \cdot 0.98 = 0.9125 = 91.25\ (\%)$$

144. The electrical power at the output of the *GG*, including the rotating system:

$$P_{\text{el}} = P_m \cdot \eta_{\text{el}}, P_{\text{el}} = 81860.552 \cdot 0.98 = 80223.341\ (\text{kW})$$

145. Power losses in the electrical rotating system:

$$P_{p_{el}} = P_m - P_{el} = 81860.552 - 80223.341 = 1637.211\ (\text{kW})$$

146. The yield of the *GG* with the *PHIT*:

$$\eta_{\text{GG}} = \frac{P_{\text{el}}}{P_{i_0}} = \frac{80223.341}{89713.235} = 0.8942 = 89.42\ (\%)$$

or

$$\eta_{\mathrm{GG}} = \eta_t \cdot \eta_{\mathrm{el}} = 0.9125 \cdot 0.98 = 0.8946 = 89.42\ (\%)$$

147. The inner radius of the *DF*:

$$r_{\mathrm{de}} = 2.6 \cdot d_2 = 2.65 \cdot 0.251 = 0.653\ (\mathrm{m})$$

148. The width of the *DF*:

$$b_d = 0.927 \cdot d_2 = 0.927 \cdot 0.251 = 0.233\ (\mathrm{m})$$

149. The turning angle of the *DF*:

$$\theta_g = \arcsin \frac{b_d}{r_{de}} = \arcsin \frac{0.233}{0.653} = 20.905^o = \frac{\pi}{180} 20.905 = 0.365\ (\mathrm{rd})$$

150. The length of the inner arc of the *DF*:

$$L_d = \theta_g \cdot r_{\mathrm{de}} = 0.365 \cdot 0.653 = 0.238\ (\mathrm{m})$$

151. The inner radius of the *DF*:

$$r_i = 0.48 \cdot d_{\mathrm{in}} = 0.48 \cdot 0.251 = 0.121\ (\mathrm{m})$$

152. The distance between the parallels of the *DF*:

$$L_h = 2 \cdot r_i = 2 \cdot 0.121 = 0.242\ (\mathrm{m})$$

153. The lower diameter of the *DF*:

$$d_{\mathrm{ed}} = 0.802 \cdot d_{\mathrm{in}} = 0.802 \cdot 0.251 = 0.201\ (\mathrm{m})$$

154. The shaft diameter of the regulation system:

$$d_{\mathrm{ep}} = 0.508 \cdot d_{\mathrm{in}} = 0.508 \cdot 0.251 = 0.128\ (\mathrm{m})$$

155. The radius external to the axis of the *DF*:

$$r_n = 0.7 \cdot d_{\text{in}} = 0.7 \cdot 0.251 = 0.172 \text{ (m)}$$

156. The diameter of the lower hole of the *DF* that connects it with the *PHIT* double adjustment system:

$$d_{\text{en}} = 0.254 \cdot d_{in} = 0.254 \cdot 0.251 = 0.064 \text{ (m)}$$

157. The diameter for fixing the *DF* rods:

$$d_m = 1.23 \cdot d_{\text{in}} = 1.23 \cdot 0.251 = 0.309 \text{ (m)}$$

158. The vertical distance between the axis of the *DF* and the axis of the *PHIT* dual adjustment system:

$$L_{m*} = 0.976 \cdot d_{\text{in}} = 0.976 \cdot 0.251 = 0.245 \text{ (m)}$$

159. The horizontal stroke of the *PHIT* dual adjustment system:

$$L_m = 0.333 \cdot d_{\text{in}} = 0.333 \cdot 0.251 = 0.084 \text{ (m)}$$

160. The vertical diameter of the rotor shaft:

$$d_e = 0.118 \cdot \sqrt[3]{\frac{g \cdot \eta_t \cdot Q_{\max} \cdot H_{\max}}{n}} = 0.118 \cdot \sqrt[3]{\frac{9.81 \cdot 0.9125 \cdot 15.02 \cdot 640}{400}}$$
$$\cong 0{,}707 \text{ (m)}$$

As:

$$d_e = 0.710 \text{ (m)} = 710 \text{ (mm)}$$

161. The maximum head with load shedding:

$$H_1 = (1 + h_s) \cdot H_{\max} = (1 + 0.1) \cdot 640 = 704 \text{ (m)}$$

162. The absolute average velocity of the jet flow when subjected to Head H_1:

$$c_{4_1} = \phi_{0;4} \cdot \sqrt{2 \cdot g \cdot H_1} = 0.98 \cdot \sqrt{2 \cdot 9.81 \cdot 704} = 115.176 \text{ (m/s)}$$

163. The maximum flow in each IN:

$$Q_{1*} = \frac{Q_{\max}}{z_j} = \frac{15.02}{4} = 3.755 \left(\text{m}^3/\text{s}\right)$$

164. The corresponding jet diameter:

$$d_{o_1} = \sqrt{\frac{4 \cdot Q_{1*}}{\pi \cdot c_{4_1}}} = \sqrt{\frac{4 \cdot 3.755}{\pi \cdot 115.176}} = 0.204 \text{ (m)}$$

165. The absolute average velocities of the flow, in the cylindrical tube sections of the *SA*:

$$c_{0;1} = \frac{4 \cdot Q_{0;1}}{\pi \cdot d_{t_{0;1}}^2} = \frac{4 \cdot 15.02}{\pi \cdot 1.552^2} = 7.940 \text{ (m/s)}$$

$$c_{2;3} = \frac{4 \cdot Q_{2;3}}{\pi \cdot d_{t_{2;3}}^2} = \frac{4 \cdot 11.265}{\pi \cdot 1.344^2} = 7.940 \text{ (m/s)}$$

$$c_{4;5} = \frac{4 \cdot Q_{4;5}}{\pi \cdot d_{t_{4;5}}^2} = \frac{4 \cdot 7.51}{\pi \cdot 1.097^2} = 7.946 \text{ (m/s)}$$

$$c_{6;7} = \frac{4 \cdot Q_{6;7}}{\pi \cdot d_{t_{6;7}}^2} = \frac{4 \cdot 3.755}{\pi \cdot 0.787^2} = 7.719 \text{ (m/s)}$$

Continue with the lowest speed, i.e.:

$$c_{6;7} = 7.719 \text{ (m/s)}$$

166. The lowest specific kinetic energy in the cylindrical tube sections:

$$E_{ct} = \frac{c_{6;7}^2}{2 \cdot g} = \frac{7.719^2}{2 \cdot 9.81} = 0.995 \cong 1.0 \text{ (mca)}$$

167. The highest internal pressure in the *SA* to calculate the thicknesses:

$$H_{1*} = (H_1 - E_{ct}) = (704 - 1) = 703 \text{ (m)}$$

168. The tube thickness of each *SA* section:

$$e_{t_i} = \left[\left(1 + \frac{1.962 \cdot H_{1*}}{\sigma_{t_{\text{adm}}} - 1.275 \cdot H_{1*}}\right)^{0.5} - 1\right] \cdot \frac{d_{t_1}}{2}$$
$$= \left[\left(1 + \frac{1.962 \cdot 703}{16744 - 1.275 \cdot 703}\right)^{0,5} - 1\right] \cdot \frac{d_i}{2} = 0.0213 \cdot d_i \rightarrow e_{t_{0;1}} = 0.0213$$
$$\cdot d_{t_{0;1}} = 0.0213 \cdot 155.2 = 3.31 \text{ (cm)} \rightarrow e_{t_{2;3}} = 0.0213 \cdot d_{t_{2;3}} = 0.0213 \cdot 134.4$$
$$= 2.86 \text{ (cm)} \rightarrow e_{t_{4;5}} = 0.0213 \cdot d_{t_{4;5}} = 0.0213 \cdot 109.7 = 2.34 \text{ (cm)} \rightarrow e_{t_{6;7}}$$
$$= 0.0213 \cdot d_{t_{6;7}} = 0.0213 \cdot 78.7 = 1.68 \text{ (cm)}$$

169. The absolute average speed at the ring gear, cylinder/needle:

$$c_{ci/a} = \frac{4 \cdot Q_{6;7}}{\pi \cdot \left(d_{ci}^2 - d_a^2\right)} = \frac{4 \cdot 3.755}{\pi \cdot \left(0.854^2 - 0.461^2\right)} = 9.251 \text{ (m/s)}$$

170. The specific kinetic energy at the crown, cylinder/needle:

$$e_{ci/a} = \frac{c_{ci/a}^2}{2 \cdot g} = \frac{9.251^2}{2 \cdot 9.81} \cong 4.4 \text{ (mca)}$$

171. The internal pressure in the crown, cylinder/needle:

$$H_{1_{ci/a}} = \left(H_{1*} - e_{ci/a}\right) = (704 - 4) = 700 \text{ (mca)}$$

172. The thickness of the cylindrical tube:

$$e_{ci} = \left[\left(1 + \frac{1.962 \cdot H_{1_{ci/a}}}{\sigma_{t_{\text{adm}}} - 1.275 \cdot H_{1_{ci/a}}}\right)^{0.5} - 1\right] \cdot \frac{d_{ci}}{2} =$$
$$= \left[\left(1 + \frac{1.962 \cdot 700}{16744 - 1.275 \cdot 700}\right)^{0.5} - 1\right] \cdot \frac{85.4}{2} = 1.81\ (\text{cm})$$

173. The thrust force of the jet in the *DF*:

$$F_j = \rho \cdot Q_1 \cdot c_{4_1} = 1000 \cdot 3.755 \cdot 115.176 = 432485.880\ (\text{N}) = 44086.226\ (\text{kgf})$$

174. The hydraulic force resulting from the flow over the *DF*:

$$F_h = 2 \cdot F_j \cdot cos\,\theta_d = 2 \cdot 432485.88 \cdot cos\,45 = 611627.397\ (\text{N})$$
$$= 62347339\ (\text{kgf})$$

175. The angle θ^o_{d*} that F_h forms with F_j:

$$\theta_{d*} = 0.5 \cdot (180 - \theta_d) = 0.5 \cdot (180 - 45) = 67.5^o$$

176. The projections of $F_h$on the orthogonal x and y axes:

$$F_{h_x} = F_h \cdot cos\,\theta_{d*} = 611627.397 \cdot cos\,67.5 = 234059.672\ (\text{N})$$

and

$$F_{h_y} = F_h \cdot sin\,\theta_{d*} = 611627.397 \cdot sin\,67.5 = 565070.034\ (\text{N}) = 57601.431(\text{kgf})$$

177. The frictional force of the flow on the inner surface of the *DF*:

$$F_{a_h} = \mu_{ad} \cdot F_h = 0.01 \cdot 611627.397 = 9116.247\ (\text{N}) = 623.473\ (\text{kgf})$$

178. The force resulting from flow and friction on the *DF*:

$$F_r = \sqrt{F_h^2 + F_{a_h}^2} = \sqrt{611627.397^2 + 6116.274^2} = 611657.978 \text{ (N)}$$
$$= 62350.456 \text{ (kgf)}$$

179. The thickness of the *DF*:

$$e_d = \frac{F_r \cdot f_s}{0.6 \cdot \sigma_{esc} \cdot L_d} \cdot \sqrt{1 + \left(\frac{3 \cdot r_i}{L_d}\right)^2} = \frac{611657.978 \cdot f_s}{0.6 \cdot 24623 \cdot 24.5} \cdot \sqrt{1 + \left(\frac{3 \cdot 10.8}{24.5}\right)^2}$$
$$= 2.802 \cdot f_s.$$

Yields:

$$e_{d_{\min}} = 2.802 \cdot 2 = 5.603 \text{ (cm)}$$

And:

$$e_{d_{\max}} = 2.802 \cdot 4 = 11.206 \text{ (cm)}$$

Store these values for the following steps, since the definitive value of e_d, will only be fixed after the laboratory tests of the *DF*. The validation, in this case, will have a sequence at:

$$e_d = e_h = e_{su} = 8 \text{ (cm)}$$

180. The *DF* head volume:

$$V_{ca} = \frac{\pi}{2} \cdot (2 \cdot r_i + e_d) \cdot e_d \cdot L_d = \frac{\pi}{2} \cdot (2 \cdot 10.8 + 8) \cdot 8 \cdot 24.5 = 9113.132 \left(\text{cm}^3\right)$$

181. The turning angle of the *DF*:

$$\theta_g = \frac{180 \cdot L_d}{\pi \cdot r_{de}} = \frac{180 \cdot 24.5}{\pi \cdot 65.3} = 21.497^o$$

182. The volume of the two stems of the *DF*, part h_1:

$$V_{h_1} = \left[\left(b_d \cdot \sin\theta_g^o + \frac{\pi \cdot \theta_g^o}{90}\right) \cdot r_{de} - \frac{\pi \cdot d_{ed}^2}{8}\right] \cdot e_h =$$
$$= \left[\left(23.3 \cdot \sin 21.497 + \frac{\pi \cdot 21.497}{90}\right) \cdot 65.3 - \frac{\pi \cdot 20.1^2}{8}\right] \cdot 8 = 3583.198\ \left(\mathrm{cm}^3\right)$$

183. The volume of the two stems of the *DF*, part h_2:

$$V_{h_2} = \pi \cdot \left[r_n^2 - \left(\frac{d_{ed}}{2}\right)^2\right] \cdot e_h = \pi \cdot \left[17.6^2 - \left(\frac{20.1}{2}\right)^2\right] \cdot 8 = 5246.648\ \left(\mathrm{cm}^3\right)$$

184. The radius r_j of the h_3 part of the *DF*:

$$r_j = 0.5 \cdot r_n + 0.25 \cdot d_{ed} = 0.5 \cdot 17.6 + 0.25 \cdot 20.1 = 13.825\ (\mathrm{cm})$$

185. The radius $r_e$of the h_3 part of the *DF*:

$$r_e = 0.5 \cdot b_d + 0.25 \cdot d_{ed} = 0.5 \cdot 23.3 + 0.25 \cdot 20.1 = 16.675\ (\mathrm{cm})$$

186. The radius $r_m$of the h_3 part of the *DF*:

$$r_{m*} = 0.5 \cdot \left(r_e + r_j\right) = 0.5 \cdot (16.675 + 13.825) = 15.250\ (\mathrm{cm})$$

187. The arc length h_p of the *DF* part, h_3:

$$h_p = 0.5 \cdot \pi \cdot r_m = 0.5 \cdot \pi \cdot 15.25 = 23.955\ (\mathrm{cm})$$

188. The largest prism area of the h_3 part of the *DF*:

$$A_B = (b_d - 0.5 \cdot d_{ed}) \cdot e_h = (23.3 - 0.5 \cdot 20.1) \cdot 8 = 106\ \left(\mathrm{cm}^2\right)$$

189. The smaller prism area of the h_3 part of the *DF*:

$$A_b = (r_n - 0.5 \cdot d_{ed}) \cdot e_h = (17.6 - 0.5 \cdot 20.1) \cdot 8 = 60.4\ \left(\mathrm{cm}^2\right)$$

190. The volume of the two rods, part h_3 of the *DF*:

$$V_{h_3} = \frac{2 \cdot h_p}{3} \cdot \left(A_B + A_b + \sqrt{A_B \cdot A_b}\right) = \frac{2 \cdot 23.955}{3} \cdot \left(106 + 60.4 + \sqrt{106 \cdot 60.4}\right) = 3935.248 \ \left(\text{cm}^3\right)$$

191. The distance between the *DF* rods:

$$L_h = 2 \cdot r_i = 2 \cdot 12 = 24.0 \ (\text{cm})$$

192. The volume of the plate between the rods:

$$V_{pd} = L_h \cdot b_d \cdot e_{su} = 24 \cdot 23.3 \cdot 8 = 4473.6 \ \left(\text{cm}^3\right)$$

193. The volume of the hollow shaft between the *DF* rods:

$$V_{ev} = \frac{\pi}{4} \cdot \left(d_m^2 - d_{ed}^2\right) \cdot L_h = \frac{\pi}{4} \cdot \left(30.9^2 - 20.1^2\right) \cdot 24 = 10382.335 \ \left(\text{cm}^3\right)$$

194. The volume of the *DF*:

$$V_{DE} = V_{ca} + V_{h_1} + V_{h_2} + V_{h_3} + V_{pd} + V_{ev} = = 9113.132 + 3583.198 + 5246.648 + 3935.248 + 4473.6 + 10382.335 = 36734.161 \ \left(\text{cm}^3\right)$$

195. The mass of the *DF*:

$$G_{DE} = \rho_m \cdot V_{DE} = 7698 \cdot 10^{-6} \cdot 36734.161 = 282.780 \ (\text{N}) \cong 28.825{,}644 \ (\text{kg})$$

196. The inner diameter of the disk:

$$d_{\text{di}} \cong 0.95 \cdot d_e = 0.95 \cdot 71.0 = 67.5 \ (\text{cm})$$

197. The bolt circle hole diameter:

$$d_p \cong 1.5 \cdot d_e = 1.5 \cdot 71.0 = 106.5 \ (\text{cm})$$

198. The diameter of the disk fixing to the shaft:

$$d_f = 1.81 \cdot d_e = 1.81 \cdot 71 = 128.5 \text{ (cm)}$$

199. The thickness of the internal disk for fixing the shells:

$$e_{cs} \cong 0.406 \cdot d_e = 0.406 \cdot 71 = 28.8 \text{ (cm)}$$

200. The maximum radius of the equivalent paraboloid of revolution:

$$R_m = 0.5 \cdot (T + 0.25 \cdot B) = 0.5 \cdot (23.3 + 0.25 \cdot 61.8) = 19.375 \text{ (cm)}$$

201. The equivalent internal volume of half a shell:

$$V_{in_{c/2}} = 0.25 \cdot \pi \cdot R_m^2 \cdot L = 0.25 \cdot \pi \cdot 19.375^2 \cdot 59.7 = 17601.417 \left(\text{cm}^2\right)$$

202. The equivalent external volume of half a shell:

$$V_{ex_{c/2}} = 0.25 \cdot \pi \cdot (R_m + e_c)^2 \cdot L = 0.25 \cdot \pi \cdot (19.375 + 8)^2 \cdot 59.7$$
$$= 35137.630 \left(\text{cm}^3\right)$$

203. The equivalent volume of a shell:

$$V_c = 2 \cdot \left(V_{ex_{c/2}} - V_{in_{c/2}}\right) = 2 \cdot (35137.630 - 17601.417) = 35072.426 \left(\text{cm}^3\right)$$

204. The weight/mass of an equivalent shell:

$$G_{co} = \rho_m \cdot V_c = 7698 \cdot 10^{-6} \cdot 35072.426 = 269.988 \text{ (kg)} = 2648.578 \text{ (N)}$$

205. The weight/mass of the equivalent shell crown:

$$G_{cc} = z_c \cdot G_{co} = 20 \cdot 269.988 = 5399.76 \text{ (kg)} = 52971.646 \text{ (N)}$$

206. The volume of the disk that connects the shell to the axis:

$$V_d = \frac{\pi}{4} \cdot \left[\left(D_{ic}^2 - d_{di}^2 \right) \cdot e_{cs} \right] = \frac{\pi}{4} \cdot \left[\left(196.2^2 - 67.5^2 \right) \cdot 28.8 \right] = 767663.772 \ \left(\text{cm}^3 \right)$$

207. The weight/mass of the disc that connects the shell to the axle:

$$G_d = \rho_m \cdot V_d = 7698 \cdot 10^{-6} \cdot 767663.772 = 5909.476 \ (\text{kg}) = 57971.957 \ (\text{N})$$

208. The weight/mass of the *RO*:

$$G_{RO} = G_{cc} + G_d = 5399.76 + 5909.476 = 11309.236 \ (\text{kg}) = 110943.605 \ (\text{N})$$

209. The coordinate:

$$x = \frac{z_j^{0,5} \cdot D}{z_c \cdot B} \cdot n_{qA_1} = \frac{4^{0.5} \cdot 2523}{20 \cdot 618} \cdot 18.179 = 7.422$$

The graph in Fig. 11.13 shows that the result is within the predicted range.

210. The diameter of the flow after exiting the brake:

$$d_f = 0.5 \cdot d_o = 0.5 \cdot 20.6 = 10.3 \ (\text{cm})$$

211. The flow flowing through the brake circuit:

$$Q_f = u \cdot \pi \cdot \frac{d_o^2}{16} = 52.844 \cdot \pi \cdot \frac{0.206^2}{16} = 0.440 \ (\text{m/s})$$

212. The inside diameter of the brake circuit tubes:

$$d_{ti_f} = \sqrt{\frac{4 \cdot Q_f}{\pi \cdot c_{0;1}}} = \sqrt{\frac{4 \cdot 0.44}{\pi \cdot 7.940}} = 0.266 \ (\text{m})$$

213. The thickness of the brake circuit tubes:

$$e_{t_f} = \left[\left(1 + \frac{1.962 \cdot H_1}{\sigma_{t_{\text{adm}}} - 1.275 \cdot H_1}\right)^{0.5} - 1\right] \cdot \frac{d_{ti_f}}{2}$$
$$= \left[\left(1 + \frac{1.962 \cdot 703}{16744 - 1.275 \cdot 703}\right)^{0.5} - 1\right] \cdot \frac{d_{ti_f}}{2} = 0.0213 \cdot d_{ti_f} = 0.0213$$
$$\cdot 0.266 = 0.0057\ (\text{m}) = 0.57\ (\text{cm})$$

With $e_{t_f} = 10$ (mm) fixed.

214. The penstock hydraulic time:

$$t_h = \frac{L_{cf} \cdot c_{0;1}}{g \cdot H_{cf}} = \frac{900 \cdot 7.940}{9.81 \cdot 620} = 1.17\ (\text{s})$$

This time indicates a need for double regulation.

215. The transient time of the *GG*:

$$t_g < 0.5056 \cdot \frac{P_{el}^{0.333}}{n^{0.15}} < 0.5056 \cdot \frac{79374.113^{0.333}}{400^{0.15}} < 8.81\ (\text{s})$$

Fix the immediate lower integer to continue i.e., $t_g = 8$ (s).

216. The time for the *DF* to fully divert the jet:

$$t_{d_{\min}} = 0.25 \cdot d_g = 0.25 \cdot 8 = 2.0\ (\text{s})$$

and

$$t_{d_{\max}} = 0.75 \cdot d_g = 0.75 \cdot 8 = 6.0\ (\text{s})$$

217. The average turning speed of the *DF*:

$$c_g = \frac{L_g}{t_{d_{\max}}} = \frac{24.5}{6} = 4.083\ (cm/s) \cong 0.041\ (\mathrm{m/s})$$

218. The GD^2_{GG} of the *GG*:

$$GD^2_{GG} = 3578256.4 \cdot \frac{P_{el} \cdot t_g}{n^2} = 3578256.4 \cdot \frac{79374.113 \cdot 8}{400^2}$$
$$= 14201046.39\ \left(\mathrm{N \cdot m^2}\right)$$

219. The *EG* rotor moment of inertia:

$$GD^2_{GE} = 1608252.9 \cdot \frac{P_{el}^{1.333}}{n^{2.15}} = 1608252.9 \cdot \frac{79374.113^{1.333}}{400^{2.15}}$$
$$= 13905842.8\ \left(\mathrm{N \cdot m^2}\right)$$

220. The moment of inertia of the rotating system of the *PHIT*:

$$GD^2_{TH} = 0.125 \cdot GD^2_{GE} = 0.125 \cdot 13905842.8 = 1738230.35\ \left(\mathrm{N \cdot m^2}\right)$$

221. The sum:

$$GD^2_{GE} + GD^2_{TH} = 13905842.8 + 1738230.35 = 15644073.15\ \left(\mathrm{N \cdot m^2}\right)$$

If this sum is greater than GD^2_{GG}, you don't need a *FW*. If the sum is smaller you do not need a *FW* with:

$$GD^2_{VO} = GD^2_{GG} - \left(GD^2_{GE} + GD^2_{TH}\right)\ \left(\mathrm{N \cdot m^2}\right)$$

Upon validation:

$$GD^2_{GE} + GD^2_{TH} = 15644073.15$$

and

$$GD_{GG}^2 = 14201046.39$$

or

$$GD_{GG}^2 = 14201046.39 = 1443026.76\ \left(\mathrm{N \cdot m^2}\right)$$

Since the value was positive, there is no need for a *FW*. If the result were negative, this value would be GD_{VO}^2 and it would be sized observing the following script, which is valid for the *FW* geometry shown in Fig. 11.19 in Chap. 11

221.1. The outer radius of the *FW* :

$$r_e = 0.156 \cdot \sqrt[5]{GD_{VO}^2} = (\mathrm{m})$$

221.2. The inner radius of the *FW* :

$$r_i = 0.74 \cdot r_e\ (\mathrm{m})$$

221.3. The thickness of the *HW* crown:

$$B = r_e - r_i (\mathrm{m})$$

221.4. The thickness of the *FW ABNT* disc:

$$b = 0.5 \cdot B\ (\mathrm{m})$$

222. Preliminary drawing of the *DF* (Fig. 12.10) having as geometric reference Fig. 11.7.

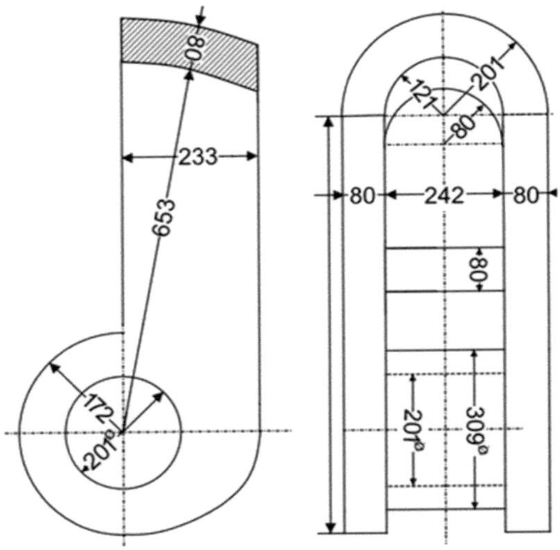

Fig. 12.10 To-scale *DF* drawing

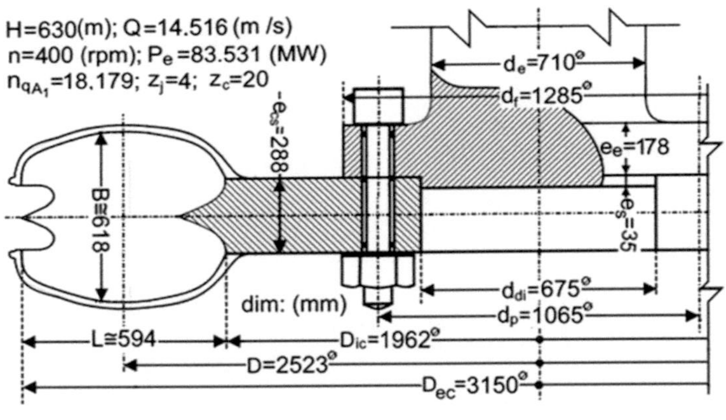

Fig. 12.11 To-scale *RO* drawing

223. Preliminary sized drawing of the *RO* longitudinal section, Fig. 12.11, referencing the geometry in Fig. 11.8
224. Referencing the longitudinal and transversal sections established and drawn, and the respective calculated and/or established thicknesses, perform the drawings. In this case, as an example, referencing Figs. 12.2, 12.3, and 12.4, the section was drawn as cross section $g*$ (Fig. 12.12) showing the basic flat section of the areas for drawing, and subsequent fabrication of the template and its movement. In Fig. 12.13 this was the longitudinal section *f*.
225. Using the cross sections and longitudinal sections of the cutaway in Fig. 12.1, make a sized drawing that will serve as a reference for machining the outside of the shells. Figure 12.14 and the script in the sequence illustrate how this can be applied to a section of the cutaway, using data from the present validation.

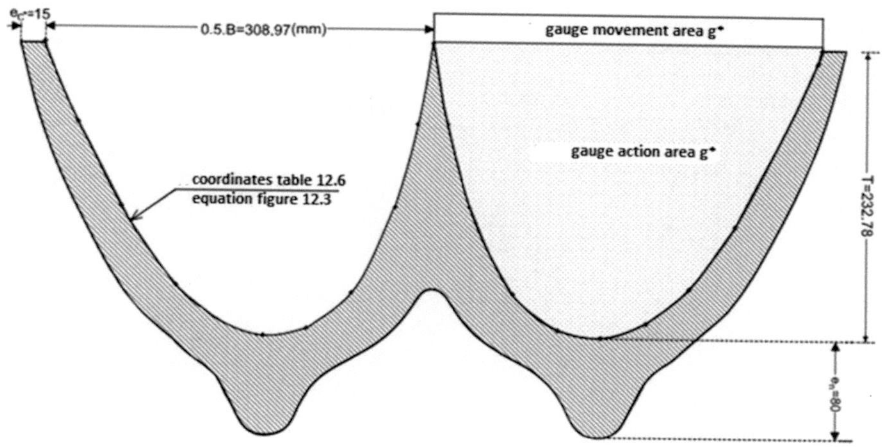

Fig. 12.12 Cross section $g*$

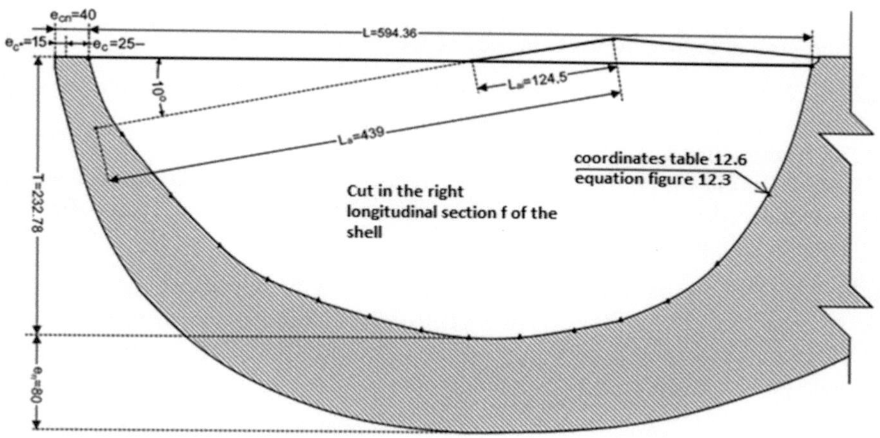

Fig. 12.13 Longitudinal section f

225.1. In the spreadsheet of the computer program, draw an orthogonal system of axes and with the center on the vertical axis and draw circumferences:

$$D_{e_c} = 3150 \text{ (mm)}$$

$$D_{e_a} = 3152 \text{ (mm)}$$

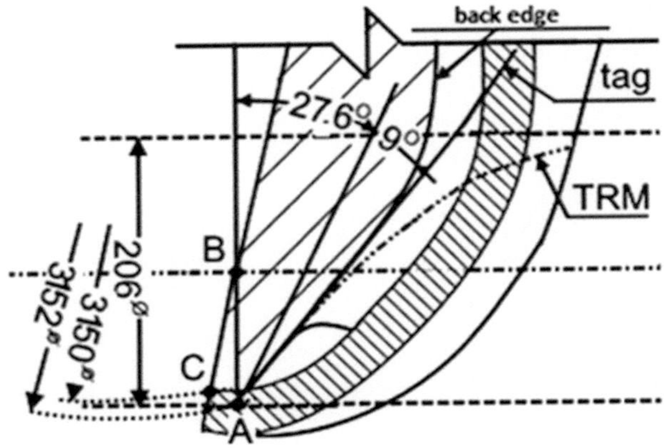

Fig. 12.14 Sized drawing of a section for machining the shell to avoid shocks to the back of the shells

$D = 2523$ (mm)

… and normal to this, point *B*, the axis of the jet and its limits.

225.2. With reference to *B*, draw a cutaway of the shell on its edge to scale referencing Fig. 12.4 section *AA* and Fig. 12.5

225.3. From point *A* draw the *RAT* and its tangent.

225.4. To avoid hitting the back of the shells, draw a projection of the line that externally delimits the shell in this section from *A*, in this case, the part corresponding to its edge.

226. With reference to the constant in Fig. 12.5, and the dimensions of the shell, a schematic of the positioning of 5 shells of the *RO* was drawn in Fig. 12.15, observing the script in sequence, using a graphical computer software program with precision/accuracy set to millimeters.

226.1. On the worksheet, draw two orthogonal axes, marking their meeting point *O*.

226.2. With center *O* draw the circumferences of the radii and diameters, in millimeters:

$$r = 487$$

$$r_{1*} = 259$$

$$D_{e_a} = 3152$$

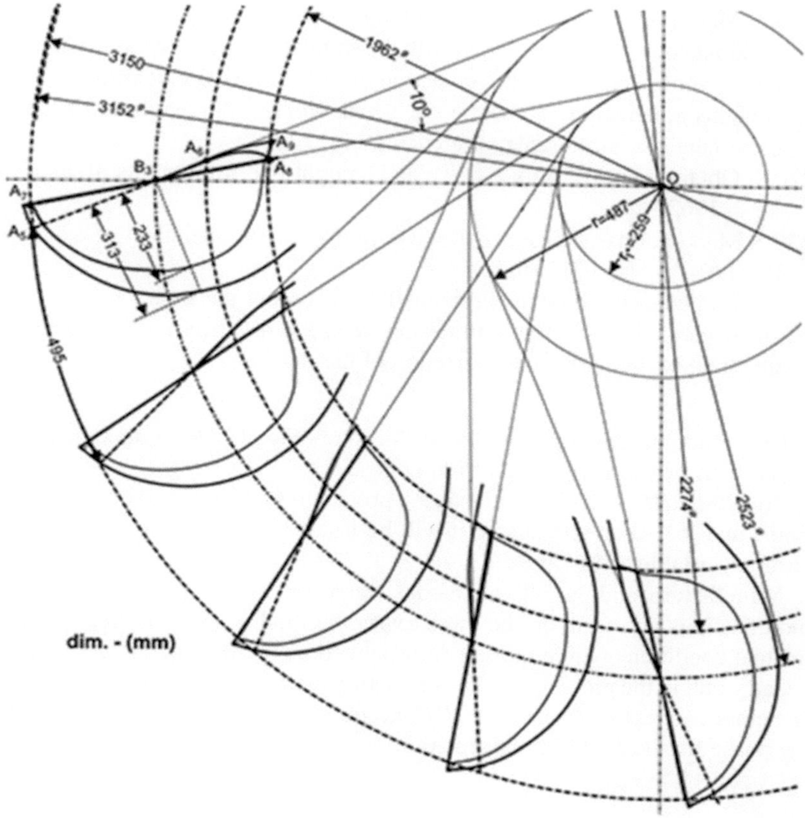

Fig. 12.15 Sized horizontal projection of 1/4 of the *RO* shells

$$D = 2523$$

$$D_{i_a} = 2274$$

$$D_{e_c} = 3150$$

$$D_{i_c} = 1962$$

226.3. Mark point B_3 on the horizontal axis on $D = 2523$ (mm), transposing the dividing edge of the shell, $A_5\bar{A}_6$ with the straight line that contains it, tangent to the circumference r and the projection of the shell $A_7\bar{A}_8$ exit with the line that contains it, tangent to the circle r_{1*}, which will result in angle of 10° between these tangents, all to scale, referencing Fig. 12.5.

226.4. Obtain point $A9$ from a curve and straight line from points $A6$ and $A8$ on $D_{i_c} = 1962$ (mm).

226.5. Mark, normally the $A_5\bar{A}_6$, $T = 233$ (mm), and the shell thickness on the rib, fixed as $T + 80 = 313$ (mm).

226.6. Preliminarily draw the horizontal projection of the shell.

226.7. Rotate this projection 4 times the step $p_c = 495$ (mm) $= 18^o$, obtaining a quadrant of the horizontal projection of the *RO*.

227. Draw the horizontal projection of the rotor, Fig. 12.16, complementing this figure with part of the *IN* and *DF* drawings.
228. Draw the *PHIT* model manufacturing project in scale within the manufacturing limits and the laboratory where it will be tested, at a 1/1 scale, with precision/accuracy established by the *PT*.
229. Manufacture and install the model, carry out preliminary tests, carry out modifications and tests on the model with the modifications to obtain the contractual conditions within the standards agreed on by the contracting parties.
230. Carry out, in the presence of the contracting parties, the model tests and prepare an authorized report so that the *PHIT* executive project can enter the manufacturing phase for the *PHIT* implementation.

12.2.4 Exit

At a minimum, generate the following tables (Tables 12.9, 12.10, 12.11, 12.12, 12.13):

12.3 Comments

12.3.1 Generalizations

- From the outset, it is important to note that the algorithm was validated using a preliminary dataset comprising approximately 24 values, corresponding to characteristics that vary across a range of intervals, some relatively narrow, others significantly wide. Consequently, the algorithm may exhibit limited average

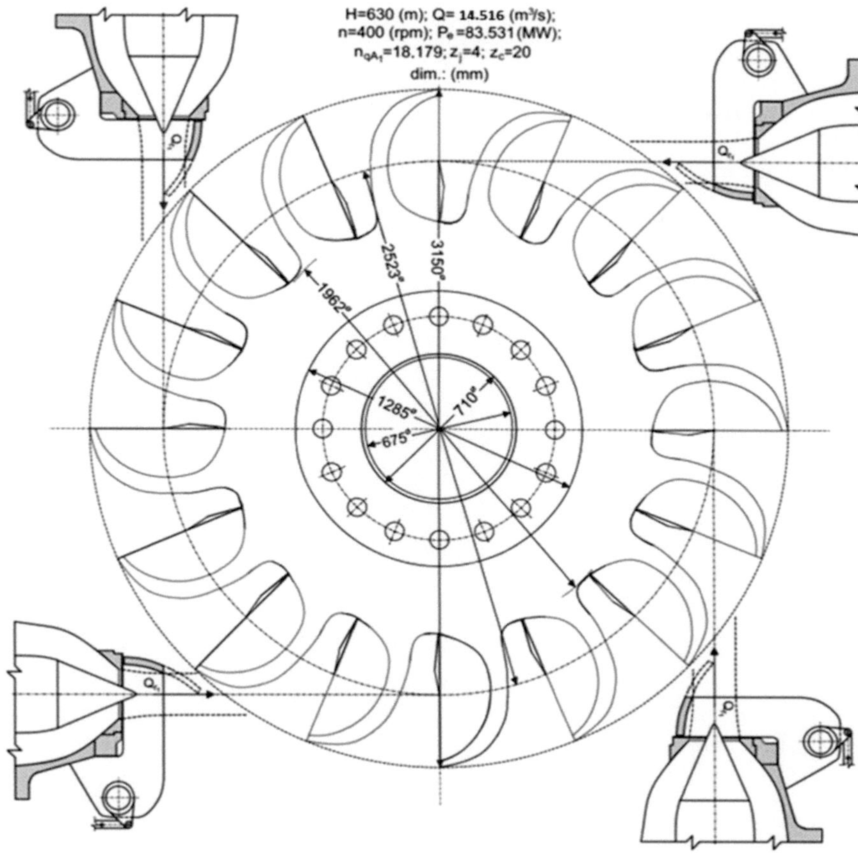

Fig. 12.16 Sized horizontal projection of *RO*, *IN*, and *DF*

accuracy and elevated uncertainty, both of which can influence the final results to varying extents.

- One should also remember that the algorithm initially addressed hydrodynamic aspects of the flow and geometric aspects of its limits, the so-called *HFP* from the cross section 0 inlet to the 5 outlet section of the *RO* shells. After this section, the algorithm calculates and fixes characteristics for calculating the external energy ventilation losses and after fixing the mechanical and electrical losses via the yields, first determining the value for the energy that will be made available to the consumer at the output, for *EG*s, this is called electrical power, and does this with uncertainties that can reach up to more than 50 (%).
- Additionally, we should note that the structural calculations observed limitations contained in Chap. 11, and that the validation was carried out with constant restrictions on the stress values.

Table 12.9 Geometry and structure of the intake system

$dt_0 = dt_1$	$dt_2 = dt_3$	$dt_4 = dt_5$	$dt_6 = dt_8$	D	D_{ea}	Le_1	Le_2	Le_3	Le_4	$e1E$
mm										
1552.00	1344.00	1097.00	776.00	2523.00	3152.00	4141.00	3933.00	3686.00	3365.00	1552.00

δe	$e_1 1=e_1 3=e_1 5$	$e_1 2=e_1 4=e_1 6$	$b_1= b_2= b_3$	L^*	L_1	L_2	L_3	L_4
degrees	mm							
50	1850.00	776.00	2626.00	944.70	1552.00	1657.00	1696.00	2798.00

Table 12.10 CCL Geometry and structure

δ^*	δ_1	δ_2	R_1	R_2	R_3	R_4	$et_{0;1}$	$et_{2;3}$	$et_{4;5}$	$et_{6;7}$
degrees			mm							
30	15	30	2840.00	2593.00	2322.00	776.00	331.00	286.00	234.00	168.00

d_{tb}	d_{ci}	d_a	d_2	L_{di}	L_{dc}	L_{cii}	L_{in}	L_{ci}	L_{cii}	L_{cu}	L_{Fea}
mm											
776.00	854.00	461.00	251.00	318.00	295.00	945.00	557.00	1240.00	1797.00	303.00	474.00

Continued and Hydraulic Brake

L_{EF}	L_{eaE}	θ_{in}	θ_{co}	e_{ci}	d_f	$d_{t\ if}$	e_{tf}
mm		degrees		mm			
2115.00	2589.00	75	45	181.00	103.00	266.00	10.00

Table 12.11 Geometry and structure of the CCR—DF

r_{de}	b_b	θ_g	t_{dmin}	c_g	L_d	r_I	L_h	d_{ed}	d_{ep}	r_n	d_{en}
mm		degrees	s	m/s	mm						
653.00	233.00	20905	6.00	0.04	238.00	121.00	242.00	201.00	128.00	172.00	64.00

d_m	L_{m*}	L_m	θ_d*	F_r	$e_d = e_h = e_{su}$	θ_g	V_{DE}	G_{DE}
mm			degrees	N	mm	degrees	cm^3	N
309.00	245.00	84.00	67.5	611,657.98	80.00	21.5	36,734.16	282.78

Table 12.12 Shell and Rotor geometry and structure

f_B	f_L	f_T	L	B	E	T	D_{ia}	D_{ea}	D_{ec}	D_{ic}	d_e	r
scale coefficients			mm									
0.96	1.11	1.13	594.36	617.94	236.66	232.78	2274.00	3152.00	3150.00	1962.00	710.00	487.00
Inner surface of shells: Tables 12.3, 12.4, 12.5 and 12.6; Figures 12.1, 12.2, 12.3, and 12.4												

r_{I*}	d_f	d_{di}	d_p	e_e	e_s	e_{cs}	R_m	e^*_{cmin}	e^*_{cmax}	e_n	e_c
259.00	1285.00	675.00	1065.00	178.00	35.00	288.00	193.75	129.00	65.00	80.00	25.00

V_c	V_d	G_{co}	G_{cc}	G_d	G_R	x
cm^3		N				-
35,072.43	637,095.87	2,648.58	52,971.65	48,111.81	101,083.46	7.42

Table 12.13 Generator set

t_h	td_{max}	td_{min}	GD^2_{GG}	GD^2_{GE}	GD^2_{TH}	GD^2_V	P_{el}	η_{GG}
s			N/m^2				kW	%
1.17	2.00	6.00	14201046.39	13905842.80	1738230.35	0.00	80223.34	89.4

- The results of the algorithm (feature values and scaled sketches of the geometry of components) are essential in beginning of the structural calculations, which can indicate recalculations, or iterative processes for making changes to the *HFP*.
- The geometry of the components, particularly those in the slow and quick control components, and the Deflector (*DF*) vary from manufacturer to manufacturer, therefore their equations and respective relationships must be considered on a case-by-case basis.

12.3.2 Specifics from the Validation

- The procedures in steps 14 to 34 referring to the inner surface of the shells, can be used by those who have already designed, built, and tested *RO* shells for a *PHIT*, whether as models and/or as prototypes, based on the method developed in Chap. 10.
- Choosing characteristics z_j and d_0 (step 4) are fundamentally important in all aspects, particularly for the manufacturer and the *PT* in establishing the manufactured product and the respective expected costs.
- If the mechanical energy transformer supplied by the *PHIT*, coupled to the axis, does not require n being constant and synchronous, e.g., when activating hydraulic or piston pumps, or using speed variators (hydrodynamic union, mechanical boxes or frequency inverters), the *PHIT* can have a constant performance curve, depending on the load variations, within the operational limits of the flows and heads.
- In step 7, we demonstrated the use of practical analytical processes for preliminary fixing of the characteristics, n_{qA_1}, r_v, D and, consequently, the possibility for calculating the tangential velocity u.
- In step 27, the graphical process developed in Chap. 9 was shown to obtain the direction of the dividing edge of the shells, imposing the condition of normality with the full jet, between two consecutive positions of each *RO* shell.
- The results in validation step 112, and the complements, showed and commented on what can happen when $\phi^*_{0,4}$ calculated $\neq \phi_{0;4}$ fixed. and the respective consequences.
- An analysis of the energy loss results in the *PHIT* study, from the entrance of the flow in section 0, to the mechanical energy available in the axis, indicated the need for concentrating forces to maximumly reduce the losses that occur in each component. However, in preliminary studies, these forces must also be observed in all other phases of the calculations and projections, i.e., the structural, model,

and laboratory tests, manufacturing controls, and factory testing, transportation, assembly, and operational tests on site, under normal load operations, and reception tests, along with maintenance of the *GG*.

- Based on the developed and validated algorithm, we recommend that contracting parties develop and validate a step-by-step algorithm for a horizontal axis *PHIT*.

Annex

Definitions of the Characteristic Quantities of Hydraulic Machines

Introduction

Flow machines are devices that convert energy between a fluid and a rotating shaft through dynamic interactions between the fluid and the machine's moving parts. These interactions allow the exchange of the fluid's kinetic and potential energy for mechanical energy, making these machines essential in industrial processes and power generation.

Examples of flow machines include hydraulic turbines, compressors, pumps, and fans. These machines either extract energy from a fluid, as in turbines, or transfer energy to the fluid, as in pumps and compressors.

The efficiency and performance of flow machines depend on key parameters, such as the flow rate, head, and rotational speed. These parameters are important for accurate design and optimal operation.

Flow Forms: Reynolds Number

The Reynolds number (Re) is a dimensionless parameter used to characterize the flow regime of a fluid. It determines whether the flow is laminar or turbulent and is defined as:

$$Re = \frac{\rho.v.L}{\mu} \tag{A.1}$$

where

© The Editor(s) (if applicable) and The Author(s), under exclusive license to Springer Nature Switzerland AG 2025

Z. de Souza, G. L. Tiago Filho, *Pelton and Crossflow Impulse Hydraulic Turbines*,
https://doi.org/10.1007/978-3-031-87086-6

- ρ = fluid density (kg/m^3)
- v = fluid velocity (m/s)
- L = characteristic length (m)
- μ = dynamic viscosity (Pa·s)
  - If Re < 2000, then the flow is defined as being laminar.
  - If Re > 4000, then the flow is defined as being turbulent.
  - 2000 < Re < 4000, then the flow is defined as being transitional.

Stream Tubes: Continuity Equation

The continuity equation represents the principle of mass conservation in fluid flow. For an incompressible fluid, the volumetric flow rate remains constant throughout a stream tube:

$$A_1.v_1 = A_2.v_2 \tag{A.2}$$

where

- A_1, A_2 = cross-sectional areas (m^2)
- v_1, v_2 = average fluid velocities (m/s)

This equation indicates that fluid velocity increases as the cross-sectional area decreases and vice versa.

Equations of Motion: Euler's Equation

Euler's equation explains the force balance in an ideal fluid and serves as the foundation for analyzing hydraulic turbines:

$$\rho.\left(\frac{dv}{dt}\right) + \nabla p.f \tag{A.3}$$

where

- dv/dt = fluid acceleration (m/s^2)
- ∇p = pressure gradient (N/m^2)
- ρ = fluid density (kg/m^3)
- f = external forces (like gravity) (N)

This equation is fundamental for understanding the behavior of moving fluids.

Flow Machines

The fundamental principle of flow machines is the conservation of energy, as described by Bernoulli's theorem. This theorem states that in the steady flow of an ideal fluid, the total energy along a streamline remains constant. It involves the conversion of potential energy (positional energy), kinetic energy (motion energy), and pressure energy into useful mechanical energy.

Flow machines are classified based on their interaction with the fluid (e.g., reaction or action) and the direction of energy conversion—either from mechanical energy to fluid energy (as in pumps and compressors) or from fluid energy to mechanical energy (as in hydraulic turbines).

Fundamental Principles for Flow Machine Equations

The theoretical foundation of flow machines is based on the *five principles* of mechanics for steady-flow systems, which are essential for understanding and formulating these machines:

1. **Mass Conservation:** The principle of flow continuity states that the total mass of fluid entering a system must equal the total mass leaving it. Therefore, the continuity equation is expressed as:

$$\sum \dot{m}_{e_i} = \sum \dot{m}_{s_j} \rightarrow \rho.\mathrm{A}_1.\mathrm{v}_1 = \rho.\mathrm{A}_2.\mathrm{v}_2 \tag{A.4}$$

where

- ρ = fluid density (kg/m^3)
- A_1 and A_2 = the areas of cross-sections 1 and 2 (m^2)
- v_1 and v_2 = the fluid velocities at sections 1 and 2 (m/s)

2. **Energy Conservation:** Based on Bernoulli's theorem, the total energy along a streamline remains constant, including kinetic, potential, and pressure energy:

$$\sum E_{e_i} = \sum E_{s_j} \rightarrow \left(\frac{\mathrm{v}^2}{2}\right) + \left(\frac{\mathrm{p}}{\rho}\right) + gz = \text{constant} \tag{A.5}$$

where

- v = velocity (m/s)
- p = pressure (N/m^2)
- ρ = density (kg/m^3)
- g = gravitational acceleration (m/s^2)

3. **Momentum Conservation:** It relates the change in linear momentum of a fluid to the external forces applied:

$$\sum \vec{F}\,dt = d(m.\overline{v}) \tag{A.6}$$

where

- F = resulting force (N)
- m = mass (kg)
- v = fluid velocity (m/s)

4. **Angular Momentum Conservation:** It states that the resulting torque applied to a fluid is related to the change in angular momentum:

$$\sum Mdt = d(I\omega) \tag{A.7}$$

where

- M = torque (N.m)
- I = moment of inertia (kg,m^2)
- ω = angular velocity (rad/s)

5. **Energy Degradation:** It represents the loss of mechanical energy due to dissipative effects, such as friction and turbulence. The head loss equation is commonly used:

$$Y_p = \lambda.\frac{v^2}{2}.\frac{L}{D} \quad [J/kg] \tag{A.8}$$

where

- Y_P = head loss
- λ = friction factor (−)
- L = length (m)
- D = diameter (m)
- g = gravitational acceleration (m/s^2)

Characteristic Quantities of Hydraulic Machines

The characteristic quantities of hydraulic machines are essential parameters for describing their operation and comparing different machines. They allow one to analyze performance, efficiency, and operational behavior under various conditions.

The main quantities include:

- **Flow Rate (Q):** It represents the volume of fluid flowing per unit time, typically measured in m^3/s. It is a fundamental parameter for determining the machine's capacity.
- **Head (H):** It represents the energy per unit weight of the fluid, expressed in meters of water column (m.w.c.), transferred by the machine. It is determined by the elevation difference the machine must overcome.
- **Power (P):** It measures the amount of energy transferred per unit time, expressed in Watts (W) or kilowatts (kW).
- **Efficiency (η):** The ratio of useful power to supplied power, indicating the machine's efficiency.
- **Specific Speed (n_s):** A dimensionless quantity that characterizes the geometry and dynamic behavior of the machine as a function of flow rate and head.

These characteristic quantities are crucial for the proper sizing and selection of a hydraulic machine, as well as for comparing performance and suitability for specific operational requirements.

Specific Speed

The specific speed (n_s) is a dimensionless quantity that identifies the type of turbine most suitable for a given height and flow rate. It is, according to the Technical System of Units, calculated using the following expression:

$$n_s = \frac{n.\sqrt{P}}{H^{(5/4)}} \tag{A.10}$$

where

- n = turbine rotation speed (rpm)
- P = power generated (kW)
- H = net head (m)

The specific speed then allows us to classify turbines, according to the following: When specific speed is:

1. $3 < n_s < 40$: Low → Pelton turbines (high head, low flow rate)
2. $30 < n_s < 100$: Medium→ Francis turbines (medium head, medium flow rate)
3. $100 < n_s < 300$:High → Kaplan turbines (low head, high flow rate)

The specific speed n_{qA}, which is another way of expressing specific speed, is shown in the Cordier diagram in Fig. A.1. According to the International System of Units, this can be expressed in the following equation:

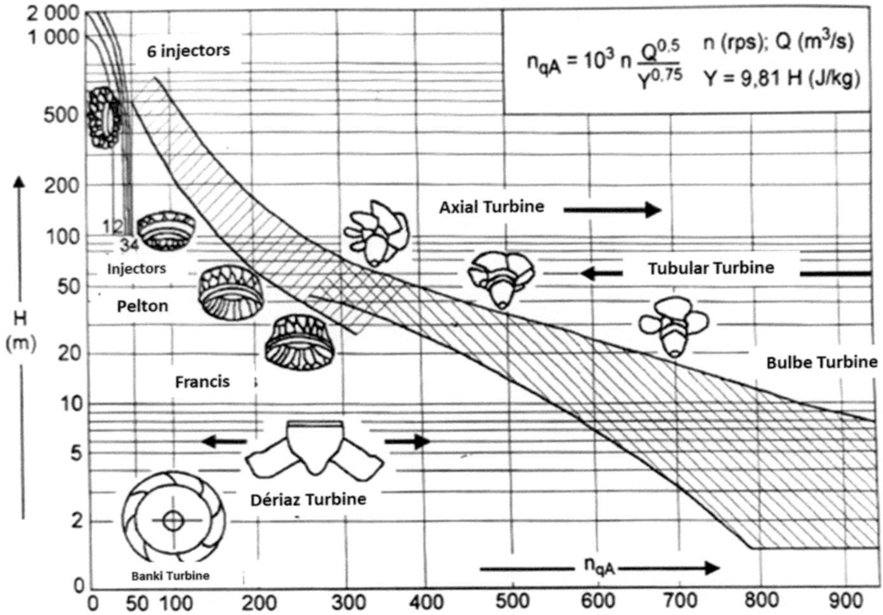

Fig. A.1 Limits of specific speeds, n_{qA}, as a function of head, H, for different types of turbines

$$n_{qA} = 1000.\frac{n.\sqrt{Q}}{(g.H)^{(3/4)}} \tag{A.11}$$

where

- $n =$ turbine rotation speed (rps)
- $H =$ net head (m)
- $Q =$ flow (m^3/s)
- $g =$ gravitational acceleration (m/s^2).

Unit Quantities of Flow Machines

Unit quantities are parameters that allow for comparisons between different flow machines, regardless of their physical size. They are normalized for the same operating conditions and include:

- **Unit Flow Rate (Qu)**: It represents the dimensionless fluid flow rate:

$$Q_1 = \frac{Q}{\sqrt{P.H}} \tag{A.12}$$

where

- $Q =$ volumetric flow rate (m^3/s)
- $H =$ net head (m)
- $P =$ the power supplied by the machine (kW)
- **Unit Speed (Nu):** It relates the turbine speed with the net head:

$$n_1 = \frac{n}{\sqrt{P.H^{(3/4)}}} \tag{A.13}$$

where

- $n =$ rotation speed (rpm)
- $H =$ net head (m)
- $P =$ power supplied by the machine (kW)

Conclusion

The use of unit quantities is essential in the study, and in the application of hydraulic machines in general, as it simplifies and generalizes their behavior by focusing on operational conditions. This approach helps in selecting and sizing the most appropriate machines for various applications.

These quantities are also crucial for developing the characteristic curves of flow machines. These curves illustrate the relationships between key variables, such as flow rate, power, head, and pressure, which directly affect the machine's performance. The curves are typically derived from experimental tests and play a critical role in determining the optimal operation of a machine in specific applications.

Therefore, parameters of the characteristic quantities of hydraulic machines—which include *the Reynolds number*, *the continuity equation*, *Euler's equation*, *unit quantities*, and *specific speed*—are all vital for understanding the behavior and performance of hydraulic machines. These parameters are fundamental for accurate sizing, specification, and operation. A thorough analysis of these parameters allows one to select the most suitable hydraulic turbine or pump based on the operational conditions it will be subjected to.

© The Editor(s) (if applicable) and The Author(s), under exclusive license to Springer Nature Switzerland AG 2025

Z. de Souza, G. L. Tiago Filho, *Pelton and Crossflow Impulse Hydraulic Turbines*,
https://doi.org/10.1007/978-3-031-87086-6

Index

© The Editor(s) (if applicable) and The Author(s), under exclusive license to Springer Nature Switzerland AG 2025
Z. de Souza, G. L. Tiago Filho, *Pelton and Crossflow Impulse Hydraulic Turbines*,
https://doi.org/10.1007/978-3-031-87086-6

MIX
Papier aus verantwortungsvollen Quellen
Paper from responsible sources
FSC® C105338

If you have any concerns about our products,
you can contact us on
ProductSafety@springernature.com

In case Publisher is established outside the EU,
the EU authorized representative is:
Springer Nature Customer Service Center GmbH
Europaplatz 3, 69115 Heidelberg, Germany

Printed by Libri Plureos GmbH
in Hamburg, Germany